Mathematics in Population Biology

PRINCETON SERIES IN THEORETICAL AND COMPUTATIONAL BIOLOGY

Simon A. Levin, Series Editor

Mathematics in Population Biology

Horst R. Thieme

PRINCETON UNIVERSITY PRESS
PRINCETON AND OXFORD

Published by Princeton University Press, 41 William Street, Princeton, New Jersey 08540

In the United Kingdom: Princeton University Press, 3 Market Place, Woodstock,
Oxfordshire OX20 1SY

Library of Congress Cataloguing-in-Publication Data

Thieme, Horst R., 1948–
Mathematics in population biology / Horst R. Thieme
p. cm. -- (Princeton series in theoretical and computational biology)
Includes bibliographical references and index.
ISBN 0-691-09290-7 (cl : alk. paper) -- ISBN 0-691-09291-5 (pbk. : alk. paper)
1. Population biology--Mathematical models. 2. Communicable diseases--Mathematical models. I. Title. II. Series.

QH352 .T45 2003
577.8′8′015118--dc21

British Library Cataloguing-in-Publication Data

A catalogue record for this book is available from the British Library.

This book has been composed in Times. Typeset by T&T Productions Ltd, London.
Printed on acid-free paper. ♾
www.pupress.princeton.edu

10 9 8 7 6 5 4 3 2 1

To Adelheid

Contents

Preface xiii

Chapter 1. Some General Remarks on Mathematical Modeling 1
Bibliographic Remarks 3

PART 1. BASIC POPULATION GROWTH MODELS 5

Chapter 2. Birth, Death, and Migration 7
2.1 The Fundamental Balance Equation of Population Dynamics 7
2.2 Birth Date Dependent Life Expectancies 9
2.3 The Probability of Lifetime Emigration 11

Chapter 3. Unconstrained Population Growth for Single Species 13
3.1 Closed Populations 13
3.1.1 The Average Intrinsic Growth Rate for Periodic Environments 14
3.1.2 The Average Intrinsic Growth Rate for Nonperiodic Environments 17
3.2 Open Populations 19
3.2.1 Nonzero Average Intrinsic Growth Rate 21
3.2.2 Zero Average Intrinsic Growth Rate 28

Chapter 4. Von Bertalanffy Growth of Body Size 33

Chapter 5. Classic Models of Density-Dependent Population Growth for Single Species 37
5.1 The Bernoulli and the Verhulst Equations 37
5.2 The Beverton–Holt and Smith Differential Equation 39
5.2.1 Derivation from a Resource–Consumer Model 40
5.2.2 Derivation from Cannibalism of Juveniles by Adults 42
5.3 The Ricker Differential Equation 45
5.4 The Gompertz Equation 47
5.5 A First Comparison of the Various Equations 47

Chapter 6. Sigmoid Growth 51
6.1 General Conditions for Sigmoid Growth 52
6.2 Fitting Sigmoid Population Data 57

Chapter 7. The Allee Effect 65

7.1 First Model Derivation: Search for a Mate 65
7.2 Second Model Derivation: Impact of a Satiating Generalist Predator 67
7.3 Model Analysis 69

Chapter 8. Nonautonomous Population Growth: Asymptotic Equality of Population Sizes 75

Chapter 9. Discrete-Time Single-Species Models 81

9.1 The Discrete Analog of the Verhulst (Logistic) and the Bernoulli Equation: the Beverton–Holt Difference Equation and Its Generalization 81
9.2 The Ricker Difference Equation 83
9.3 Some Analytic Results for Scalar Difference Equations 84
9.4 Some Remarks Concerning the Quadratic Difference Equation 99
Bibliographic Remarks 104

Chapter 10. Dynamics of an Aquatic Population Interacting with a Polluted Environment 107

10.1 Modeling Toxicant and Population Dynamics 108
10.2 Open Loop Toxicant Input 114
10.3 Feedback Loop Toxicant Input 117
10.4 Extinction and Persistence Equilibria and a Threshold Condition for Population Extinction 120
10.5 Stability of Equilibria and Global Behavior of Solutions 125
10.6 Multiple Extinction Equilibria, Bistability and Periodic Oscillations 135
10.7 Linear Dose Response 139
Bibliographic Remarks 149

Chapter 11. Population Growth Under Basic Stage Structure 151

11.1 A Most Basic Stage-Structured Model 151
11.2 Well-Posedness and Dissipativity 153
11.3 Equilibria and Reproduction Ratios 155
11.4 Basic Reproduction Ratios and Threshold Conditions for Extinction versus Persistence 156
11.5 Weakly Density-Dependent Stage-Transition Rates and Global Stability of Nontrivial Equilibria 157
11.6 The Number and Nature of Possible Multiple Nontrivial Equilibria 160
11.7 Strongly Density-Dependent Stage-Transition Rates and Periodic Oscillations 162
11.8 An Example for Multiple Periodic Orbits and Both Supercritical and Subcritical Hopf Bifurcation 166
11.9 Multiple Interior Equilibria, Bistability, and Many Bifurcations for Pure Intrastage Competition 168
Bibliographic Remarks 181

PART 2. STAGE TRANSITIONS AND DEMOGRAPHICS 183

Chapter 12. The Transition Through a Stage 185

12.1 The Sojourn Function 185
12.2 Mean Sojourn Time, Expected Exit Age, and Expectation of Life 187
12.3 The Variance of the Sojourn Time, Moments and Central Moments 189
12.4 Remaining Sojourn Time and Its Expectation 190
12.5 Fixed Stage Durations 197
12.6 Per Capita Exit Rates (Mortality Rates) 199
12.7 Exponentially Distributed Stage Durations 201
12.8 Log-Normally Distributed Stage Durations 202
12.9 A Stochastic Interpretation of Stage Transition 206
Bibliographic Remarks 209

Chapter 13. Stage Dynamics with Given Input 211

13.1 Input and Stage-Age Density 211
13.2 The Partial Differential Equation Formulation 212
13.3 Stage Content and Average Stage Duration 217
13.4 Average Stage Age 219
13.5 Stage Exit Rates 221
13.5.1 The Fundamental Balance Equation of Stage Dynamics 222
13.5.2 Average Age at Stage Exit 224
13.6 Stage Outputs 226
13.7 Which Recruitment Curves Can Be Explained by Cannibalism of Newborns by Adults? 230
Bibliographic Remarks 237

Chapter 14. Demographics in an Unlimiting Constant Environment 239

14.1 The Renewal Equation 240
14.2 Balanced Exponential Growth 241
14.3 The Renewal Theorem: Approach to Balanced Exponential Growth 244

Chapter 15. Some Demographic Lessons from Balanced Exponential Growth 255

15.1 Inequalities and Estimates for the Malthusian Parameter 255
15.2 Average Age and Average Age at Death in a Population at Balanced Exponential Growth. Average Per Capita Death Rate 262
15.3 Ratio of Population Size and Birth Rate 266
15.4 Consequences of an Abrupt Shift in Maternity: Momentum of Population Growth 267
Bibliographic Remarks 270

Chapter 16. Some Nonlinear Demographics 273

16.1 A Demographic Model with a Juvenile and an Adult Stage 274
16.2 A Differential Delay Equation 277
Bibliographic Remarks 279

PART 3. HOST–PARASITE POPULATION GROWTH: EPIDEMIOLOGY OF INFECTIOUS DISEASES 281

Chapter 17. Background 283

17.1 Impact of Infectious Diseases in Past and Present Time 284
17.2 Epidemiological Terms and Principles 289
Bibliographic Remarks 291

Chapter 18. The Simplified Kermack–McKendrick Epidemic Model 293

18.1 A Model with Mass-Action Incidence 293
18.2 Phase-Plane Analysis of the Model Equations. The Epidemic Threshold Theorem 295
18.3 The Final Size of the Epidemic. Alternative Formulation of the Threshold Theorem 297

Chapter 19. Generalization of the Mass-Action Law of Infection 305

19.1 Population-Size Dependent Contact Rates 305
19.2 Model Modification 306
19.3 The Generalized Epidemic Threshold Theorem 307

Chapter 20. The Kermack–McKendrick Epidemic Model with Variable Infectivity 311

20.1 A Stage-Age Structured Model 311
20.2 Reduction to a Scalar Integral Equation 313
Bibliographic Remarks 316

Chapter 21. SEIR (→ S) Type Endemic Models for "Childhood Diseases" 317

21.1 The Model and Its Well-Posedness 318
21.2 Equilibrium States and the Basic Replacement Ratio 321
21.3 The Disease Dynamics in the Vicinities of the Disease-Free and the Endemic Equilibrium: Local Stability and the Interepidemic Period 325
21.4 Some Global Results: Extinction, Persistence of the Disease; Conditions for Attraction to the Endemic Equilibrium 332
Bibliographic Remarks 339

Chapter 22. Age-Structured Models for Endemic Diseases and Optimal Vaccination Strategies 341

22.1 A Model with Chronological Age-Structure 341
22.2 Disease-Free and Endemic Equilibrium: the Replacement Ratio 348
22.3 The Net Replacement Ratio, and Disease Extinction and Persistence 351
22.4 Cost of Vaccinations and Optimal Age Schedules 358

22.5 Estimating the Net Replacement Ratio: Average Duration of Susceptibility and Average Age at Infection. Optimal Vaccination Schedules Revisited 366
Bibliographic Remarks 381

Chapter 23. Endemic Models with Multiple Groups or Populations 383

23.1 The Model 384
23.2 Equilibrium Solutions 388
23.3 Local Asymptotic Stability of Strongly Endemic Equilibria 394
23.4 Extinction or Persistence of the Disease? 399
23.5 The Basic Replacement Matrix, Alias Next-Generation Matrix 404
23.6 The Basic Replacement Ratio as Spectral Radius of the Basic Replacement Matrix 406
23.7 Some Special Cases of Mixing 411
Bibliographic Remarks 416

PART 4. TOOLBOX 419

Appendix A Ordinary Differential Equations 421

A.1 Conservation of Positivity and Boundedness 421
A.2 Planar Ordinary Differential Equation Systems 424
A.3 The Method of Fluctuations 428
A.4 Behavior in the Vicinity of an Equilibrium 433
A.5 Elements of Persistence Theory 436
Bibliographic Remarks 441
A.6 Global Stability of a Compact Minimal Set 442
A.7 Hopf Bifurcation 444
A.8 Perron–Frobenius Theory of Positive Matrices and Associated Linear Dynamical Systems 446
Bibliographic Remarks 451

Appendix B Integration, Integral Equations, and Some Convex Analysis 453

B.1 The Stieltjes Integral of Regulated Functions 453
B.2 Some Elements from Measure Theory 465
B.3 Some Elements from Convex Analysis 472
B.4 Lebesgue–Stieltjes Integration 475
B.5 Jensen's Inequality and Related Material 483
B.6 Volterra Integral Equations 486
B.7 Critical and Regular Values of a Function 490
Bibliographic Remarks 491

Appendix C Some MAPLE Worksheets with Comments for Part 1 493

C.1 Fitting the Growth of the World Population (Figure 3.1) 493
C.2 Periodic Modulation of Exponential Growth in Closed Populations (Figures 3.2 and 3.3) 496

C.3 Fitting Sigmoid Population-Growth Curves (Figures 6.1 and 6.2) 498
C.4 Fitting Bernoulli's Equation to Population Data of Sweden (Figure 6.3) 507
C.5 Illustrating the Allee Effect (Figures 7.2–7.4) 510
C.6 Dynamics of an Aquatic Population Interacting with a Polluted Environment (Figure 10.3E) 513

References 519

Index 537

Preface

Now we only see models, like reflections in a mirror; but then we shall see face to face. Now I only know partially; but then I shall know as fully as I am myself known.

St Paul, 1st letter to the Corinthians, 13:12

Plato's allegory of the cave suggests that we only see the shadows of reality, St Paul declares our knowledge patchwork, and Immanuel Kant distinguishes between the phenomenon we can recognize and the underlying noumenon (Ding an sich), which remains concealed. In more modern terms, we only see reality through models, first through the spontaneous models created by our senses, then through the deliberate models of science and arts, and finally through the meta-models of philosophy and theology. In this scenario, mathematics has its place as a form of symbolic modeling, namely, representation and analysis of reality through mathematical symbols and concepts. Biology, the science of life, has developed its own (nonmathematical) models, but recently the formulation of the dynamics of biological populations in mathematical equations, the analysis of these equations, and the reinterpretation of the results in biological terms has become a valuable source of insight, a selected sample of which has been collected in this book. As with any form of human cognition, mathematical modeling is imperfect, and its strengths and weaknesses must always be accounted for; they are discussed in more detail at the outset of our exposition.

A treatise in mathematical biology can be organized according to biological topics or mathematical concepts and tools. We have chosen the first organizing principle; we start in Part 1 with unstructured single-species population models and then add the most rudimentary stage structure with variable stage duration. In Part 2, we develop the theme of stage transition more fully and apply it to a particular stage, human life, leading to the age-structured models of demographics. In Part 3, we consider the dynamic interplay of host and parasite populations, i.e., epidemic and endemic models. The theme of stage structure recurs in the form of the different stages of infection, and age-structure again plays a role in optimizing vaccination strategies.

While the details can easily be seen from the table of contents, let us highlight some special features of this book.

Realistic models should take account of seasonal and stochastic effects, but the incorporation of either of them makes it much more difficult to obtain mathematical results with a succinct biological interpretation. For that reason they are not dealt with in many mathematical biology books. Stochastic models (as well as statistical methods) are not included in this book either, but seasonality is at least incorporated in our treatment of single-species population dynamics. We develop concepts like birth-date dependent life expectancies and emigration probabilities. For an environment that does not limit population growth, we study the impact of time-periodic immigration on the development of a population the size of which would otherwise decrease. We also touch on the topic of seasonality for density-dependent growth, introducing the concept of asymptotic equality for population sizes.

A great deal of care is spent in Part 1 of this book in deriving the classical single-species models linked to the names of von Bertalanffy, Verhulst, Beverton–Holt, Smith, Ricker, Gompertz, and Allee in the framework of continuous-time models. They are of fundamental importance because they often serve as building blocks in complex models, and knowing their origins is paramount for their correct use. This point is elaborated on when exploring the interaction of an aquatic population with a toxicant, where revisiting the Beverton–Holt growth function allows us to specify the mechanisms by which the toxicant affects individual growth. I prefer to discuss this planar model (and the beautiful and powerful phase-plane methods that come with it) in place of the traditional predator–prey and competition models; excellent treatments of the latter have been and are still being written by people who are more expert in this topic than me. As a second paradigm, a single-species model with two stages is investigated. There is a considerable amount of literature on stage-structured models with variable stage length, but apparently the most elementary model consisting of two ordinary differential equations has been overlooked. Surprisingly, it comes with bifurcation phenomena as complex as they can get in two space dimensions. The two planar models are qualitatively different, as the first can display multiple extinction equilibria (or, mathematically speaking, boundary steady states), while the second can have multiple persistence equilibria (interior steady states).

The second part of the book fully develops the theme of stage structure in an age-dependent context, and demographic concepts such as life expectancy, expectation of remaining life, variance of life length, and their dynamic consequences get a mathematically rigorous treatment. They will later be linked to age-dependent models for infectious childhood diseases.

Reflecting my personal research history, this book contains a relatively large part on host–parasite dynamics, i.e., epidemics and endemics of infectious diseases. A section on the Kermack–McKendrick epidemic model derives estimates from below and above for the final size of an epidemic that cannot be found elsewhere; they are sharp for basic replacement ratios that are either close to 1 or very large. For childhood diseases we derive the formula for the length of the interepidemic period. The relation between the basic replacement ratio and the average age at infection (or, more correctly, average duration of susceptibility) is established without any restrictions on the survivorship. Sexual-disease transmission has highlighted the importance of population heterogeneity and the importance of multi-group models; in this book the relation between overall basic replacement ratio on the one hand and group-to-group basic replacement ratios on the other hand is interpreted in the light of the Perron–Frobenius theory for positive matrices. The interaction between population structure and dynamics has always fascinated me. This book concentrates on age-structure, stage structure and multiple diverse groups; the theory of general physiological structure is still under development and is not included.

This text has grown from a series of courses I taught at Arizona State University over the years, ranging from the junior undergraduate to the advanced graduate level. So, hidden underneath the biological organization is an arrangement of the material according to mathematical difficulty, which the table of contents may not reveal clearly. The first four chapters of Part 1 mostly require one-variable calculus. A course in elementary differential equations may be helpful, but the methods of integrating factors, variation of parameters, separation of variables, and transformation of variables are developed from scratch and applied to more examples and practiced in more exercises than in a typical textbook for elementary differential equations. In Chapters 7–9, some advanced calculus may be needed; Chapter 9, on difference equations, is actually a celebration of the intermediate value theorem. Chapters 10 and 11 require an intermediate or advanced course in ordinary differential equations, notably phase-plane analysis and Hopf bifurcation in the plane. The required theorems are collected in a "toolbox."

Part 2, on stage transition, again starts at the one-variable calculus level and provides ample material to revisit the definition of the integral, guided by a biological interpretation, and its properties. In a few instances, the concept of Stieltjes integration is needed; a full exposition of the Stieltjes integral of regulated functions is presented in the toolbox. In the last section, advanced calculus is again helpful.

The epidemics division of Part 3 (Chapters 17–20) returns to the calculus level; Chapter 21 on childhood diseases requires an intermediate course on ordinary differential equations. Chapter 22, which approaches the question of optimal age-dependent vaccination strategies via linear optimization, has the most demanding background material, some measure theory and Lebesgue integration and elements of convex analysis which are also included in the toolbox. The background material for Chapter 23 (multi-group endemic models) is less demanding, but also less likely to be taught in a graduate course, namely the Perron–Frobenius theory of quasi-positive matrices and their role in systems of ordinary differential equations, the importance of which cannot be overestimated for models with an extended population structure.

The more advanced last sections of both Part 1 and Part 3 are permeated by persistence theory, a dynamical systems method that allows us to determine which parts of an ecological or epidemic system do not become extinct on the one hand and remain bounded on the other hand. In spite of its abstract nature, the concept of a semiflow is used because it is the most natural in this context. Apart from this, the presentation in the toolbox is kept as elementary as possible by using uniform weak persistence as an intermediate step. Uniform weak persistence can often be checked by easy ad hoc considerations and implies the stronger property of uniform strong persistence under natural extra conditions.

Another recurrent technique is the use of time-scale methods, which come in the form of quasi-steady-state simplifications, power expansions, and other approximations. They allow us to arrive at simpler models after starting from more complex ones which are formulated from first principles. They also help us to obtain simple approximation formulas which have a suggestive biological interpretation, in the context of not only differential equations but also of integral equations. I feel that scaling methods can best be learned by ad hoc applications, the paradigms in this being: the derivation of the Beverton–Holt, Ricker, and Allee population-growth functions; the discussion of the kind of recruitment functions that can be explained by cannibalism; the derivation of the formulas for the interval between outbreaks in childhood diseases; the relation between basic replacement ratio and average duration of the susceptible stage; and the complex formation approach to modeling infectious contacts in a multi-group population.

Each chapter has exercises at its end, a few of them with solutions. Bibliographic remarks which provide further background information or make suggestions for further reading are added at the end of a chapter or after several shorter chapters. Chapter C in the toolbox consists of a collection of commented MAPLE worksheets which I used for producing some of the figures in the text.

Acknowledgments

This text grew from courses I taught at Arizona State University (ASU) in six semesters from Fall 1989 to Spring 1999. I thank the students who attended these courses and contributed to the development of the material in various ways, among them Zhilan Feng, Tim Lant, Irakli Loladze, Jimmy Mopecha, Natalia Navarova, and Jinling Yang. The only material never taught in a course at ASU is that in Chapter 11; it evolved from one of the lectures I gave at the Third Winterschool on Population Dynamics in Woudschoten (the Netherlands) in January 2000, which was organized by Odo Diekmann, Hans Heesterbeek, Fleur Kelpin and Bob Kooi.

This book could not have been accomplished without the support of my colleagues in the Department of Mathematics and Statistics, starting with Tom Trotter, chairman at that time, who allowed me to teach one of the earlier courses with only two students, so that the project could get off the ground. The members of our Mathematical Biology group, Steve Baer, Frank Hoppensteadt, Yang Kuang, Hal Smith, and the late Betty Tang, created the nourishing and inspiring environment in which such an endeavor could flourish. Steve Baer spent many hours with me in performing the AUTO calculations for Figures 11.5 and 11.6 in Chapter 11, and Hal Smith read parts of the typescript. John McDonald helped with the convex optimization material in Toolbox B.3, Matt Hassett and Dennis Young with the probability distributions in Chapter 13, and Joe Rody was always willing to answer my MAPLE questions. Renate Mittelmann and Jialong He and their student assistants tended to the various computers which were used over the years in producing this book.

I would like to mention a few things that influenced this book without leaving direct evidence in the contents: the papers by Frank Hoppensteadt and Paul Waltman (1970, 1971) on an epidemic model suggested by Ken Cooke (1967) with which my "Doktorvater", Willi Jäger, made me acquainted as preprints and which got me started as a mathematical biologist; the book by Krasnosel'skii, *Positive Solutions of Operator Equations* (1964), the methods of which made my Ph.D. dissertation fly (they actually play a role in Chapter 23); the discussions on mathematical biology in Willi Jäger's research group, particularly with Wolfgang Alt and Konrad Schumacher; the joint work with Odo Diekmann, Mats Gyllenberg, Henk Heijmans, and Hans Metz on physiologically structured population models stretching over two decades; and, at ASU, the collaboration with Hal Smith.

I am grateful for the many detailed and constructive reviews the typescript has obtained, from Reinhard Bürger, Odo Diekmann, Herb Hethcote, and anonymous reviewers. Many others have helped by commenting on the typescript, providing important references, or through their general support, among them Fred Brauer, Carlos Castillo-Chavéz, John Chance, Jim Collins, Jim Cushing, Patrick DeLeenheer, Klaus Dietz, Zhilan Feng, Bob Kooi, Bas Kooijman, Karl Hadeler, Mimmo Iannelli, Ulrich Krause, Torsten Lindström, Maia Martcheva, Fabio Milner, Roger Nisbet, Diana Thomas, and Pauline van den Driessche. As part of their honors projects, Michael Belisle, Carrie Eggen, and Sumir Mathur read some sections and worked some exercises and tested some of the Maple worksheets for snags.

During most of the years I worked on this book, I had continuing partial support from the Division of Mathematical Sciences at the National Science Foundation, under the program directorate of Mike Steuerwalt.

It must have been in the early 1990s that I first mentioned my germinating book project to Simon Levin and showed him a few sections of an early version. He immediately became supportive, with a view to later publication. I warned him, though, that it might take quite a few more years (it turned out to be more than 10) and asked him when he would lose interest. He answered "probably never," and he never did, eventually guiding the project to Princeton University Press. Thanks also to David Ireland and his colleagues at Princeton University Press, and Sam Clark, copy-editor and typesetter at T&T Productions Ltd, who lent their experience and hard work to the task of bringing the typescript to hard- and soft-cover reality.

I dedicate this book to my wife Adelheid, for her love and her support. During the years in which this book took shape, she kept the family and the house running (along with her own professional career), and shielded me from the time-consuming and annoying disturbances of everyday life. She also checked the typescript for nonmathematical mistakes and typos; all remaining errors are my responsibility, though. I thank my daughters, Ruth and Clara, for their tolerance.

Last, above all, praise to the Holy Spirit, source of knowledge and persistence, for granting the completion of this book.

Mathematics in Population Biology

Chapter One

Some General Remarks on Mathematical Modeling

ἐκ μέρους γὰρ γινώσκομεν

St Paul, 1st letter to the Corinthians, 13:9

Modeling is an attempt to see the wood for the trees. A model is a simplification or abstraction of reality (whatever that is), separating the important from the irrelevant. Actually, modeling is a part of our existence. If we want to be philosophical, we could say that we do not perceive reality as it is, but only realize a model our mind has designed from sensory stimuli and their interpretation. It seems that certain animal species perceive different models of reality which, compared to ours, are based more on hearing and smell than on sight. Many philosophers have had much deeper thoughts on this problem than I present here; following Plato's famous allegory of the cave we may say that we only see the shadows of reality, or, following Kant, that we see the phenomena rather than the noumena. St Paul (1st letter to the Corinthians, 13:9), puts it this way: *We obtain our knowledge in parts, and we prophesy in parts*. St Paul, of course, had a broader and deeper reality in mind than we are concerned with here, but if we apply his statement to a more restricted reality, we may (rather freely) paraphrase it as follows: *We obtain our knowledge from models, and we make our predictions on the basis of models.*

Since, in a wide sense, we are modeling anyway, modeling in the strict sense is the purposeful attempt to replace one model (the so-called "real world," which we typically accept without questioning) by another, deliberate, model which may give us more insight. Essentially, there seem to be two incentives for doing so: either the "real-world" model is too complex to obtain the desired insight and so is replaced by a simpler or more abstract one. Or the "real-world" model does not allow certain experiments for ethical, practical or other reasons and is replaced by a model in which all kinds of changes can be readily made and their consequences studied without causing harm.

The word *model* presumably traces back to the Latin word *modulus*, which means "little measure" (Merriam-Webster, 1994), alluding to a small-scale physical representation of a large object (e.g., a model airplane).

Mathematical modeling uses symbolic rather than physical representations, unleashing the power of mathematical analysis to increase scientific understanding. It can conveniently be divided into three stages (cf. Lin, Segel, 1974, 1988).

(i) *Model formulation*: the translation of the scientific problem into mathematical terms.

(ii) *Model analysis*: the mathematical solution of the model thus created.

(iii) *Model interpretation and verification*: the interpretation of the solution and its empirical verification in terms of the original problem.

All three steps are important and useful. Already the first step—model formulation—can lead to considerable insight. For building a mathematical model, one needs clear-cut assumptions about the operating mechanisms, and it often turns out to be an unpleasant surprise that the old model—the "real world"—is far less understood than one thought. In many cases the modeling procedure—at least if one chooses parameters that are meaningful—already teaches what further knowledge is needed in order to apply the mathematical model successfully; the model analysis and its interpretation help to determine to what extent and precision new information and new data have to be collected.

Analytic and numerical tools allow the extrapolation of present states of the mathematical model into the future and, sometimes, into the past. Assumptions, initial states, and parameters can easily be changed and the different outcomes compared. So, models can be used to identify trends or to estimate uncertainties in forecasts. Different detection, prevention and control strategies can be tested and evaluated.

While the model analysis may require sophisticated analytic or computational methods, mathematical modeling ideally leads to conceptual insight, which can be expressed without elaborate mathematics.

A model is, purposefully, a simplification or abstraction; very often it is an oversimplification or overabstraction. Insight obtained from a model should be checked against empirical evidence and common sense. It can also be checked against insight from other models: how much does the model's behavior depend on the degree of complexity, on the form of the model equations, on the choice of the parameters? Dealing with a concrete problem, a modeler should work with a whole scale of models starting from one which is as simple as possible to obtain some (hopefully) basic insight and then adding complexity and checking whether the insight is confirmed. Starting with a complex model has the risk that the point will be missed because it is obscured by all the details.

The use of a whole range of models also educates the modeler on how critically qualitative and quantitative results depend on the assumptions one has made. A mathematical modeler should become a guardian keeping him/herself and others from jumping to premature conclusions.

When modeling concrete phenomena, there is typically a dilemma between incorporating enough complexity (or realism) on the one hand and keeping the model tractable on the other. In some cases the modeler feels forced to add a lot of detail to her/his model in order to grasp all the important features and ends up with a huge number of unknown parameters, which may never be known even approximately and cannot be determined by fitting procedures either. Such a mathematical model will be of limited value for quantitative and maybe even qualitative forecasts, but still has the other benefits described above.

Mathematical modeling has its place in all sciences. This book, according to the interests of the author, concerns mathematical models in the biosciences, or rather a very small area within the biosciences, namely population biology. Another restriction is the one to deterministic models (as opposed to stochastic models), which neglect the influence of random events. To some degree one can dispute whether stochastic models are more realistic than deterministic models; there is still the possibility that everything is deterministic, but just incredibly complex. In this case, stochasticity would simply be a certain way to deal with the fact that there are many factors we do not know. Stochastic models are theoretically more satisfying as they count individuals with integer numbers, while deterministic models, usually differential or difference equations, have to allow population sizes that are not integers. (This actually is of no concern if population size is modeled as biomass and not as number of individuals, but the latter is not always appropriate.) While a typical tool of deterministic-model analysis consists of discussing large-time limits, stochastic models take account of the truism that nothing lasts forever and make it possible to analyze the expected time until extinction—a concept that has no counterpart in deterministic models. In many cases, deterministic models can theoretically be justified as approximations of stochastic models for large populations sizes; however, the population size needed to make the approximation good enough may be unrealistically large. Nevertheless, deterministic models have the values which we described above, as long as one keeps their limitations in mind. The latter particularly concerns predictions, which are of very limited use in this uncertain world if no confidence intervals for the predicted phenomena are provided.

Bibliographic Remarks

Other discussions of mathematical modeling can be found in Lin, Segel (1974, 1988), Hethcote, Van Ark (1992), Kooijman (2000), and Kirkilionis et al. (2002). Section 1.1 in Lin and Segel's *On the Nature of Applied Mathematics* more generally speaks about the scope, purpose, and practice of applied mathematics (also in comparison with pure mathematics and theoretical science), while Section 1.6 in Hethcote, Van Ark (1992), *Purposes and limitations of epidemiological modeling*, refers to the modeling of infectious diseases and of the

AIDS epidemic in particular. Their respective remarks, however, easily specialize or generalize to mathematical modeling. A "playful" approach to the role of modeling and mathematics in the sciences can be found in Sigmund (1993, Chapters 1, 3, and 9). The idea that a concrete problem should be addressed by a chain of models has been realized in a textbook by Mesterton-Gibbons (1992), who calls this the layered approach. Section 1.2 in Kooijman (2000) shares the philosophical touch of this chapter; the outlooks are rather similar for modeling itself, but quite different in the meta-modeling aspects. For a discussion of deterministic versus stochastic models see the papers by Nåsell (n.d.) and Section 1.6 in the book by Gurney, Nisbet (1998). Genetic and evolutionary aspects of population biology are not touched on at all in this book, and I refer to the books by Bürger (2000), Charlesworth (1980, 1994), Farkas (2001), and Hofbauer, Sigmund (1988, 1998) as sources for their mathematical modeling and analysis.

PART 1

Basic Population Growth Models

Chapter Two

Birth, Death, and Migration

In a natural environment, the dynamics of one species can hardly be isolated from those of other species. The particular species we are interested in may depend on others as resources or may serve as a resource itself (prey–predator relation), and may compete or cooperate with other species in the quest for food and habitat. Nevertheless, dividing to conquer, mathematical modeling first concentrates on one population and expresses the change of its size in terms of concepts like birth rates, mortality rates, and emigration and immigration rates which take the form of either population rates or per capita rates.

2.1 The Fundamental Balance Equation of Population Dynamics

As a first goal, mathematical models try to describe the size of a population as a function of time. A second goal may be to describe the distribution of the population over space or according to important individual characteristics like age, body size, protein content, etc., but we ignore population structure at this point. So, let $N(t)$ denote the size of the single-species population we are focusing on at time t. Depending on the species, the population size may be the number of individuals or the collective biomass of the population. It can also be the average spatial density of the population or, for aquatic populations, the average concentration per water unit.

The rate of change of the population size is given by the time derivative $N'(t) = \mathrm{d}N(t)/\mathrm{d}t$. If we consider a species of mammal, we may like to have the population size to be given in numbers of individuals. Then the size can change by four processes: births, deaths, emigration, and immigration. If there is no emigration or immigration, we speak of a *closed population*, otherwise of an *open population*. The *fundamental balance equation of population dynamics* takes these four processes into account,

$$N'(t) = B(t) - D(t) + I(t) - E(t), \tag{2.1}$$

where, at time t, $B(t)$ denotes the population birth rate, $D(t)$ the population death rate, $I(t)$ the immigration rate, and $E(t)$ the emigration rate. More precisely, $B(t)$ is the number of births per time unit, at time t, $D(t)$ is the number of deaths per unit of time, etc. This balance equation can equivalently be written as

$$N(t)-N(r)=\int_r^t B(s)\,\mathrm{d}s-\int_r^t D(s)\,\mathrm{d}s+\int_r^t I(s)\,\mathrm{d}s-\int_r^t E(s)\,\mathrm{d}s,\qquad t>r.$$

As an illustration, we give the example of Sweden in 1988 (Preston et al., 2001, Box 1.1, p. 3). There t is January 1, 1989, and r is January 1, 1988, and

$$N(t)=8\,461\,554,\qquad N(r)=8\,416\,599.$$

The number of births and deaths in 1988 are

$$\int_r^t B(s)\,\mathrm{d}s=112\,080,\qquad \int_r^t D(s)\,\mathrm{d}s=96\,756,$$

while the number of immigrants and emigrants are

$$\int_r^t I(s)\,\mathrm{d}s=51\,092,\qquad \int_r^t E(s)\,\mathrm{d}s=21\,461.$$

If the individuals are very small, it may be better to describe the population size in terms of biomass. Then B would be the biomass acquisition rate through food uptake and D the biomass loss rate due not only to death but also to nonfatal starvation.

As for births, deaths, and emigration, it also makes sense to speak about per capita rates. The per capita birth rate, $\beta(t)$, is the average number of offspring born to one typical individual per unit of time, at time t. At this point, however, the per capita birth rate is just a theoretical construct. In sexually reproducing populations, for example, only females give birth, and this only in a certain age-window, and only after mating. So, in order to determine the per capita birth rate, one would need to know the sex ratio, the age-distribution of the female population, and the mating probability in order to determine the per capita birth rate as defined above. To complicate matters even more, all these ingredients may change with time.

As for the per capita death or mortality rate, $\mu(t)$, even the definition takes substantial effort. Let $\Pi(t,r)$, $t\geqslant r$, denote the probability of surviving from time r to time t, $\Pi(r,r)=1$, $0\leqslant\Pi(t,r)\leqslant 1$. The probability of dying from time r to time t is $1-\Pi(t,r)$. The *per capita mortality rate* $\mu(t)$ is defined as

$$\mu(t)=\lim_{h\to 0}\frac{1-\Pi(t+h,t)}{h},$$

assuming that these limits exist. The number of people dying from time t to time $t+h$ is $N(t)[1-\Pi(t+h,t)]$, so the population death rate at time t is

$$D(t)=\lim_{h\to 0}\frac{N(t)[1-\Pi(t+h,t)]}{h}=\mu(t)N(t).$$

Π can be recovered from μ if the probability of surviving from r to t is independent from the past up to time r, in particular if mortality does not depend on age. Then the following multiplicative rule holds:

$$\Pi(t,r) = \Pi(t,s)\Pi(s,r), \quad t \geqslant s \geqslant r, \quad \Pi(r,r) = 1.$$

This implies

$$\frac{\Pi(t+h,r) - \Pi(t,r)}{h} = -\frac{1-\Pi(t+h,t)}{h}\Pi(t,r),$$

and the following differential equation holds for Π,

$$\frac{\mathrm{d}}{\mathrm{d}t}\Pi(t,r) = -\mu(t)\Pi(t,r), \quad \Pi(r,r) = 1.$$

We fix r and set $p(t) = \Pi(t,r)$. In terms of p, the differential equation can be solved by separation of variables,

$$-\mu(t) = \frac{p'(t)}{p(t)} = \frac{\mathrm{d}}{\mathrm{d}t}\ln p(t), \quad p(r) = 1.$$

We integrate both sides and use the fundamental theorem of calculus,

$$-\int_r^t \mu(s)\,\mathrm{d}s = \ln p(t) - \ln p(r) = \ln p(t) - \ln 1.$$

Exponentiating both sides provides

$$\Pi(t,r) = p(t) = \exp\left(-\int_r^t \mu(s)\,\mathrm{d}s\right). \tag{2.2}$$

2.2 Birth Date Dependent Life Expectancies

For later use, let us calculate the life expectancy, $L(r)$, of an individual born at time r. Then $r + L(r)$ is the expected time of death. We first assume that there is a maximum finite life span strictly less than some number c. This means that the individual born at time r will be dead by time $r+c$, i.e., $p(r+c) = \Pi(r+c,r) = 0$. We choose a partition $r = t_1 < \cdots < t_{n+1} = c + r$ of the interval $[r, r+c]$ such that $t_{j+1} - t_j$ is very small for all j. Observe that $p_j = p(t_{j+1}) - p(t_j) = \Pi(t_{j+1},r) - \Pi(t_j,r)$ is the probability that the individual dies between t_j and t_{j+1}. In this case, the time of death is some number s_j between t_j and t_{j+1}. The expected time of death is given as the limit of Riemann–Stieltjes sums,

$$L(r) + r = \lim \sum_{j=1}^{n} s_j p_j = \lim \sum_{j=1}^{n} s_j[p(t_j) - p(t_{j+1})],$$

where the lengths of all intervals $[t_j, t_{j+1}]$ tend to 0. Taking the sums apart and changing the index of summation we obtain

$$L(r) + r = \lim\left(\sum_{j=1}^{n} s_j p(t_j) - \sum_{j=2}^{n+1} s_{j-1} p(t_j)\right).$$

We set $s_0 = r$ and $s_{n+1} = r + c$. Then $r = s_0 < \cdots < s_{n+1} = r + c$ is a partition of $r + c$ and $L(r) + r$ can be expressed as a limit of Riemann sums, i.e., as a Riemann integral,

$$\begin{aligned} L(r) + r &= \lim \sum_{j=1}^{n+1} (s_j - s_{j-1}) p(t_j) - s_{n+1} p(t_{n+1}) + s_0 p(t_1) \\ &= \int_r^{r+c} p(t)\,\mathrm{d}t - (r+c)p(r+c) + rp(r). \end{aligned}$$

Since $p(r) = 1$ and $p(r + c) = 0$, by substituting $t = s + r$,

$$L(r) = \int_0^c p(s+r)\,\mathrm{d}s = \int_0^c \Pi(s+r, r)\,\mathrm{d}s.$$

If there is no finite maximum life span, we let $c \to \infty$, and $L(r)$ is the improper Riemann integral

$$L(r) = \int_0^\infty \Pi(s+r, r)\,\mathrm{d}s.$$

In terms of per capita mortality rates μ, by (2.2),

$$L(r) = \int_0^\infty \exp\left(-\int_r^{r+s} \mu(t)\,\mathrm{d}t\right)\mathrm{d}s = \int_0^\infty \exp\left(-\int_0^s \mu(t+r)\,\mathrm{d}t\right)\mathrm{d}s.$$

In the important special case that the per capita mortality rate is constant, $L(r)$ does not depend on r and the life expectancy is given by

$$L = \int_0^\infty \mathrm{e}^{-\mu s}\,\mathrm{d}s = \frac{1}{\mu}. \tag{2.3}$$

Mathematically, there is not much difference between leaving the population by death or by emigration. The per capita emigration rates are denoted by $\eta(t)$ and can be defined from the probabilities $\Xi(t, r)$ of not emigrating from time r to time t. A derivation similar to the one above gives that the mean sojourn time of an individual, born at time r, in this particular population (death neglected) is given by

$$\int_0^\infty \exp\left(-\int_0^s \eta(t+r)\,\mathrm{d}t\right)\mathrm{d}s,$$

and by $1/\eta$ if the per capita emigration rate is constant. If both death and emigration are regarded and the per capita mortality and emigration rates are constant, the mean sojourn time in the population is then $1/(\mu + \eta)$, because $\mu + \eta$ is the rate of leaving the population by either death or emigration.

2.3 The Probability of Lifetime Emigration

Again for later use, let us determine the probability that an individual born at time r will emigrate during its lifetime. Fix the birth date r and let $\xi(t) = \Xi(t, r)$ denote the probability of not having emigrated until time t. The probability of emigrating in the time interval between t_{j-1} and t_j, where $t_j > t_{j-1} \geqslant r$, is $\xi(t_{j-1}) - \xi(t_j)$. So, the unconditional probability of emigrating between those times is $[\xi(t_{j-1}) - \xi(t_j)]p(s_j)$ with some s_j between t_{j-1} and t_j. Here $p(s)$ again denotes the probability to be still alive at time s. We assume that ξ is continuously differentiable as is the case when we have a continuous emigration rate $\eta = \xi'$. By the mean value theorem, there exists some $\tilde{s}_j$ between t_{j-1} and t_j such that the probability of emigrating between these times is $-\xi'(\tilde{s}_j)(t_j - t_{j-1})p(s_j)$. The probability of emigrating before some time $T > r$ is obtained by choosing a partition $r = t_0 < \cdots < t_n = T$, by summing these probabilities, and letting the number of partition points tend to infinity:

$$\lim_{n\to\infty} \sum_{j=1}^{n} (-1)\xi'(\tilde{s}_j)p(s_j)(t_j - t_{j-1}) = -\int_r^T \xi'(t)p(t)\,\mathrm{d}t.$$

The probability of emigrating during one's lifetime, $P_{\text{out}}(r)$, is obtained by taking the limit $T \to \infty$. If we have continuous per capita mortality and emigration rates, $\mu(\cdot)$ and $\eta(\cdot)$, then

$$p(t) = \exp\left(-\int_r^t \mu(s)\,\mathrm{d}s\right) \quad \text{and} \quad \xi(t) = \exp\left(-\int_r^t \eta(s)\,\mathrm{d}s\right)$$

by (2.2). So, the probability of emigrating during one's lifetime is

$$P_{\text{out}}(r) = \int_r^\infty \eta(t)\exp\left(-\int_r^t [\eta(s) + \mu(s)]\,\mathrm{d}s\right)\mathrm{d}t.$$

If η and μ are constant, so is this probability, and

$$P_{\text{out}} = \frac{\eta}{\eta + \mu} \tag{2.4}$$

is the product of the mean sojourn time in the population and the constant per capita emigration rate.

The following relations hold between population and per capita rates:

$$B(t) = \beta(t)N(t), \qquad D(t) = \mu(t)N(t), \qquad E(t) = \eta(t)N(t).$$

It makes little sense in this context to speak about a per capita immigration rate; for it may well happen that $N(t) = 0$ and $I(t) > 0$, if a species invades a formerly empty habitat.

We mention that these considerations also apply if we only consider the part of a population in a certain stage (e.g., larvae). The birth rate has then to be interpreted as a stage influx rate (stage input, the hatching rate for larvae, for example) and the emigration rate as the stage exit rate (stage output, the pupation rate for larvae, for example). For constant per capita rates, $1/\eta$ gives the average stage length (deaths disregarded), and $1/(\mu + \eta)$ the stage sojourn time, while $\eta/(\eta + \mu)$ is the probability of surviving the stage and $\mu/(\eta + \mu)$ the probability of dying during the stage (cf. Chapter 11).

Chapter Three

Unconstrained Population Growth for Single Species

If we look at a single species exclusively, the population rates in the fundamental balance equation of population dynamics would only depend on the size of the respective population, N, and on the time variable. Although in most cases the dynamics of this population cannot be isolated from the dynamics of others, it is still useful to look at single-species dynamics for a number of reasons:

- Single-species dynamics occur in laboratory environments.
- Sometimes it may be possible to blend other species into the environment.
- Understanding one-species dynamics helps us to understand the more complex multi-species dynamics.

There are two separate cases to consider: unconstrained population growth, where the per capita rates depend only on environmental conditions, i.e., they are functions of time only or are just constant; and density-dependent population growth, where the per capita rates are affected by the population size, N. The first case occurs if a species has unlimited resources and is not subject to predators or competitors; the per capita rates are then just functions of time t or may even be constant.

3.1 Closed Populations

If there is neither immigration nor emigration, equation (2.1) takes the form of a linear ordinary differential equation,

$$\left.\begin{aligned} N' &= \rho(t)N, \\ \rho(t) &= \beta(t) - \mu(t), \\ N(t_0) &= N_0, \end{aligned}\right\} \tag{3.1}$$

where N_0 is the population size at some time t_0 (e.g., the time when we start observing the population). N_0 and t_0 are called the *initial data* of our model. We assume that the per capita birth rate β and per capita mortality rate μ are continuous. Their difference ρ is sometimes called the *intrinsic* (per capita) *growth rate*.

Equation (3.1) can be solved by a method called *separation of variables*, where we collect all terms containing the dependent variable N on one side of the equation:

$$\rho(t) = \frac{N'(t)}{N(t)} = \frac{\mathrm{d}}{\mathrm{d}t} \ln N(t).$$

We integrate both sides and use the fundamental theorem of calculus,

$$\ln N(t) - \ln N(t_0) = \int_{t_0}^{t} \rho(s)\,\mathrm{d}s = \int_0^t \rho(s)\,\mathrm{d}s - \int_0^{t_0} \rho(s)\,\mathrm{d}s.$$

Exponentiating both sides provides

$$\left.\begin{aligned} N(t) &= N_0 \frac{Q(t)}{Q(t_0)}, \\ Q(t) &= \exp\left(\int_0^t \rho(s)\,\mathrm{d}s\right). \end{aligned}\right\} \tag{3.2}$$

Q is called a *fundamental solution* of (3.1).

Notice that, in the formula for Q, the lower integration limit 0 can be replaced by a fixed, but arbitrary, number r_0, which can be chosen at convenience. Actually, Q can be chosen as the exponential of any anti-derivative of ρ.

If the per capita birth and death rates are constant, so is $\rho(t) = \bar{\rho}$, and we obtain

$$N(t) = N(t_0)\mathrm{e}^{\bar{\rho}(t-t_0)}.$$

This is the notorious exponential growth, in which a population doubles (or halves) its size in a fixed amount of time (see Exercise 3.1) if $\bar{\rho} > 0$ (or $\bar{\rho} < 0$, respectively). The British economist Thomas Malthus (1798) was one of the first to be concerned about its consequences, and the parameter $\bar{\rho}$ is often called the *Malthusian parameter* in his honor. Other names for $\bar{\rho}$ are *intrinsic rate of natural increase* or *intrinsic growth constant*. In Figure 3.1 we have fitted a line and an exponential curve to the world population data from 1950 to 1985 taken from Keyfitz, Flieger (1990, p. 105). The doubling time for the exponential curve is 36.34 years. For details see the MAPLE worksheet in Toolbox C.

3.1.1 The Average Intrinsic Growth Rate for Periodic Environments

Constant per capita birth and mortality rates are highly unlikely for most natural populations; rather they are usually subject to seasonal fluctuations. A more realistic, though still rather idealistic, assumption is a periodic time-dependence where $\beta(t)$ and $\mu(t)$, and so $\rho(t)$, are periodic functions of t (with the same period), i.e.,

$$\rho(t) = \rho(t+p), \quad t \in \mathbb{R}.$$

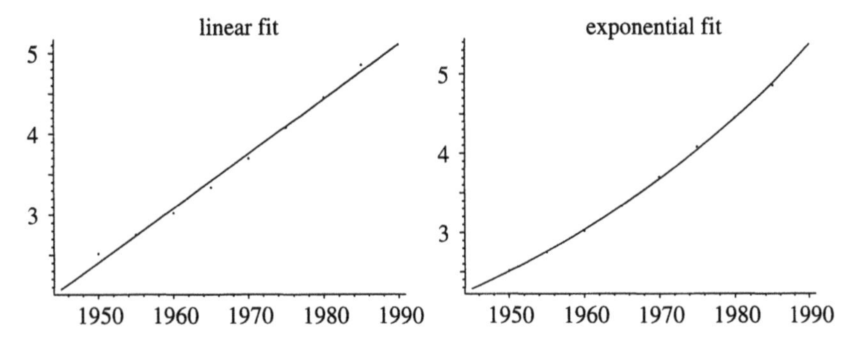

Figure 3.1. Linear and exponential fits to the growth of the world population (in billions) from 1950 to 1985 according to data from Keyfitz, Flieger (1990, p. 105).

The fixed number p is the period of β, μ, and ρ. In many applications, p will be one year.

Example. Let $\rho(t) = 2 + \sin t$.

We could find a solution of (3.1) by substituting ρ into (3.2). But let us instead do it from scratch. Separating the variables we obtain

$$2 + \sin t = \frac{N'}{N} = \frac{\mathrm{d}}{\mathrm{d}t} \ln N.$$

Integrating both sides we have

$$\ln N(t) = 2t - \cos t + c$$

with an integration constant c. Exponentiating,

$$N(t) = k\mathrm{e}^{2t}\mathrm{e}^{-\cos t}, \qquad k = \mathrm{e}^{c}.$$

So, the population size exhibits exponential growth with rate $\bar{\rho} = 2$, modulated by the periodic function $\mathrm{e}^{-\cos t}$. N, given in this form with an arbitrary constant k, is called a *general solution*. The constant k can be adjusted to make N satisfy the appropriate initial data. For instance, if $N(1) = 5$, then, by substitution into the general solution, $5 = k\mathrm{e}^{2}\mathrm{e}^{-\cos 1}$. Solving for k and fitting it into the general solution,

$$N(t) = 5\mathrm{e}^{2(t-1)}\mathrm{e}^{\cos 1 - \cos t}.$$

The behavior of the population size in this example, periodically modulated exponential growth, makes us wonder whether this is a general phenomenon. More precisely, we ask ourselves the following question.

If ρ is periodic, is there a number $\bar{\rho}$ and a periodic function q (with same period p) such that

$$N(t) = N(t_0)\mathrm{e}^{(t-t_0)\bar{\rho}}\left(\frac{q(t)}{q(t_0)}\right)?$$

By (3.2), $N(t) = N(t_0)(Q(t)/Q(t_0))$. We look for $\bar{\rho} \in \mathbb{R}$ such that $q(t) = e^{-\bar{\rho}t} Q(t)$ is p-periodic. This is equivalent to $\sigma(t) = \ln q(t)$ being p-periodic. By the form of Q in (3.2),

$$\sigma(t) = \int_0^t \rho(r)\, dr - \bar{\rho}t.$$

Obviously, $\sigma(0) = 0$. A minimum requirement for σ to be p-periodic is $\sigma(p) = 0$ as well. So, we set

$$\bar{\rho} = \frac{1}{p} \int_0^p \rho(r)\, dr.$$

With this choice, σ inherits p-periodicity from ρ and so does $q(t) = e^{\sigma(t)}$. Indeed, $\sigma'(t) = \rho(t) - \bar{\rho}$, so $\sigma'(t+p) - \sigma'(t) = 0$ for all t. Thus $\sigma(t+p) - \sigma(t)$ is a constant function of t and $\sigma(t+p) - \sigma(t) = \sigma(p) - \sigma(0) = 0$ for all t.

Notice that $q(t) = e^{\sigma(t)}$ is a bounded, strictly positive function.

Let us summarize. For periodic ρ, we define $\bar{\rho}$ to be the *time average* of ρ,

$$\bar{\rho} = \frac{1}{p} \int_0^p \rho(s)\, ds. \tag{3.3}$$

$\bar{\rho}$ is the *average intrinsic growth rate* for periodic ρ. We have the following result.

Theorem 3.1. *Consider a closed population, $N' = \rho(t)N$, where ρ is a p-periodic function and $\bar{\rho}$ its time average. Then there exists some p-periodic function q such that the solution of N of (3.1) satisfies*

$$N(t) = N(t_0) e^{(t-t_0)\bar{\rho}} \frac{q(t)}{q(t_0)}.$$

Remark 3.2. In other words, the fundamental solution Q in (3.2) has the form $Q(t) = e^{\bar{\rho}t} q(t)$ with a p-periodic function q. q is continuously differentiable and strictly positive.

With some justification one can again call $\bar{\rho}$ the *Malthusian parameter* if ρ is periodic. Notice the following relation from Theorem 3.1,

$$N(t_0 + mp) = N(t_0) e^{m\bar{\rho}}, \quad m \in \mathbb{Z},$$

which actually holds for all times t_0. This means that the population grows geometrically from season to season with growth factor $e^{\bar{\rho}}$. Theorem 3.1 is a special case of *Floquet theory* in the context of which $\bar{\rho}$ is called a Floquet exponent and $e^{\bar{\rho}}$ a *Floquet multiplier*. We call

$$Q(t) = e^{\bar{\rho}t} q(t)$$

the *Floquet representation* of the fundamental solution Q.

Illustration

We consider a population subject to a periodic environment with a period of one year. We want to mimic the situation in which the per capita birth rate has a much more pronounced seasonality than the per capita mortality rate; in other words, the reproductive season is relatively short. Furthermore, we want the birth rate to be at its maximum when the mortality rate is at its minimum and vice versa,

$$\tilde{\beta}(t) = (1 + \cos(2\pi t))^8, \qquad \tilde{\mu}(t) = 1 + 0.5\sin(2\pi t - \pi/2)$$

(see Figure 3.2A). Both per capita rates have been normalized such that their averages are 1.

We combine these normalized per capita birth and death rates into an intrinsic growth rate ρ in three different ways as follows.

(A) $\rho_{\mathrm{A}}(t) = \tilde{\beta}(t) - \tilde{\mu}(t), \qquad \bar{\rho}_{\mathrm{A}} = 0.$

(B) $\rho_{\mathrm{B}}(t) = \tilde{\beta}(t) - 0.7\tilde{\mu}(t), \quad \bar{\rho}_{\mathrm{B}} = 0.3.$

(C) $\rho_{\mathrm{C}}(t) = \tilde{\beta}(t) - 1.9\tilde{\mu}(t), \quad \bar{\rho}_{\mathrm{C}} = -0.9.$

Figure 3.3 illustrates Theorem 3.1 for these various choices of ρ. If $\bar{\rho} = 0$, the population size remains in periodic motion; for $\bar{\rho} \neq 0$ the population increases or decreases exponentially with periodic modulation. The MAPLE worksheet used to produce Figures 3.2 and 3.3 can be found in Toolbox C.

3.1.2 The Average Intrinsic Growth Rate for Nonperiodic Environments

If ρ is not necessarily periodic, we may consider the asymptotic time average

$$\bar{\rho} = \lim_{t\to\infty} \frac{1}{t} \int_0^t \rho(s)\,\mathrm{d}s \tag{3.4}$$

provided that this limit exists. If ρ is periodic, this definition is consistent with our previous one. If $t > p$, choose $m \in \mathbb{N}$ such that $t \in [mp, (m+1)p)$. Obviously, $t \to \infty$ if and only if $m \to \infty$. Let $\bar{\rho}$ be defined by (3.3). We split up the integral,

$$\frac{1}{t}\int_0^t \rho(s)\,\mathrm{d}s = \frac{1}{t}\sum_{j=0}^{m-1}\int_{jp}^{(j+1)p} \rho(s)\,\mathrm{d}s + \frac{1}{t}\int_{mp}^{t} \rho(s)\,\mathrm{d}s.$$

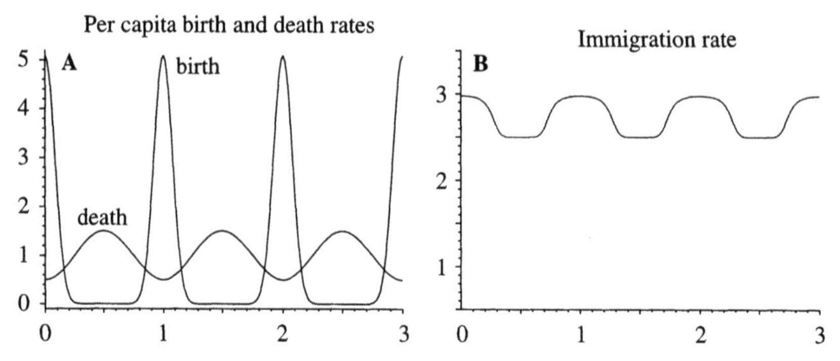

Figure 3.2. Fictitious periodic per capita birth, death, and immigration rates for a population with short reproductive season. The period is one year with the year beginning at the time when the birth rate is at its maximum. The immigration rate has its maximum at the same time, while the death rate is at its minimum. The birth and death rates have been normalized such that their averages are 1.

Using substitution and the periodicity of ρ,

$$\frac{1}{t}\int_0^t \rho(s)\,\mathrm{d}s = \frac{1}{t}\sum_{j=0}^{m-1}\int_0^p \rho(s+jp)\,\mathrm{d}s + \frac{1}{t}\int_0^{t-mp} \rho(s+mp)\,\mathrm{d}s$$
$$= \frac{mp}{t}\bar{\rho} + \frac{1}{t}\int_0^{t-mp} \rho(s)\,\mathrm{d}s.$$

Since $mp \leqslant t < mp + p$, we have $(mp/t) \to 1$ and $(1/t)\int_0^{t-mp} \rho(s)\,\mathrm{d}s \to 0$ as $t \to \infty$, which shows that $\bar{\rho}$ in (3.4) is the same as in (3.3). So, we have proved the following result.

Proposition 3.3. *The asymptotic time average of a periodic function equals its time average.*

If the limit in formula (3.4) exists, we call $\bar{\rho}$ the *average intrinsic growth rate.*

Theorem 3.4. *Assume that the average intrinsic growth rate $\bar{\rho}$ in (3.4) is defined. Then the population increases exponentially if $\bar{\rho} > 0$, and decreases exponentially if $\bar{\rho} < 0$.*

Proof. Let $\bar{\rho} > 0$. Then there exists some $r > 0$ such that

$$\frac{1}{t}\int_0^t \rho(s)\,\mathrm{d}s \geqslant \tfrac{1}{2}\bar{\rho}, \quad t \geqslant r.$$

By (3.2),

$$Q(t) \geqslant \mathrm{e}^{\bar{\rho}t/2}, \quad t \geqslant r,$$

and $N(t)$ increases exponentially. The case $\bar{\rho} < 0$ is left as an exercise. □

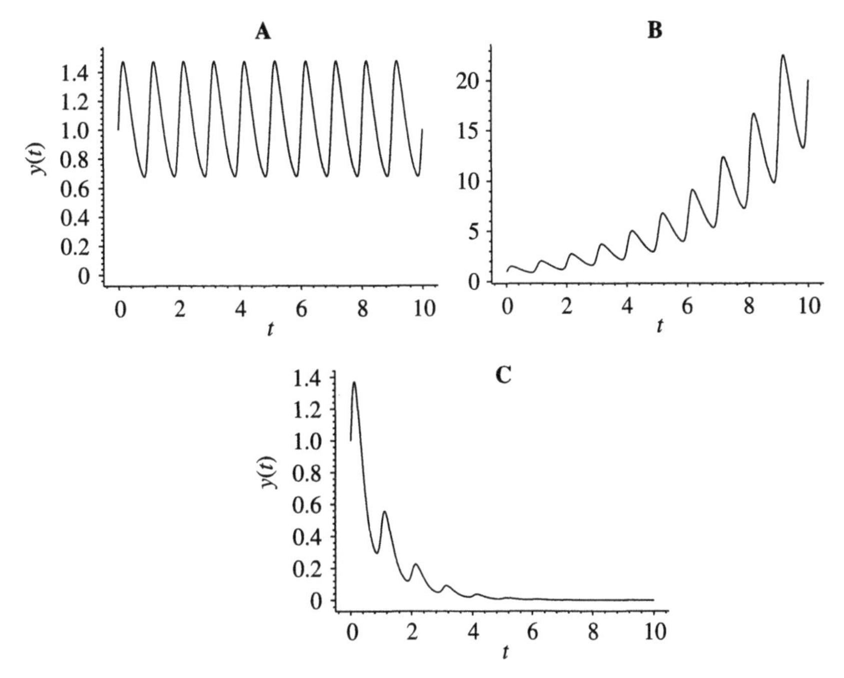

Figure 3.3. Dynamics of closed populations under various combinations of the per capita birth and death rates of Figure 3.2. (**A**) $\bar{\rho} = 0$. (**B**) $\bar{\rho} = 0.3$. (**C**) $\bar{\rho} = -0.9$.

3.2 Open Populations

If immigration and emigration are included, we have

$$\left.\begin{aligned} N'(t) &= \rho(t)N + I(t), \\ \rho(t) &= \beta(t) - \mu(t) - \eta(t). \end{aligned}\right\} \tag{3.5}$$

The differential equation in (3.5) is called a *nonhomogeneous linear* equation, while (3.1) is called a *homogeneous linear* equation.

The equations in (3.5) can be solved by using an *integrating factor*. We collect the N terms on one side of the equation and multiply by some function $F(t)$ (the integrating factor):

$$F(t)N'(t) - \rho(t)F(t)N(t) = F(t)I(t).$$

If

$$F'(t) = -\rho(t)F(t),$$

then the product rule allows us to write this equation as

$$\frac{\mathrm{d}}{\mathrm{d}t}(F(t)N(t)) = F(t)I(t).$$

The first equation is of the form (3.1), so a special solution is

$$F(t) = \exp\left(-\int_0^t \rho(s)\,\mathrm{d}s\right) = \frac{1}{Q(t)}.$$

Integrating the second equation gives

$$F(t)N(t) - F(t_0)N(t_0) = \int_{t_0}^t F(s)I(s)\,\mathrm{d}s.$$

We solve for $N(t)$:

$$\left.\begin{aligned} N(t) &= N_0\frac{Q(t)}{Q(t_0)} + \int_{t_0}^t I(s)\frac{Q(t)}{Q(s)}\,\mathrm{d}s, \\ Q(t) &= \exp\left(\int_0^t \rho(r)\,\mathrm{d}r\right). \end{aligned}\right\} \tag{3.6}$$

Again, in the formula for the fundamental solution Q, the lower integration limit 0 can be replaced by a fixed, but arbitrary, number r_0 which can be chosen at convenience. In other words, we can choose Q as the exponential of any anti-derivative of ρ.

Formula (3.6) is often called the *variation of parameters formula* or the *variation of constants formula* because it can also be found by a method called variation of parameters. The phenomenon that the nonhomogeneous equation can be solved in this way, in terms of the fundamental solution of the homogeneous equation and the nonhomogeneous term I, is also called *Duhamel's principle*.

We notice from formula (3.6) that a solution to (3.5) is uniquely determined by its initial data $N = N_0$ at $t = t_0$.

Example.

$$N' = -\frac{1}{t^2}N + \frac{1}{t^3}, \qquad N(1) = N_0.$$

Here the immigration rate tends to 0 as $t \to \infty$, and the per capita death and emigration rates together are larger than the per capita birth rate, though the difference tends to 0 as well.

As $\rho(t) = -1/t^2$ is not defined at $t = 0$, we do not use the formula for Q in (3.6), but take Q as the exponential of a convenient anti-derivative of ρ, $Q(t) = \mathrm{e}^{1/t}$. Then

$$N(t) = N_0\mathrm{e}^{(1/t)-1} + \mathrm{e}^{1/t}\int_1^t \frac{1}{s^3}\mathrm{e}^{-1/s}\,\mathrm{d}s.$$

Substituting $s = -1/r$,

$$N(t) = N_0\mathrm{e}^{(1/t)-1} + \mathrm{e}^{1/t}\int_{-1}^{-1/t}(-1)r\mathrm{e}^r\,\mathrm{d}s.$$

Finding an anti-derivative for re^r (integrating by parts if we cannot guess it),

$$N(t) = N_0 e^{(1/t)-1} + e^{1/t}[-re^r + e^r]_{r=-1}^{r=-1/t}$$
$$= N_0 e^{(1/t)-1} + e^{1/t}\left[\frac{1}{t}e^{-1/t} + e^{-1/t} - e^{-1} - e^{-1}\right].$$

Simplifying,

$$N(t) = N_0 e^{(1/t)-1} + \frac{1}{t} + 1 - 2e^{(1/t)-1}.$$

We notice that $N(t) \to (N_0 - 2)e^{-1} + 1$ as $t \to \infty$, i.e., the population size approaches a positive constant after a sufficiently long time.

3.2.1 Nonzero Average Intrinsic Growth Rate

We now consider an open population in which the per capita birth rate is smaller than the per capita death and emigration rates together, in the asymptotic time average. It is in this situation that immigration has the biggest impact on the population development. An example is a fish pond in which the fish population cannot survive on its own and so is stocked regularly. We show that the solutions of (3.5) "forget" their initial data if the asymptotic time average $\bar{\rho}$ in (3.4) is negative.

Theorem 3.5. *Assume that $\bar{\rho} < 0$ and consider two solutions N and M of (3.5) satisfying different initial data. Then $N(t) - M(t)$ is a solution of (3.1) which converges to 0 as $t \to \infty$.*

Proof. Let $\tilde{N} = N - M$ be the difference of the two solutions. Then

$$\tilde{N}' = \rho(t)\tilde{N}$$

and $\tilde{N}(t)$ decreases exponentially to 0 by Theorem 3.4. □

We now return to the scenario in which the various per capita rates are periodic and assume that the immigration rate is also periodic with the same period p.

Example. Let $\rho(t) = \cos(t) - 1$, $I(t) = 1 - \cos(t)$.

Rather than using the variation of parameters formula, we go through the integrating procedure again. The equation for the integrating factor F specializes to

$$F' = -(\cos(t) - 1)F.$$

Separating the variables we find the special solution

$$F(t) = e^{t-\sin(t)}.$$

With this integrating factor the original differential equation becomes

$$\frac{\mathrm{d}}{\mathrm{d}t}(F(t)N(t)) = F(t)I(t) = \mathrm{e}^{t-\sin t}(1-\cos(t)) = \frac{\mathrm{d}}{\mathrm{d}t}\mathrm{e}^{t-\sin(t)}.$$

Integrating both sides,

$$F(t)N(t) = \mathrm{e}^{t-\sin(t)} + c,$$

and dividing by $F(t)$,

$$N(t) = 1 + c\mathrm{e}^{\sin(t)-t}.$$

We see that $N(t) \to 1$ as $t \to \infty$.

This asymptotic behavior is more regular than we can generally expect, as we will see in an example below. It is due to the fact that this example is of the form

$$N' = \rho(t)N - \rho(t)A$$

with a positive constant A and that $M(t) = A$ is a special solution. If the asymptotic time average $\bar{\rho}$ exists and is strictly negative, it follows from Theorem 3.5 that $N(t) - M(t) \to 0$ as $t \to \infty$ and we have the following result. Notice that ρ does not need to be periodic.

Corollary 3.6. *Consider $N' = \rho(t)N - \rho(t)A$ with a constant A and assume that $\bar{\rho} < 0$.*

Then $N(t) \to A$ as $t \to \infty$.

In particular, if $N' = -\alpha N + A$ with positive constants α and A, then $N(t) \to A/\alpha$ as $t \to \infty$.

We will see a less regular asymptotic behavior in the next example.

Example. Let $\rho(t) = \bar{\rho} = -2$ and again $I(t) = 1 - \cos t$.

We could specialize the general formula (3.6) or redo the integration factor procedure from scratch. In this special case, another approach called *the method of undetermined coefficients* is more effective. As we have seen in Theorem 3.5, the difference of two solutions of the nonhomogeneous equation (3.5) is a solution of the homogeneous equation (3.1). In turn, the sum of a solution of (3.5) and a solution of (3.1) is again a solution of (3.5). So, it is enough to find a single solution of (3.5) and the general solution of (3.1) and to add them in order to get the general solution of (3.5). In this context the single solution of (3.5) is often referred to as a *particular solution*, while the general solution of (3.1) is called the *complementary solution*.

This so-called *principle of superposition* generally holds for linear equations (and only for linear equations).

The method of undetermined coefficients looks for a special solution of (3.5) which has the same terms as $I(t)$ and its derivatives. So, we try

$$M(t) = A + B\cos t + C\sin t.$$

A, B, C are the undetermined coefficients which give this method its name. Substituting this "ansatz" into (3.5) yields

$$-B\sin t + C\cos t = -2(A + B\cos t + C\sin t) + 1 - \cos t.$$

Comparing the constant terms and the terms associated with $\sin t$ and $\cos t$, we see that this equality holds if

$$0 = -2A + 1, \qquad -B = -2C, \qquad C = -2B - 1.$$

So, $A = \frac{1}{2}$, $C = -\frac{1}{5}$, $B = -\frac{2}{5}$, and we have found a special solution of (3.5),

$$M(t) = \tfrac{1}{2} - \tfrac{2}{5}\cos t - \tfrac{1}{5}\sin t.$$

The general solution of (3.1) is given by $N(t) = k\mathrm{e}^{-2t}$ and the general solution of (3.5) by

$$N(t) = M(t) + k\mathrm{e}^{-2t}.$$

We notice that N, while not approaching a constant, approaches the periodic function M as $t \to \infty$. This is a phenomenon that holds in general for our scenario.

Theorem 3.7. *Consider* $N' = \rho(t)N + I(t)$, $N(t_0) = N_0$ *and assume that* ρ *is periodic with period* p, $\bar{\rho} \neq 0$.

(a) *If* $\bar{\rho} < 0$ *and* I *is bounded on the interval* $[t_0, \infty)$, *so is* N.

(b) *If* I *is periodic, with same period* p, *then there exists some* p*-periodic function* N_∞ *such that*

$$N(t) = [N(t_0) - N_\infty(t_0)]\mathrm{e}^{(t-t_0)\bar{\rho}}\frac{q(t)}{q(t_0)} + N_\infty(t),$$

$$N_\infty(t) = \frac{q(t)}{1 - \mathrm{e}^{\bar{\rho}p}}\int_0^p \mathrm{e}^{\bar{\rho}s}\frac{I(t-s)}{q(t-s)}\,\mathrm{d}s,$$

with q *being the* p*-periodic function from the Floquet representation of the fundamental solution.*

For $\bar{\rho} < 0$, $N(t) - N_\infty(t) \to 0$ *as* $t \to \infty$. *If* I *is nonnegative and not identically* 0, *then* N_∞ *is bounded away from* 0 *and has the opposite sign of* ρ.

So, if a fish pond in which the fish population cannot survive on its own is stocked periodically, the population size will eventually vary periodically as well (Figure 3.4C). In this context N_∞ is called the *persistent* solution and $N - N_\infty$ the *transient* solution of the nonhomogeneous linear differential equation. However, if the population already grows on its own, then the stocking adds to the growth, which is exponential with periodic modulation (Figure 3.4B).

Proof. (a) From the variation of constants formula (3.6) and Remark 3.2,

$$N(t) = N(t_0)e^{(t-t_0)\bar{\rho}}\frac{q(t)}{q(t_0)} + \int_{t_0}^{t} e^{(t-s)\bar{\rho}}\frac{I(s)}{q(s)}q(t)\,ds.$$

Since q is periodic and strictly positive, it is bounded and bounded away from 0 on $\mathbb{R}$. So, there exists a constant $c > 0$ such that

$$\frac{q(t)}{q(s)} \leqslant c \quad \text{for all } t, s \in \mathbb{R}.$$

So, for $t \geqslant t_0$,

$$N(t) \leqslant cN(t_0) + c\int_{t_0}^{t} e^{(t-s)\bar{\rho}} I(s)\,ds.$$

Furthermore, since I is bounded on $[t_0, \infty)$, there exists some $k > 0$ such that $I(s) \leqslant k$ for all $s \geqslant t_0$, thus

$$N(t) \leqslant cN(t_0) + ck\frac{1}{\bar{\rho}}(e^{(t-t_0)\bar{\rho}} - 1).$$

Since $\bar{\rho} < 0$,

$$N(t) \leqslant cN(t_0) - \frac{ck}{\bar{\rho}}.$$

(b) We make the "ansatz"

$$N_\infty(t) = \alpha\int_{t-p}^{t} \frac{Q(t)}{Q(s)} I(s)\,ds.$$

We will determine α such that N_∞ satisfies $N_\infty' = \rho(t)N_\infty + I(t)$ and show that N_∞ has the form of part (b) of our theorem. Part (b) then follows from the fact that $\tilde{N} = N - N_\infty$ satisfies $\tilde{N}' = \rho(t)\tilde{N}$ and from Theorem 3.5.

Using the product rule (remember $Q' = \rho(t)Q$) and the fundamental theorem of calculus, we see that

$$N_\infty'(t) = \rho(t)N_\infty(t) + \alpha\left(I(t) - \frac{Q(t)}{Q(t-p)}I(t-p)\right).$$

By Remark 3.2 and the periodicity of I,

$$N_\infty'(t) = \rho(t)N_\infty(t) + \alpha I(t)\left(1 - \frac{e^{\bar{\rho}t}q(t)}{e^{\bar{\rho}(t-p)}q(t-p)}\right).$$

Since q is p-periodic,

$$N_\infty'(t) = \rho(t)N_\infty(t) + \alpha I(t)(1 - e^{\bar{\rho}p}).$$

So, we choose $\alpha = (1 - e^{\bar{\rho}p})^{-1}$. Notice that α has the opposite sign of $\bar{\rho}$. Using Remark 3.2 in our ansatz,

$$N_\infty(t) = \alpha \int_{t-p}^{t} e^{(t-s)\bar{\rho}} \frac{q(t)}{q(s)} I(s)\, ds.$$

Substituting $t - s = r$ we obtain the form in Theorem 3.7 (b), from which we realize that N_∞ is p-periodic. Furthermore, we have the estimate

$$\frac{N_\infty(t)}{\alpha} \geqslant q(t) e^{\bar{\rho}p} w(t)$$

with

$$w(t) = \int_{t-p}^{t} \frac{I(s)}{q(s)}\, ds.$$

By the fundamental theorem of calculus,

$$w'(t) = \frac{I(t)}{q(t)} - \frac{I(t-p)}{q(t-p)} = 0,$$

because I and q are p-periodic. This implies that w is constant and

$$w(t) = w(p) = \int_0^p \frac{I(s)}{q(s)}\, ds > 0.$$

Since q is strictly positive (Remark 3.2), so is N_∞/α. □

More information can be obtained if $\rho = \bar{\rho} = -\alpha$ is constant (and negative). Then we can rewrite (3.5) as

$$N' = \alpha(-N(t) + f(t)),$$

where f is a periodic nonnegative function that is not identically 0, and use Fourier series. After scaling time we can assume that the period of f is 2π. If f is twice continuously differentiable, we can represent f by a Fourier series

$$f(t) = \sum_{k=-\infty}^{\infty} a_k e^{ikt}$$

with

$$\sum_{k=-\infty}^{\infty} |a_k| < \infty.$$

Recall that

$$a_0 = \frac{1}{2\pi} \int_0^{2\pi} f(t)\, dt = \bar{f}$$

is the time average of f and the other Fourier coefficients are given by

$$a_j = \frac{1}{2\pi}\int_0^{2\pi} f(t)\mathrm{e}^{-\mathrm{i}jt}\,\mathrm{d}t.$$

We look for $N = N_1 + N_2$ with N_1 being a general solution of $N' = -\alpha N$ and N_2 a particular solution in the form

$$N_2 = \sum_{k=-\infty}^{\infty} a_k M_k,$$
$$M_k' = \alpha(-M_k + \mathrm{e}^{\mathrm{i}kt}).$$

We make the following "ansatz" for M_k,

$$M_k = \gamma_k \mathrm{e}^{\mathrm{i}k(t-t_k)}.$$

Substituting the ansatz into the differential equation yields after simplification that

$$\gamma_k \mathrm{i}k\mathrm{e}^{-\mathrm{i}kt_k} = -\alpha\gamma_k\mathrm{e}^{-\mathrm{i}kt_k} + \alpha.$$

After reorganization,

$$\mathrm{e}^{\mathrm{i}kt_k} = \frac{\alpha + \mathrm{i}k}{\alpha}\gamma_k.$$

The left-hand side has absolute value 1, so

$$\gamma_k = \frac{\alpha}{\sqrt{\alpha^2+k^2}} \quad \text{and} \quad \mathrm{e}^{\mathrm{i}kt_k} = \frac{\alpha + \mathrm{i}k}{\sqrt{\alpha^2+k^2}}.$$

Separating real and imaginary parts (recall $\mathrm{e}^{\mathrm{i}\theta} = \cos\theta + \mathrm{i}\sin\theta$),

$$\tan(kt_k) = \frac{k}{\alpha},$$

which, for every k, has a unique solution $t_k \in [0, \pi/(2k))$. Moreover, for every $k \neq 0$,

$$\begin{aligned} t_k &\to 0, & \alpha &\to \infty, \\ t_k &\to \pi/(2k), & \alpha &\to 0. \end{aligned}$$

Since

$$N_2(t) = \sum_{k=-\infty}^{\infty} a_k \frac{\alpha}{\sqrt{\alpha^2+k^2}}\mathrm{e}^{\mathrm{i}k(t-t_k)}$$

and also the sum of the respective derivatives converge absolutely and uniformly, N_2 is a solution of the differential equation. We see that the higher-frequency terms of N_2 as compared with those of f are damped and that every term except the constant term for $k = 0$ lags behind that of f with a delay between 0 and a quarter of a period. The delays decrease with k.

Let us consider the case where α is large. Since $t_k \approx 0$ except for large k,

$$N_2 \approx f.$$

Since the complementary solution is of the form $N_1(t) = c\mathrm{e}^{-\alpha t}$, the population size N tracks the immigration rate.

We turn to the case where $\alpha > 0$ is small. As f is nonnegative and not identically 0, $a_0 > 0$ and

$$N_2(t) = a_0 + \alpha \sum_{k \neq 0} \frac{a_k}{\sqrt{\alpha^2 + k^2}} \mathrm{e}^{\mathrm{i}k(t - t_k)}.$$

For small α, since $kt_k \approx \pi/2$ except for large k,

$$N_2(t) \approx a_0 + \alpha \sum_{k \neq 0} \frac{a_k}{k} \mathrm{e}^{\mathrm{i}(kt - \pi/2)}.$$

In this case, the population size exhibits small-amplitude oscillations of the order of α about its mean value with the fundamental mode of oscillation ($k = 1$) lagging behind that of the immigration rate by about one-quarter of a period. The population cannot react fast enough for its size to follow the oscillations of the immigration rate. Also

$$N_2 \approx a_0 + \alpha \frac{1}{\mathrm{i}} \sum_{k \neq 0} \frac{a_k}{k} \mathrm{e}^{\mathrm{i}kt} = \bar{f} + \alpha \left(\int_0^t [f(s) - \bar{f}] \, \mathrm{d}s - c \right)$$

for some constant c. This follows from the fact that

$$\frac{1}{\mathrm{i}} \sum_{k \neq 0} \frac{a_k}{k} \mathrm{e}^{\mathrm{i}kt}$$

is an anti-derivative for $f - \bar{f}$. To determine c, we observe from the differential equation for N_2, $N_2' = \alpha(-N_2 + f)$, and the periodicity of N_2, that N_2 and f have identical time averages. This implies that

$$c = \frac{1}{2\pi} \int_0^{2\pi} \int_0^t [f(s) - \bar{f}] \, \mathrm{d}s \, \mathrm{d}t.$$

Changing the order of integration

$$c = \frac{1}{2\pi} \int_0^{2\pi} (2\pi - s)[f(s) - \bar{f}] \, \mathrm{d}s.$$

Recalling the definition of $\bar{f}$,

$$c = \pi \bar{f} - \frac{1}{2\pi} \int_0^{2\pi} s f(s) \, \mathrm{d}s.$$

Substituting this back into the equation for N_2,

$$N_2 \approx \bar{f} + \alpha\left(\int_0^t f(s)\,\mathrm{d}s + (\pi - t)\bar{f} - \frac{1}{2\pi}\int_0^{2\pi} sf(s)\,\mathrm{d}s\right).$$

3.2.2 *Zero Average Intrinsic Growth Rate*

Let us briefly turn to the case where ρ is periodic with $\bar{\rho} = 0$ (i.e., on average over time, the per capita birth rate balances the per capita death and emigration rates) and that the immigration rate is periodic with the same period. Then

$$N(t) = N_0\frac{Q(t)}{Q(t_0)} + Q(t)\int_{t_0}^t \frac{I(s)}{Q(s)}\,\mathrm{d}s.$$

By Theorem 3.1 and Remark 3.2, Q is a periodic function with period p and so is the product I/Q. By Proposition 3.3,

$$\frac{1}{t}\int_{t_0}^t \frac{I(s)}{Q(s)}\,\mathrm{d}s \to \frac{1}{p}\int_0^p \frac{I(s)}{Q(s)}\,\mathrm{d}s =: \kappa, \quad t \to \infty.$$

So, $N(t)/(tQ(t)) \to \kappa$ as $t \to \infty$, in more suggestive formalism

$$N(t) \approx \kappa t Q(t), \quad t \to \infty.$$

This means that, for large times, the population size exhibits a linear increase in time which is modulated by a periodic function.

Let us summarize the impact of immigration under periodicity. For an illustration, compare Figures 3.3 and 3.4, where we use the normalized per capita birth and death rates, and the immigration rate in Figure 3.2. If the average intrinsic growth rate $\bar{\rho}$ is negative (case C in Figures 3.3 and 3.4), immigration turns a population that would become extinct otherwise into an eventually periodic population. If the time average $\bar{\rho}$ is zero (case A), immigration converts a periodic population into one that grows linearly with periodic modulation. If $\bar{\rho} > 0$ (case B), the population size increases exponentially with exponent $\bar{\rho}$ and periodic modulation, independently of whether or not there is immigration, but immigration adds to the growth.

Exercises

3.1. Consider unrestricted population growth in a closed population with constant per capita birth and death rates. Calculate the doubling time of the population in terms of $\bar{\rho}$ if $\bar{\rho} > 0$.

Remark. The doubling time T is characterized by $N(t_0 + T) = 2N(t_0)$.

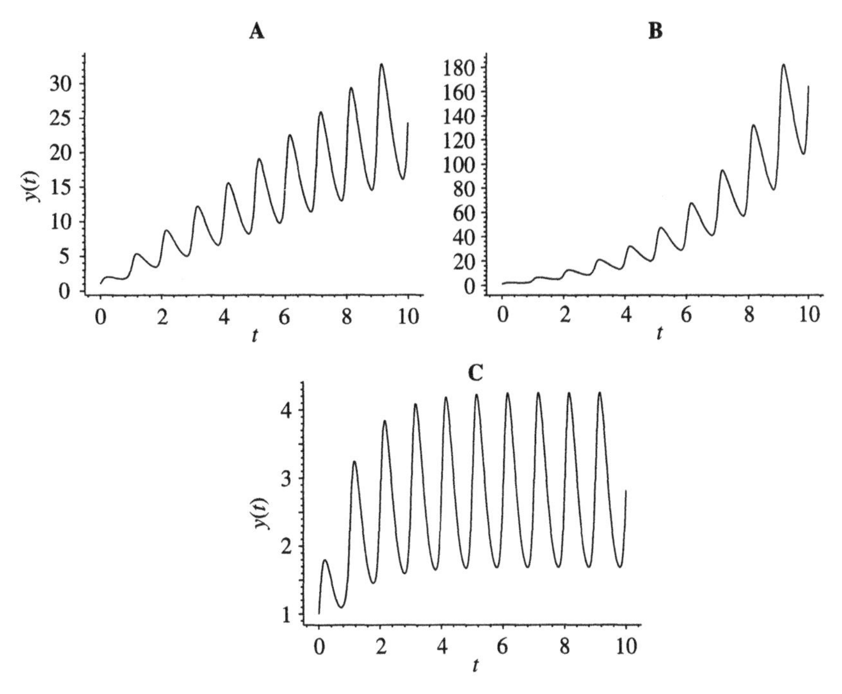

Figure 3.4. Dynamics of open populations under various combinations of the per capita birth and death rates of Figure 3.2 and the immigration rate of Figure 3.2. The averages of the intrinsic growth constant have been chosen as in Figure 3.3. (**A**) $\bar{\rho} = 0$. (**B**) $\bar{\rho} = 0.3$. (**C**) $\bar{\rho} = -0.9$.

3.2. If the per capita mortality rate μ is constant, then $L = 1/\mu$ is the life expectancy (expected age at death) of the individuals in the population (see (2.3)). If the per capita birth rate β is also constant, then $\mathcal{R}_0 := \beta L = \beta/\mu$ is the average number of offspring that one typical individual produces during its lifetime. $\mathcal{R}_0$ is called the *basic reproduction ratio* of the population. In a population with two sexes, one would divide the average number of offspring by two or only count the average number of daughters per female (in other words one would only consider the female part of the population).

Express the doubling time of the population in terms of $\mathcal{R}_0$ and L. Calculate the doubling time if the life expectancy L is 70 years and $\mathcal{R}_0 = 1.5$ (i.e., in a human population, every female would have an average number of 1.5 daughters during her lifetime).

3.3. Determine the solution of

$$N' = \tan(t)N, \qquad N(0) = N_0.$$

Where is the solution defined? How does it behave? Where does it make biological sense?

3.4. (a) Solve $N' = -\alpha N + A$ with positive constants α and A and initial data $N(t_0) = N_0$.

(b) Using (a), show that $N(t) \to A/\alpha$ as $t \to \infty$.

3.5. Determine the solution of

$$N' = \rho N + I(t), \qquad N(0) = N_0$$

with constant ρ and immigration rate $I(t) = 1 + \sin t$.

What can you say about the behavior for large times?

3.6. Consider

$$N' = \sin(t)N + 1, \qquad N(0) = N_0.$$

What can you tell about the growth of the population as $t \to \infty$ from the theory developed in the text. Try the problem on a computer for various initial values N_0.

3.7. (a) Solve

$$N' = \frac{a}{t}N + bt^{\gamma}, \quad t > 0,$$
$$N(t_0) = N_0,$$

with constants $a \in \mathbb{R}, \gamma \in \mathbb{R}, b > 0$ and initial data $N_0 \geqslant 0, t_0 > 0$.

(b) Discuss the behavior of the solution as $t \to \infty$.

3.8. Solve

$$N' = \rho(t)N + I(t), \qquad N(t_0) = N_0$$

for $\rho(t) = -\sin t - 1$, $I(t) = a(1 + \sin t)$ with some constant $a > 0$.

Discuss the behavior as $t \to \infty$.

3.9. Prove the second statement in Theorem 3.4 that the size of a closed population decreases exponentially if the asymptotic time average of ρ exists and $\bar{\rho} < 0$.

3.10. A continuous function $f : \mathbb{R} \to \mathbb{R}$ is called *quasi-periodic* if it is the finite sum of periodic functions with different periods.

Show that the asymptotic time average exists for quasi-periodic functions.

3.11. A continuous function $f : [0, \infty) \to [0, \infty)$ is called *asymptotically almost periodic* if for every $\epsilon > 0$ there exists a quasi-periodic function g and some $t > 0$ such that

$$|f(s) - g(s)| \leqslant \epsilon, \quad \text{for all } s \geqslant t.$$

Show that the asymptotic time average exists for asymptotically almost-periodic functions.

Hint: consider the sequence

$$x_n = \frac{1}{n} \int_0^n f(s)\,\mathrm{d}s$$

and show that it is a Cauchy sequence.

Chapter Four

Von Bertalanffy Growth of Body Size

Here we do not consider the growth of a population, but the size growth of a single individual. We include this topic for two reasons. On the one hand, we can consider the body to be a population of cells so that this topic fits under the umbrella of population growth. On the other hand, growth models for the body sizes of individuals are the basis for size-structured population models, which stratify a population along the body sizes of its constituents (Metz, Diekmann, 1986; Kooijman, 1993, 2000). The von Bertalanffy model (1934) is built on the assumption that the acquisition of food is proportional to the surface of the body, S, while maintenance costs for the body are proportional to the volume of the body, V:

$$V' = F(t)S - \nu(t)V.$$

$F(t)$ is the quantity of food taken up per unit surface area and $\nu(t)$ denotes the maintenance costs per unit volume. Since

$$S = \gamma V^{2/3}$$

with a proportionality constant γ that depends on the shape of the body, we have

$$V' = \gamma F(t)V^{2/3} - \nu(t)V. \tag{4.1}$$

Since γ does not depend on time, we have implicitly assumed that, though growing, the body does not change its shape. In order to transform this equation into a linear one, we introduce the length of the body, L. Since $V = (\alpha L)^3$ with some proportionality constant α depending on the shape of the body, we have $V' = 3\alpha^3 L^2 L'$. Thus we obtain the following ordinary differential equation for L:

$$3\alpha^3 L^2 L' = \gamma F(t)\alpha^2 L^2 - \nu(t)\alpha^3 L^3.$$

Hence

$$L' = f(t) - \mu(t)L, \tag{4.2}$$

with $f(t) = (\gamma/3)\alpha^{-1}F(t)$, $\mu(t) = \nu(t)/3$. Equation (4.2) can now be solved by formula (3.6). In a laboratory environment, one can assume that the organism under question is fed with a period of one day. Furthermore, since the maintenance costs often depend on temperature, μ may have the same period. From Theorem 3.7 and $V = (\alpha L)^3$ we immediately obtain the following result.

Theorem 4.1. *Let F and ν be p-periodic nonnegative continuous functions that are not identically 0. Then there exists a strictly positive periodic function V_∞ such that $V(t) - V_\infty(t) \to 0$ as $t \to \infty$.*

Let F and ν, and so f and μ, be constant. If L_0 is the length at time $t = 0$, the length at time t is given by

$$\left.\begin{aligned} L(t) &= L_0 e^{-\mu t} + L_\infty(1 - e^{-\mu t}) = (L_0 - L_\infty)e^{-\mu t} + L_\infty, \\ L_\infty &= \frac{f}{\mu}. \end{aligned}\right\} \tag{4.3}$$

This shows that the length converges to $L_\infty = f/\mu$ as $t \to \infty$, which can be interpreted as the final length under constant conditions. If we go backwards in time and if $L_0 < L_\infty$, we notice that L becomes negative at

$$t_- = \frac{1}{\mu} \ln \frac{L_\infty - L_0}{L_\infty},$$

and the solution stops making sense. If we go back to body volume, however, we find that the following is also a solution to (4.1) which is defined for all times:

$$V(t) = \begin{cases} (L_0 e^{-\mu t} + L_\infty(1 - e^{-\mu t}))^3, & t \geqslant t_-, \\ 0, & t \leqslant t_-. \end{cases} \tag{4.4}$$

This is, of course, related to the fact that (4.1) can have more than one solution in a neighborhood of $V = 0$, because the vector field fails to be Lipschitz there.

While the graph of the solution L in (4.3) is concave, the graph of V in (4.4) is sigmoid. We will see more of this later.

The von Bertalanffy model is an example of modeling from first principles. The principles are that the per capita food intake scales with body surface, while the maintenance costs scale with body volume. For small creatures it may be difficult to verify these principles directly, and it is even possible that they may not tell the whole story. So, it is imperative to check whether the above formulas provide results that agree with real data of body sizes. To this end one needs to know the parameters in (4.3). We now interpret t as age and L_0 as length at birth, which occurs at $t = 0$.

By observing for long enough, it is possible to get a good idea of the final body length L_∞, while it may be difficult to determine the length at birth (the individuals may yet be too fragile to be measured). The parameter μ, which is related to maintenance costs, is very difficult to determine directly. The idea is to find L_0 and μ such that the length growth data and the outputs of equation (4.3) deviate from each other as little as possible. Rewriting (4.3) as

$$L(t) - L_\infty = (L_0 - L_\infty)e^{-\mu t},$$

we can take logarithms,

$$\ln(L_\infty - L(t)) = \ln(L_\infty - L_0) - \mu t.$$

The right-hand side is a line with slope $-\mu$.

Set $b(t) = \ln(L_\infty - L(t))$, $b_0 = \ln(L_\infty - L_0)$ and assume that we have n measurements $b(t_j) = b_j$ at ages $t_1, \ldots, t_n$. This gives us the linear system $b_j = b_0 - \mu t_j$ for $j = 1, \ldots, n$, which takes the matrix form $Ax = b$ with $b = (b_1, \ldots, b_n)^{\mathrm{T}}$, $x = (b_0, -\mu)^{\mathrm{T}}$ and

$$A^{\mathrm{T}} = \begin{pmatrix} 1 & \cdots & 1 \\ t_1 & \cdots & t_n \end{pmatrix}.$$

The superscript "T" denotes the transpose of a matrix. For $n > 2$ this is an overdetermined system which typically has no solution. However, it can be shown by projection methods (Edwards, Penny, 1988, Sections 5.2, 5.3, for example) that the system $A^{\mathrm{T}}Ax^* = A^{\mathrm{T}}b$ has a solution x^* and that the Euclidean distance $\|Ax^* - b\|$ is the minimum of all distances $\|Ax - b\|$, $x \in \mathbb{R}^2$. x^* is called the *least-squares solution* of $Ax = b$ and the respective procedure *least-squares fitting*.

Once we have extracted L_0 and μ from x^*, we can solve

$$L' = \mu(L_\infty - L), \qquad L(0) = L_0$$

numerically and see whether the plotted solution graph comes sufficiently close to the measurements.

Typically, the von Bertalanffy model fits individual growth curves under constant food regimes quite well. See Metz, Diekmann (1986, A.I.3.1), Kooijman, Metz (1984) and Kooijman (1986). For refinements of this model and more growth curves see the book by Kooijman (1993, 2000).

Exercises

4.1. Determine whether the following growth data of body length resemble von Bertalanffy growth under constant food regime:

age (days)	4	8	9.5	12	16
length (mm)	1.8	2.3	2.5	2.6	2.8

The final body length is assumed to be $L_\infty = 3$ mm.

4.2 Consider $V' = \nu(t)(AV^{2/3} - V)$ with a positive constant A and a nonnegative, continuous function ν defined on $\mathbb{R}$. Assume that the asymptotic time average $\bar{\nu}$ exists and $\bar{\nu} > 0$. What will be the final body volume (with proof)?

Chapter Five

Classic Models of Density-Dependent Population Growth for Single Species

For many natural populations, the available resources are limited and, directly or indirectly, the size of the population couples back to the per capita birth and/or mortality rates. The simplest way to incorporate this feedback consists in assuming that the per capita birth and mortality rates at time t depend on the population size at that very moment t:

$$B(t) = \beta(t, N(t))N(t), \qquad D(t) = \mu(t, N(t))N(t).$$

For a closed population, equation (2.1), with migration $E = I = 0$, takes the form

$$N' = N[\beta(t, N) - \mu(t, N)]. \tag{5.1}$$

We will present the classical equations by Bernoulli and Verhulst (the second being a special case of the first), by Ricker, Beverton–Holt, Smith, and Gompertz, and analyze them as special cases of a certain class of nonlinear scalar ordinary differential equations. We will treat them in some detail because they not only model the one-species situation, but often occur also as modules in multi-species models.

5.1 The Bernoulli and the Verhulst Equations

A first class of models assumes that the population density only affects the per capita mortality rate,

$$\mu(t, N) = m_0(t) + m(t)N^\theta \tag{5.2}$$

where θ is a positive constant. Here m_0 is the density-independent component of the per capita mortality rate. Then

$$N' = \rho(t)N - m(t)N^{\theta+1} \tag{5.3}$$

with

$$\rho(t) = \beta(t) - m_0(t). \tag{5.4}$$

This is (Jakob) Bernoulli's equation. Leibnitz (1696) suggested the following transformation to reduce it to a linear equation (see the problems in Section 2.2

of Boyce, DiPrima, 1992):

$$x = N^{-\theta}.$$

Then

$$x' = -\theta N^{-\theta-1} N' = -\theta N^{-\theta-1}[\rho(t)N - m(t)N^{\theta+1}]$$
$$= -\theta\rho(t)x + \theta m(t).$$

This equation is of the form (3.5), and a solution x can be found from formula (3.6). In turn, $N = x^{-1/\theta}$ is a solution of (5.4).

Again we may like to model seasonal influences on the birth and mortality rates by periodic functions (with the period being one year, for example). Then we obtain the following result from Theorem 3.7 (b).

Theorem 5.1. *Let $\theta > 0$ and let $\rho(t), m(t)$ be continuous p-periodic functions. Assume that*

$$\bar{\rho} = \frac{1}{p}\int_0^p \rho(t)\,\mathrm{d}t > 0$$

and that m is nonnegative and not identically 0. Then there exists a strictly positive p-periodic function N_∞ such that $N(t) - N_\infty(t) \to 0$ as $t \to \infty$ for every solution N of (5.3).

If ρ and m are time independent and positive, it is convenient to rewrite (5.3) as

$$N' = \rho N(1 - (N/K)^\theta) \tag{5.5}$$

with

$$K = (\rho/m)^{1/\theta}. \tag{5.6}$$

K is called the *carrying capacity* for the population, because $N(t) \to K$ as $t \to \infty$. (This follows from Corollary 3.6 applied to x.) In Chapter 6 we will see that this is also related to the fact that K is the unique positive zero of the right-hand side of (5.5).

For the case $\theta = 1$, equation (5.5) has become known as Verhulst's equation, the Verhulst–Pearl equation, or the *logistic equation* as Verhulst (1845) called it himself. Gilpin and Ayala (1973) have rediscovered (5.5) for $\theta \neq 1$, and theoretical ecologists sometimes refer to it as the asymmetric or generalized logistic equation rather than Bernoulli's equation (Hutchinson, 1978, Chapter 1).

As we will see later in a more general context, the solution graph of the Verhulst equation, for any initial value between 0 and K, has an S-shape or sigmoid shape, and connects 0 (for $t \to -\infty$) with K (for $t \to \infty$), and has an inflection point at $\frac{1}{2}K$. But the symmetry of the solution graph with respect to $\frac{1}{2}K$ (half the carrying capacity) is even more striking.

Proposition 5.2. *Let N be the solution of the Verhulst equation $N' = \rho N(1 - N/K)$ with $\rho, K > 0$. Without restriction of generality let $N(0) = \frac{1}{2}K$. Then $N(t) - \frac{1}{2}K = \frac{1}{2}K - N(-t)$.*

Proof. We set

$$x(t) = N(t) - \tfrac{1}{2}K, \qquad y(t) = \tfrac{1}{2}K - N(-t).$$

Obviously, $x(0) = 0 = y(0)$. We shall show that x and y satisfy the same differential equation with a locally Lipschitz continuous vector field. Then $x = y$ by the uniqueness theorem for ordinary differential equations (see Hale (1980), for example). Indeed,

$$x' = N' = \rho(x + \tfrac{1}{2}K)\left(1 - \frac{x + \frac{1}{2}K}{K}\right) = \rho\left(x + \frac{K}{2}\right)\left(\frac{1}{2} - \frac{x}{K}\right),$$

$$\begin{aligned} y'(t) = N'(-t) &= \rho\left(\frac{K}{2} - y(t)\right)\left(1 - \frac{\frac{1}{2}K - y(t)}{K}\right) \\ &= \rho\left(\frac{1}{2} - \frac{y}{K}\right)\left(\frac{K}{2} + y\right). \end{aligned}$$

□

It was this striking symmetry that led Gilpin and Ayala (1973) to consider the asymmetric logistic alias Bernoulli equation. Let us explore the symmetry properties of solutions N to (5.5) for $\theta > 0$.

Set $z = N^\theta$. Then

$$z' = \theta N^{\theta-1}N' = \theta\rho z\left(1 - \frac{z}{K^\theta}\right).$$

Hence z is a solution of the Verhulst equation with carrying capacity K^θ. By Proposition 5.2, if $z(0) = K^\theta/2$,

$$z(t) - K^\theta/2 = K^\theta/2 - z(-t).$$

Going back to N we have the following result.

Proposition 5.3. *Let N be a solution to (5.5) with $\rho, K, \theta > 0$. Let $N(0) = K2^{-1/\theta}$. Then*

$$(N(t))^\theta + (N(-t))^\theta = K^\theta \quad \forall t \geqslant 0.$$

5.2 The Beverton–Holt and Smith Differential Equation

Other models assume that the population density affects the (effective) per capita birth rate rather than the per capita mortality rate. The effective per capita birth rate takes account of higher than usual mortality at or immediately after birth. For fishery models, though in a somewhat different context, Ricker (1954, 1975) suggested

$$\beta(N) = \beta_0 e^{-\alpha N},$$

while Beverton, Holt (1957) suggested

$$\beta(N) = \frac{\beta_0}{1 + \alpha N}.$$

As we will see, both functional forms can be explained by heavy cannibalization of juveniles by adults of the same species, during a very short period immediately after birth. The different forms originate from different assumptions concerning the length distribution of the juvenile period; in the Ricker equation the juvenile period has a fixed length, while in the Beverton–Holt equation its length is exponentially distributed, i.e., individuals make the transition from juvenile to adult stage at a fixed constant rate. In a way, it is disconcerting that different assumptions about the length distribution of the juvenile period lead to these dramatically different forms (see Section 13.7 for an elaboration of this theme).

Other explanations than cannibalism are possible, in particular for the Beverton–Holt reproduction function, which can also be obtained from a resource–consumer model via a time scale argument. We will deal with this explanation first and turn to cannibalism later.

5.2.1 Derivation from a Resource–Consumer Model

Let M be the biomass of the population under consideration (the consumers), and F the biomass of the food (resource) on which the population lives:

$$F' = \Lambda - \nu F - \frac{b}{\gamma} F M, \qquad M' = bFM - \mu M.$$

The food is provided at the constant rate Λ and degrades at a constant per unit rate ν. Food consumption follows the *law of mass action*, i.e., the consumption rate is proportional to both the biomass of food available and the biomass of consumers; the proportionality constant b gives the per unit consumer biomass gained from one unit biomass of food per unit of time. The *yield constant* γ describes the conversion of consumed food biomass into consumer biomass, i.e., γ is the per unit consumer biomass gained from one unit of consumed food. The quotient b/γ is the per unit food biomass consumed per unit of consumer biomass and per unit of time.

We mention that the biomass gain of the consumers is not only due to the increase of individuals in weight or volume, but also to the production of offspring.

Assuming that the food dynamics are much faster than the consumer dynamics, we want to make a *quasi-steady-state approach* to the food equation. In order to be able to compare the speeds of the different dynamics we introduce new, dimensionless variables. We start with the dependent variables. Without the consuming population being around, the food would converge to Λ/ν, so we introduce

$$F = \frac{\Lambda}{\nu} z.$$

Substituting this expression into the differential equations,

$$\frac{\Lambda}{\nu} z' = \Lambda - \Lambda z - \frac{b}{\gamma}\frac{\Lambda}{\nu} z M,$$
$$M' = \mu M\left(\frac{b}{\mu}\frac{\Lambda}{\nu} z - 1\right).$$

Simplifying and introducing the *basic biomass production ratio*,

$$\mathcal{R}_0 = \frac{\Lambda}{\nu}\frac{b}{\mu}, \tag{5.7}$$

yields

$$z' = \nu\left(1 - z - \frac{b}{\gamma\nu} z M\right),$$
$$M' = \mu M(\mathcal{R}_0 z - 1).$$

The form of $\mathcal{R}_0$ motivates us to define

$$M = \frac{\Lambda\gamma}{\mu} x,$$

and obtain

$$z' = \nu(1 - z - \mathcal{R}_0 z x),$$
$$x' = \mu x(\mathcal{R}_0 z - 1).$$

To interpret $\mathcal{R}_0$, recall that Λ/ν is the large-time limit of the food concentration if there are no consumers around. Since b is the per unit/unit biomass gain rate and $1/\mu$ is the life expectancy of the consumer, we see that $\mathcal{R}_0$ is the ratio by which consumer biomass grows over a consumer's lifetime, without competition from conspecifics.

Finally, we introduce the dimensionless time τ, $\tau = \mu t$; in this way our time unit becomes the life expectancy of the consumer, $1/\mu$. Let $\dot{z}$ and $\dot{x}$ denote the derivatives with respect to the new time τ, by the chain rule $\dot{z} = \mu z'$ and $\dot{x} = \mu z'$. Dividing the first equation by ν and the second by μ we obtain

$$\epsilon\dot{z} = 1 - z - \mathcal{R}_0 z x,$$
$$\dot{x} = x(\mathcal{R}_0 z - 1),$$

with

$$\epsilon = \mu/\nu.$$

The assumption that the food dynamics are fast compared to the population dynamics can now be rephrased more precisely as $\epsilon \ll 1$ and $\mathcal{R}_0$ being not too small compared to 1. We expect this system to behave similarly to the system

$$\begin{aligned} 0 &= 1 - z - \mathcal{R}_0 z x, \\ \dot{x} &= x(\mathcal{R}_0 z - 1). \end{aligned}$$

This formal argument can be made rigorous using singular perturbation theory (Tikhonov et al., 1985; Hoppensteadt, 1974). From the first equation we get the *quasi-steady-state*

$$z = \frac{1}{1 + \mathcal{R}_0 x},$$

which we substitute into the second equation:

$$\dot{x} = x\left(\frac{\mathcal{R}_0}{1 + \mathcal{R}_0 x} - 1\right). \tag{5.8}$$

We return to unscaled time and consumer biomass and obtain the Beverton–Holt equation,

$$\begin{aligned} M' &= \mu M\left(\frac{\mathcal{R}_0}{1 + (\mathcal{R}_0 \mu / \Lambda\gamma) M} - 1\right) \\ &= M\left(\frac{\beta}{1 + \alpha M} - \mu\right), \end{aligned} \tag{5.9}$$

with

$$\beta = \mu \mathcal{R}_0 = \frac{\Lambda}{\nu} b, \qquad \alpha = \frac{\beta}{\Lambda\gamma}.$$

5.2.2 *Derivation from Cannibalism of Juveniles by Adults*

In order to derive the Beverton–Holt reproduction function from adults cannibalizing juveniles, let $N(t)$ denote the size of the adult population and $J(t)$ the size of the juvenile population. Then

$$J' = \beta_0 N - \nu_0 J - \kappa J N - \gamma J, \qquad N' = \gamma J - \mu N.$$

Here β_0 is the per capita birth rate of the population, ν_0 and μ are the natural per capita mortality rates of juveniles and adults, respectively, γ is the per capita rate of a juvenile turning into an adult, and κ is the per capita rate of juveniles being cannibalized by adults per capita of adults per unit of time. The cannibalism rate is modeled according to the law of mass action, i.e., it is proportional to both the size of the adult and the juvenile populations.

$1/\gamma$ is the average length of the juvenile period (death disregarded), and we assume that it is very short compared to the life expectancy of an adult, $1/\mu$, i.e.,

$$\mu = \epsilon\gamma,$$

with $\epsilon > 0$ being a very small dimensionless parameter. We also assume that cannibalism occurs on a similar time scale to leaving the juvenile stage, i.e.,

$$\epsilon\kappa = \tilde{\alpha}$$

for some parameter $\tilde{\alpha}$. Let $J = \epsilon x N$. Then

$$\epsilon x' + \epsilon x\mu(x-1) = \beta_0 - \nu_0\epsilon x - \tilde{\alpha}xN - \mu x, \qquad N' = \mu N(x-1).$$

Finally, we introduce the dimensionless time τ, $\tau = \mu t$; in this way our time unit becomes the life expectancy of the adult population, $1/\mu$. Let $\dot{z}$ and $\dot{x}$ denote the derivatives with respect to the new time τ, by the chain rule $\dot{z} = \mu z'$ and $\dot{x} = \mu z'$. Dividing both equations by μ we obtain

$$\epsilon\dot{x} + \epsilon x(x-1) = \frac{\beta_0}{\mu} - \frac{\nu_0}{\mu}\epsilon x - \frac{\tilde{\alpha}}{\mu}xN - x, \qquad \dot{N} = N(x-1).$$

Since ϵ is very small, we assume the solution of the first equation to behave as if $\epsilon = 0$,

$$0 = \beta_0 - \tilde{\alpha}xN - \mu x.$$

Solving this equation, we obtain the quasi-steady-state

$$x = \frac{\beta_0}{\mu + \tilde{\alpha}N},$$

the substitution of which into the equation $N' = \mu(x-1)$ again yields equation (5.9) with $\alpha = \tilde{\alpha}/\mu$.

If $\beta > \mu$, we can give equation (5.9) the following form suggested by Smith (1963) (via a different derivation (see also Hutchinson, 1978, Chapter 1)):

$$M' = \rho M \frac{1 - (M/K)}{1 + \alpha M} \tag{5.10}$$

with

$$\rho = \beta - \mu, \qquad K = \frac{\beta - \mu}{\alpha\mu}.$$

Notice that we formally recover the Verhulst equation (also known as the logistic equation) from (5.10) setting $\alpha = 0$. We realize that all positive solutions remain bounded, since $M' < 0$ for $M > K$; in fact $M(t) \leqslant \max\{M(0), K\}$ for all $t \geqslant 0$.

Let us try to find a solution of the Beverton–Holt equation in the form of Smith's equation. Equation (5.10) has two constant solutions, $M \equiv 0$ and $M \equiv K$. To obtain other solutions we separate the variables,

$$\rho = M' \frac{1 + \alpha M}{M[1 - (M/K)]} = M' \frac{1 - (M/K) + (\alpha + (1/K))M}{M[1 - (M/K)]}$$
$$= \frac{M'}{M} + \left(\alpha + \frac{1}{K}\right) \frac{M'}{1 - (M/K)} = \frac{(M/K)'}{(M/K)} + (1 + \alpha K) \frac{(M/K)'}{1 - (M/K)}.$$

Integrating,

$$\rho t + k = \ln \frac{|M|}{K} - (1 + \alpha K) \ln \left| 1 - \frac{M}{K} \right|, \tag{5.11}$$

with an integration constant k. It seems impossible to find an explicit solution unless $\alpha = 0$, in which case Smith's equation reduces to the logistic equation. But we see from this equation that $M(t) \to K$ as $t \to \infty$, because this is the only way in which the right-hand side can approach infinity (recall that all positive solutions are bounded). So, as with the logistic equation, K plays the role of the carrying capacity.

We can also use relation (5.11) to test whether given population data can be fitted by Smith's equation, provided that we can observe the population for long enough to get a good idea what the carrying capacity K may be.

Let $x = (k, \rho, 1 + \alpha K)^{\mathrm{T}}$ be the vector that contains the parameters we want to determine and $M_j = M(t_j)$, $j = 1, \ldots, n$, be measurements taken at times $t_1, \ldots, t_n$. Then we have the linear system

$$x_1 + t_j x_2 + \ln |1 - (M_j/K)| x_3 = \ln |M_j/K|, \quad j = 1, \ldots, n,$$

in matrix form $Ax = b$ with $b^{\mathrm{T}} = (\ln |M_1/K|, \ldots, \ln |M_n/K|)$,

$$A^{\mathrm{T}} = \begin{pmatrix} 1 & \cdots & 1 \\ t_1 & \cdots & t_n \\ \ln |1 - (M_1/K)| & \cdots & \ln |1 - (M_n/K)| \end{pmatrix}.$$

If $n > 3$, this is an overdetermined system which typically has no solution; its least-squares solution can be found by solving the linear system $A^{\mathrm{T}} A x = A^{\mathrm{T}} b$ (cf. Chapter 4). Fortunately this has already been programmed into packages like MAPLE, MATHEMATICA or EXCEL.

If we restrict our fitting exercise to the Verhulst equation, $\alpha = 0$, we obtain the relation

$$x_1 + t_j x_2 = \ln \left| \frac{M_j}{K - M_j} \right|, \quad j = 1, \ldots, n.$$

By (5.11), $K - M$ always has the same sign; most data are collected for the case in which M takes values between 0 and K,

$$x_1 + t_j x_2 = \ln \frac{M_j}{K - M_j}, \quad j = 1, \ldots, n.$$

If the data are collected at equidistant times $t_j = t_1 + (j-1)h$, we can try to use the following procedure for estimating K, which has already been suggested by Verhulst (1845, Section 8). By subtracting the last equation from the analogous one for $j+1$, we obtain

$$hx_2 = \ln \frac{M_{j+1}}{K - M_{j+1}} - \ln \frac{M_j}{K - M_j}, \qquad j = 1, \dots, n-1.$$

Equating the right-hand sides and rearranging,

$$\ln \frac{M_{j+2}}{K - M_{j+2}} + \ln \frac{M_j}{K - M_j} = 2 \ln \frac{M_{j+1}}{K - M_{j+1}}, \qquad j = 1, \dots, n-2.$$

We exponentiate and rearrange,

$$M_{j+1}^2 (K - M_{j+2})(K - M_j) = M_{j+2} M_j (K - M_{j+1})^2.$$

We expand this expression and solve for K,

$$K = M_{j+1} \frac{M_{j+2}M_{j+1} + M_{j+1}M_j - 2M_{j+2}M_j}{M_{j+1}^2 - M_{j+2}M_j}, \qquad j = 1, \dots, n-2.$$

For real data, we will obtain different values for K as $j = 1, \dots, n-2$, so we could take their average. If a population grows less than exponentially, one has

$$\frac{M_{j+2}}{M_{j+1}} < \frac{M_{j+1}}{M_j},$$

i.e., the denominator in the expression for K is positive. If the data are taken from a regime, however, where the growth is still close to being exponential, then the denominator is close to 0. The numerator can be rewritten as

$$2M_{j+1} \left(\frac{M_{j+2} + M_j}{2} - \sqrt{M_{j+2}M_j} + \frac{\sqrt{M_{j+2}M_j}}{M_{j+1}} \left[M_{j+1} - \sqrt{M_{j+2}M_j} \right] \right).$$

Since the arithmetic mean is larger than the geometric mean, this expression is positive and safely bounded away from 0. This means that our formula for K is very sensitive to errors in measurement and hardly practical, as long as the population growth is still almost exponential. In fact, it can easily happen that one obtains negative values if one uses real data. But even for data which level off for large times, I have found this procedure impractical because the values for K that one obtains vary too widely.

5.3 The Ricker Differential Equation

While we have assumed before that juveniles turn into adults at a constant rate, we now assume that the juvenile stage has a fixed length, τ, which is very short compared to the life expectancy of adults, $1/\mu$. So, individuals that enter the adult population at time t have been born at time $t - \tau$, and we obtain the equation

$$N'(t) = \beta_0 N(t-\tau) P(t, t-\tau) - \mu N.$$

As before, β_0 is the per capita birth rate and μ the per capita mortality rate of adults. $P(t,r)$ denotes the probability of a juvenile to survive from time r to time t, $t \geqslant r$. A similar derivation as for (2.2) expresses P in terms of a per capita juvenile mortality rate ν,

$$P(t,r) = \exp\left(- \int_r^t \nu(s)\, \mathrm{d}s \right).$$

We assume that the per capita juvenile mortality rate splits into a natural mortality rate and a mortality rate due to cannibalism,

$$\nu(t) = \nu_0 + \kappa N(t).$$

Here ν_0 is the natural per capita mortality rate, and κ is the per capita rate at which juveniles are cannibalized by adults, per capita of adults. In other words, $1/\kappa$ is the average time it takes a juvenile to be killed by one unit of adults. As in the previous section, we assume that $1/\kappa$ is on the same time scale as the length of the juvenile period,

$$\kappa = \alpha/\tau.$$

By substitution into the formula for P,

$$P(t, t-\tau) = \mathrm{e}^{-\nu_0 \tau} \exp\left(-\frac{\alpha}{\tau} \int_{t-\tau}^t N(s)\, \mathrm{d}s \right).$$

Taking the limit for $\tau \to 0$,

$$P(t, t-) = \mathrm{e}^{-\alpha N(t)}.$$

Taking also the limit $\tau \to 0$ in the original differential equation for N, we obtain the Ricker equation

$$N' = \beta_0 N \mathrm{e}^{-\alpha N} - \mu N. \tag{5.12}$$

Let us rewrite (5.12) in a form that displays the carrying capacity. Factoring out μ,

$$N' = \mu N \left(\frac{\beta_0}{\mu} \mathrm{e}^{-\alpha N} - 1 \right).$$

Set

$$K = \frac{1}{\alpha} \ln \frac{\beta_0}{\mu}.$$

Then

$$N' = \mu N (\mathrm{e}^{\alpha K (1 - (N/K))} - 1).$$

Now set

$$\rho = \mu(e^{\alpha K} - 1), \qquad \gamma = \alpha K.$$

Then

$$\rho = \beta_0 - \mu, \qquad \gamma = \ln\left(\frac{\beta_0}{\mu}\right),$$

and

$$N' = \rho N \frac{e^{\gamma(1-(N/K))} - 1}{e^{\gamma} - 1}.$$

Notice that we recover the logistic equation as the limit of this equation for $\gamma \to 0$ (apply l'Hôpital's rule). K again plays the role of the carrying capacity as the right-hand side of this equation is 0 for $N = K$.

5.4 The Gompertz Equation

The equation

$$N' = rN \ln(K/N),$$

with positive constants $r, K > 0$, which ultimately goes back to Gompertz (1825), does not fit into the fundamental balance equation of population dynamics, equation (2.1), with per capita birth and mortality rates. It has been very successful in fitting data of tumor growth, with N representing the number of tumor cells. This equation can be solved explicitly. Let

$$x = \ln(N/K) = \ln N - \ln K.$$

Then

$$x' = N'/N = -rx.$$

Thus $\ln|x| = -rt + k$ and

$$\ln(|\ln(N/K)|) = k - rt.$$

From this equation we see that $N(t) \to K$ as $t \to \infty$ and $N(t) \to 0$ as $t \to -\infty$ if $N(0) < K$, while $N(t) \to \infty$ as $t \to -\infty$ if $N(0) > K$. As in the logistic equation, we can estimate K from the large-time behavior of the data and r and k by least-squares fitting as in Chapter 4. This is a further reason for the popularity of the Gompertz equation.

5.5 A First Comparison of the Various Equations

In order to compare these various differential equations in the time-autonomous case, we write them in the respective forms that display the carrying capacity,

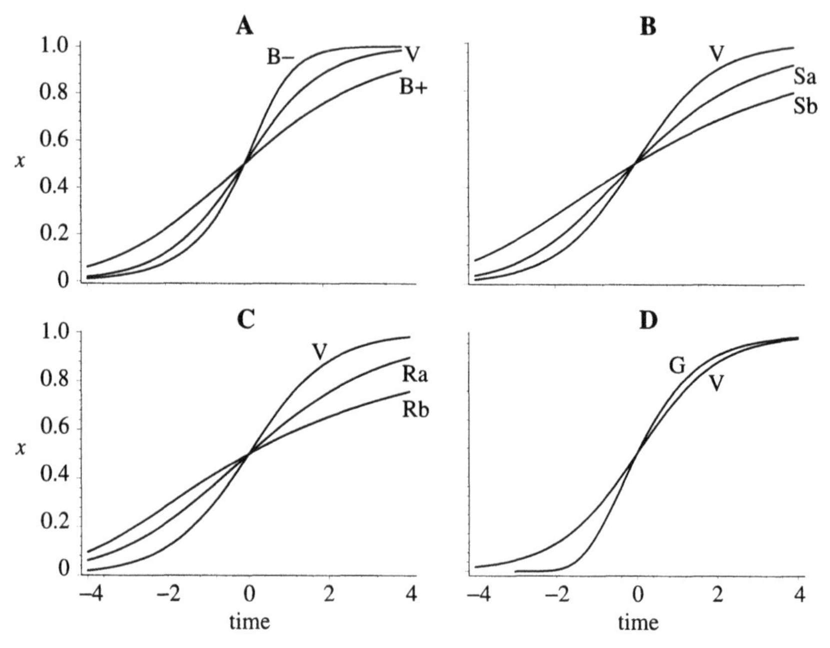

Figure 5.1. Comparison of the Verhulst (logistic) equation (labelled "V" and shown in all parts of the figure), (**A**) the Bernoulli equation, (**B**) the Smith equation, (**C**) the Ricker equation, and (**D**) the Gompertz equation. (**A**) V, $\theta = 1$; B−, $\theta = 0.5$; B+, $\theta = 2$. (**B**) V, $\alpha = 0$; Sa, $\alpha = 1$; Sb, $\alpha = 3$. (**C**) V, $\gamma \to 0$; Ra, $\gamma = 1$; Rb, $\gamma = 3$. See the text for further explanation.

K. Scaling the population size, we can achieve that $K = 1$. Scaling time, we can make an additional parameter equal to 1, $\rho = 1$ or $r = 1$, respectively:

$$x' = x(1 - x) \qquad \text{(Verhulst, logistic)},$$

$$x' = x(1 - x^{\theta}), \qquad \theta > 0 \qquad \text{(Bernoulli)},$$

$$x' = x\frac{1 - x}{1 + \alpha x}, \qquad \alpha \geqslant 0 \qquad \text{(Smith, Beverton–Holt)},$$

$$x' = x\frac{\mathrm{e}^{\gamma(1-x)} - 1}{\mathrm{e}^{\gamma} - 1}, \qquad \gamma \geqslant 0 \qquad \text{(Ricker)},$$

$$x' = -x \ln x \qquad \text{(Gompertz)}.$$

As we have noticed before, the Verhulst equation is a special case of the Bernoulli equation with $\theta = 1$, and a limit case of the Smith equation with $\alpha = 0$ and of the Ricker equation with $\gamma \to 0$. In Figure 5.1 we have plotted solutions for the Bernoulli equation with $\theta = 1$ (Verhulst), $\theta = 0.5$, and $\theta = 2$; for the Smith equation with $\alpha = 0$ (Verhulst), $\alpha = 1$, and $\alpha = 3$; and for the Ricker equation with $\gamma \to 0$ (Verhulst), $\gamma = 1$, and $\gamma = 3$.

All solutions show an S-shaped growth, often called *sigmoid growth*. They increase from one equilibrium state, $x = 0$, to another one, $x = 1$, first in a convex and then in a concave manner. Notice that in all other examples the symmetry of the Verhulst equation is lost; in particular, the inflection point is no longer at $x = 0.5$. Only the Bernoulli equation is flexible enough to have the inflection point at both $x > 0.5$ (Figure 5.1A, B− for $\theta = 0.5$) and $x < 0.5$ (Figure 5.1A, B+ for $\theta = 2$). In the Smith, Ricker and Gompertz equations, the inflection point is at $x < 0.5$. In Chapter 6, we will find that these are general features that do not depend on the particular choices of the parameters we have made. See the remark between Lemma 6.2 and Theorem 6.4.

Chapter Six

Sigmoid Growth

Rather than analyzing the many differential equation models we derived in Chapter 5 individually, we try a unified approach. First we notice that the qualitative behavior of solutions of scalar ordinary differential equations is quite restricted.

Theorem 6.1. *Let $f : (0, \infty) \to \mathbb{R}$ be continuously differentiable, $N_0 \in (0, \infty)$. Then there exists an interval (a, b), $a \in [-\infty, 0)$, $b \in (0, \infty]$, and a unique strictly positive solution N of*

$$N' = f(N) \quad \text{on } (a, b), \quad N(0) = N_0,$$

with the following properties:

(i) *N is either constant (if $f(N_0) = 0$), or strictly monotone increasing on (a, b) (if $f(N_0) > 0$), or strictly monotone decreasing (if $f(N_0) < 0$).*

(ii) *If $c \in \{a, b\}$ is finite, then $N(t) \to \infty$ or $N(t) \to 0$ as $t \to c$, $t \in (a, b)$.*

(iii) *If $c \in \{a, b\}$ is not finite, then $N(t) \to \infty$, or $N(t) \to 0$, or $N(t) \to K \in (0, \infty)$ with $f(K) = 0$, as $t \to c$.*

Proof. The local existence and uniqueness theorem for ordinary differential equations provides a unique strictly positive solution N with $N(0) = N_0$ on an interval (a, b) with $a < 0 < b$.

If $f(N_0) = 0$, then N is identically equal to N_0 on (a, b) and the solution can be extended to $\mathbb{R}$.

Let us assume that $f(N_0) > 0$. Then $f(N(t)) > 0$ for all $t \in (a, b)$. Otherwise, by the intermediate value theorem, there were $t_0 \in (a, b)$, $t_0 \neq 0$, such that $f(N(t_0)) = 0$, and the uniqueness theorem would imply that N is identically equal to $N(t_0)$ on (a, b), which is different from N_0, a contradiction. So, $N' = f(N)$ is strictly positive on (a, b), and N is strictly increasing on (a, b), and the limits of $N(t)$ as $t \to b$ or $t \to a$ exist in $[0, \infty]$. If these limits are in the interval $(0, \infty)$, the local existence and uniqueness theorem allows us to extend the solution beyond a or b, respectively. Taking the union of all open intervals around 0 on which solutions of $N' = f(N)$, $N(0) = N_0$ exist, we arrive at an interval (a, b) and a unique solution N that cannot be extended to a larger interval. So, a and b are either not finite or the limits of N at these points are 0 or ∞. Now assume that $b = \infty$ and that the limit of $N(t)$, $t \to \infty$, is a number $K \in (0, \infty)$. Since $f(N(t)) > 0$ for all $t \in (a, \infty)$, $f(K) \geqslant 0$. Assume $f(K) > 0$. Then there exists some $r > a$ and some $\epsilon \in (0, f(K))$ such that $N'(t) = f(N(t)) \geqslant \epsilon$ for all $t \geqslant r$. So, for all $t \geqslant r$,

$$N(t) \geqslant N(r) + \int_r^t N'(s)\,ds \geqslant N(r) + \epsilon(t - r),$$

which implies that $N(t) \to \infty$ as $t \to \infty$, a contradiction. Similarly, we deal with the case $a = -\infty$. If $f(N_0) < 0$, the proof is the same, except that N is strictly decreasing. □

6.1 General Conditions for Sigmoid Growth

In the examples we derived in Chapter 5, the solutions have more specific properties. Rather than exploring them separately for each example, we treat them as special cases of the equation

$$N' = f(N),$$

with f having the following properties:

(1) f is continuous on $[0, \infty)$ and continuously differentiable on $(0, \infty)$.

(2) $f(0) = 0$.

(3) $f(N) < 0$ for some $N > 0$.

(4) There is at most one $N > 0$ such that $f'(N) = 0$.

(5) $\rho := \lim_{x \to 0+}(f(x)/x)$ exists, but is possibly infinite, and $f(N)/N < \rho$ for all $N > 0$.

In the Gompertz equation (Section 5.4), f is not defined for $N = 0$, but can be continuously extended by setting $f(0) = 0$. Property (5) holds in particular if $f(x)/x$ is strictly decreasing in $x > 0$, which is actually the case in all our previous examples. If ρ is finite, it is the right derivative of f at 0.

The parameter ρ is sometimes called the *intrinsic rate of natural increase*, the *innate capacity for increase*, or the *potential rate of increase*. If ρ exists but $\rho < f(N)/N$ for some $N > 0$, these names do not make sense anymore and the population is said to be subject to an Allee effect (cf. Chapter 7). If (5) holds (together with the other properties, as we assume throughout this section), there are two cases to consider.

Lemma 6.2.

(a) *If $\rho \leqslant 0$, then $f(N) < 0$ for all $N > 0$. If $\rho < 0$, then also $f'(N) < 0$ for all $N > 0$, with the possible exception of one point.*

(b) *Let $\rho > 0$. Then there exist uniquely determined numbers*

$$K > L > 0 \quad \textit{such that } f(K) = 0, \; f'(L) = 0. \tag{6.1}$$

Moreover, f is strictly positive on $(0, K)$ and strictly negative on (K, ∞), while f' is strictly positive on $(0, L)$ and strictly negative on (L, ∞).

Again K is called the *carrying capacity* of the environment or the *saturation level* of the population. The function f has a global (and local) maximum at L, so we call L the *population size with maximum growth.*

Proof. The proof of (a) is similar to (b) and left to the reader.

(b) Since $\rho > 0$, $f(N) > 0$ for small $N > 0$. By (3) and the intermediate value theorem, there exists some $K > 0$ such that $f(K) = 0$. By the mean value theorem, there exists some $L \in (0, K)$ such that $f'(L) = 0$. L is uniquely determined by (4) as is K. Indeed, if there were $0 < N \neq K$ such that $f(N) = 0$, then f' would have two positive zeros by the mean value theorem. Since f has only one sign on $(0, K)$ and (K, ∞), respectively, we conclude that f is strictly positive on $(0, K)$ and strictly negative on (K, ∞).

In order to show that f is strictly increasing on $[0, L]$, we observe that $0 < f(L) = f(L) - f(0) = f'(N)L$ for some $N \in (0, L)$ by the mean value theorem. So, $f'(N) > 0$ for some $N \in (0, L)$. Since L is the only zero of f', we have $f'(N) > 0$ for all $N \in (0, L)$. So, f is strictly increasing on $[0, L]$.

Since f' has one sign on (L, ∞), f' is either strictly positive or strictly negative throughout (L, ∞). If f' were strictly positive on (L, ∞), f would be strictly increasing and $f(N) > f(L) > 0$ for all $N > L$. So, f would be nonnegative on $[0, \infty)$, contradicting (3) that $f(N) < 0$ for some $N > 0$. Hence f' is strictly negative throughout (L, ∞) and f is strictly decreasing on (L, ∞). □

Remark 6.3. If we choose to model one-species growth by the Bernoulli equation, we automatically introduce the restriction $1 < K/L < \mathrm{e}$, with $K/L = 2$ for the special case of the Verhulst equation, while choosing the Beverton–Holt or the Ricker equation involves the restriction $K/L > 2$. For the Gompertz equation, one has $K/L = \mathrm{e}$ (see Exercise 6.2).

The next theorem confirms the S-shaped or *sigmoid* growth pattern often observed in laboratory experiments.

Theorem 6.4. *Consider $N' = f(N)$ with f satisfying the properties (1)–(5), $\rho > 0$. Then the following statements hold:*

(a) *Let $N(0) \in (0, K)$. Then there exists a unique nonnegative solution N defined on $\mathbb{R}$. It has the following properties:*

$$N(t) \to 0, \quad t \to -\infty, \qquad N(t) \to K, \quad t \to \infty.$$

Furthermore, $N(t)$ is a nondecreasing function of t which is strictly increasing where it is strictly positive. Furthermore, $N(t)$ is strictly convex as long as $0 < N(t) < L$ and strictly concave as soon as $L < N(t) < K$.

(b) *Let $N(0) > K$. Then there exists a unique solution defined on an interval (a, ∞) with $a \in [-\infty, 0)$ such that*

$$N(t) \to \infty, \quad t \searrow a, \qquad N(t) \to K, \quad t \to \infty.$$

Furthermore, $N(t)$ is a strictly decreasing convex function of t.

Proof. Let $N(0) > 0$. By Theorem 6.1, there exists a unique strictly positive and strictly monotone solution $N(t)$, defined on an interval (a, b) with $a \in [-\infty, 0)$, $b \in (0, \infty]$ with the behavior of N at the endpoints as described therein.

(a) Let $N(0) \in (0, K)$. Since the constant function K is a solution of $N' = f(N)$ and f is continuously differentiable in a neighborhood of K, uniqueness of solutions implies that $N(t) < K$ for all $t \in (a, b)$. Since f is strictly positive on $(0, K)$ and $N' = f(N)$, N is strictly monotone increasing and bounded above by K on (a, b). It follows from Theorem 6.1 that $b = \infty$ and that $N(t) \to K$ as $t \to \infty$.

If $a = -\infty$, a similar argument shows that $N(t) \to 0$ as $t \to -\infty$.

If $a > -\infty$, then $N(a+) = 0$ by Theorem 6.1 and the fact that N is strictly increasing. We extend the solution N by setting $N(t) = 0$ for $t \leqslant a$. One easily checks that N is continuously differentiable on $\mathbb{R}$ and satisfies the differential equation.

Finally, we have that

$$N'' = (f(N))' = f'(N)N' = f'(N)f(N).$$

Remember that $f(N) > 0$ for $N \in (0, K)$ and that, by Lemma 6.2, f' is strictly positive on $(0, L)$ and strictly negative on (L, ∞) with $L < K$. So, $N'' > 0$ as long as $N \in (0, L)$, and $N'' < 0$ as soon as $N \in (L, K)$.

(b) Let $N(0) > K$. Then, by uniqueness, $N(t) > K$ for all $t \in (a, b)$. Thus $f(N(t)) < 0$ for all $t \in (a, b)$ and $N(t)$ is strictly decreasing. By Theorem 6.1, this implies that $b = \infty$. An argument similar to the one in part (a) shows that $N(t) \to K$ as $t \to \infty$. Furthermore,

$$N'' = f'(N)f(N) > 0,$$

because $N > K > L$ and thus $f(N) < 0$ and $f'(N) < 0$. So, N is strictly convex. If $a > -\infty$, $N(t) \to \infty$ as $t \downarrow a$, by Theorem 6.1. If $a = -\infty$, $N(t) \to \infty$ as $t \to -\infty$, because N is not only strictly decreasing but also strictly convex on $\mathbb{R}$. $\square$

Remark 6.5. In part (a) of Theorem 6.4, as we have seen from the proof above, the solution N exists on an interval (a, ∞), $a < 0$, such that N is strictly positive on (a, ∞) and $N(t) \to 0$ as $t \downarrow a$. In order to determine a, we separate the variables and obtain

$$\int_{N(0)}^{N(t)} \frac{1}{f(x)} \,\mathrm{d}x = \int_0^t \frac{N'(s)}{f(N(s))} \,\mathrm{d}s = t, \qquad a < t < 0.$$

Taking the limit for $t \to a$,

$$a = \int_{N(0)}^{0} \frac{1}{f(x)} \,\mathrm{d}x. \tag{6.2}$$

a is finite if the improper integral in (6.2) is finite. If the integral in (6.2) is infinity, N is strictly positive throughout $\mathbb{R}$. In order to be able to discriminate between the two situations, we introduce the following terminology.

We call $1/f(x)$ *integrable at* 0 if there exists some $M > 0$ such that f has one sign on $(0, M)$ and

$$\int_0^M \frac{\mathrm{d}x}{|f(x)|} < \infty. \tag{6.3}$$

$1/f(x)$ is called *not integrable at* 0 if M exists as before, but the integral in (6.3) is infinity. So, we have the following alternative for solutions between 0 and the carrying capacity K.

If $1/f(x)$ is not integrable at 0, all these solutions are positive throughout $\mathbb{R}$.

If $1/f(x)$ is integrable at 0, they are all 0 on some interval $(-\infty, a)$ with a depending on the initial data via (6.2).

In part (b) of Theorem 6.4, the solution is defined on an interval (a, ∞) with $a < 0$ and $N(t) \to \infty$ as $t \downarrow a$. As before we have

$$\int_{N(0)}^{N(t)} \frac{\mathrm{d}x}{f(x)} = t, \qquad a < t < 0.$$

Taking the limit for $t \to a$,

$$a = \int_{N(0)}^{\infty} \frac{\mathrm{d}x}{f(x)}. \tag{6.4}$$

We have "blow up" at the finite backward time a if the integral in (6.4) is finite. If this integral is not finite, N is defined throughout $\mathbb{R}$. In order to be able to discriminate between the two situations, we introduce a similar terminology as before.

We call $1/f(x)$ *integrable at* ∞ if there exists some $M > 0$ such that f has one sign on (M, ∞) and

$$\int_M^{\infty} \frac{\mathrm{d}x}{|f(x)|} < \infty. \tag{6.5}$$

$1/f(x)$ is called *not integrable at* ∞ if M exists as before, but the integral in (6.5) is infinity. So, we have the following alternative for solutions with initial data larger than K.

If $1/f(x)$ is integrable at infinity, all these solutions blow up in finite backward time, i.e., there exists a finite time $a < 0$, determined by (6.4), such that $N(t) \to \infty$ as $t \downarrow a$.

If $1/f(x)$ is not integrable at infinity, all these solutions exist throughout $\mathbb{R}$ and $N(t) \to \infty$ as $t \to -\infty$.

There is an easy condition for $1/f(x)$ not to be integrable at 0.

Lemma 6.6. *If ρ in (5) is finite, $1/f(x)$ is not integrable at* 0.

Proof. Since $|f(x)|/x \to |\rho|$ as $x \searrow 0$, there exists some $M > 0$ such that f has one sign on $(0, M)$ and $|f(x)|/x \leqslant |\rho| + 1$ for all $x \in (0, M)$. Then, for all $\delta \in (0, M)$,

$$\int_\delta^M \frac{1}{|f(x)|}\,dx \geqslant \frac{1}{|\rho|+1}\int_\delta^M \frac{1}{x}\,dx = \frac{1}{|\rho|+1}(\ln M - \ln \delta).$$

Taking the limit for $\delta > 0$, we see that the integral on the left-hand side of the inequality is infinite.

If $\rho = \infty$, $1/f(x)$ may or may not be integrable at 0. The von Bertalanffy equation is an example for the first case and the Gompertz equation is an example for the second case (see Exercise 6.4).

So, if $\rho > 0$ and $N(0) < K$, the von Bertalanffy equation is the only one among the equations we have discussed where N is 0 on some interval $(-\infty, a]$. The endpoint a can be determined from (6.2) and coincides with the number t_- in Chapter 4, introduced shortly before (4.4).

If $\rho > 0$ and $N(0) > K$, the Bernoulli equation and the Verhulst equation as its special case display blow up in finite backward time a which can be determined from (6.4).

The solutions of the Beverton–Holt (and Smith), Ricker, and Gompertz equations are, in this situation, defined for all backward times, and $N(t) \to \infty$ as $t \to -\infty$ (see Exercise 6.5).

Actually, there is a more general pattern. If large populations sizes N make the per capita mortality rate $\mu(N)$ grow to infinity faster than a positive power of N, there is blow up in finite backward time. However, if $\mu(N)$ only grows like the logarithm of N, every solution is defined for all backward times. Similarly, if large population sizes do not affect the per capita mortality rate, but make the per capita birth rate arbitrarily small, then the solutions are defined for all backward times (see exercises 6.6–6.8).

The less interesting case $\rho \leqslant 0$ (see Lemma 6.2(a)) is stated without proof and left for Exercise 6.3. □

Theorem 6.7. *Consider $N' = f(N)$ with f satisfying (1)–(5). Let $\rho \leqslant 0$ and $N(0) > 0$. Then there exists a unique solution defined on an interval (a, ∞) with $a \in [-\infty, 0)$ such that*

$$N(t) \to \infty, \quad t \downarrow a, \qquad N(t) \to 0, \quad t \to \infty.$$

Furthermore, $N(t)$ is a strictly decreasing function of t. If $\rho < 0$, N is strictly convex. Finally, $a > -\infty$ if and only if $1/f(x)$ is integrable at ∞.

6.2 Fitting Sigmoid Population Data

Many laboratory populations show sigmoid growth—the question is whether their growth data can be fitted well to any of the models we have discussed. One of their features that I have not yet mentioned is the following. Each of the models can be rewritten in the form

$$N' = Ng(N), \quad N > 0,$$

with $g(N) = f(N)/N$ being strictly decreasing for $N > 0$. This means that $(\mathrm{d}/\mathrm{d}t) \ln N(t)$ is strictly decreasing, i.e., $\ln N(t)$ is a strictly concave (or convex down) function. Equivalently, $N(t+h)/N(t)$ is a strictly decreasing function of t for every $h > 0$. The logarithmic convexity of N provides a test as to whether we can expect to obtain a reasonable fit to our data. As an example let us revisit a classical laboratory experiment for the population growth of the bacterium *Escherichia coli* (McKendrick, Kesava Pai, 1911; cf. Hutchinson, 1978, p. 21, p. 23, figure 8). The results for McKendrick and Kesava Pai's curve 5 follow.

time (h)	0	0.5	1	2	3	4	5	6	7	8
number (millions)	0.176	0.280	0.608	3.87	28.2	74.2	127	150	149	154

As I understand it, McKendrick and Kesava Pai give the number of bacteria in the test tube, while Hutchinson (figure 8) uses millions of cells per milliliter.

When we plot the natural logarithm of these data, we see that the curve is slightly convex rather than concave in the early stage of bacteria multiplication; McKendrick and Kesava Pai call this a period of "latency," in which growth is slower than would be expected. But the deviation is slight enough not to destroy our hopes.

The tricky part consists of obtaining a good guess for the carrying capacity K. We will try 154.2. (Here I am cheating; I have tried various values for K and this choice has given me the most satisfactory results.) When we plot the data, it appears that the value L at which the growth curve switches from convexity to concavity is not far from $\frac{1}{2}K$, so first we try the logistic equation $N' = \rho N(1 - N/K)$.

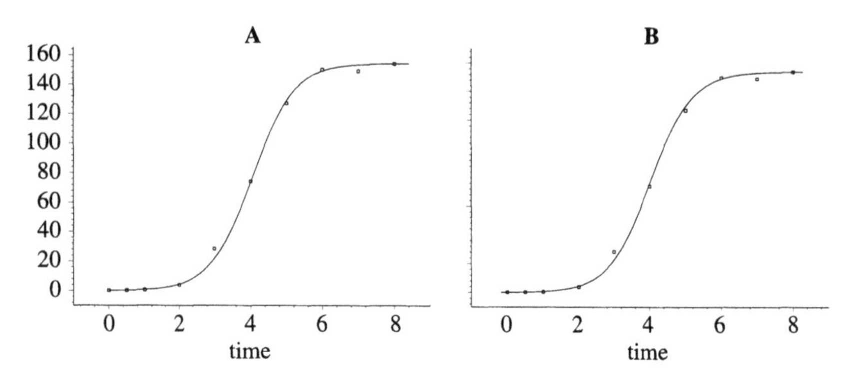

Figure 6.1. Two least-squares fits to *E. coli* data from McKendrick, Kesava Pai (1911). (**A**) Using the logistic equation. (**B**) Using Smith's equation. Since Smith's equation has one more degree of freedom, the fit in (**B**) is slightly better.

Fitting a cubic through the data points one finds that its inflection point has the value 77.104 455 80 as N-coordinate, but this may be a coincidence.

We normalize over K, $y = N/K$, and obtain $y' = \rho y(1-y)$. Separating the variables and using partial fractions,

$$\rho = \frac{y'}{y} + \frac{y'}{1-y}.$$

Integrating,

$$k + \rho t = \ln y - \ln(1-y).$$

This means that substituting the normalized data into the right-hand side and plotting the resulting values against time should give us a straight line. If we do so, the result is not perfect, but tolerable. In particular, it shows that we should ignore the last but one data point, (7,149), in any fitting attempt. A least-squares procedure provides the approximations $k \approx -7.004\,380\,749$ and $\rho \approx 1.727\,284\,027$. The resulting curve, together with the original data, are shown in Figure 6.1A.

To improve the fit, we try the Smith equation (which is equivalent to the Beverton–Holt equation), with the same carrying capacity, $K = 154.2$. After normalization over K,

$$k + \rho t + a\ln(1-y) = \ln y,$$

with $a = 1 + \alpha K$ (see Section 5.2). A least-squares fit now provides $k \approx -7.057\,657\,711$, $\rho = 1.765\,405\,962$, $a = 1.050\,099\,717$. We reverse the normalization,

$$k + \rho t + a\ln\left(1 - \frac{N}{K}\right) = \ln\left(\frac{N}{K}\right),$$

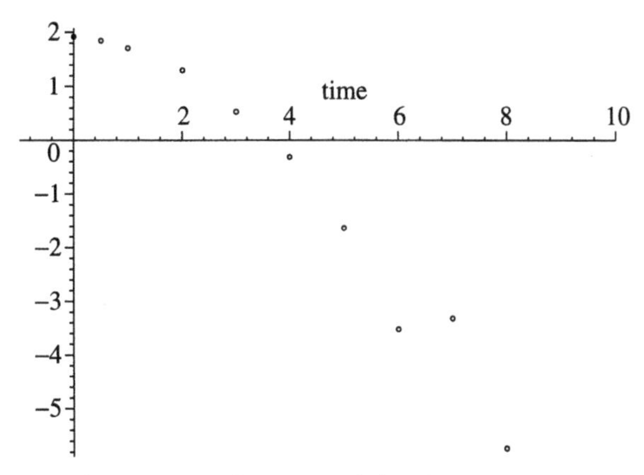

Figure 6.2. Plot of $\ln(\ln|N(t)|)$ for the *E. coli* data from McKendrick, Kesava (1911). Since there is no chance of fitting a line, Gompertz's equation is not a good model in this case.

and solve for t in terms of M. The solution is plotted together with the original data in the right-hand side of Figure 6.1. A MAPLE worksheet with detailed instructions can be found in Toolbox C.

We could also try to fit the Gompertz equation to the data. Plotting $\ln(\ln(K/N(t)))$ shows that this is not promising because the graph is apparently not close to a straight line (see Figure 6.2).

Keyfitz, Flieger (1990, p. 96) report the following population data for Sweden from 1950 to 1985 in millions.

year	1950	1955	1960	1965	1667	1970	1975	1980	1985
estimate	7.017	7.262	7.480	7.734	7.868	8.043	8.193	8.310	8.350

Plotting these data with a device like MAPLE shows that we are in the fortunate situation of being able to make reasonable guesses about the carrying capacity, K, and the point of inflection, L. See Theorem 6.4 and the data points in Figure 6.3A. We choose

$$K = 8.4, \qquad L = 7.8.$$

Since $K/L = 1.076\,923\,077$, Remark 6.3 suggests that we try the Bernoulli equation, which has the form (5.5) in terms of the carrying capacity,

$$N' = \rho N(1 - (N/K)^{\theta}) =: f(N).$$

L satisfies $f'(L) = 0$ by Lemma 6.2; this provides the relation

$$1 = (1+\theta)\left(\frac{L}{K}\right)^{\theta}.$$

Solving this equation, e.g., with MAPLE, yields

$$\theta := 54.098\,363\,52.$$

We have already observed in Section 5.1 that we can transform the Bernoulli equation to a logistic equation, namely

$$y = (N/K)^{\theta}, \qquad y' = \tilde{\rho} y(1-y), \qquad \tilde{\rho} = \theta\rho.$$

Separating the variables and using partial fractions

$$\tilde{\rho} = \frac{y'}{y(1-y)} = \frac{y'}{y} + \frac{y'}{1-y}.$$

Integrating,

$$k + \tilde{\rho} t = \ln y - \ln(1-y) = \ln\left(\frac{y}{1-y}\right).$$

This is a linear equation for k and $\tilde{\rho}$, and k and $\tilde{\rho}$ can be determined by a least-squares procedure after one has transformed the population data as well. It will be convenient to shift the time such that the year 1950 becomes 0. We obtain the following estimates,

$$k = -9.305\,258\,210, \qquad \tilde{\rho} = 0.317\,193\,374\,8.$$

Solving for y and reversing the shift in time and the transformation in the dependent variable, $N = Ky^{1/\theta}$, gives a solution for N which we can plot and compare with the original data (Figure 6.3B).

At some point we should look back critically. The Bernoulli equation has the original form

$$N' = N(\beta - m_0 - mN^{\theta}),$$

where β and m_0 are the density-independent components of the per capita birth and mortality rates. In Section 5.1 we chose a combined per capita mortality rate $\mu(N) = m_0 + mN^{\theta}$ for the simple reason that the combination $\beta(N) = \beta - mN^{\theta}$ would give us negative per capita birth rates for large N. A look at the data in Keyfitz, Flieger (1990, p. 278) shows, however, that the life expectancy of both males and females increased in Sweden from 1950 to 1985, which means that the per capita mortality rate decreased, and that the number of births also decreased. From this point of view, the Beverton–Holt equation or the Ricker equation should work better, but in practice they do not because K/L is too close to 1.

Our fitting exercise was very crude in so far as it neglected the age-structure of the population (see Part 2). Keyfitz, Flieger (1990, p. 278) took this into consideration and predicted that, starting from 1990, the Swedish population would actually decrease, while we assume that it would level off at 8.4 mil-

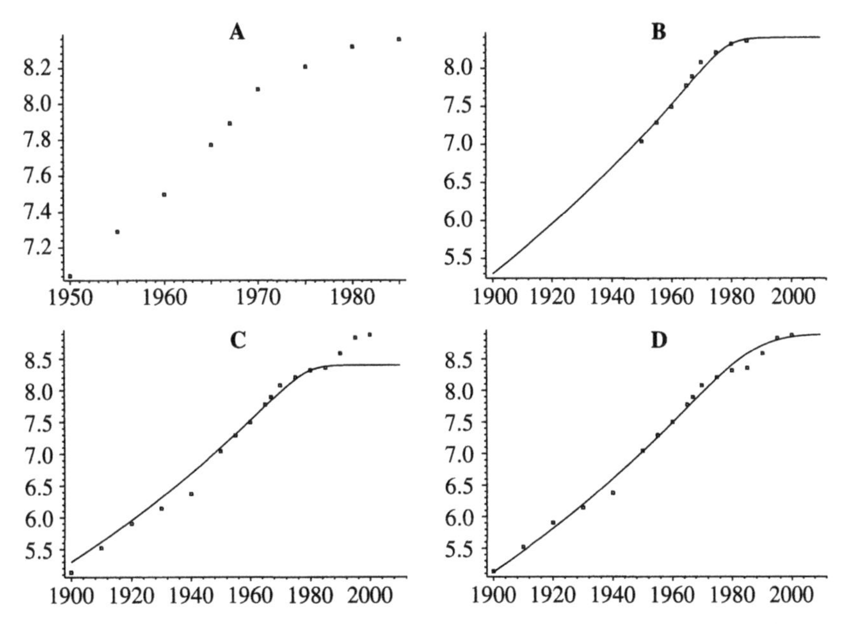

Figure 6.3. Two fits of the Bernoulli equation to the growth of the population of Sweden. (**A**) Plot of data from Keyfitz, Fieger (1990, p. 96). (**B**) Fit to the data in (**A**). (**C**) Comparison of the fit in (**B**) to a second data set which is both more recent and goes further back into the past (see text for the source). (**D**) Fit to the second data set.

lion. Today, population data are just a mouse (or track-ball) click away provided one knows the right internet address (http://www.scb.se/eng/befovalfard/befolkning/befstor/befarlig/befarligutvtab.asp), which was kindly provided to me by Torsten Lindström. There we see that, in 2000, the Swedish population was at almost 8.883 million. The reason that both we and Keyfitz, Flieger were wrong is that immigration picked up considerably after 1985. Since our nonlinear models assume a closed population, immigration and emigration are not included at all. Looking at the past, we notice that our fit overestimates the population number before 1950 (Figure 6.3C). If we redo our fitting exercise with the internet data (notice that the 1950–1985 internet data are slightly different from the mid-year data that Keyfitz and Flieger used), we end up with a guess of $K = 8.9$, $L = 7.9$. From there, we calculate $\theta = 24.569\,569\,68$, and estimate $k = -5.781\,584\,566$ and $\tilde{\rho} = 0.155\,701\,102\,9$. The associated curve can be found in Figure 6.3D.

Exercises

6.1. Consider all the special cases: von Bertalanffy, Bernoulli (both with constant coefficients), Ricker, Beverton–Holt, and Gompertz.

Show that they fit into the following framework:

$$N' = f(N)$$

with f having the properties (1)–(5).

6.2. Consider the Bernoulli, Ricker, and Beverton–Holt (in the form of Smith) equations.

(a) Investigate the flexibility of these equations: given any triple ρ, K, L with $\rho > 0$, $K > L > 0$, can you choose the parameters in the equations in such a way that $f'(0) = \rho$, $f(K) = 0$, $f'(L) = 0$? If restrictions apply, spell them out.

(b) Are the parameters in these equations uniquely determined by K, L, and $\rho = f'(0)$.

Hint: you may like to solve (a) and (b) simultaneously for each equation.

6.3. Prove Theorem 6.7.

6.4. (a) For von Bertalanffy show that $\rho = \infty$ and $1/f(x)$ is integrable at 0.

(b) For Gompertz show that $\rho = \infty$ and $1/f(x)$ is not integrable at 0.

6.5. Investigate whether or not the von Bertalanffy, Bernoulli (both with constant coefficients), Ricker, Beverton–Holt, and Gompertz equations are integrable at infinity.

6.6. Let $f(N) = N(\beta - \mu(N))$ with $\beta > 0$ being constant. Assume that there are $\epsilon, \theta > 0$ such that $\mu(N) \geqslant \epsilon N^{\theta}$. Then $1/f(N)$ is integrable at infinity.

6.7. Let $f(N) = N(\beta(N) - \mu(N))$ with $\beta(N)$ being a strictly positive and decreasing function of N and μ such that $\mu(N)/\ln N$ has a finite strictly positive limit as $N \to \infty$. Then $1/f(N)$ is not integrable at infinity.

6.8. Let $f(N) = N(\beta(N) - \mu)$ with $\mu > 0$ being constant, β nonnegative, and $\beta(N) \to 0$ as $N \to \infty$. Then $1/f(N)$ is not integrable at infinity.

6.9. Run numerical experiments with the Verhulst, Bernoulli, Ricker, and Smith equations and plot and explore the solution graphs in the case $\rho > 0, 0 < N(0) < K$. In order to compare the solution graphs choose your parameters such that K and ρ are the same.

6.10. Fit the Bernoulli equation to the following population data for Sweden (population figures are in millions) (http://www.scb.se/eng/befovalfard/befolkning/befstor/befarlig/befarligutvtab.asp).

year	1900	1910	1920	1930	1940	1950	1955	1960	1965
population	5.136	5.522	5.904	6.142	6.371	7.042	7.290	7.498	7.7725

year	1967	1970	1975	1980	1985	1990	1995	2000
population	7.893	8.081	8.208	8.318	8.358	8.591	8.837	8.883

Hint: modify the appropriate MAPLE worksheet in Toolbox C.

Chapter Seven

The Allee Effect

The term *Allee effect* seems to originate from the work of Allee (1931, 1951), though it has developed a life of its own. While the models considered in Chapters 5 and 6 assume that an increase in population density always has a negative effect on the reproduction and survival of a single individual, Allee type models only assume this for high densities, while for low densities an increase in density may be beneficial.

Possible mechanisms may be satiation of a generalist predator, group defense against a predator, or the necessity of finding a mate for reproduction.

We start from the Verhulst model and add an extra mortality term that decreases with increasing density,

$$N' = \phi N\left(1 - \frac{N}{K} - \frac{\eta}{1+\gamma N}\right). \tag{7.1}$$

Alternatively, we could start from the Beverton–Holt model, or Smith's version of it. We will present two derivations of equation (7.1): one modeling the search for a mate, the other the impact of a satiating generalist predator.

7.1 First Model Derivation: Search for a Mate

Let N denote the size of the female population. We assume a one-to-one sex ratio such that we do not need to model the males explicitly. We split the female population into two parts: one part consists of females that have recently mated and can therefore reproduce—the size of this part is denoted by N_1; and the second part consists of females who have not mated recently and so cannot reproduce at this point and are therefore looking for a mate—the size of this part is denoted by N_2. We have

$$N = N_1 + N_2.$$

The per capita reproduction rate for recently mated females (in terms of female offspring) is β, the per capita mortality rate of recently mated females is $\mu + \nu N$, while the per capita mortality rate of females looking for a mate is $\mu + \lambda + \nu N$. Here λ is the extra mortality rate which a searching female may experience because she needs to leave her well-sheltered home range and expose herself to a higher risk of predation or accidental death.

We assume that the females switch from the reproductive stage to the search stage at a constant rate σ. This means that the sojourn in the reproductive stage is exponentially distributed and that the mean sojourn time is $1/\sigma$ (cf. equation (2.3) and its derivation). When this period is over, the female starts looking for a mate. If there are N females, then there are also N males because we assume a one-to-one sex ratio. So, we assume that the per capita rate at which a searching female finds one out of N potential mates is ξN. These assumptions result in the following system:

$$\begin{aligned} N_1' &= \beta N_1 - (\mu + \nu N)N_1 - \sigma N_1 + \xi N N_2, \\ N_2' &= -(\mu + \lambda + \nu N)N_2 + \sigma N_1 - \xi N N_2. \end{aligned}$$

We rewrite the system in terms of N and N_2:

$$\begin{aligned} N' &= \beta N - (\mu + \nu N)N - (\beta + \lambda)N_2, \\ N_2' &= -(\mu + \lambda + \nu N)N_2 + \sigma(N - N_2) - \xi N N_2. \end{aligned}$$

We rewrite the equation for the size of the total female population, N, in terms of the carrying capacity, K, of the underlying logistic model,

$$K = \frac{\nu}{\beta - \mu},$$

assuming that $\beta - \mu > 0$,

$$\begin{aligned} N' &= (\beta - \mu)N\left(1 - \frac{N}{K}\right) - (\beta + \lambda)N_2, \\ N_2' &= -(\mu + \lambda + \nu N)N_2 + \sigma(N - N_2) - \xi N N_2. \end{aligned}$$

We do this because we want to justify taking the quasi-steady-state for the N_2 equation by a time scale argument. To do this properly, we want to express the system in dimensionless variables and parameters. The carrying capacity K serves as our reference population size as we introduce new, dimensionless variables

$$N = xK, \qquad N_2 = yK.$$

Then

$$\begin{aligned} x' &= (\beta - \mu)x(1 - x) - (\beta + \lambda)y, \\ y' &= -(\mu + \lambda)y - (\nu + \xi)Kxy + \sigma(x - y). \end{aligned}$$

We make $1/\mu$ our reference time (the life expectancy under the ideal, but unrealistic, circumstances that one neither experiences negative density effects nor additional mortality while searching for a mate) by introducing the new dimensionless time $s = \mu t$. Let $\dot{x}$, $\dot{y}$ denote the derivatives with respect to this new time. Then, setting

$$\epsilon = \frac{\mu}{\sigma}, \qquad \mathcal{R}_0 = \frac{\beta}{\mu},$$

we obtain

$$\dot{x} = (\mathcal{R}_0 - 1)x(1 - x) - \frac{\beta + \lambda}{\mu}y,$$

$$\epsilon y' = -\epsilon y - \frac{\lambda}{\sigma}y - \frac{\nu + \xi}{\sigma}Kxy + x - y.$$

We now assume that the average time $1/\sigma$ to be reproductive after mating and the average time needed to find a mate when the population is at carrying capacity, $1/(\xi K)$, are very short compared to the average life expectancy of a female, i.e., σ and ξK on one side are much larger than μ on the other side. We make no specific assumptions for νK and λ, which may either have the order of magnitude of σ or of μ; in the second case they can be neglected. We further assume that $\mathcal{R}_0 = \beta/\mu$, the number of female offspring produced by one average female during her lifetime, is not too large.

By these assumptions, $0 < \epsilon \ll 1$. Hence we make the quasi-steady-state approximation $\epsilon = 0$, i.e.,

$$\dot{x} = (\mathcal{R}_0 - 1)x(1 - x) - \frac{\beta + \lambda}{\mu}y,$$

$$0 = x - y - \frac{\lambda}{\sigma}y - \frac{(\nu + \xi)K}{\sigma}xy.$$

This approximation can be justified by singular perturbation theory (Hoppensteadt, 1974; Tikhonov et al., 1985); here we only argue formally. One can also analyze the planar system and will obtain similar results at a greater mathematical expense.

We solve the second equation for y,

$$y = \frac{x}{1 + (\lambda/\sigma) + (\nu + \xi)(K/\sigma)x},$$

and substitute the result into the differential equation for x,

$$\dot{x} = (\mathcal{R}_0 - 1)x(1 - x) - \frac{\beta + \lambda}{\mu}\frac{x}{1 + (\lambda/\sigma) + ((\nu + \xi)K/\sigma)x}.$$

Going back to N and the original time, we obtain (7.1) with

$$\phi = \beta - \mu, \qquad \eta = \frac{\beta + \lambda}{\mu}\frac{1}{1 + (\lambda/\sigma)} = \frac{\beta + \lambda}{\mu + \epsilon\lambda}, \qquad \gamma = \frac{\nu + \xi}{\sigma + \lambda}.$$

7.2 Second Model Derivation: Impact of a Satiating Generalist Predator

A *generalist* predator (as opposed to a *specialist* predator) preys not only on the species we are interested in, but also on other species, to the degree that its own population dynamics are not affected by the abundance of this particular prey. Let N be the population size of the species under consideration and P the size of the generalist predator population, which we assume to be constant. A predator may be hungry or satiated, so

$$P = H(t) + S(t)$$

with $H(t)$ being the size of the hungry part and $S(t)$ the size of the satiated part of the predator population, at time t. We assume that our population follows logistic (or Verhulst) growth in the absence of the predator. We have the following system of equations:

$$N' = \phi N\left(1 - \frac{N}{K}\right) - \xi NH,$$
$$H' = \sigma S - \xi NH - \zeta H = \sigma(P - H) - \xi NH - \zeta H.$$

σ is the rate at which a satiated predator becomes hungry again; in other words, $1/\sigma$ is the average duration of satiation (cf. formula (2.3)). The predation process is modeled by the law of mass action, the rate of prey killed, ξNH, is proportional to both the numbers of predators and prey. A predator that kills the prey enters the class of satiated predators. ζ is the per capita rate at which predators kill alternative prey. We introduce the dimensionless variables

$$N = xK, \qquad H = yP$$

in which the system takes the form

$$x' = \phi x(1 - x) - \xi Pxy,$$
$$y' = \sigma(1 - y) - \xi Kxy - \zeta y.$$

We choose $1/\phi$ as our reference time,

$$\frac{1}{\phi} = \frac{1}{\beta - \mu} = \frac{L}{\mathcal{R}_0 - 1}$$

with $L = 1/\mu$ being the life expectancy of the prey in the absence of intraspecies competition and of predators and $\mathcal{R}_0 = (\beta/\mu)$ the basic reproduction ratio of the prey population, i.e., the average number of offspring an average individual produces during the lifetime L. We introduce the dimensionless time $s = t\phi$ and let $\dot{x} = \phi x'$ and $\dot{y} = \phi y'$ be the derivatives with respect to the new time, by the chain rule. The system takes the form

$$\dot{x} = x(1 - x) - \frac{\xi P}{\phi}xy,$$
$$\epsilon\dot{y} = 1 - y - \frac{\xi K}{\sigma}xy - \frac{\zeta}{\sigma}y,$$

with $\epsilon = \phi/\sigma$. We assume that the duration of satiation of the predator, $1/\sigma$, is very short compared with the life expectancy of the prey under ideal conditions, $1/L$, and that the basic reproduction ratio is not very large. This implies $0 < \epsilon \ll 1$. Moreover, we assume that $\xi P/\phi$, $\xi K/\sigma$, and ζ/σ are of the order of magnitude of 1. Since $\phi = \epsilon\sigma$ with ϵ very small, this assumption has the consequence that the carrying capacity of the prey population, K, must be much larger than the number of predators, P. Setting $\epsilon = 0$, and solving for the quasi-steady-state of y, we obtain

$$y = \frac{1}{1 + (\zeta/\sigma) + (\xi K/\sigma)x}.$$

Substituting this expression into the x equation,

$$\dot{x} = x(1-x) - \frac{\xi P}{\phi} \frac{x}{1 + (\zeta/\sigma) + (\xi K/\sigma)x}.$$

Returning to the unscaled variable N, we have

$$\dot{N} = N\left(1 - \frac{N}{K}\right) - \frac{\xi P}{\phi} \frac{N}{1 + (\zeta/\sigma) + (\xi/\sigma)N}.$$

We obtain (7.1), remember $\dot{N} = \phi N'$, by setting

$$\eta = \frac{\xi}{\phi} \frac{1}{1 + (\zeta/\sigma)} = \frac{\xi}{\phi + \epsilon\zeta}, \qquad \gamma = \frac{\xi}{\sigma} \frac{1}{1 + (\zeta/\sigma)} = \frac{\xi}{\sigma + \zeta}.$$

7.3 Model Analysis

Before we start analyzing equation (7.1), we simplify it by scaling. We introduce a new dependent variable

$$x = \gamma N.$$

Then the equation takes the form

$$x' = \phi x\left(1 - \frac{x}{\gamma K} - \frac{\eta}{1 + x}\right).$$

Furthermore, we scale the independent variable by introducing a new time setting $z(s) = x(s/\phi)$. Actually, we keep the variable x but indicate differentiation with respect to the new time by $\dot{x}$. Then

$$\dot{x} = x\left(1 - \frac{x}{M} - \frac{\eta}{1 + x}\right) =: f(x), \tag{7.2}$$

where $M = \gamma K$.

From the analysis of sigmoid growth in the last section, we have seen that equilibria (or equilibrium points, critical points, steady states) play a crucial role in the analysis of the qualitative behavior of the solutions. The equilibria are constant solutions of the differential equations and are those numbers for which the right-hand side of (7.2) is 0:

$$0 = 1 - \frac{x}{M} - \frac{\eta}{1+x}.$$

Whether this equation has solutions $x > 0$ (the biologically relevant ones) will depend on η. Actually, it is more efficient to study η as a function of x than vice versa,

$$\eta = -\frac{x}{M}(1+x) + x + 1 = \frac{1}{M}(M-x)(x+1).$$

η, as a function of x, is a concave parabola. The value of η for $x = 0$, $\eta_0 = 1$, is of particular importance, because it is at this value for η where a so-called branch of positive equilibria bifurcates from the trivial equilibrium 0 which represents the extinction state of the population. See Figure 7.1, where M has been chosen to be 4. In order to find out where the parabola takes its maximum, we differentiate η with respect to x:

$$\frac{d\eta}{dx} = -\frac{2x}{M} + 1 - \frac{1}{M}.$$

Hence the maximum is attained at

$$x_\diamond = \tfrac{1}{2}(M-1),$$

and the value of η at this point is

$$\eta_\diamond = \frac{1}{M}(\tfrac{1}{2}(M+1))^2.$$

The interesting case is the one where $M > 1$. Otherwise the qualitative behavior is not much different from that studied in the previous section, because, for all $\eta \geqslant 0$, there would be at most one positive equilibrium. We change the point of view and consider how the equilibria depend on the parameter η (cf. Figure 7.1).

Lemma 7.1. *Let $M > 1$. If $\eta > \eta_\diamond$, then there exists no positive equilibrium. If $\eta = \eta_\diamond$, there exists exactly one positive equilibrium, namely $x_\diamond$. If $1 < \eta < \eta_\diamond$, there exist two positive equilibria. If $0 \leqslant \eta \leqslant 1$, there exists exactly one nontrivial equilibrium.*

The point $(1, 0)$ in the (η, x) plane is called a *bifurcation point* since a branch of nontrivial solutions bifurcates from the branch of trivial solutions $(\eta, 0)$. For obvious reasons, the point $(\eta_\diamond, x_\diamond)$ is called a *turning point* because the branch of nontrivial solutions changes direction at this point. Notice that, in the notation of the previous section, $\rho = f'(0)$ has the same sign as $1 - \eta$.

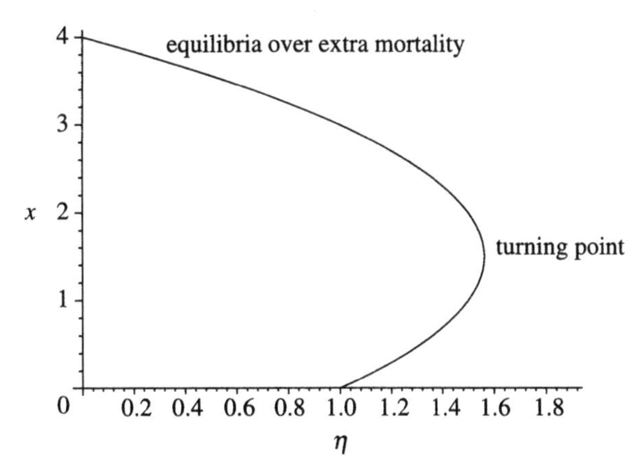

Figure 7.1. Equilibria plotted against the extra per capita mortality rate η for $M = 4$. The turning point has the coordinates $(\eta_\diamond, 1.5)$ with $\eta_\diamond = 1.5625$. Two positive equilibria exist for $\eta \in (1, \eta_\diamond)$. The lower equilibrium is the unstable watershed equilibrium, the upper one is stable.

Let us analyze the qualitative behavior of solutions in the case that $1 < \eta < \eta_\diamond$, i.e., $\rho = f'(0) < 0$. Let x_1 and x_2 be the two positive equilibria that exist according to Lemma 7.1, $0 < x_1 < x_\diamond < x_2$. We rewrite (7.2) as

$$\begin{aligned} \dot{x} &= \frac{x}{M(1+x)}((M-x)(1+x) - M\eta) \\ &= \frac{x}{1+x} \frac{(x_2 - x)(x - x_1)}{1 + x_1 + x_2}. \end{aligned} \tag{7.3}$$

The constants M and η are related to the equilibria by

$$M = 1 + x_1 + x_2, \qquad \eta = 1 + \frac{x_1 x_2}{1 + x_2 + x_2}.$$

The right-hand side of (7.3) is strictly negative for $x \in (0, x_1)$, strictly positive for $x \in (x_1, x_2)$ and strictly negative for $x > x_2$. Since the right-hand side of (7.3) is locally Lipschitz continuous, solutions cannot cross the equilibrium solutions 0, x_1, and x_2. So, as $t \to \infty$, solutions starting in $(0, x_1)$ approach 0, while solutions starting in (x_1, ∞) approach x_2. The equilibrium solution x_1 acts as a so-called *separatrix*. Solutions starting in $(0, x_2)$ approach x_1 for $t \to -\infty$ (cf. Theorem 6.1). x_1 is sometimes called the *watershed equilibrium*. If the population were a pest one would like to eradicate, x_1 is also called the *breakpoint density*. In this situation one speaks about *bi-stability*, because there are two stable equilibria, 0 and x_2, and it depends on the initial data to which of the two a solution converges as time tends to infinity.

A similar analysis as in Theorem 6.4 shows that solutions starting in (x_1, x_2) have a sigmoid or S-shaped increasing graph, while solutions starting in $(0, x_1)$

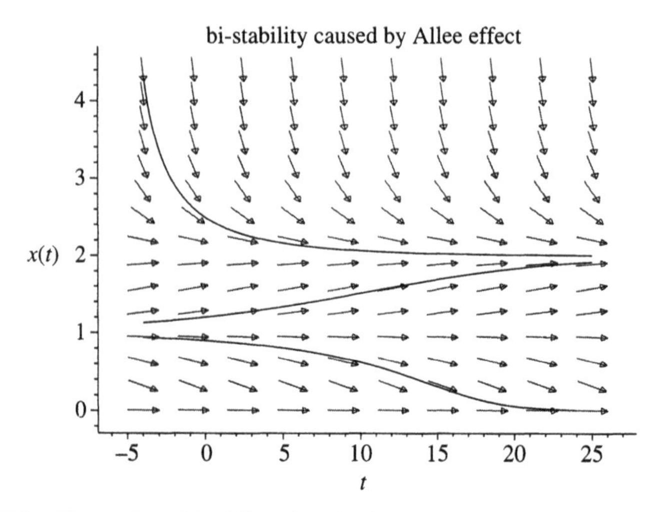

Figure 7.2. Illustration of the Allee effect for the constant extra mortality rate $\eta = 1.5$. The stable equilibria are 0 and $x_2 = 2$. $x_1 = 1$ is the watershed equilibrium which separates their domains of attraction.

have a sigmoid decreasing graph. Solutions starting in (x_2, ∞) are convex and decreasing. See Figure 7.2 for an illustration with $x_1 = 1$ and $x_2 = 2$, i.e., $M = 4$ and $\eta = 1.5$. For this value of M, $\eta_\diamond = 1.5625$.

Similarly, we analyze the other cases. If $\eta > \eta_\diamond$, all solutions converge to 0 as $t \to \infty$.

If $\eta = \eta_\diamond$, there is exactly one positive equilibrium which coincides with the turning point $x_\diamond$. All solutions which start in $[x_\diamond, \infty)$ approach $x_\diamond$ as $t \to \infty$, and all solutions starting in $[0, x_\diamond)$ approach 0 as $t \to \infty$ and $x_\diamond$ as $t \to -\infty$. In this case, $x_\diamond$ is a so-called *semistable* equilibrium.

If $0 \leqslant \eta \leqslant 1$, all solutions starting in $(0, \infty)$ approach the unique nontrivial equilibrium.

The Allee effect is related to the following phenomena, which are of particular concern to species conservation and reestablishment.

First, if the population is in the scenario with two positive equilibria, overhunting or overfishing may drive the population size into the interval $[0, x_1)$ and the population will become extinct even if the overhunting stops.

Secondly, fragmentation of the habitat may lead to a gradual, but steady, increase of the extra mortality rate η experienced during mate search or from a specialist predator.

For fixed η, the solutions converge towards an equilibrium. If the increase of η is slow, the solutions will stay close to the equilibrium which slowly changes with η. If $\eta < 1$, they will be close to the unique positive equilibrium. When η enters the interval $(1, \eta_\diamond)$, the solutions stay close to the stable upper branch of the positive equilibria. Finally, when η enters the interval $(\eta_\diamond, \infty)$,

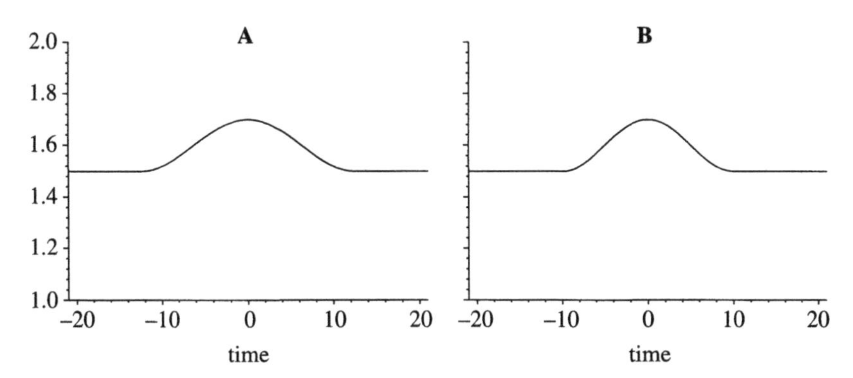

Figure 7.3. Two different time-dependent extra mortality regimes. **(A)** $\eta_1(t) = \eta_0(0.08t)$, **(B)** $\eta_2(t) = \eta_0(0.1t)$, with η_0 as in the text.

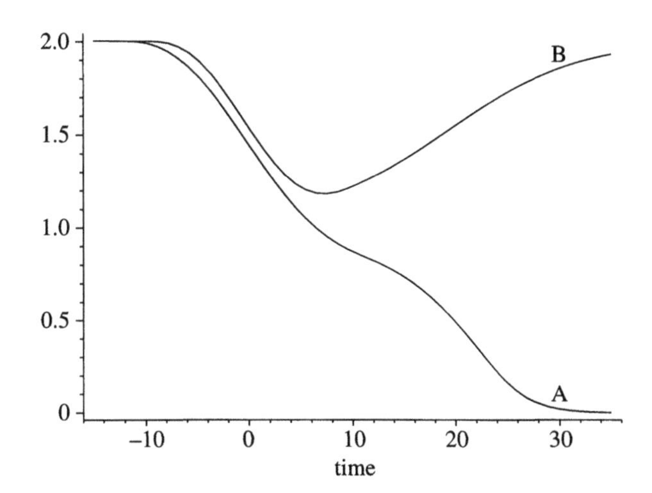

Figure 7.4. Population development under the extra mortality regimes (A) and (B) from Figure 7.2. Recall that the equilibria for constant $\eta = 1.5$ are 0, 1, and 2. The initial condition is $x(-20) = 2.01$ in both cases.

the population collapses out of the blue. Unless the fragmentation of the habitat is reversed fast, the population size continues to decrease and drops to such a level that it will die out, even if the original state is restored after a while.

We illustrate this effect by modifying the example of Figure 7.2 with a time-dependent η and $M = 4$. We start from

$$\eta_0(t) = \begin{cases} 1.5, & t \neq (-1, 1), \\ 1.5 + 0.1(1 + \cos(\pi t)), & t \in [-1, 1]. \end{cases}$$

Recall that 1.5 is the constant value of η in Figure 7.2. Furthermore, $\eta_0(0) = 1.7$, which is slightly larger than $\eta_\diamond = 1.5625$. We consider two versions of η_0 which are scaled differently in time,

$$\eta_1(t) = \eta_0(0.08t), \qquad \eta_2(t) = \eta_0(0.1t).$$

The raised levels of η_1 and η_2 at intermediate times reflect an increased extra mortality experienced during mate search or by predators (see Figure 7.3). Figure 7.4 shows the solutions $x_1(t)$ and $x_2(t)$ of (7.2) with $\eta = \eta_1(t)$ and $\eta = \eta_2(t)$, respectively. When the extra mortality is raised for a long time, the population dies out; when it is raised for a short time only, the population recovers.

If a species that is subject to an Allee effect has locally become extinct and is to be reintroduced into its former habitat, it is necessary to start with sufficiently many individuals, because the population will will die out again otherwise.

Exercises

7.1. A model with mate search that is based on the Beverton–Holt rather than the Verhulst equation takes the form

$$\begin{aligned} N_1' &= \frac{\beta N_1}{1+\alpha N} - \mu N_1 - \sigma N_1 + \xi N N_2, \\ N_2' &= -(\mu + \lambda) N_2 + \sigma N_1 - \xi N N_2. \end{aligned}$$

(a) Explain the different structure of this system compared with the analogous system based on the Verhulst equation.

(b) Derive an Allee type model for $N = N_1 + N_2$ using the appropriate scaling arguments.

7.2. (a) Derive a model with a satiating generalist predator based on the Beverton–Holt equation rather than the Verhulst equation.

(b) Derive an Allee type model for N using the appropriate scaling arguments.

7.3. Analyze the Allee type models obtained in Exercises 7.1 and 7.2 and compare whether you obtain the same qualitative results as for the models based on the Verhulst equation.

Chapter Eight

Nonautonomous Population Growth: Asymptotic Equality of Population Sizes

If the parameters in the Ricker equation (or any other of the equations we have considered) depend on time, the population size satisfies a nonautonomous equation of the form

$$N' = f(t, N). \tag{8.1}$$

It is difficult to make precise statements concerning the behavior of $N(t)$ for $t \to \infty$ because constant solutions like $N(t) = K$ no longer exist. However, since only one species is involved and the model formulation as an ordinary differential equation excludes retardation effects, one can still discover the existence of growth patterns that are identical for $t \to \infty$.

Let us first state that the nonautonomous versions of our examples, if we exclude the Allee effect, fit into the following framework.

($\diamondsuit_1$) $f : [0, \infty)^2 \to \mathbb{R}$ is continuous and the partial derivatives of $f(t, N)$ with respect to N exist and are continuous on $[0, \infty) \times (0, \infty)$.

($\diamondsuit_2$) $f(t, 0) = 0$ for all $t \geqslant 0$.

($\diamondsuit_3$) Assume that the function

$$g(t, N) = f(t, N)/N, \quad N > 0,$$

is nonincreasing in $N > 0$ for every $t \geqslant 0$.

Now consider two solutions N, M that exist and are strictly positive on some interval $[0, b)$. Without restricting the generality, let us assume $0 < M(0) < N(0)$. The assumptions for f imply that the uniqueness theorem holds. This implies that $M(t) < N(t)$ for all $t \in [0, b)$. Consider the quotient

$$x(t) = \frac{M(t)}{N(t)}.$$

Then $0 < x(t) < 1$ for all $t \in [0, b)$. By the quotient rule,

$$x' = \frac{M'}{N} - x\frac{N'}{N} = \frac{f(t, M)}{N} - x\frac{f(t, N)}{N} = x(g(t, M) - g(t, N)).$$

Since $g(t, \cdot)$ is nonincreasing and $0 < M < N$, $x > 0$, we see that $x' \geqslant 0$. So, x is increasing and we have the estimates

$$N(t) > M(t) = x(t)N(t) \geqslant x(0)N(t) = \frac{M(0)}{N(0)}N(t).$$

Theorem 4.1 in Hale (1980, I.2) provides the extra information that a solution N on a finite maximal interval $[0, b)$ of existence and positivity is either unbounded or not bounded away from 0.

So, the estimate above implies that M, N have the same maximal interval $[0, b)$ on which they exist and are strictly positive. Since x is nondecreasing and bounded, $\xi = \lim_{t\to\infty} x(t)$ exists. If we remove the assumption that $M(0) < N(0)$, we can formulate this result as follows.

Theorem 8.1. *Let f satisfy the properties $(\diamond_1)$–$(\diamond_3)$. Then there exists some $b \in (0, \infty]$ such that, for any solution N of (8.1) with $N(0) > 0$, $[0, b)$ is the maximal interval on which N exists and is strictly positive. Furthermore, for any pair M, N of solutions of (8.1) with (different!) positive initial data at 0, the quotient $M(t)/N(t)$ is monotone in $t \in [0, b)$ and converges towards a finite strictly positive limit as $t \to b$.*

If $[0, b)$ is the maximal interval for a solution N to exist and to be strictly positive and $b < \infty$, then N is either unbounded or not bounded away from 0 on $[0, b)$. If N is bounded on $[0, b)$, so is $N' = f(t, N)$ and N is uniformly continuous on $[0, b)$, so the left limit $N(b+)$ exists. Since in this case N is strictly positive but not bounded away from 0 on $[0, b)$, $N(b+) = 0$. This means that extending N beyond b by setting $N = 0$ on (b, ∞) yields a solution that exists on $[0, \infty)$. Since our assumptions do not enforce the uniqueness of solutions at $N = 0$, there may be other extensions which we ignore for biological reasons. Recall that we do not include immigration, so a population size that has dropped to 0 should remain 0. With this convention in mind, Theorem 8.1 has the following consequence.

Trichotomy of population growth.

(i) There exists some $b \in (0, \infty]$ such that all solutions of (8.1) with positive initial data exist but are unbounded on $[0, b)$.

(ii) All solutions of (8.1) with positive initial data exist and are bounded on $[0, \infty)$, but are not bounded away from 0 on $[0, \infty)$.

(iii) All solutions of (8.1) with positive initial data exist on $[0, \infty)$ and are both bounded and bounded away from 0 on $[0, \infty)$.

We want to investigate the third case somewhat further. Then, for two solutions M and N with $M(0) < N(0)$ as before, the quotient x exists and is nondecreasing on $[0, \infty)$, and is bounded by 1. So, the limit $\xi = \lim_{t\to\infty} x(t)$

not only exists but also satisfies $\xi \in (0, 1]$. We would like to find conditions for $\xi = 1$ to hold, i.e., $M(t)/N(t) \to 1$ as $t \to \infty$. So, we suppose that $\xi \in (0, 1)$. Since x is nondecreasing and so $x(t) \leqslant \xi$ for all $t \geqslant 0$ and g is a nonincreasing function of population size,

$$\frac{x'(t)}{x(t)} = g(t, xN) - g(t, N) \geqslant g(t, \xi N) - g(t, N).$$

By the mean value theorem, there exist numbers $u(t) \in (\xi N(t), N(t))$ such that

$$\frac{x'(t)}{x(t)} = g_N(t, u(t))(\xi N(t) - N(t)) = |g_N(t, u(t))| N(t)(1 - \xi).$$

Here g_N denotes the partial derivative of $g(t, N)$ with respect to N. Remember $g_N \leqslant 0$ by ($\diamond_3$).

Since N is bounded and bounded away from 0 on $[0, \infty)$, there exist $c > \epsilon > 0$ such that

$$\epsilon \leqslant \xi N(t) \leqslant N(t) \leqslant c \quad \forall t \geqslant 0.$$

Now let $\phi : [0, \infty) \to \mathbb{R}$ satisfy $|g_N(t, u)| \geqslant \phi(t)$ for all $t \geqslant 0$, $\epsilon \leqslant u \leqslant c$. Then

$$\frac{x'(t)}{x(t)} \geqslant \phi(t)\epsilon(1 - \xi).$$

If the time averages of ϕ are bounded away from 0,

$$\underline{\phi} = \liminf_{t\to\infty} \frac{1}{t} \int_0^t \phi(s)\, ds > 0,$$

the same proof as in Theorem 3.4 yields that x grows exponentially, a contradiction to x being bounded by 1. So, $\xi = 1$. Finally, we can conclude

$$0 \leqslant N(t) - M(t) = N(t)(1 - x(t)) \to 0, \quad t \to \infty,$$

because N is bounded and $x(t) \to 1$, $t \to \infty$. So, we make the following assumption.

Assumption 8.2. *For every pair of real numbers $c > \epsilon > 0$ there exists an integrable function $\phi : [0, \infty) \to [0, \infty)$ such that*

$$|g_N(t, N)| \geqslant \phi(t) \quad \text{whenever } t \geqslant 0,\ \epsilon \leqslant N \leqslant c,$$

and the time averages of ϕ are bounded away from 0,

$$\liminf_{t\to\infty} \frac{1}{t} \int_0^t \phi(s)\, ds > 0.$$

We obtain that any two solutions are asymptotically equal. (See Exercises 8.2 and 8.3.)

Theorem 8.3. *If Assumption 8.2 holds in addition to the properties* $(\diamond_1)$–$(\diamond_3)$, *we have in case (iii) of the trichotomy that any pair of solutions* M, N *(associated with different initial data at* $t = 0$*) satisfies* $M(t) - N(t) \to 0$ *as* $t \to \infty$.

An ad hoc condition for excluding possibility (i) in the trichotomy of population growth consists of postulating the existence of some $M > 0$ such that $f(t, N) < 0$ for all $t \geqslant 0$, $N > M$ (Exercise 8.1). This is actually not a bad assumption because, at very high population sizes, deaths should outweigh births even in very good seasons. An analogous ad hoc assumption for excluding (ii), the existence of some $\epsilon > 0$ such that $f(t, N) > 0$ for all $t \geqslant 0$, $N \in (0, \epsilon)$, does not capture the realistic case of a strict seasonality of the birth rate, which is 0 for parts of the year, while the mortality rate is still positive. To also exclude scenario (ii) in the growth trichotomy in such a situation, one can use nonautonomous persistence theory (Thieme, 2000), which is, however, beyond our scope.

Exercises

8.1. Assume that there exists some $M > 0$ such that $f(t, N) < 0$ for all $N \geqslant M$. Then all solutions N of (8.1) with positive initial data exist on $[0, \infty)$ and are bounded.

Hint: for a solution N, first consider the case that $N(r) < M$ for some $r \geqslant 0$, assume that the statement is false for N, and obtain a contradiction by looking at the time $t > r$ such that $N(s) < M$ for all $s \in [r, t)$, $N(t) = M$. (Of course, you must first show that such a time t exists. What is the sign of $N'(t)$?) Then consider the case $N(r) \geqslant M$ for all $r \geqslant 0$.

8.2. Consider a nonautonomous version of the Ricker equation $N' = N(\beta_0(t)\mathrm{e}^{-\alpha N} - \mu(t))$ with nonnegative continuous functions β_0 and μ and a constant $\alpha > 0$.

Show that Assumption 8.2 is satisfied if

$$\liminf_{t\to\infty} \frac{1}{t} \int_0^t \beta_0(s)\,\mathrm{d}s > 0.$$

(If $\liminf_{t\to\infty}$ is not familiar to you, replace it by $\lim_{t\to\infty}$ assuming that the limit exists.)

8.3. Consider a nonautonomous version of the Bernoulli equation $N' = N(\rho(t) - m(t)N^\theta)$ with continuous nonnegative functions ρ and m and a constant $\theta > 0$.

Show that Assumption 8.2 is satisfied if

$$\liminf_{t\to\infty} \frac{1}{t} \int_0^t m(s)\, \mathrm{d}s > 0.$$

(If $\liminf_{t\to\infty}$ is not familiar to you, replace it by $\lim_{t\to\infty}$ assuming that the limit exists.)

Chapter Nine

Discrete-Time Single-Species Models

The Ricker and Beverton–Holt equations were originally introduced as discrete-time (or difference) equations rather than differential equations (Beverton, Holt, 1957; Ricker, 1954, 1975). Autonomous difference equations are useful simplifications of time-periodic differential equation models if reproduction occurs only once a year during a very short season (compared to the rest of the year), like the spawning season for many fish species, and if all newborns are adults (i.e., reproductive) after one year.

9.1 The Discrete Analog of the Verhulst (Logistic) and the Bernoulli Equation: the Beverton–Holt Difference Equation and Its Generalization

Let P_n be the population size in the nth year immediately after the reproduction season, which is assumed to be of negligible length. During the year, outside the reproductive season, the population is only subjected to mortality, but not to births, and satisfies the differential equation,

$$N' = -\mu(N)N, \quad 0 \leqslant t \leqslant 1, \qquad N(0) = P_n, \tag{9.1}$$

with t denoting the time that has elapsed since the end of the reproductive season. In (9.1), we have chosen one year as time unit. $\mu(N)$ is the per capita mortality rate at population size N and is assumed to be a strictly positive continuous functions of $N \geqslant 0$.

$N(1)$ gives the population size after one year, immediately before the reproductive season. Then the population size in the year $n + 1$, immediately after the reproductive season, is

$$P_{n+1} = (p + b)N(1), \tag{9.2}$$

where p is the probability of surviving the reproductive season and b the per capita number of (viable) offspring.

In order to obtain explicit formulas in the following, we have assumed in (9.1) that the per capita mortality does not depend on the season. The qualitative features are preserved for seasonally dependent rates $\mu(t, N)$ as one can see by applying the techniques developed in Chapter 8 (see Exercise 9.7). If μ does not depend on time, we can separate the variables in (9.1),

$$1 = -\int_0^1 \frac{N'(s)\,\mathrm{d}s}{N(s)\mu(N(s))} = \int_{N(1)}^{N(0)} \frac{\mathrm{d}x}{x\mu(x)}.$$

For every $N(0)$ this equation has a solution $N(1) \in (0, N(0))$ because

$$\int_0^M \frac{\mathrm{d}x}{x\mu(x)} = \infty \quad \forall M > 0.$$

Since the integrand is strictly positive, we see that $N(1)$ is uniquely determined by $N(0)$ and depends in a strictly monotone increasing way on $N(0)$. We also see that $N(1)$ continuously depends on $N(0)$. Setting

$$f(N(0)) = (p+b)N(1)$$

and recalling $N(0) = P_n$, we obtain

$$P_{n+1} = f(P_n), \quad n \in \mathbb{N}. \tag{9.3}$$

We summarize our considerations (cf. Clark, 1976, Section 7.2; Gyllenberg et al., 1997).

Lemma 9.1. *The model (9.1) and (9.2) leads to the difference equation (9.3) with strictly increasing continuous* $f : (0, \infty) \to (0, \infty)$.

Example. Let the mortality rate be of the same form as in the Bernoulli equation (cf. Section 5.1),

$$\mu(N) = \mu_0 + mN^\theta.$$

Then

$$1 = \int_{N(1)}^{N(0)} \frac{\mathrm{d}x}{x(\mu_0 + mx^\theta)}.$$

Using the Leibnitz substitution $x^{-\theta} = y$, we have

$$1 = -\frac{1}{\theta}\int_{N(1)^{-\theta}}^{N(0)^{-\theta}} \frac{\mathrm{d}y}{\mu_0 y + m} = \frac{1}{\theta\mu_0}\ln\left(\frac{\mu_0 N(1)^{-\theta} + m}{\mu_0 N(0)^{-\theta} + m}\right).$$

Exponentiating both sides,

$$\frac{\mu_0 N(1)^{-\theta} + m}{\mu_0 N(0)^{-\theta} + m} = \mathrm{e}^{\theta\mu_0}.$$

Solving for $N(1)$,

$$N(1) = \frac{N(0)}{[\mathrm{e}^{\theta\mu_0} + (m/\mu_0)(\mathrm{e}^{\theta\mu_0} - 1)N(0)^\theta]^{1/\theta}}.$$

This yields the difference equation (9.3) with f of the form

$$f(P) = (b+p)q\frac{P}{(1+cP^\theta)^{1/\theta}},$$

where

$$q = \mathrm{e}^{-\mu_0}, \qquad c = \frac{m}{\mu_0}(1-q^\theta).$$

q is the probability of surviving the year from one reproductive season to the other if there are no crowding effects. If we take the mortality rate of the Verhulst (or logistic) equation, $\theta = 1$, we obtain the Beverton–Holt equation:

$$P_{n+1} = (b+p)q\frac{P_n}{1+cP_n}.$$

So, in this framework, the discrete analog of the continuous logistic equation is the discrete Beverton–Holt equation. Notice that the discrete Bernoulli equation and its special case, the discrete Beverton–Holt equation, have $f(P)/P$ strictly decreasing in $P > 0$. We will see later that this is always true if $\mu(P)$ is a strictly increasing function of P.

9.2 The Ricker Difference Equation

The Ricker equation cannot be derived in this way unless we make the per capita number of births depend on population size (see Gyllenberg et al., 1997). Remember that, in the Ricker differential equation with a time-independent birth rate, cannibalism of young by adults leads to the characteristic functional form. If the juvenile period lasts less than a year, we can make a similar derivation in the case of a very short reproductive season. In contrast to the derivation of the differential equation (Section 5.3), we do not need to assume that the age-window during which juveniles are cannibalized by adults is very short, while the cannibalism rate is very high.

In this context it is more convenient to look at the size of the population in year n immediately before the reproductive season, Q_n. Let $J(t)$ denote the juvenile part of the population and $A(t)$ the adult part of the population with t being the time that has elapsed since the last reproductive season. Notice that t is also the age of the juveniles. Then

$$J' = -\nu(t)J - \kappa(t)A(t)J(t), \quad J(0) = bQ_n, \quad A' = -\mu A, \quad A(0) = pQ_n.$$

Again p is the probability of surviving the reproductive season and b the per capita number of offspring. μ is the per capita mortality rate of adults, while $\nu(t)$ is the per capita mortality rate of juveniles at age t (without cannibalism) and $\kappa(t)$ the per capita rate at time t at which juveniles are cannibalized by adults per capita of adults per unit of time. Finally,

$$Q_{n+1} = A(1) + J(1).$$

Integrating the differential equations (see Section 3.1),

$$A(t) = pQ_n e^{-\mu t},$$

$$J(t) = bQ_n \exp\left(-\int_0^t \nu(s)\, ds - \int_0^t \kappa(s)A(s)\, ds\right).$$

The exponential factors give the respective probabilities of surviving from the end of the reproductive season to time t (cf. (2.2)). Substituting the expression for $A(t)$ into the one for $J(t)$, we obtain the discrete Ricker equation,

$$Q_{n+1} = Q_n[pe^{-\mu} + be^{-a-cQ_n}],$$

with

$$a = \int_0^1 \nu(s)\, ds, \qquad c = p\int_0^1 \kappa(s)e^{-\mu s}\, ds.$$

Introducing $x_n = cQ_n$, this can be rewritten as

$$x_{n+1} = x_n(q + \gamma e^{-x_n}) = f(x_n)$$

with

$$q = pe^{-\mu}$$

being an adult's probability of surviving one year including the reproductive season and

$$\gamma = be^{-a}$$

being the number of per capita offspring still alive after one year if there is no cannibalism. Again we notice that $f(x)/x$ is a strictly decreasing function of $x > 0$.

9.3 Some Analytic Results for Scalar Difference Equations

As we have seen, the development of a population with a very short reproductive season can be described as solutions (x_n) of a *difference equation*

$$x_{n+1} = f(x_n), \quad n \in \mathbb{N},$$

with a continuous function $f : (0, \infty) \to (0, \infty)$. x_0 is the *initial value* of the solution. If f is a monotone increasing function, as we obtained from our first derivation above (Lemma 9.1), the solutions behave in a very straightforward way: if $f(x_0) = x_0$, i.e., if the initial value is a *fixed point of* f, we have $x_n = x_0$ for all $n \in \mathbb{N}$. If $f(x_0) > x_0$, we have $x_1 > x_0$, which implies by induction that $x_{n+1} \geqslant x_n$ for all $n \in \mathbb{N}$, and the increasing sequence x_n either tends to infinity or converges to a finite limit x^*, which is a fixed point of f. Indeed,

$$f(x^*) = \lim_{n\to\infty} f(x_n) = \lim_{n\to\infty} x_{n+1} = x^*.$$

Similarly, if $f(x_0) < x_0$, the solution of the difference equation is a decreasing sequence which converges to 0 or to a nonzero fixed point of f. The bottom line is that scalar difference equations with an increasing function have their solutions behave exactly like the solutions of scalar differential equations, with the single exception that they do not blow up in finite time (see Theorem 6.1). This can change dramatically if f is not increasing, as in the discrete Ricker equation, and the solutions can show very complicated, even chaotic, behavior.

For an important class of difference equations, the convergence of solutions is linked to the nonexistence of so-called period-two points of f, i.e., fixed points of f^2 that are not fixed points of f. Here $f^2 = f \circ f$ is the double iterate of f, $f^2(x) = f(f(x))$. For if y^* is a fixed point of f^2, but not of f, and $z^* := f(y^*)$, we have $f(z^*) = y^*$; so, every solution starting at y^* alternates between z^* and y^* and, since $y^* \neq z^*$, does not converge. Surprisingly, the converse is also true under certain circumstances. If x^* is a unique fixed point of f and f^2 has no fixed points different from x^*, then all solutions of the difference equation will converge to x^*.

In order to prove this, we will use the following result.

Proposition 9.2. *Let f be a continuous function mapping an interval $[a, b]$, $a < b$, onto itself.*

Then either

- *f has at least two fixed points, or*
- *f has at least one fixed point, x^*, and f^2 has at least one fixed point y^* such that $y^* < x^* < f(y^*)$ (with $f(y^*)$ also being a fixed point of f^2).*

Proof. Since $f(a) \geqslant a$ and $f(b) \leqslant b$, f has at least one fixed point, x^*, by the intermediate value theorem. Let us assume that x^* is the only fixed point of f.

Step 1. $a < x^* < b$.

For if $x^* = a$, i.e., $f(a) = a$, we have $f(b) < b$ and $f(x) = b > x$ for some x. Then, by the intermediate value theorem, f has a second fixed point in (x, b). Similarly, b cannot be a fixed point of f.

Step 2. $[a, x^*] \subseteq f([x^*, b])$ and $[x^*, b] \subseteq f([a, x^*])$.

By Step 1, $f(a) > a$ and $f(b) < b$. Since x^* is the only fixed point of f, $f(x) > x > a$ for all $x \in [a, x^*)$. So, there exists some $y \in (x^*, b]$ such that $f(y) = a$. Since $f([x^*, b])$ is an interval and contains a and x^*, we have that $[a, x^*] \subseteq f([x^*, b])$. A similar argument shows that $[x^*, b] \subseteq f([a, x^*])$.

Step 3. Existence of a period-two point.

By Step 2, there exists some $y \in (x^*, b]$ such that $a = f(y)$ and some $z \in [a, x^*)$ such that $y = f(z)$. Hence $f^2(z) = a < z$. Since $f^2(a) \geqslant a$, the intermediate value theorem, applied to f^2, provides some $y^* \in [a, x^*)$ such that

$f^2(y^*) = y^*$. In order to show that $f(y^*) > x^*$, let us consider two points u, v such that $f(u) = v$ and $f(v) = u$. Let $u \leqslant v$. Then $f(u) \geqslant u$ and $f(v) \leqslant v$ and there exists a fixed point of f between u and v. Since x^* is the only fixed point of f, $u \leqslant x^* \leqslant v$. If we apply this consideration to $\{u, v\} = \{y^*, f(y^*)\}$ and use the fact that $y^* < x^*$, it follows that $y^* < x^* < f(y^*)$. □

Proposition 9.3. *Let f be a continuous function mapping an interval $[a_0, b_0]$, $a_0 < b_0$, into itself. Let $[c, d] \subseteq [a_0, b_0] \cap f([c, d])$, $c < d$.*
Then either

- *f has at least two fixed points, or*
- *f has at least one fixed point x^* and f^2 has at least one fixed point y^* such that $y^* < x^* < f(y^*)$ (with $f(y^*)$ also being a fixed point of f^2).*

Proof. Define

$$I = \bigcup_{n=0}^{\infty} f^n([c, d])$$

with f^0 being the identity map, $f^0(x) = x$. Since $[c, d] \subseteq f([c, d])$, I is the union of increasing intervals contained in $[a_0, b_0]$, and so I is an interval itself, $[c, d] \subseteq I \subseteq [a_0, b_0]$. Now

$$f(I) = \bigcup_{n=0}^{\infty} f(f^n([c, d])) = \bigcup_{n=1}^{\infty} f^n([c, d]) = \bigcup_{n=0}^{\infty} f^n(f([c, d])).$$

Since $[c, d] \subseteq f([c, d])$, it follows that $f(I) = I$. Let $\bar{I} = [a, b]$ with a and b being the endpoints of I. $\bar{I}$ inherits the properties of I,

$$[c, d] \subseteq \bar{I} = f(\bar{I}) \subseteq [a_0, b_0].$$

We are now in the same situation as in Proposition 9.2 and the assertion follows. □

Many of the functions f in the scalar difference equations of population dynamics have the following properties.

Assumption 9.4. *$f : (0, \infty) \to (0, \infty)$ is continuous, has a unique fixed point $x^* > 0$, and is bounded on $(0, x^*]$. Furthermore, there exist x_1 and x_2, $0 < x_1 < x^* < x_2 < \infty$, such that $f(x_1) > x_1$ and $f(x_2) < x_2$.*

Proposition 9.5. *Let Assumption 9.4 hold and $b_0 > \sup f((0, x^*])$ and $0 < a_0 < \inf f([x^*, b_0])$. Then $f([a_0, b_0]) \subseteq [a_0, b_0]$, and every solution of the difference equation associated with f enters $[a_0, b_0]$ without ever leaving it again.*

Proof. Since x^* is the only fixed point of f, the intermediate value theorem implies that $f(x) > x$ for all $x \in (0, x^*)$ and $f(x) < x$ for all $x > x^*$. Let $\tilde{x} = \sup f((0, x^*])$ and let $b_0 > \tilde{x} \geqslant x^*$.

Step 1. $f((0, b_0]) \subseteq (0, b_0]$ and every solution (x_n) of the difference equation enters $(0, b_0]$ and stays in this interval.

Indeed, if $x \in (0, x^*]$, then $f(x) \leqslant \tilde{x} < b_0$. If $x \in (x^*, b_0]$, $f(x) < x \leqslant b_0$. Now assume that a solution (x_n) satisfies $x_n \geqslant b_0$ for all $n \in \mathbb{N}$. Then $x_{n+1} = f(x_n) < x_n$ for all $n \in \mathbb{N}$ and the sequence is strictly decreasing. Since it is bounded below by b_0, it has a limit $x \geqslant b_0$ which is a fixed point different from x^*, a contradiction. Once the solution has entered the interval $(0, b_0]$, it must stay there because f maps the interval into itself.

Step 2. Let $0 < a_0 < \inf f([x^*, b_0])$. Then $f([a_0, b_0]) \subseteq [a_0, b_0]$ and all solutions of the difference equation enter $[a_0, b_0]$ and stay in this interval.

If $x \in [x^*, b_0]$, then $f(x) \geqslant a_0$ by the choice of a_0. If $x \in [a_0, x^*)$, $f(x) > x \geqslant a_0$. So, $f([a_0, b_0] \subseteq [a_0, \infty)$, and, together with Step 1, $f([a_0, b_0]) \subseteq [a_0, b_0]$. Now consider a solution (x_n) of the difference equation. By Step 1, there exists some $m \in \mathbb{N}$ such that $x_n \leqslant b_0$ for all $n \geqslant m$. Suppose that $x_n \leqslant a_0$ for all $n \geqslant m$. Since $a_0 \leqslant x^*$ we have $x_{n+1} = f(x_n) \geqslant x_n$ for all $n \geqslant m$. So, the sequence x_n is increasing for $n \geqslant m$ and bounded above by a_0. Thus it converges towards a limit which, as in the proof of Step 1, is a second fixed point of f, a contradiction. So, any solution of the difference equation finally enters the interval $[a_0, b_0]$, and stays therein because $f([a_0, b_0]) \subseteq [a_0, b_0]$. □

We combine these considerations and obtain the convergence result we announced before.

Theorem 9.6. *Assume that Assumption 9.4 holds and that there is no fixed point of f^2 different from the unique fixed point x^*. Then all solutions of the difference equation associated with f converge to x^* as $n \to \infty$.*

Proof. Choose b_0 and a_0 as in Proposition 9.5. Notice that $\inf f([x^*, b_0]) > 0$, so we can choose $a_0 > 0$. Furthermore, $a_0 \leqslant x^* \leqslant b_0$.

Let (x_n) be a solution of the difference equation, and $x_\infty = \liminf_{n\to\infty} x_n$, $x^\infty = \limsup_{n\to\infty} x_n$ the smallest and largest accumulations points of (x_n).

By Proposition 9.5, (x_n) enters and then stays in $[a_0, b_0]$, so $[x_\infty, x^\infty] \subseteq [a_0, b_0]$. There exists a subsequence n_k such that

$$x^\infty = \lim_{k\to\infty} x_{n_k} = \lim_{t\to\infty} f(x_{n_k-1}).$$

By the Bolzano–Weierstraß theorem, the sequence (x_{n_k-1}) has a convergent subsequence with limit in $[x_\infty, x^\infty]$, hence $x^\infty \in f([x_\infty, x^\infty])$. A similar argument shows that $x_\infty \in f([x_\infty, x^\infty])$. Since $f([x_\infty, x^\infty])$ is an interval, $[x_\infty, x^\infty] \subseteq f([x_\infty, x^\infty])$.

If the sequence (x_n) does not converge, $x_\infty < x^\infty$ and we are in the situation of Proposition 9.3 which implies that there is a fixed point of f^2 different from x^*, which is excluded by assumption. So, (x_n) converges. □

If f and f^2 have at most one positive fixed point, one generically (i.e., except in a rare, degenerate situation) has a complete classification of the large-time behavior.

Theorem 9.7. *Let $f : (0, \infty) \to (0, \infty)$ be continuous. Assume that f and f^2 have at most one fixed point.*

(a) *One of the following six mutually exclusive scenarios hold:*

(i) *All solutions of the difference equation converge to ∞.*

(ii) *All solutions of the difference equation converge to 0.*

(iii) *There exists a unique fixed point of f which attracts all solutions of the difference equation.*

(iv) *There exists a unique fixed point x^* of f such that all solutions starting in $(0, x^*)$ converge to 0 while all solutions starting in (x^*, ∞) converge to ∞.*

(v) *There exists a unique fixed point x^* of f such that all solutions starting in (x^*, ∞) converge to ∞, while all solutions starting in $(0, x^*)$ converge to x^* or to ∞.*

(vi) *There exists a unique fixed point x^* of f such that all solutions starting in $(0, x^*)$ converge to 0, while all solutions starting in (x^*, ∞) converge to x^* or to 0.*

(b) *If $f(x)/x$ is strictly decreasing in $x > 0$ (which implies that there is at most one fixed point of f), we have the limit trichotomy that exactly one of (i), (ii), or (iii) in part (a) holds. Moreover*

$$\text{(i)} \iff \lim_{x\to\infty} \frac{f(x)}{x} \geqslant 1,$$

$$\text{(ii)} \iff \lim_{x\to 0} \frac{f(x)}{x} \leqslant 1,$$

$$\text{(iii)} \iff \lim_{x\to\infty} \frac{f(x)}{x} < 1 < \lim_{x\to 0} \frac{f(x)}{x}.$$

Proof. (a) Assume that (i) and (ii) do not hold.

Step 1. Then f has a fixed point.

Case 1. There exists an unbounded solution sequence that does not converge to ∞.

This implies that the sequence is not eventually monotone. So, there exist natural numbers m, n such that $x_{n+1} \geqslant x_n$ and $x_{m+1} \leqslant x_m$, and f has a fixed point by the intermediate value theorem.

Case 2. There exists a solution sequence that is not bounded away from 0 but does not converge to 0.

The same argument as in case 1 provides a fixed point.

Case 3. There exists a bounded solution sequence (x_n) and there exists a solution sequence (y_n) that is bounded away from 0.

As before let x^∞ and y_∞ be the limit superior of (x_n) and the limit inferior of (y_n). Choose a subsequence x_{n_k} such that $x_{n_k} \to x^\infty$ as $k \to \infty$. Then

$$f(x^\infty) = \lim_{k\to\infty} f(x_{n_k}) = \lim_{k\to\infty} x_{n_k+1} \leqslant x^\infty.$$

Similarly, $f(y_\infty) \geqslant y_\infty > 0$ and f has a positive fixed point by the intermediate value theorem.

These three cases exhaust all the possibilities and Step 1 is complete.

Step 2. Proof of part (a).

By Step 1, we can assume that f has a unique fixed point x^*.
By the intermediate value theorem the following cases can occur.

(iii) Assumptions 9.4 hold, and all solutions converge to x^* by Theorem 9.6.

(iv) $f(x) < x$ for all $x \in (0, x^*)$ and $f(x) > x$ for all $x > x^*$. Then every solution sequence starting in $(0, x^*)$ is decreasing and every solution sequence starting in (x^*, ∞) is increasing. This case cannot occur if $f(x)/x$ is strictly decreasing.

(v) $f(x) > x$ for all $x \neq x^*$. Then all solution sequences are increasing. Those starting in (x^*, ∞) converge to ∞. Those starting $(0, x^*]$ either stay in this interval and converge to x^*, or they enter (x^*, ∞) and converge to ∞. This case cannot occur if $f(x)/x$ decreases strictly.

(vi) $f(x) < x$ for all $x \neq x^*$. All solution sequences are decreasing. Those starting in $(0, x^*)$ converge to 0. If a solution starts in $[x^*, \infty)$, it either stays in that interval and converges to x^*, or it enters $(0, x^*)$ and converges to 0. This case cannot occur if $f(x)/x$ decreases strictly.

Step 3. Proof of part (b).

Since $f(x)/x$ strictly decreases in $x > 0$, only (i), (ii), or (iii) can occur, as we have seen in the proof of part (a). This proves the limit trichotomy in part (b). If $\lim_{x\to\infty}(f(x)/x) \geqslant 1$, then $f(x) > x$ for all $x > 0$. This means that every solution sequence is strictly increasing and converges to ∞ because there is no fixed point. A similar argument settles the characterization of (ii). If

$$\lim_{x\to\infty} \frac{f(x)}{x} < 1 < \lim_{x\to 0} \frac{f(x)}{x},$$

then f has a unique fixed point and (iii) holds. □

If the nonlinear function f has zeros, the picture becomes more complicated.

Theorem 9.8. *Let $f : [0, \infty) \to [0, \infty)$ be continuous, $f(0) = 0$. Assume that f has a unique positive fixed point x^* and that the iterate f^2 has no positive fixed point other than x^*. Furthermore, assume that there exist $x_\diamond$ and $\hat{x}$ such that $0 < x_\diamond < x^* < \hat{x} < \infty$ and $f(x_\diamond) > x_\diamond$, while $f(x) > 0$ for all $x \in (0, \hat{x})$ and $f(\hat{x}) = 0$.*

Then every solution of the difference equation $x_{n+1} = f(x_n)$ with $0 < x_0 < \hat{x}$ converges to x^* as $n \to \infty$. If $x_0 > \hat{x}$, then either $x_n = 0$ for all but finitely many n or x_n converges to x^* as $n \to \infty$.

Proof. We claim that there is no $\tilde{x} \in [0, \hat{x}]$ such that $f(\tilde{x}) = \hat{x}$.

Suppose there is such an $\tilde{x}$. By the intermediate value theorem, $f(x) < x$ for all $x > x^*$, hence $\tilde{x} \in (0, x^*)$. Also $f(x) > x$ for all $x \in (0, x^*)$, so $f^2(x) > x$ if $x > 0$ is small enough. But also $f^2(\tilde{x}) = 0 < \tilde{x} < x^*$. So, f^2 has a fixed point in $(0, x^*)$, a contradiction.

Again by the intermediate value theorem, there exists some $\check{x} \in (x^*, \hat{x})$ such that $f((0, \hat{x}]) \subseteq (0, \check{x})$. If the sequence starts in $(0, \hat{x})$, it immediately enters $(0, \check{x})$, so we can assume that $x_0 \in (0, \check{x})$. Then $x_n \in (0, \check{x})$ for all n.

Consider a sequence that starts in $(\hat{x}, \infty)$ and does not become 0 after a couple of iterations. If the sequence does not eventually enter $(0, x^*)$, it is decreasing and so converges towards the unique positive fixed point of f, x^*.

This means that we can restrict our consideration to sequences (x_n) which start in $(0, \check{x})$ and so have all their terms in $(0, \check{x})$.

We now modify f on $[\check{x}, \infty)$ by setting $\tilde{f}(x) = f(\check{x})$ for $x \geqslant \check{x}$, $\tilde{f} = f$ on $[0, \check{x}]$. Then $\tilde{f}$ satisfies Assumption 9.4 and (x_n) satisfies also the difference equation with $\tilde{f}$ replacing f. By Theorem 9.6, x_n converges towards the unique fixed point of $\tilde{f}$, which is x^*. □

While the condition that f^2 has no other positive fixed points than the unique positive fixed point of f is not only sufficient, but also necessary, it may be difficult to check for concrete functions f.

Corollary 9.9. *Let $f : (0, \infty) \to (0, \infty)$ have the unique fixed point $x^* > 0$.*

(a) *The iterate f^2 then has no fixed point other than x^* if at least one of the following three conditions is satisfied.*

 (i) *$f'(x) \geqslant -1$ for all $x \in (0, \infty)$, with strict inequality for all $x \in (0, x^*)$ or all $x \in (x^*, \infty)$.*

 (ii) *$xf(x)$ is a strictly monotone function of $x > 0$.*

 (iii) *f is nondecreasing on $(0, x^*]$ or on $[x^*, \infty)$.*

(b) *If f satisfies Assumption 9.4 and one of the conditions (i), (ii), or (iii) in part (a), then all solutions of the difference equation associated with f converge to x^* as $n \to \infty$.*

Similarly, Corollary 9.9 (a) can be combined with Theorem 9.8.

Remark 9.10.

(a) It is sufficient for (i) to hold for $x < \sup f((0, x^*])$. In (ii), it is sufficient to assume that $yf(y) \neq zf(z)$ whenever $0 < y < x^* < z \leqslant \sup f((0, x^*])$. In (iii), it is sufficient to assume that f is nondecreasing on $[x^*, \sup f((0, x^*])]$.

(b) Assume that $xf(x)$ strictly increases and $f(x)/x$ strictly decreases in $x > 0$. Then f and f^2 have at most one positive fixed point. By Theorem 9.7 (b) we have the limit trichotomy that one of the mutually exclusive scenarios (i), (ii), or (iii) holds. If f is differentiable, the two monotonicity assumptions for f can elegantly be combined into

$$f(x) > x|f'(x)| \quad \text{for almost all } x > 0.$$

Proof of Corollary 9.9 and Remark 9.10 (a). If we can rule out fixed points of f^2 different from x^*, then the assertion follows from Theorem 9.6. So, we assume that there exist $y, z \neq x^*$, $y \neq z$, such that $f(z) = y$ and $f(y) = z$. By the intermediate value theorem, x^* lies between y and z. So, $y, z \leqslant \sup f((0, x^*))$.

To prove (a) (i), we consider three cases.

Case 1. $|z - x^*| < |y - x^*|$. By the mean value theorem, $y - x^* = f(z) - f(x^*) = f'(\xi)(z - x^*)$ for some ξ strictly between x^* and z. Since x^* lies between y and z, $f'(\xi) < 0$. Furthermore, since $|z - x^*| < |y - x^*|$, $f'(\xi) < -1$, a contradiction.

Case 2. $|y - x^*| < |z - x^*|$. The proof is analogous to case 1 by switching the roles of y and z.

Case 3. $|z - x^*| = |y - x^*|$. Let us assume that $y < x^* < z$ and $f'(x) > -1$ for all $x \in (0, x^*)$. Then $z - x^* = f'(\xi)(y - x^*)$ for some $\xi \in (y, x^*)$ and $f'(\xi) = -1$, a contradiction. If $f'(x) > -1$ for all $x \in (x^*, \infty)$, we switch the roles of y and z.

(ii) and (iii). Without loss of generality we can assume that $0 < y < x^* < z \leqslant \sup f((0, x^*])$. Furthermore, $yf(y) = yz = zf(z)$, which contradicts condition (ii). See also part (a) of the remark.

If f is nondecreasing on $(0, x^*]$, then $z = f(y) \leqslant f(x^*) = x^*$, contradicting $z > x^*$. If f is nondecreasing on $[x^*, \sup f((0, x^*])]$, then $y = f(z) \geqslant f(x^*) = x^*$, contradicting $y < x^*$. □

In the following we show that the condition $f'(x^*) \geqslant -1$ is necessary for all solutions of the difference equations to converge to x^*. In fact, we will see that, if $f'(x^*) < -1$, all but countably many solutions will not converge to x^*. That some solutions will still converge to x^* has to do with the following.

Since, in general, f is not one to one, there often is an $x \in (0, \infty)$ such that $x \neq x^* = f(x)$. More generally, let us consider those x which are mapped onto x^* by some iterate of f,

$$M_-(x^*) = \bigcup_{n=1}^{\infty} \{x;\ f^n(x) = x^*\}.$$

Obviously, every solution of the difference equation that starts in M converges to x^*. Under reasonable assumptions, M is a thin set.

Proposition 9.11. *Assume that $f^{-1}(\{x\})$ is countable for every $x \in (0, \infty)$. Then $M_-(y)$ is countable for every point $y \in (0, \infty)$.*

Proof. Set $M_n = \{x;\ f^n(x) = y\}$. Since countable unions of countable sets are countable, it is sufficient to show that M_n is countable for each $n \geqslant 1$. This follows by induction. Indeed, $M_1 = f^{-1}(\{y\})$ is countable by assumption; furthermore,

$$M_{n+1} = \bigcup_{x \in M_n} f^{-1}(\{x\}),$$

so M_{n+1} is countable whenever M_n is. □

Theorem 9.12. *Let $f : (0, \infty) \to (0, \infty)$ be a continuously differentiable function and x^* a fixed point of f with $|f'(x^*)| > 1$. Then only solutions that start in $M_-(x^*)$ converge to x^* as $n \to \infty$. If Assumption 9.4 is satisfied, there exists a fixed point of f^2 different from x^*.*

Proof. Let $x_0 \notin M_-(x^*)$ and assume that the solution with initial value x_0 converges towards x. Since f' is continuous, there exists some $\epsilon \in (0, x^*)$ such that $|f'(x)| > 1$ for all $x \in (x^* - \epsilon, x^* + \epsilon)$. Furthermore, there exists some

$m \in \mathbb{N}$ such that $|x_n - x^*| < \epsilon$ for all $n \geqslant m$. Since $x_0 \notin M_-(x^*)$, $x_n \neq x^*$. By the mean value theorem, for all $n \geqslant m$,

$$|x_{n+1} - x^*| = |f(x_n) - f(x^*)| = |f'(\xi_n)||x_n - x^*| > |x_n - x^*| > 0,$$

with some ξ_n between x^* and x_n. This contradicts the convergence of x_n towards x^*.

Choose $b_0 > a_0 > 0$ as in Proposition 9.5 and let $\epsilon > 0$ be small enough such that $[x^* - \epsilon, x^* + \epsilon] \subseteq [a_0, b_0]$ and $|f'(x)| \geqslant -1$, whenever $|x - x^*| \leqslant \epsilon$. We consider the case that is more relevant in this context (see Remark 9.13), namely that $f'(x) \leqslant -1$ whenever $|x - x^*| \leqslant \epsilon$. The other case, $f'(x) \geqslant 1$, is done similarly. By the mean value theorem, $f(x^* + \epsilon) = f(x^*) + f'(\xi)\epsilon \leqslant x^* - \epsilon$, where ξ lies between x^* and $x^* + \epsilon$. Similarly, $f(x^* - \epsilon) \geqslant x^* + \epsilon$. Since $f([x^* - \epsilon, x^* + \epsilon])$ is an interval, $f([x^* - \epsilon, x^* + \epsilon]) \supseteq [f(x^* + \epsilon), f(x^* - \epsilon)] \supseteq [x^* - \epsilon, x^* + \epsilon]$. The existence of a fixed point of f^2 that is not a fixed point of f now follows from Proposition 9.3. □

Remark 9.13. If $f : (0, \infty) \to (0, \infty)$ is differentiable and Assumption 9.4 holds, then $f'(x^*) \leqslant 1$. So, $|f'(x^*)| > 1$ if and only if $f'(x^*) < -1$.

Proof. We have seen that $f(x) < x$ for all $x > x^*$. Hence $f(x^* + h) - f(x^*) < h$ for all $h > 0$. Dividing by h and taking the limit $h \to 0$ implies the assertion. □

There is a possible gap between Corollary 9.9 (i) and Theorem 9.12. There is also the possibility that $|f'(x^*)| = 1$, but we do not consider it here. It is convenient to introduce the following notion.

We say that a fixed point x^* of f *attracts a point* x_0 if the solution of the difference equation with initial value x_0 converges to x^*. We say that x^* *pointwise attracts a set* $M \subseteq (0, \infty)$ if x^* attracts every point of M, i.e., if every solution of the difference equation starting in M converges to x^*. The set $W^s(x^*)$ of all points that are attracted to x^* is called the *stable set* or the *domain of attraction* of x^*.

Theorem 9.14. *Let $f : (0, \infty) \to (0, \infty)$ be continuously differentiable and x^* a fixed point of f with $|f'(x^*)| < 1$. Then one of the following possibilities occurs.*

(i) *x^* pointwise attracts $(0, \infty)$.*

(ii) *x^* pointwise attracts an interval $(0, z^*)$ with $z^* > x^*$ being another fixed point of f.*

(iii) *x^* pointwise attracts an interval (y^*, ∞) with $y^* < x^*$ being another fixed point of f.*

(iv) *x^* pointwise attracts an interval (y^*, z^*), $y^* < x^* < z^*$, with at least one of y^*, z^* being another fixed point of f and $f(\{y^*, z^*\}) = \{y^*, z^*\}$.*

(v) *x^* pointwise attracts an interval (y^*, z^*) with $y^* < x^* < z^*$ being two fixed points of f^2, $f(y^*) = z^*$, $f(z^*) = y^*$.*

If x^* is the unique fixed point of f, then only possibilities (i) and (v) can occur.

First we prove the following lemma.

Lemma 9.15. *Let $f : (0, \infty) \to (0, \infty)$ be continuous and x^* a fixed point of f with $|f'(x^*)| < 1$.*

(a) *There then exists $\epsilon \in (0, x^*)$ such that x^* pointwise attracts the interval $I = (x^* - \epsilon, x^* + \epsilon)$. Furthermore, any solution of the difference equation with initial value $x_0 \in I$ satisfies $|x_n - x^*| \leqslant |x_0 - x^*|$ for all $n \in \mathbb{N}$.*

(b) *The stable set of x^*, $W^s(x^*)$, is open, i.e., every point x in $W^s(x^*)$ is contained in an open interval $J \subseteq W^s(x^*)$.*

Actually, the assumptions in Theorem 9.14 can be replaced by assuming that f is continuous and attracts an open interval containing x^*.

Proof. (a) Since f' is continuous, there exists some $\epsilon \in (0, x^*)$ such that $|f'(x)| < \delta < 1$ whenever $|x - x^*| < \epsilon$. By the mean value theorem, we have

$$|f(x) - x^*| = |f(x) - f(x^*)| = |f'(\xi)|\,|x - x^*| \leqslant \delta|x - x^*|.$$

If (x_n) is a solution of the difference equation with $x_0 \in I$, then, by induction,

$$|x_n - x^*| \leqslant \delta^n |x_0 - x^*| \quad \forall n \in \mathbb{N}.$$

This implies (a).

(b) Let $x \in W^s(x^*)$. Then there exists $m \in \mathbb{N}$ such that $f^m(x) \in I = (x^* - \epsilon, x^* + \epsilon)$ with ϵ from (a). Since f^m is continuous, there exists an open interval J such that $f^m(J) \subseteq I$. Since every solution starting in I is attracted to x^*, $f^n(y) \to x^*$, $n \to \infty$, for every $y \in J$. □

Proof of Theorem 9.14. If x^* does not attract $(0, \infty)$, there are the following cases to consider.

Case 1. x^* attracts $(0, x^*]$, but not $[x^*, \infty)$.

Case 2. x^* attracts $[x^*, \infty)$, but not $(0, x^*]$.

Case 3. x^* attracts neither $(0, x^*]$ nor $[x^*, \infty)$.

We will only deal with case 3; the others are handled analogously.

Let b be the largest number such that x^* pointwise attracts $[x^*, b)$ and $a > 0$ the smallest number such that x^* attracts $(a, x^*]$. Since we are in case 3 and since x^* pointwise attracts an open interval around itself by Lemma 9.15 (a), $0 < a < x^* < b < \infty$. By Lemma 9.15 (b), x^* does not attract a or b. Obviously, if x is attracted by x^*, so is $f(x)$. So, $f((a, b))$ is an interval which is also attracted by x^* and has nonempty intersection with (a, b) because both contain x^*. So, the choice of a and b implies $f((a, b)) \subseteq (a, b)$.

Since f is continuous, $f([a, b]) \subseteq [a, b]$. Since x^* does not attract a or b, $f(\{a, b\}) \subseteq \{a, b\}$. So, either (iv) or (v) holds. □

Application to the Generalized Beverton–Holt Difference Equation

Recall the model in Section 9.1,

$$N' = -\mu(N)N, \quad 0 \leqslant t \leqslant 1, \qquad N(0) = P_n,$$
$$P_{n+1} = (p + b)N(1),$$

with $\mu : [0, \infty) \to (0, \infty)$ being a strictly increasing continuous function, $\mu(0) > 0$. We have already learned in Section 9.1 that $P_{n+1} = f(P_n)$ with $f : (0, \infty) \to (0, \infty)$ being strictly increasing. We want to apply Theorem 9.7 (b). Making the change of variables $N(t) = xy(t)$, $x = N(0)$, we observe that $f(x) = (p + b)xg(x)$ with

$$g(x) = y(1),$$
$$y' = -\mu(xy)y, \quad 0 \leqslant t \leqslant 1, \qquad y(0) = 1.$$

Separating the variables,

$$1 = \int_{g(x)}^{1} \frac{\mathrm{d}y}{y\mu(xy)}.$$

Since μ is strictly increasing, g is strictly decreasing. Taking the limit for $x \to 0$,

$$1 = \int_{g(0+)}^{1} \frac{\mathrm{d}y}{y\mu(0)} = -\frac{\ln(g(0+)}{\mu(0)},$$

i.e.,

$$g(0+) = \mathrm{e}^{-\mu(0)}.$$

Let us first assume that $\mu(\infty) = \lim_{N\to\infty} \mu(N) < \infty$. Then $y' \geqslant -\mu(\infty)y$ and $y(t) \geqslant \mathrm{e}^{-\mu(\infty)t}$ for all $x \geqslant 0$. So, $g(x) \geqslant \mathrm{e}^{-\mu(\infty)} > 0$. Taking the limit $x \to \infty$ in the integral equation above,

$$1 = \int_{g(\infty)}^{1} \frac{\mathrm{d}y}{y\mu(\infty)} = -\frac{\ln(g(\infty)}{\mu(\infty)},$$

i.e.,

$$g(\infty) = \mathrm{e}^{-\mu(\infty)}.$$

We now assume that $\mu(\infty) = \infty$ and claim that $g(\infty) = 0$. Suppose that $g(\infty) > 0$. Then

$$1 \leqslant \int_{g(\infty)}^{1} \frac{\mathrm{d}y}{y\mu(xy)} \leqslant \int_{g(\infty)}^{1} \frac{\mathrm{d}y}{y\mu(xg(\infty))} = -\frac{\ln g(\infty)}{\mu(xg(\infty))} \to 0, \quad x \to \infty,$$

a contradiction. So, in either case, $g(\infty) = \mathrm{e}^{-\mu(\infty)}$.

The following **limit trichotomy** follows from Theorem 9.7(b).

(i) $(p+b)\mathrm{e}^{-\mu(\infty)} \geqslant 1$. All solution sequences (P_n) converge to ∞. The population grows monotonically and without bound.

(ii) $(p+b)\mathrm{e}^{\mu(0)} \leqslant 1$. All solution sequences (P_n) converge to 0. The population becomes extinct.

(iii) $(p+b)\mathrm{e}^{-\mu(\infty)} < 1 < (p+b)\mathrm{e}^{-\mu(0)}$. There exists a unique fixed point of f which attracts all solution sequences (P_n). The population converges towards a uniquely determined equilibrium state.

Application to the Ricker Difference Equation

Recall that the Ricker difference equation can be written as

$$x_{n+1} = x_n(q + \gamma \mathrm{e}^{-x_n})$$

with $0 \leqslant q < 1$. By Theorem 9.7 (b), all solutions (x_n) converge to 0 if $q + \gamma \leqslant 1$. If $q + \gamma > 1$, there exists a unique fixed point $x^* > 0$. It is convenient to normalize x^* to 1; to do this we observe that $1 = q + \gamma \mathrm{e}^{-x^*}$. So,

$$x_{n+1} = x_n(q + (1-q)\mathrm{e}^{x^* - x_n}).$$

Setting $x_n = x^* y_n$ and $r = x^*$,

$$y_{n+1} = y_n(q + (1-q)\mathrm{e}^{r(1-y_n)}) =: f(y_n).$$

So, the fixed point of the previous equation has become a parameter $r = x^*$ in the new equation. We easily observe that 1 is the unique nonzero fixed point of f. We will prove the following result.

Theorem 9.16.

(a) *If*

$$r = x^* \leqslant 2 + \ln\left(1 + \frac{2q}{1-q}\right),$$

all solutions of the Ricker difference equation with strictly positive initial value converge to the unique positive fixed point.

(b) *If*

$$r = x^* > \frac{2}{1-q},$$

then the set of initial data for which the solutions converge to the positive fixed point is countable. Moreover, f^2 has a positive fixed point different from the fixed point of f.

Note that the result is sharp for $q = 0$, i.e., when adults do not survive the reproductive season. Otherwise there is a gap for r between

$$2 + \ln\left(1 + \frac{2q}{1-q}\right) \quad \text{and} \quad 2 + \frac{2q}{1-q}.$$

The gap is larger for larger q. To prove Theorem 9.16, we let $r > 0$ and notice that

$$\frac{f(y)}{y} \to q + (1-q)\mathrm{e}^r > 1, \quad 0 < y \to 0, \qquad \frac{f(y)}{y} \to q < 1, \quad y \to \infty.$$

So, f satisfies Assumption 9.4 if $r > 0$. Notice that

$$f'(y) = q + (1-q)(1-ry)\mathrm{e}^{r(1-y)},$$
$$f'(1) = 1 - (1-q)r.$$

We see that $f'(0) > 1$ and $\lim_{y\to\infty} f'(y) = q$. In order to find the infimum of f', we differentiate once more,

$$f''(y) = [1-q]\left(-r\mathrm{e}^{r(1-y)} - r(1-ry)\mathrm{e}^{r(1-y)}\right) = r[1-q]\mathrm{e}^{r(1-y)}(ry-2).$$

We see that $f''(y) = 0$ if and only if $y = 2/r$, and that f'' is positive on $(2/r, \infty)$ and negative on $(0, 2/r)$. So, the infimum of f' is given by

$$f'(2/r) = q - (1-q)\mathrm{e}^{r-2}.$$

This is a strict infimum. By Corollary 9.9 (ii), the fixed point 1 attracts all solutions if $f'(2/r) \geqslant -1$, i.e.,

$$r \leqslant 2 + \ln\left(1 + \frac{2q}{1-q}\right).$$

This proves part (a). As for part (b), we observe that $f(0) = 0$ and that f'' has exactly one zero. Therefore, f is either monotone increasing or has exactly one local maximum and at most one local minimum. In particular, $f^{-1}(\{x\})$ has at most three points. From Theorem 9.12 and Proposition 9.11 we conclude that 1 attracts only a countable set of initial data if $f'(1) < -1$, i.e., if

$$r > \frac{2}{1-q} = 2 + \frac{2q}{1-q}.$$

As we already mentioned, Theorem 9.16 leaves a gap for r between

$$2+\ln\left(1+\frac{2q}{1-q}\right) \quad \text{and} \quad 2+\frac{2q}{1-q}.$$

Plotting the graph of f^2 in this critical region for r seems to indicate, however, that there are no fixed points of f^2 other than 1, so Theorem 9.6 suggests that for these r also, all solutions converge to 1. See Figure 9.1, where q is chosen as $q=\frac{1}{3}$. For this choice of q, $f'(1)=-1$ if and only if $r=3$.

A famous theorem by Sharkovskii (1964) states that if f^3 has fixed points different from the fixed point of f, f has periodic points of all periods, i.e., for every n there exists a fixed point of f^n which is not a fixed point of f^m for all $m<n$. Looking at the graph of f^3, one sees that f^3 can have fixed points different from 1 if r is large enough. If $q=\frac{1}{3}$, this appears to happen for r between 5.4 and 5.5 (see Figure 9.2). Numerical computation with MAPLE provides the following fixed points of f^3 different from 0 and 1 for $r=5.5$: 0.566 177 250 9, 0.585 683 409 2, 1.335 928 641, 1.430 566 417, 4.007 785 398, 4.291 699 134.

Figure 9.3 shows solutions of the Ricker equation for $q=\frac{1}{3}$ and the initial value $y_0=1.01$. For $r=4$, the solutions eventually oscillate between the two fixed points of f^2 that are different from 1. For $r=6$, we have a much more complicated behavior, though some kind of pattern seems to exist.

Application to a Combination of Models by Hassell and Maynard Smith

The following difference equation,

$$x_{n+1}=\frac{ax_n}{(1+bx_n^{\xi})^{\zeta}},$$

with $a>1$, $b,\xi,\zeta>0$, has been proposed as a model for single-species population dynamics for $\xi=1$ by Hassell (1975) and $\zeta=1$ by Maynard Smith (1974). Denoting the right-hand side by $f(x_n)$, f has the following properties: $f(x)/x$ is strictly decreasing and converges to $a>1$ as $x\to 0$ and to 0 as $x\to\infty$. The intermediate value theorem implies that Assumption 9.4 is satisfied. Let us check under which conditions assumption (ii) of Corollary 9.9 is satisfied. Differentiating the function $x^2/(1+b^{\xi})^{\zeta}$ with respect to x, we find that it is strictly increasing if $\xi\zeta\leqslant 2$. We conclude from Corollary 9.9 (b) that, for $\xi\zeta\leqslant 2$, all solutions of the difference equation starting from $x_0>0$ converge towards the unique fixed-point theorem point

$$x^*=\left(\frac{a^{1/\zeta}-1}{b}\right)^{1/\xi}.$$

For the cases $\xi=1$, $\zeta>2$ and $\xi>2$, $\zeta=1$ see Exercises 9.3 and 9.4.

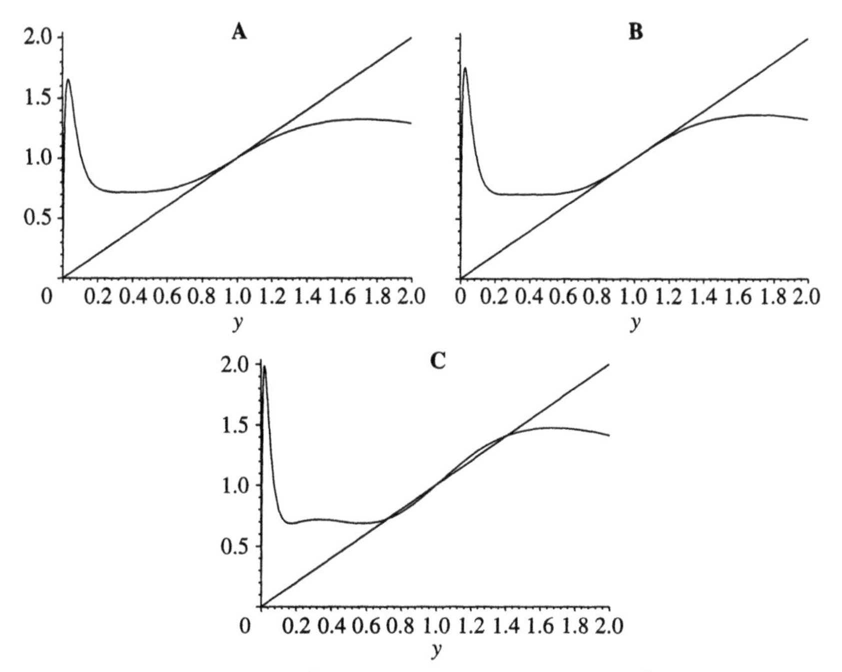

Figure 9.1. Graphs of f^2 for the Ricker equation with $q = \frac{1}{3}$. (**A**) $r = 2.9$. (**B**) $r = 3$. (**C**) $r = 3.2$.

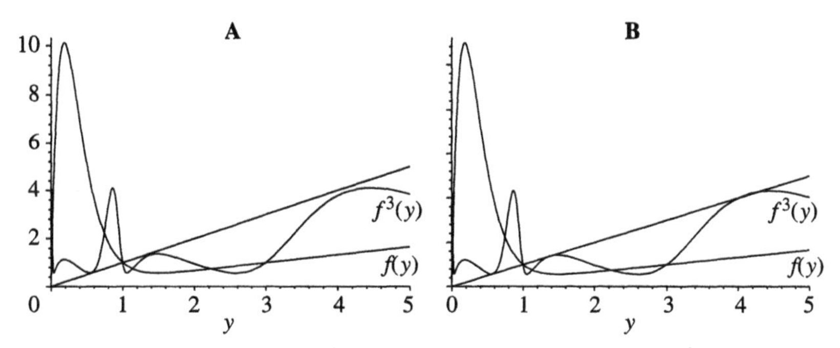

Figure 9.2. Graphs of f and f^3 for the Ricker equation with $q = \frac{1}{3}$. (**A**) $r = 5.4$. (**B**) $r = 5.5$. In (**A**) it looks as if there might be fixed points y of f^3, but then $f(y)$ and $f^2(y)$ would be fixed points of f^3 as well, which is not the case.

9.4 Some Remarks Concerning the Quadratic Difference Equation

Making a second-order Taylor expansion at 0 of the difference equation $x_{n+1} = f(x_n)$, $f(0) = 0$, yields the quadratic difference equation

$$y_{n+1} = ay_n(1 - by_n)$$

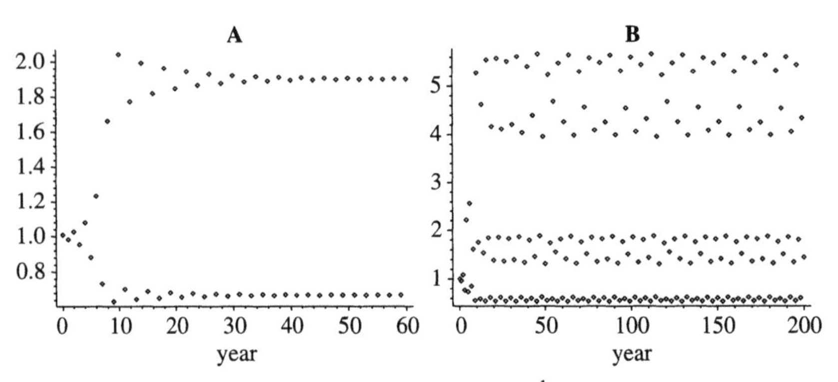

Figure 9.3. Solution of the Ricker equation for $q = \frac{1}{3}$ and initial value $y_0 = 1.01$. **(A)** $r = 4$. The solution eventually oscillates between the two fixed points of f^2 that are different from 1, 0.668 334 483 4 and 1.901 832 013. **(B)** $r = 6$. There seems to be some kind of pattern, but it is not clear whether it is periodic.

with $a, b > 0$, provided we assume a negative feedback from large population sizes, $f''(0) < 0$. The quadratic equation has gained notoriety because it illustrates the route to chaos in the most direct way and, in contrast to the discrete Ricker equation, for example, often allows explicit calculations. However, the parameters are difficult to interpret biologically. Its form reminds us of the Verhulst equation, the continuous logistic equation, and can formally be obtained by an explicit Euler discretization of the Verhulst equation. It is therefore sometimes called the discrete logistic equation, though this is not justified as we will show below. The biological interpretation of the Verhulst equation is only valid for its discretization, as long as the solutions behave in the same way, i.e., converge towards an equilibrium or fixed point. As we have seen above, the biologically meaningful discrete analog of the Verhulst equation, in the case of a very short reproductive season, is the Beverton–Holt difference equation. Actually, this even holds without assuming a very short reproductive season. Consider the logistic equation, or more generally the Bernoulli equation,

$$N' = \rho(t)N - m(t)N^{\theta+1}$$

with $\theta > 0$ and periodic functions ρ and m with the same period $p > 0$. The Leibnitz transformation $x = N^{-\theta}$ yields $x' = -\theta\rho(t)x + \theta m(t)$. Set $x_n = x(np)$. This means that we take periodically timed snapshots of the solution. By Theorem 3.7,

$$x_{n+1} = (x_n - b)\mathrm{e}^{-\theta\bar{\rho}} + b$$

with an appropriate number b with $\bar{\rho}b < 0$. So,

$$x_{n+1} = \alpha x_n + c$$

with appropriate constant $\alpha, c > 0$. Inverting the Leibnitz transformation,

$$N_{n+1} = \frac{N_n}{(\alpha + cN_n^\theta)^{1/\theta}}.$$

For $\theta = 1$, this is the discrete Beverton–Holt equation.

Out of curiosity, let us look at various discrete approximations of the logistic equation, in its simplest form,

$$y' = y(1 - y).$$

Replacing the derivative by the right difference quotient (explicit Euler discretization) we obtain

$$y(t + h) = y(t) + hy(t)(1 - y(t)) = y(t)(1 + h - hy(t)),$$

the quadratic difference equation. Instead we can replace the derivative by the left difference quotient (implicit Euler discretization) and obtain

$$h^{-1}(y(t) - y(t - h)) = y(t)(1 - y(t)).$$

Solving for $y(t)$ gives

$$y(t) = -\frac{h^{-1} - 1}{2} \pm \sqrt{\frac{h^{-1} - 1}{2} + h^{-1}y(t - h)}.$$

Obviously, we are only interested in the positive square root, in which case we get a difference equation with an increasing function: all solutions converge towards a (unique) fixed point, in complete analogy to the Verhulst equation. For $h > 1$, this equation is not meaningful, though, because then the right-hand side is not defined for small $y(t - h)$ because the radicand is negative.

We can also mix explicit and implicit discretization,

$$h^{-1}(y(t + h) - y(t)) = y(t)(1 - y(t + h)),$$

which gives us

$$y(t + h) = (1 + h)\frac{y(t)}{1 + hy(t)},$$

the discrete Beverton–Holt equation, the solutions of which again behave in complete analogy to the logistic equation as they converge towards a unique fixed point.

Finally, we can use a logarithmic discretization. We write the logistic equation as $(x'/x) = 1 - x$ and notice that

$$\frac{x'(t)}{x(t)} = \lim_{h \to 0} \frac{\ln x(t + h) - \ln x(t)}{h}.$$

This yields the approximation $\ln x(t+h) = \ln x(t) + h[1 - x(t)]$. Exponentiating,

$$x(t+h) = x(t)e^{h[1-x(t)]},$$

a special case of the discrete Ricker equation. While, as we have seen, the Ricker equation makes perfect biological sense, it can behave very differently from the continuous logistic equation for certain parameters.

The bottom line of Section 9.4 is that the quadratic difference equation, the discrete Beverton–Holt equation, and the discrete Ricker equation can all be considered discrete approximations of the continuous logistic equation.

Exercises

9.1. Consider the difference equation $x_{n+1} = f(x_n)$,

$$f(x) = \frac{ax}{1 + bx + cx^2},$$

with $a > 1$, $b, c \geqslant 0$, $b + c > 0$.

Show that f has a unique fixed point $x^* > 0$ and that all solutions of the difference equation with $x_0 > 0$ converge towards x^* as $n \to \infty$.

9.2. Consider the difference equation $x_{n+1} = f(x_n)$,

$$f(x) = \frac{x}{\ln(1 + x^2)}.$$

Show that f has a unique positive fixed point x^* and that all solutions of the difference equation converge to x^*.

9.3. Consider the difference equation $x_{n+1} = f(x_n)$ with

$$f(x) = \frac{ax}{1 + bx^k}$$

with $a > 1$, $b > 0$, $k \in \mathbb{N}$, $k > 2$.

(a) Show that there exists a unique positive fixed point of f.

(b) For given $k \in \mathbb{N}$ determine $b > 1$ as large as possible such that you can still show that all solutions of the difference equation with $x_0 > 0$ converge towards the positive fixed point.

(c) For given $k \in \mathbb{N}$ determine $b > 1$ as small as possible such that the solutions do not converge towards the unique positive fixed point for all but countably many initial data.

9.4. Consider the difference equation $x_{n+1} = f(x_n)$ with

$$f(x) = \frac{ax}{(1+bx)^k}$$

with $a > 1$, $b > 0$, $k \in \mathbb{N}$, $k > 2$.

(a) Show that there exists a unique positive fixed point of f.

(b) For given $k \in \mathbb{N}$ determine $b > 1$ as large as possible such that you can still show that all solutions of the difference equation with $x_0 > 0$ converge towards the positive fixed point.

(c) For given $k \in \mathbb{N}$ determine $b > 1$ as small as possible such that the solutions do not converge towards the unique positive fixed point for all but countably many initial data $x_0 > 0$.

9.5. Consider the Ricker equation $y_{n+1} = \frac{1}{2}y_n(1 + e^{r-y_n}) = f(y_n)$.

(a) For which $r > 0$ do our theorems guarantee that all solutions converge to 1?

(b) For which $r > 0$ do the solutions, for all but countably many initial data, not converge to 1?

(c) Explore the window between the r in (a) and the r in (b) by checking graphically for three different r in the window where f^2, $f^2(y) = f(f(y))$, has no fixed point different from 1. What conjecture do you conclude from this observation?

9.6. Consider the Ricker equation $y_{n+1} = \frac{1}{2}y_n(1 + e^{r-y_n}) = f(y_n)$. Find graphically (using appropriate computer software) an approximate value for the smallest r such that f has a period-three point.

9.7. Revisit the derivation of the discrete Beverton–Holt equation with a seasonally dependent per capita mortality rate; in particular, consider the following nonautonomous version of (9.1),

$$N' = -\mu(t, N)N, \quad 0 \leqslant t \leqslant 1, \qquad N(0) = P_n,$$

with a nonnegative continuous function $\mu : [0, 1] \times (0, \infty) \to [0, \infty)$ with continuous partial derivatives in N. Show that we still obtain $P_{n+1} = f(P_n)$, $n \in \mathbb{N}$, with a strictly increasing function f, and that $f(P)/P$ (strictly) decreases in $P > 0$ if $\mu(t, N)$ is (strictly) increasing in $N > 0$ for every $t \in [0, 1]$.

Hint: apply the considerations preceding Theorem 8.1 in Chapter 8, with $g(t, N) = -\mu(t, N)$.

Bibliographic Remarks

An exposition of Floquet theory for general systems of linear ordinary differential equations can be found in Hale (1980). In spite of their obvious importance, few textbooks in mathematical biology consider fluctuating environments because of their mathematical intricacies; an exception is Nisbet, Gurney (1982).

Kooijman, Metz (1984) and Kooijman (1986) show growth curves for body lengths of *Daphnia magna* that have been fitted by the von Bertalanffy equation. Kooijman (2000) also shows body-length growth curves for other organisms (see figures 1.1, 2.4, 2.5, 3.5 in his book) and one weight growth curve (figure 2.4 in his book).

Hutchinson (1978, Chapter 1) exhibits many sigmoid growth curves for various populations that have been fitted by the logistic equation (see also Hallam, 1986). This work also contains interesting remarks about Bernoulli's equation as a generalized or asymmetric logistic equation and about Smith's equation. Last but not least one finds interesting background information on Verhulst, including a portrait of him. I could not find any reference to what, if anything, Jakob Bernoulli intended to model with his equation. In most textbooks on elementary ordinary differential equations, Bernoulli's equation has been banished to the exercises, while more prominence is given to treating its special case, the logistic equation, as an example of a separable equation. In comparison, Leibnitz's transformation is just a trick, but a powerful one, because it allows us to deal with time-varying environments. The scientific history of the logistic equation, in particular an explanation of why it is also called the Verhulst–Pearl equation, can be found in Kingsland (1982, 1995).

A more mathematical exposition of the Allee effect is given by Edelstein-Keshet (1988). If the function f in the population equation $N' = f(N)$ is analytic in $N \geqslant 0$, it can be expanded into a Taylor series

$$f(N) = \sum_{j=0}^{\infty} a_j N^j.$$

The condition $f(0) = 0$ (no immigration) implies $a_0 = 0$. If the Taylor series is truncated after $n = 2$ and $a_2 < 0 < a_1$, we have the logistic (Verhulst) equation $N' = N(a_1 + a_2 N)$. Truncation after $n = 3$ with $a_3 < 0 < a_2$ results in an equation with Allee effect, $N' = N(a_1 + a_2 N + a_3 N^2)$.

Similar derivations and analyses of our model with Allee effect can be found in Dennis, Patil (1984) and Dennis (1989) and the references mentioned there. A nice review of Allee effect models for single species has been given by Boukal, Berec (n.d.), and the role of spatial dynamics has been investigated by Berec et al. (2002).

Our presentation of elementary population models has been restricted to self-reproducing populations, where the influx of new individuals (through birth) is proportional to the population size. The model for a population of red blood cells after peripheral stem cell transplantation by Merrill, Murphy (2002), for example, assumes that the influx of new cells is constant. This, in conjunction with an auto-catalytic effect, leads to hyperbolic rather than sigmoid growth patterns.

If seasonal effects are included in the models, it is very difficult to extract useful information in other than special situations. The results in Chapter 8 no longer hold if several species are considered or if delay effects are incorporated, unless very restrictive assumptions are made (Inaba, 1989; Nussbaum, 1990; Thieme, 1988a) which are automatically satisfied and hardly noticeable in Chapter 8. In this more general context, the analog of Theorem 8.1 is called a *(nonlinear) weak ergodic theorem* and the solutions are said to be *asymptotically proportional* to each other.

For a comparison of continuous versus discrete-time models and more references on the subject, see Gyllenberg et al. (1997). The derivation of the Beverton–Holt and Ricker difference equations is very similar to the one in Clark (1976), who calls this modeling approach discrete-time metered models; see also Brauer, Castillo-Chávez (2001). An alternative derivation of the Ricker equation is given by Royama (1992, Section 4.2.5), who also presents the Ricker equation as a logarithmic discretization of the continuous logistic equation (Section 4.2.4). Theorems 9.6 and 9.7 have been proved by Cull (1986) under an additional assumption. Cull also gives some other sufficient assumptions for global stability. The condition $f(x) > x|f'(x)|$ in Remark 9.10 can be generalized to multidimensional difference equations (Krause, 1997) and even to infinite-dimensional equations (Takáč, 1996). A proof of Sharkovkii's theorem and much more on difference equations can be found in Devaney (1987) or Robinson (1995); for more on biological aspects see Hoppensteadt (1982), Edelstein-Keshet (1988, Chapter 2), Murray (1989, Chapter 2), Busenberg, Cooke (1993, Chapter 3), and Brauer, Castillo-Chávez (2001, Chapter 2). Deterministic chaos has been verified both mathematically and experimentally for discrete-time models of the population dynamics of the flour beetle (Cushing et al., 1996; Cushing, 1998). These models, systems of difference equations, contain functions of Ricker type as essential building blocks.

Chapter Ten

Dynamics of an Aquatic Population Interacting with a Polluted Environment

So far we have considered elementary models for the dynamics of one population, which are formed by one scalar differential or difference equation. Naturally, more complex models are needed when populations of several species are involved, unless a reduction by a time scale argument is possible, as in the derivation of the Allee effect generated by a satiating generalist predator in Section 7.2. More complex models are also needed, when the population is subject to external or internal structure. External structure can be provided by spatial distribution or by interaction with the environment. Internal structure is induced by development stages, age, body, etc. In the following two chapters, we investigate planar (i.e., two-dimensional) differential equation models for a single-species population interacting with its environment and a single-species population with a juvenile and an adult stage. At the same time, we explore the beautiful mathematical theory available for the analysis of planar systems, like the Poincaré–Bendixson theory (Toolbox A.2).

In the previous chapters, we have considered the dynamics of one population in a given environment, be it constant or time dependent. The environment has an impact on the development of the population, but the population does not change its environment. In this chapter, we model a population which interacts with an environment that is subject to chemical pollution.

This interaction has several facets: the individuals take up the toxicant from the water and egest (excrete it with the feces, for example) or depurate (detoxify) it through metabolic processes. The toxicant in turn affects the food uptake and biomass conversion and the mortality of the population. Toxicant uptake can occur via food and environmental pathways. This exposition will be restricted to environmental pathways, because a proper modeling of the food pathway should involve the interaction between food resource and toxicant, a complication we want to avoid.

We will briefly consider an open-loop toxicant input where the pollutant is introduced into the environment at a given, though possibly time-dependent, rate. Our main emphasis, however, will be on feedback-loop toxicant inputs, which steer the environmental concentration of the toxicant to a given target value. The model then reduces to a planar system of ordinary differential equations which can be made subject to the powerful methods of phase plane analysis

(Toolbox A.2): the Poincaré–Bendixson theory above all. In the next chapter we will discuss another planar population model that describes the dynamics of one population with two stages: juveniles and adults. Interesting dynamics in the toxicant model will be driven by multiple boundary equilibria, more precisely by several equilibrium solutions where the population is extinct but the extinction happens at different internal toxicant concentrations. In the two-stage model, interesting dynamics are due to multiple interior equilibria, i.e., several equilibrium solutions, none of which lies on one of the coordinate axes.

10.1 Modeling Toxicant and Population Dynamics

Since we consider an aquatic environment, we formulate the model in terms of concentrations of population biomass and of toxicant in the environment and not in terms of amounts or numbers. We use the following dependent variables:

$M(t)$, concentration of biomass of the population at time t,

$C_{\mathrm{E}}(t)$, concentration of toxicant in the environment at time t,

$C_{\mathrm{I}}(t)$, amount of toxicant in the population biomass at time t,

$y(t)$, concentration of toxicant per unit population biomass at time t (*body burden*), $y = C_{\mathrm{I}}/M$.

The subscript "E" stands for "environmental," the subscript "I" for "internal." A simple model for the interaction between population and toxicant can be formulated in these terms as follows:

$$\left.\begin{aligned} M' &= M[\beta(M, y) - \mu], \\ C_{\mathrm{I}}' &= aC_{\mathrm{E}}M - [\mu + \eta + \kappa]C_{\mathrm{I}}, \\ C_{\mathrm{E}}' &= -aC_{\mathrm{E}}M + [p\mu + \eta]C_{\mathrm{I}} - \theta C_{\mathrm{E}} + u(t), \\ y &= C_{\mathrm{I}}/M, \end{aligned}\right\} \tag{10.1}$$

with appropriate initial conditions. The first equation presents a generic description of the growth of the population under the influence of the toxicant, while the second and third equations are balance equations for the concentrations of the toxicant contained in the individuals of the population and of the toxicant dissolved in the aquatic environment.

The positive constant μ represents the per unit biomass loss rate of the population due to death, and $\beta(M, y)$ denotes the per unit rate of biomass change of the population. We assume that the toxicant and the population concentrations only affect the per unit biomass change, but not the per unit biomass death rate.

The death of an individual not only leads to a loss of population biomass, but also to a loss of internal toxicant concentration, at rate μC_{I}. This leads to the term $-\mu C_{\mathrm{I}}$ in the second equation. A fixed fraction of the internal toxicant, denoted by p, is recycled into the environment, $p \in [0, 1]$; this gives rise to the term $p\mu C_{\mathrm{I}}$ in the third equation.

The positive constants η and κ are per unit rates of toxicant egestion and depuration due to the metabolic processes of the population. Internal toxicant that is egested reappears as environmental toxicant such that we have a term $-\eta C_{\mathrm{I}}$ in the second equation and a term $+\eta C_{\mathrm{I}}$ in the third equation. Internal toxicant that is depurated just vanishes from the whole system such that there is only a term $-\kappa C_{\mathrm{I}}$ in the second equation.

The toxicant uptake rate by the population from the environment, $aC_{\mathrm{E}}M$, is modeled according to the *law of mass action* and is proportional to both the concentration of toxicant in the environment and the concentration of population biomass. Environmental toxicant in water can be decomposed by sunlight (photolysis) or by reaction with water (hydrolysis), or can be vaporized (volatilization) or be detoxified in other ways; the parameter $\theta > 0$ denotes the per unit rate of environmental detoxification. The time-dependent function u in the third equation stands for the exogenous input of toxicant into the environment.

It is instructive to write down the equation for the body burden $y = C_{\mathrm{I}}/M$. Since

$$y' = \frac{C_{\mathrm{I}}'}{M} - y\frac{M'}{M} = aC_{\mathrm{E}} - [\mu + \eta + \kappa]y - y[\beta(M, y) - \mu],$$

we obtain

$$y' = aC_{\mathrm{E}} - [\eta + \kappa + \beta(M, y)]y. \tag{10.2}$$

Here the term $\beta(M, y)$ plays the role of a dilution rate of per unit concentration of toxicant, with the dilution being caused by biomass gain. This reflects our assumption that the food resource of the population is not contaminated by the toxicant.

We also notice that there is no conservation of the total concentration of toxicant, $C = C_{\mathrm{E}} + C_{\mathrm{I}}$, because toxicant is introduced into the environment at rate u, and degraded by both environmental and population internal processes. Indeed, by adding the equations for C_{E} and C_{I},

$$\begin{aligned} C' &= u(t) - [\kappa + (1 - p)\mu]C_{\mathrm{I}} - \theta C_{\mathrm{E}} \\ &= u(t) - [\kappa + (1 - p)\mu](C - C_{\mathrm{E}}) - \theta C_{\mathrm{E}}. \end{aligned} \tag{10.3}$$

The following assumptions (1)–(4) for β seem appropriate.

(1) $\beta : [0, \infty)^2 \to \mathbb{R}$ is locally Lipschitz continuous, i.e., for every constant $c > 0$ there is a Lipschitz constant $L > 0$ such that

$$|\beta(M, y) - \beta(N, z)| \leqslant L(|M - N| + |y - z|) \quad \text{whenever } 0 \leqslant y, z, M, N \leqslant c.$$

(2) $\beta(M, y)$ is monotone decreasing in both arguments. If $\beta(0, y) > 0$, then the decrease of $\beta(\cdot, y)$ is strict. For fixed, but arbitrary, $M \geqslant 0$, $\beta(M, \cdot)$ is a bounded function on $[0, \infty)$.

(3) $\beta(0, 0) > 0$, $\beta(0, y) \geqslant 0$ for all $y > 0$, and $\lim_{M \to \infty} \beta(M, 0) < \mu$.

(4) For every $\epsilon > 0$ there exists some $\delta > 0$ such that $\beta(M, y) \geqslant -\epsilon$ whenever $M \leqslant \delta$, $y \geqslant 0$.

Let us explain the assumptions.

Hypothesis (1) is a technical mathematical assumption which guarantees that the solutions to the model are unique. It does not restrict biological generality.

As for (2), the decrease of β as a function of y reflects that the toxicant, to a degree related to its per unit biomass concentration, hampers food uptake and/or food conversion into population biomass. β also decreases as a function of M, because increasing M means decreasing the per unit share of resources. Requiring that the decrease in M is strict if $\beta(0, y) > 0$ definitely has a technical flavor. We do this in order to include the important linear dose response,

$$\beta(M, y) = \tilde{\beta}(M)[1 - by]_+,$$

where $\tilde{\beta}$ is a strictly decreasing and strictly positive function and $[\cdot]_+$ denotes the positive part of a number.

The second assumption in (3) states that, for very large population biomass, the per unit biomass change rate even drops below the per unit loss rate by death, while the first assumption maintains that it is strictly positive under optimal conditions (no pollution, no competition). If M is very small, one can assume that the biomass per individual is also very small such that there are hardly any starvation effects even if the body burden of toxicant is high, so $\beta(0, y) \geqslant 0$ and (4).

To underpin these assumptions we derive a concrete expression for the per unit rate of biomass change, β. For our general results this concrete expression is not needed, but our examples will be based on it.

Intermezzo. Derivation of a functional form for β

Let us revisit our derivation of the continuous Beverton–Holt equation (Section 5.2). We prefer the Beverton–Holt formulation over the equivalent Smith formulation because it allows us to differentiate between the impacts of the toxicant on food uptake, food conversion, and biomass gain. For the same reason we prefer to build on the Beverton–Holt equation rather than on the logistic equation (Section 5.1). We repeat the derivation in Section 5.2 because the impact of the body burden creates an additional difficulty since it varies with time.

With F denoting the concentration of food on which the population lives, we have the equations

$$F' = \Lambda - \nu F - \frac{b(y)}{\gamma(y)} FM, \qquad M' = b(y)FM - \mu M.$$

The food is provided at the constant rate Λ and degrades at a constant per unit rate ν.

Food consumption also follows the *law of mass action*, i.e., the consumption rate is proportional to both the biomass of food available and the biomass of consumers. We neglect that there may be biomass loss due to starvation other than by death; so, the per unit death rate, μ, coincides with the per unit biomass loss rate. The interpretation of the parameters is the same as in Section 5.2, but the biomass gain rate b and the biomass conversion rate γ now depend on the body burden y.

Assuming that the food dynamics are much faster than the consumer dynamics, we want to make a *quasi-steady-state approach* to the food equation. In order to be able to compare the speeds of the different dynamics, we introduce new, dimensionless variables. We start with the dependent variables,

$$F = \frac{\Lambda}{\nu} z, \qquad M = \frac{\Lambda \gamma(y)}{\mu} x,$$

and introduce the *basic biomass production rate*,

$$\rho(y) = \frac{\Lambda}{\nu} b(y), \tag{10.4}$$

and the *basic biomass production ratio*,

$$\mathcal{R}_0(y) = \frac{\Lambda}{\nu} \frac{b(y)}{\mu}. \tag{10.5}$$

In these new variables, the food resource and population biomass system takes the form

$$\frac{\Lambda}{\nu} z' = \Lambda - \Lambda z - \frac{b(y)}{\gamma(y)} \frac{\Lambda}{\nu} z \frac{\Lambda \gamma(y)}{\mu} x,$$

$$\frac{\Lambda \gamma(y)}{\mu} x' + \frac{\Lambda (\gamma(y))'}{\mu} x = \left(b(y) \frac{\Lambda}{\nu} z - \mu \right) \frac{\Lambda \gamma(y)}{\mu} x.$$

After simplification, using (10.4) and (10.5),

$$z' = \nu(1 - z - \mathcal{R}_0 z x),$$

$$x' = \mu x(\mathcal{R}_0 z - 1) - \frac{\gamma'}{\gamma} x y'.$$

Λ/ν is the large-time limit of the food concentration if there are no consumers around. Recalling that $b(y)$ is the per unit/unit biomass gain rate and μ is the per unit consumer biomass loss rate, we see that $\rho(y)$ is the per unit biomass production rate and $\mathcal{R}_0(y)$ is the per unit biomass production ratio in situations without competing conspecifics. (In both cases, biomass gain includes the biomass invested into offspring.)

We also introduce the dimensionless time τ, $\tau = \mu t$, and let $\dot{z}$ and $\dot{x}$ denote the derivative with respect to the new time τ, $\dot{z} = \mu z'$ and $\dot{x} = \mu z'$. Dividing the first equation by ν and the second by μ we obtain

$$\epsilon \dot{z} = 1 - z - \mathcal{R}_0 z x,$$
$$\dot{x} = x(\mathcal{R}_0 z - 1) - \frac{\gamma'(y)}{\gamma(y)} x \dot{y},$$

with

$$\epsilon = \mu/\nu.$$

The assumption that the food dynamics are fast compared to the population dynamics can now be rephrased more precisely as $\epsilon \ll 1$ and $\mathcal{R}_0$ being not too small compared to 1. Setting $\epsilon = 0$ and solving for z, we obtain

$$z = \frac{1}{1 + \mathcal{R}_0 x}.$$

Substituting this into the original equation yields

$$\dot{x} = x\left(\frac{\mathcal{R}_0}{1 + \mathcal{R}_0 x} - 1\right) - \frac{\gamma'(y)}{\gamma(y)} x \dot{y}.$$

We return to unscaled time and consumer biomass,

$$M' = \mu M\left(\frac{\mathcal{R}_0}{1 + (\mathcal{R}_0 \mu/\Lambda\gamma)M} - 1\right)$$
$$= M\left(\frac{\rho}{1 + (\rho/\Lambda\gamma)M} - \mu\right).$$

The dependence of this model equation on the body burden y can be most easily seen by rewriting the equation as

$$M' = M\left(\frac{1}{(1/\rho(y)) + (1/\Lambda\gamma(y))M} - \mu\right).$$

It is reasonable to assume that the basic production rate ρ and the yield constant γ depend on the body burden in a monotone decreasing way. So, we can write the equation for M as

$$M' = M[\beta(M, y) - \mu],$$

with a nonnegative function β which is decreasing in both M and y,

$$\beta(M, y) = \frac{\rho(y)}{1 + (\rho(y)/\Lambda\gamma(y))M} = \frac{\Lambda\gamma(y)\rho(y)}{\Lambda\gamma(y) + \rho(y)M}. \tag{10.6}$$

From the formula for ρ in (10.4) it follows that

$$\frac{\rho(y)}{\Lambda\gamma(y)} = \frac{1}{\nu}\frac{b(y)}{\gamma(y)}$$

is the per unit/unit food consumption rate $\tilde{b}(y) = b(y)/\gamma(y)$ times the life expectancy $1/\nu$ of the food. We assume that $\tilde{b}$ is a strictly positive continuously differentiable function on $[0, \infty)$, while we will allow $b(y)$ and so $\rho(y)$ to be zero for large body burdens y.

It follows from (10.6) and the remarks immediately thereafter that β has a particularly easy form if the toxicant concentration does not affect the per unit/unit food consumption rate $\tilde{b}$ but only the food conversion rate η, namely that

$$\beta(M, y) = \frac{\rho(y)}{1 + cM}, \tag{10.7}$$

with a monotone nonincreasing function ρ and a positive constant c. As a special case, we want to include a linear dose response for the per unit biomass production rate, $\rho(y) = \alpha_1[1 - \alpha_2 y]_+$ with positive constants α_i and $[r]_+ = \max\{0, r\}$ being the positive part of a real number r (Hallam, De Luna, 1984). This motivates the technicalities in the assumptions for β.

Exercises

10.1. Consider the following model for several species competing for one food resource:

$$F' = \Lambda - \nu F - \sum_{j=1}^{n} \frac{b_j}{\gamma_j} F M_j,$$
$$M_j' = b_j F M_j - \mu_j M_j, \quad j = 1, \ldots, m.$$

Remark: there is no pollution in this model.

(a) Explain the model. (What is the meaning of the variables and parameters?)

(b) Use scaling (introduction of dimensionless variables) and a quasi-steady-state approach to derive a model with m rather than $m + 1$ equations by eliminating F.

Hint: use $\bar{\mu} = (1/m)\sum_{j=1}^{m} \mu_j$ and $\epsilon = \bar{\mu}/\nu$. Assume $\epsilon \ll 1$. Explain what this assumption means.

10.2. (*Chemostat.*) Consider the same model as in (1), but assume that $\mu_j = \nu$ for all $j = 1, \ldots, m$. Do the same scaling as in (1). Obviously, under these different assumptions, you cannot get rid of the equation for F.

(a) Derive a differential equation for

$$S = z + \sum_{j=1}^{m} x_j,$$

where z is the scaled F and x_j are the scaled M_j.

(b) From the differential equation for S, what can you tell about the asymptotic behavior of S?

10.2 Open Loop Toxicant Input

We first consider the case in which the toxicant input, u, is some prescribed function and derive an estimate for the total toxicant concentration. Our tool is fluctuation theory (see Toolbox A.3).

The biological interpretation suggests that C_{E}, the concentration of toxicant in the environment, and C_{I}, the amount of toxicant in the population biomass, are nonnegative, and it can be shown that this is indeed the case if it holds for the initial data. The proof is omitted because we will have a similar proof later (Theorem 10.5).

Theorem 10.1. *Let the toxicant input u be a given continuous nonnegative function on $[0, \infty)$. Then, for given nonnegative initial data, $M(0) > 0$, there exists a unique solution M, C_{E}, C_{I} of (10.1) on $[0, \infty)$ which is nonnegative, $M(t) > 0$ for all $t \geqslant 0$. M is bounded on $[0, \infty)$.*

Corollary 10.2. *If u satisfies the assumptions of Theorem 10.1 and is bounded on $[0, \infty)$, so are C_{E} and C_{I}. For $C = C_{\mathrm{E}} + C_{\mathrm{I}}$ we have the inequality*

$$C(t) \leqslant \max\left\{C(0), \frac{\tilde{u}}{\xi_\diamond}\right\},$$

where

$$\tilde{u} = \sup_{[0,\infty)} u, \qquad \xi_\diamond = \min\{\kappa + (1-p)\mu, \theta\}.$$

Proof. We apply Theorem A.4 from the toolbox. First we check the assumptions concerning the vector field. For instance, let us consider the right-hand side of the equation for C_{E} in (10.1). If $C_{\mathrm{E}} = 0$ and $C_{\mathrm{I}} \geqslant 0$, we have

$$-aC_{\mathrm{E}}M + [p\mu + \eta]C_{\mathrm{I}} - \theta C_{\mathrm{E}} + u(t) = [p\mu + \eta]C_{\mathrm{I}} + u(t) \geqslant 0.$$

We make similar considerations for the other equations. Now consider a non-negative solution on an interval $[0, b)$. It follows from (10.3) that

$$C' \leqslant u(t) - \xi_\diamond C,$$

with

$$\xi_\diamond = \min\{\kappa + (1-p)\mu, \theta\}.$$

Let $t \in (0, b)$ and $\bar{C} = \max_{[0,t]} C$, $\bar{u} = \sup_{[0,b)} u$. By Lemma A.6 in the toolbox, $\bar{C} = C(0)$ or there exists some $s \in [0, t]$ such that $C(s) = \bar{C}$, $C'(s) \geqslant 0$. In the second case, from the first inequality for C' above, we obtain $0 \leqslant \bar{u} - \xi_\diamond \bar{C}$ which implies that $\bar{C} \leqslant \bar{u}/\xi_\diamond$. Combining the two cases,

$$\bar{C} \leqslant \max\left\{C(0), \frac{\bar{u}}{\xi_\diamond}\right\}.$$

We notice that this estimate does not depend on $t \in (0, b)$, so C is bounded on $[0, b)$ and the estimate also holds for $\bar{C} = \sup_{[0,b)} C$. Since C_{E} and C_{I} are nonnegative and $C = C_{\mathrm{E}} + C_{\mathrm{I}}$, C_{E} and C_{I} are bounded on $[0, b)$ as well.

By assumption (2), the biomass concentration satisfies the differential inequality

$$M' \leqslant M[\beta(M, 0) - \mu].$$

Let $t \in (0, b)$ and $\bar{M} = \max_{[0,t]} M$. Again by Lemma A.6 in the toolbox, $\bar{M} = M(0)$ or

$$0 \leqslant \bar{M}[\beta(\bar{M}, 0) - \mu].$$

If $\bar{M} > 0$,

$$\beta(\bar{M}, 0) \geqslant \mu.$$

By assumption (3), there exists some $M^\sharp > 0$ such that $\beta(M, 0) < \mu$ for all $M > M^\sharp$. So, $\bar{M} \leqslant M^\sharp$. Since this estimate does not depend on t, M is bounded on $[0, b)$ and $M(t) \leqslant \max\{M(0), M^\sharp\}$ for all $t \in [0, b)$. By Theorem A.4 in the toolbox, the solutions are defined on $[0, \infty)$. The bound for M does not depend on b, so M is bounded on $[0, \infty)$ with the same upper bound. If u is bounded on $[0, \infty)$, the estimates for C do not depend on b either, so C_{I} and C_{E} are bounded on $[0, \infty)$. □

From the fluctuation lemma we obtain the following large-time estimates for the total toxicant concentration.

Proposition 10.3. *Let $u : [0, \infty) \to \mathbb{R}$ be bounded. Then*

$$\limsup_{t\to\infty} C(t) =: C^\infty \leqslant \frac{u^\infty}{\xi_\diamond}, \qquad \liminf_{t\to\infty} C(t) =: C_\infty \geqslant \frac{u_\infty}{\xi^\diamond}.$$

Proof. By Proposition A.22, we choose a sequence $t_n \to \infty$ such that $C(t_n) \to C^\infty$ and $C'(t_n) \to 0$ as $n \to \infty$. By the inequalities, derived from (10.3),

$$C' \leqslant u(t) - \xi_\diamond C, \qquad C' \geqslant u(t) - \xi^\diamond C,$$

with

$$\xi_\diamond = \min\{\kappa + (1-p)\mu, \theta\}, \qquad \xi^\diamond = \max\{\kappa + (1-p)\mu, \theta\},$$

we have

$$0 \leqslant \limsup_{n\to\infty} u(t_n) - \xi C^\infty \leqslant u^\infty - \xi C^\infty.$$

This implies the first estimate. The second is proved similarly. □

It is an immediate consequence that the total toxicant concentration tends to 0 if the toxicant input tends to 0. This is due to our assumption that the toxicant is degraded both in the environment and within the biomass of the species.

Corollary 10.4. *If $u(t) = 0$ for all $t \geqslant 0$, then $C_E(t) + C_I(t) = C(t) \to 0$ as $t \to \infty$. Then the body burden $y(t)$ converges to 0 as $t \to \infty$. If $0 < \beta(0,0) \leqslant \mu$, the biomass $M(t)$ converges to 0 as $t \to \infty$, while $M(t)$ converges to the unique number $\bar{M}$ characterized by $\beta(\bar{M}, 0) = \mu$ if $\beta(0,0) > \mu$.*

Proof. Since $u_\infty = u^\infty = 0$, it follows from Proposition 10.3 that $0 \leqslant C_\infty \leqslant C^\infty \leqslant 0$, so $C_\infty = C^\infty = 0$. This means that $C(t)$ converges as $t \to \infty$ and the limit is 0.

Let $\beta(0,0) \leqslant \mu$. By Proposition A.22, $0 \leqslant M^\infty[\beta(M^\infty, 0) - \mu]$. So, $M^\infty = 0$, or $M^\infty > 0$ and $\beta(M^\infty, 0) \geqslant \mu$. Since $\beta(0,0) \leqslant \mu$ and $\beta(\cdot, 0)$ is strictly increasing by our assumptions, $M^\infty = 0$, i.e., $M(t) \to 0$ as $t \to \infty$.

By (**4**), there exists some $\xi > 0$ such that $\eta + \kappa + \beta(M(t), y(t)) \geqslant \xi$ for large t. Equation (10.2) implies that y is bounded, and, by Theorem A.23, $0 = -[\eta + \kappa + \beta(0, y^\infty)]y^\infty$. This implies $y^\infty = 0$.

Now let $\beta(0,0) > \mu$. Since $M' \leqslant M[\beta(M,0) - \mu]$, we have $M^\infty \leqslant \bar{M}$ for every solution.

First we show that there is some $\delta > 0$ such that $M^\infty \geqslant \delta$ for all solutions starting with $M(0) > 0$, $y(0) \geqslant 0$. By (**4**), we find some $\delta, \xi > 0$ such that $\eta + \kappa + \beta(\delta, y) \geqslant \xi > 0$ for all $y \geqslant 0$ and $\beta(\delta, \delta) > \mu + \xi$. So, if there is a solution with $M^\infty < \delta$, we have $y(t) \to 0$ as $t \to \infty$ by (10.2). Then, for large t, $M'(t) \geqslant M(t)\xi$ and M grows exponentially, a contradiction because M is bounded.

If $u = 0$, the solutions of (10.1) induce a continuous semiflow on $\mathbb{R}^3_+$ by Corollary A.30. We apply Theorem A.32 to $X = \{(M, C_I, C_E);\ M > 0,\ C_I \geqslant 0,\ C_E \geqslant 0\}$ with $\rho(M, C_I, C_E) = M$. Then $\sigma(\Phi(t,x)) = M(t)$. It follows that there exists some $\delta > 0$ such that $M_\infty > \delta$ for all solutions with $M(0) > 0$. Then

$$y(t) = \frac{C_I(t)}{M(t)} \to 0 \quad \text{as } t \to \infty.$$

M satisfies

$$M'(t) - M(t)[\beta(M(t),0) - \mu] \to 0, \quad t \to \infty.$$

Since we assume $\beta(0,0) > 0$, $\beta(\cdot,0)$ is strictly increasing on $[0,\infty)$ by (**2**). By Theorem A.26, $M(t)$ converges to some $M^* \geqslant M_\infty$ with $M^*[\beta(M^*,0) - \mu] = 0$. So, $\beta(M^*,0) = \mu$. Since $\beta(0,0) > \mu$, there is a unique $\bar{M} > 0$ such that $\beta(\bar{M},0) = \mu$, again by (2). $M^* = \bar{M}$, again by the strict decrease of $\beta(\cdot,0)$.

Corollary 10.4 also holds if we assume that $u(t) \to 0$ as $t \to \infty$ rather than $u(t) = 0$ for all $t \geqslant 0$. In the proof, one needs to replace Theorem A.32 by a nonautonomous generalization (Thieme, 2000). □

10.3 Feedback Loop Toxicant Input

In our model equations,

$$\begin{aligned}
M' &= M[\beta(M,y) - \mu],\\
C_I' &= aC_EM - [\mu + \eta + \kappa]C_I,\\
C_E' &= -aC_EM + [p\mu + \eta]C_I - \theta C_E + u(t),\\
y &= \frac{C_I}{M},
\end{aligned}$$

the function u represents the toxicant input. We want to construct this input in such a way that the environmental concentration of toxicant, C_E, converges to a prescribed target value T. This can be achieved by letting u neutralize the other terms in the C_E equation and add a term that has the sign of $T - C_E$,

$$u(t) = aC_EM - [p\mu + \eta]C_I + \theta C_E - v\left(\frac{C_E}{T} - 1\right),$$

where $v : \mathbb{R} \to \mathbb{R}$ is continuous, $zv(z) \geqslant 0$ for all $z \in \mathbb{R}$, and $v(0) = 0$. Notice that $u(t)$ can possibly become negative every now and then, which means that the environment needs to be cleaned from time to time in order to meet the target value. With this choice of u, the C_E equation takes the form,

$$C_E' = -v\left(\frac{C_E}{T} - 1\right).$$

Set $C_E = T(1+z)$ and rename aT as a. Let us also use (10.2) instead of the third equation in (10.1) and set $\sigma = \eta + \kappa$. Then we obtain the system

$$\left.\begin{aligned}
M' &= M[\beta(M,y) - \mu],\\
y' &= a(1+z) - [\sigma + \beta(M,y)]y,\\
z' &= -v(z).
\end{aligned}\right\} \tag{10.8}$$

Even though we have not assumed that v is Lipschitz continuous, one can easily see that 0 is the only solution of $z' = -v(z), z(0) = 0$ and that $z(t)$ has the same sign for all $t \geqslant 0$ as $z(0)$. This implies that solutions are unique. For simplicity we assume v to be odd and nonnegative, $v(-z) = -v(z) \leqslant 0$ for $z \geqslant 0$, so the solutions are symmetric with respect to 0. More specifically we choose

$$v(z) = -\alpha z^{\zeta}, \quad z \geqslant 0,$$

with some constant $\zeta > 0$. For $\zeta = 1$, we obtain $z(t) = z(0)e^{-\alpha t}$. If $\zeta \neq 1$, $z(0) > 0$, then

$$z(t) = [z(0)^{1-\zeta} + \alpha(\zeta - 1)t]_+^{1/(1-\zeta)}.$$

In any case we have $z(t) \to 0$ as $t \to \infty$. For $\zeta = 1$, the target is reached exponentially fast, but not in finite time, for $\zeta > 1$ it is reached even more slowly. For $\zeta \in (0, 1)$, however, it is reached in finite time, the length of which depends on how far off we are initially. Rather than the full system (10.8), we study the two-dimensional limiting system

$$\left.\begin{aligned} M' &= M[\beta(M, y) - \mu], \\ y' &= a - [\sigma + \beta(M, y)]y. \end{aligned}\right\} \tag{10.9}$$

Very often the large-time behavior of solutions to the limiting system is similar to the large-time behavior of the solutions to the original system, though differences may occur, for $\zeta > 1$ more frequently than for $\zeta = 1$. For specifics consult the theory of asymptotically autonomous differential equations (Benaïm, Hirsch, 1996; Smith, Waltman, 1995, Appendix F; Thieme, 1992, 1994a,b). For $\zeta \in (0, 1)$, the systems (10.8) and (10.9) are identical after finite time, and so the solutions show the same type of large-time behavior.

Theorem 10.5. *All solutions of (10.9) that have nonnegative initial data at time 0 exist for all forward times and are nonnegative and bounded.*

If $M(0) = 0$, then $M(t) = 0$ for all $t \geqslant 0$.

If $M(0) > 0$, then $M(t) > 0$ for all $t \geqslant 0$.

If $y(0) \geqslant 0$, then $y(t) > 0$ for all $t > 0$.

Furthermore, we have the following estimates:

$$y_\infty \geqslant \frac{a}{\sigma + \beta(0, 0)}$$

and

$$M(t) \leqslant M(0)e^{(\beta(0,0)-\mu)t} \quad \forall t \geqslant 0.$$

Moreover, if $\beta(0,0) \geqslant \mu$, *the following estimate holds for the large-time behavior of* M,

$$M^{\infty} \leqslant \bar{M},$$

where $\bar{M}$ *is characterized by* $\beta(\bar{M},0) = \mu$.

Finally, there exists some $c_0 > 0$ *such that* $y^{\infty} \leqslant c_0$ *for all solutions starting from initial data.*

In other words, the system (10.9) is *dissipative*, i.e., there exists some constant $c > 0$ such that for all nonnegative solutions of (10.9) there is some time $r > 0$ with $\|(M(t), y(t))\| \leqslant c$ for all $t \geqslant r$. Here $\|\cdot\|$ is any norm of $\mathbb{R}^2$.

Proof. Consider a solution of (10.9) defined on an interval $[0,\alpha)$. It is readily checked that the vector field in (10.9) satisfies the assumptions of Theorem A.4. So, all solutions starting with nonnegative initial values remain nonnegative.

Since β is decreasing in both arguments,

$$M' \leqslant [\beta(0,0) - \mu]M.$$

Integrating these differential inequalities provides the exponential estimate for M. In particular, M is bounded on $[0,\alpha)$ if $\alpha < \infty$, let us say $M(t) \leqslant \tilde{M}$. By assumption (**2**), $\beta(M(t), y(t)) \geqslant \breve{\beta}$ on $[0,\alpha)$ with some $\breve{\beta} \in \mathbb{R}$. So,

$$y' \leqslant a - \breve{\beta}y,$$

which implies that y is bounded on $[0,\alpha)$ if $\alpha < \infty$. By Theorem A.4, the solutions of (10.9) are defined on $[0,\infty)$.

Notice that the first equation in (10.9) satisfies the assumptions in Lemma A.2. This proves the relation between the signs of $M(0)$ and $M(t)$.

From the second equation in (10.9) we obtain the differential inequalities

$$y' \geqslant a - [\sigma + \beta(0,0)]y.$$

By Proposition A.22,

$$a - [\sigma + \beta(0,0)]y_{\infty} \leqslant 0.$$

This provides the estimate for y_{∞}.

M satisfies the differential inequality

$$M' \leqslant M[\beta(M,0) - \mu].$$

Assume $\beta(0,0) > \mu$. By assumption (**3**), $\beta(M,0) < \mu$ for sufficiently large $M > 0$, so the intermediate value theorem provides a $\bar{M} > 0$ such that $\beta(\bar{M},0) = \mu$, which is unique by assumption (**2**). By Proposition A.22,

$$0 \leqslant M^{\infty}[\beta(M^{\infty},0) - \mu].$$

So, $M^\infty = 0$ or $\beta(M^\infty, 0) \geqslant \mu$. In either case, $M^\infty \leqslant \bar{M}$ because $\beta(\cdot, 0)$ is strictly decreasing.

First we show that there is some $c_1 > 0$ such that $y_\infty \leqslant c_1$ for all solutions.

Let $c_1 > 0$ and assume that there is a solution with $y_\infty > c_1$. By choosing c_1 large enough, we will eventually obtain a contradiction. First choose $c_1 > 0$ so large that $\beta(0, c_1) < \mu$. Then, for large t, $\beta(M(t), y(t)) \leqslant \beta(0, c_1) < \mu$. This implies that $M(t) \to 0$ as $t \to \infty$. By assumption (4), choose $\delta > 0$ such that $\beta(\delta, y) > -\sigma/2$ for all $y \geqslant 0$. Then, for large t,

$$y'(t) \leqslant a - [\sigma + \beta(\delta, y(t))]y(t) \leqslant a - \tfrac{1}{2}\sigma y(t).$$

This implies that $y_\infty \leqslant (2a/\sigma)$. Choosing $c_1 > (2a/\sigma)$ provides a contradiction.

Together with our results for M, this implies that there is some $c_2 > 0$ such that $\liminf_{t\to\infty} \|(M(t), y(t))\| < c_2$ for every solution. By Corollary A.30, the solutions of (10.9) induce a semiflow on $\mathbb{R}^2_+$. By Corollary A.33, this implies that there is some $c_0 > 0$ such that $\limsup_{t\to\infty} \|(M(t), y(t))\| < c_0$ for all solutions. □

Exercises

10.3. Assume that $\beta(M, y)$ is decreasing in both $M \geqslant 0$ and $y \geqslant 0$, that $\beta(M, 0)$ is strictly decreasing in $M \geqslant 0$ and that $\beta(0, 0) \leqslant \mu$. Show that any solution of (10.9) with nonnegative initial data satisfies $M(t) \to 0$ as $t \to \infty$.

10.4 Extinction and Persistence Equilibria and a Threshold Condition for Population Extinction

We make the following additional assumptions from now on.

$\beta(M, y)$ is continuously differentiable on those points $M, y \geqslant 0$, where $\beta(M, y) > 0$ and the partial derivatives are strictly negative at these points.

We write $\beta_M = (\partial\beta/\partial M)$ and $\beta_y = (\partial\beta/\partial y)$ for the partial derivatives with respect to M and y, respectively. In order to explore the equilibria, we set the vector field in (10.9) equal to 0 and obtain the equilibrium equations

$$0 = M[\beta(M, y) - \mu],$$
$$0 = a - [\sigma + \beta(M, y)]y.$$

From the first equation $M = 0$ or $\beta(M, y) = \mu$. $M = 0$ will give rise to so-called *boundary equilibria* $x^\diamond = (M^\diamond, y^\diamond) = (0, y^\diamond)$. In this context, one could also call them *extinction equilibria*. If we substitute $M = 0 = M^\diamond$ into the second equilibrium equation, we obtain

$$a = [\sigma + \beta(0, y^\diamond)]y^\diamond =: \phi(y^\diamond).$$

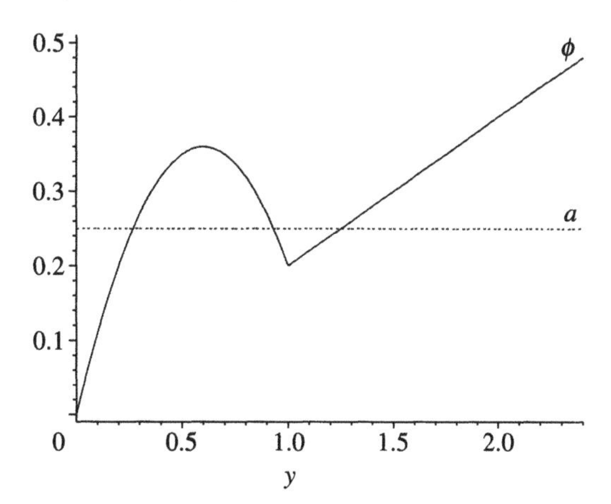

Figure 10.1. Graph of $\phi(y) = y(\sigma + \beta(0, y))$ for linear dose response, $\beta(0, y) = [1 - y]_+$, with $\sigma = 0.25$. The equation $\phi(y^\diamond) = a$ with $a = 0.25$ has three roots $y_1^\diamond \approx 0.27$, $y_2^\diamond \approx 0.93$, and $y_3^\diamond = 1.25$, corresponding to three boundary equilibria.

We notice that $\phi(0) = 0$ and $\phi(y) \to \infty$ as $y \to \infty$. So there is always a positive solution to this equation and we will have at least one boundary equilibrium. As we will see, we may have several (Figure 10.1). If ϕ equals a on a nondegenerate interval, we even have an uncountable continuum of boundary equilibria.

Let us turn to the alternative $M > 0$ and $\beta(M, y) = \mu$. This will give rise to a so-called *interior equilibrium*, which could also be called a *survival equilibrium* in this context. Substituting $\beta(M, y) = \mu$ into the second equilibrium equation leads to

$$y^* = \frac{a}{\sigma + \mu}.$$

Substituting back we obtain $\beta(M, y^*) = \mu$. The intermediate value theorem, in conjunction with assumption (**3**), provides a positive solution $M = M^*$ if and only if the *threshold condition*

$$\beta(0, y^*) > \mu$$

is satisfied. If this threshold condition is not satisfied, the population will die out. To prove this, we use the method of fluctuation in Toolbox A.3.

Theorem 10.6. *Let $\beta(0, y^*) \leqslant \mu$, i.e., there is no interior equilibrium. Then $M(t) \to 0$ as $t \to \infty$ for every solution (M, y) of (10.9), while $\{0\} \times [y_\infty, y^\infty]$ consists of boundary equilibria. If all boundary equilibria are isolated, $(M(t), y(t))$ converges to a boundary equilibrium.*

Proof. Without restricting the generality, we can assume that $M(0) > 0$ and so $M(t) > 0$ for all $t \geqslant 0$. First suppose that $M(t)$ does not converge as $t \to \infty$.

By the fluctuation Lemma A.20, there exists a sequence $s_k \to \infty$ such that $M(s_k) \to M_\infty$ and $M'(s_k) = 0$. Hence

$$0 = M(s_k)(\beta(M(s_k), y(s_k)) - \mu).$$

Since $M(s_k) > 0$, $\beta(M(s_k), y(s_k)) = \mu$. Since β is assumed to be differentiable in every point where it is strictly positive, $y(t)$ is twice differentiable at s_k and, by Lemma A.20 and by $M'(s_k) = 0$,

$$0 \leqslant M''(s_k) = M(s_k)\beta_y(M(s_k), y(s_k))y'(s_k).$$

Since $\beta_y(M(s_k), y(s_k)) < 0$ and $M(s_k) > 0$,

$$0 \geqslant y'(s_k) = a - [\sigma + \beta(M(s_k), y(s_k))]y(s_k) = a - [\sigma + \mu]y(s_k).$$

This implies $y(s_k) \geqslant y^*$. Since β is decreasing in both arguments and $\beta_M(M(s_k), y(s_k)) < 0$, we have that

$$\mu = \beta(M(s_k), y(s_k)) < \beta(0, y(s_k)) \leqslant \beta(0, y^*) \leqslant \mu,$$

a contradiction.

So, we can assume that $M(t)$ converges as $t \to \infty$. Since M, y are bounded and uniformly continuous on $[0, \infty)$ and β is uniformly continuous on every bounded subset of the first quadrant, M' is uniformly continuous by the differential equation. So, we can apply Barbalat's lemma, Corollary A.18: $M'(t) \to 0$ as $t \to \infty$. Assume that $M^\bullet := \lim_{t\to\infty} M(t) \neq 0$. Then

$$\beta(M(t), y(t)) \to \mu, \quad t \to \infty.$$

Since β is uniformly continuous on every bounded subset of the first quadrant,

$$\beta(M^\bullet, y(t)) \to \mu, \quad t \to \infty.$$

Since y is bounded on $[0, \infty)$, there exists a sequence $t_k \to \infty$ such that $y(t_k) \to y^\infty$ (Proposition A.22). So, $\beta(M^\bullet, y^\infty) = \mu$. Similarly, $\beta(M^\bullet, y_\infty) = \mu$. Since $\beta(M^\bullet, \cdot)$ is strictly decreasing as long as it is strictly positive, we have $y^\infty = y_\infty$. So, $y(t) \to y^\bullet$ as $t \to \infty$ for some $y^\bullet$. Together with $M(t) \to M^\bullet > 0$, this implies that $(M^\bullet, y^\bullet)$ is an interior equilibrium, contradicting the fact that such an equilibrium does not exist. So, we have shown that $M(t) \to 0$ as $t \to \infty$.

Since y is bounded,

$$y'(t) - a + [\sigma + \beta(0, y(t))]y(t) \to 0, \quad t \to 0.$$

It follows from Proposition A.22 that both y^∞ and y_∞ are coordinates of boundary equilibria. Let $y_\infty < y^\bullet < y^\infty$. By Theorem A.26, $a - [\sigma + \beta(0, y^\bullet)]y^\bullet = 0$. So, $y^\bullet$ is the coordinate of a boundary equilibrium. So, if $y_\infty < y^\infty$, there is a nondegenerate interval formed of coordinates of boundary equilibria. If every boundary equilibrium is isolated, this cannot happen, so $y_\infty = y^\infty$ and $(M(t), y(t))$ converges towards a boundary equilibrium. □

As we will see later, this extinction result does not exclude that there may be transient solutions which show a considerable build up of population biomass which finally vanishes again.

If an interior equilibrium exists, the y-coordinate of which is necessarily y^*, no boundary equilibrium $(0, y^\diamond)$ with $y^\diamond < y^*$ is the limit of a solution that starts in the interior of the first quadrant. Actually, a stronger result holds.

Proposition 10.7.

(a) *If* $(0, y^\diamond)$ *is a boundary equilibrium, then the region* $\{M \geqslant 0,\ y \geqslant y^\diamond\}$ *is forward invariant.*

(b) *If* (M, y) *is a solution with nonnegative initial data, then there exists a boundary equilibrium* $(0, y^\diamond)$ *such that* $y_\infty \geqslant y^\diamond$.

(c) *Let* $(0, y^\diamond)$ *be a boundary equilibrium and* $y^\diamond < y^* = a/(\sigma + \mu)$. *Then there exists some* $\epsilon > 0$ *such that every solution of (10.9) starting with* $M(0) > 0$, $y(0) \geqslant 0$ *satisfies*

$$\liminf_{t\to\infty}(M(t))^2 + [y(t) - y^\diamond]_+^2) > \epsilon^2.$$

Here $[r]_+ = \max\{0, r\}$ again is the positive part of a real number r. A set is called *forward invariant* if all solutions which start in this set stay in it for all future (or forward) times. It is called *invariant* if all solutions which start in this set stay in it for all forward and backward times (cf. Section A.6). Geometrically, Proposition 10.7 means that every solution, where the population is not initially extinct, in the long run keeps a distance of at least ϵ from the segment on the $M = 0$ axis that stretches from 0 to $y^\diamond$.

Proof. For part (a) the following cases apply.

Case 1. $\beta(0, y^\diamond) > 0$. Then $\beta(\cdot, y^\diamond)$ is strictly decreasing by assumption (**2**). So, if $t \geqslant 0$ and $y(t) = y^\diamond$, $M(t) > 0$, then

$$y'(t) = a - y^\diamond[\sigma + \beta(M(t), y^\diamond)] > a - \phi(y^\diamond) = 0.$$

Therefore, y cannot decrease through a point $(M, y^\diamond)$ where $M > 0$. If $y(t) = y^\diamond$, $M(t) = 0$, then $y(s) = y^\diamond$, $M(s) = 0$ for all $s \geqslant 0$. This follows from the uniqueness of solutions, because $(0, y^\diamond)$ is an equilibrium. Hence y cannot decrease through $(0, y^\diamond)$ either.

Case 2. $\beta(0, y^\diamond) = 0$. Then $y^\diamond = a/\sigma$.

Let $M^\diamond \in [0, \infty]$ such that $\beta(M, y^\diamond) = 0$ for $M \leqslant M^\diamond$ and $\beta(M, y^\diamond) < 0$ for $M > M^\diamond$. By the same argument as above, y cannot decrease through a point $(M, y^\diamond)$ with $M > M^\diamond$. Let $M_0 \in [0, M^\diamond]$. Notice that every solution of

$$M' = -\mu M, \quad y' = 0, \qquad y(0) = y^\diamond, \quad M(0) = M_0,$$

is also a solution of our system. By uniqueness of solutions, every solution that crosses the line segment $y = y^\diamond$, $0 \leqslant M \leqslant M^\diamond$, stays in this segment. So, for no solution, $y(t)$ can decrease through a point $(M, y^\diamond)$ with $M \geqslant 0$.

(b) By Proposition A.22 in the toolbox, there exists a sequence $s_n \to \infty$ as $n \to \infty$ such that $y(s_n) \to y_\infty$ and $y'(s_n) \to 0$. So,

$$\begin{aligned} 0 &= \lim_{n\to\infty} (a - y(s_n)[\sigma + \beta(M(s_n), y(s_n)]) \\ &= \lim_{n\to\infty} (a - y_\infty[\sigma + \beta(M(s_n), y_\infty)]) \geqslant a - \phi(y_\infty). \end{aligned}$$

So, $\phi(y_\infty) \geqslant a$. Since $\phi(0) = 0 < a$, there exists some $y^\diamond \in (0, y_\infty]$ such that $\phi(y^\diamond) = a$ by the intermediate value theorem.

(c) First we give a weaker statement for the limit superior rather than the limit inferior. From $a = [\sigma + \beta(0, y^\diamond)]y^\diamond$ we obtain

$$\beta(0, y^\diamond) - \mu = \frac{a}{y^\diamond} - \sigma - \mu = a\left(\frac{1}{y^\diamond} - \frac{1}{y^*}\right) > 0.$$

Because β is continuous and nonincreasing in both variables, there exists some $\epsilon, \delta > 0$ such that

$$\beta(M, y) - \mu \geqslant \delta, \quad M \in [0, \epsilon], \quad y \in [0, y^\diamond + \epsilon].$$

Now suppose that there is a solution with $M(0) > 0$, $y(0) \geqslant 0$ and

$$\limsup_{t\to\infty} \left((M(t))^2 + [y(t) - y^\diamond]_+^2\right) < \epsilon^2.$$

Then there exists some $r \geqslant 0$ such that

$$M(t) \in [0, \epsilon), \quad y(t) \in [0, y^\diamond + \epsilon], \quad t \geqslant r.$$

Hence $M'(t) \geqslant \delta M(t)$ for all $t \geqslant r$. Integrating this differential inequality we obtain $M(t) \geqslant M(r)e^{\delta(t-r)}$ for all $t \geqslant r$ with $M(r) > 0$ by Theorem 10.5. So, M increases exponentially, a contradiction. In order to obtain the stronger statement for the limit inferior we apply the persistence result in Theorem A.32. By Corollary A.30, the solutions of (10.9) induce a continuous semiflow on $\mathbb{R}_+^2$. The compactness condition (C) follows from Theorem 10.5. We can choose B as the compact set $[0, \bar{M}] \times [0, c_0]$. We define

$$\rho(M, y) = (M^2 + [y - y^\diamond]_+^2)^{1/2},$$

then $\sigma^2(t,x) = (M(t))^2 + [y(t) - y]_+^2$ for $x = (M(0), y(0))$. In the language of Toolbox A.5 we have shown above that the semiflow is uniformly weakly ρ-persistent. We still need to show that $\rho(M(0), y(0)) > 0$ implies $\sigma(t,x) > 0$ for all $t > 0$. If $\rho(M(0), y(0)) > 0$, then $M(0) > 0$, or $M(0) = 0$ and $y(0) > y^\diamond$. If $M(0) > 0$, then $M(t) > 0$ and $\sigma(t,x) > 0$ for all $t > 0$ by Theorem 10.5. If $M(0) = 0$ and $y(0) > y^\diamond$, then $M(t) = 0$ for all $t \geqslant 0$ and $y^\diamond$ and $y(\cdot)$ are solutions of the same scalar ordinary differential equation $y' = a - [\sigma + \beta(0,y)]y$, but with different initial data. The uniqueness theorem for ordinary differential equations with locally Lipschitz continuous vector fields implies that $y(t) > y^\diamond$ and $\sigma(t,x) > 0$ for all $t > 0$. Since the solution semiflow is uniformly weakly ρ-persistent, it is uniformly strongly ρ-persistent by Theorem A.32. This translates to the statement of this proposition. □

10.5 Stability of Equilibria and Global Behavior of Solutions

In order to determine the behavior of solutions of (10.9) in the neighborhood of an equilibrium (linearized stability) we write the vector field F in the following form:

$$F(M,y) = \begin{pmatrix} M[\beta(M,y) - \mu] \\ a - [\sigma + \beta(M,y)]y \end{pmatrix}$$

(see Toolbox A.4). We assume in this section that β is continuously differentiable on $(0,\infty)^2$ and that the partial derivatives are strictly negative and bounded on every bounded subset of $(0,\infty)^2$. We also assume that the partial derivatives $\beta_y(0,y)$ exist and are continuous in y.

We will see that boundary equilibria with $M = 0$ will play a crucial role in the dynamics of our system. So, we extend β to $\mathbb{R}^2$ by $\beta(M,y) = \beta(M_+, y_+)$, where $M_+ = \max\{M, 0\}$ is the positive part of M. Then β is locally Lipschitz on $\mathbb{R}^2$ by assumption (**1**). Moreover, we have

$$\frac{\partial}{\partial M}(M\beta(M,y)) = \begin{cases} \beta(M,y) + M\beta_M(M,y), & M > 0, \\ \beta(M,y), & M \leqslant 0. \end{cases}$$

So, with this extension, F is continuously differentiable on $\mathbb{R} \times (0,\infty)$. By Theorem 10.5, every solution starting at nonnegative initial data eventually enters and then stays in a compact subset of $[0,\infty) \times (0,\infty)$. The Jacobian matrix of F at $M \geqslant 0$, $y > 0$ is given by

$$DF(M,y) = \begin{pmatrix} \beta(M,y) - \mu + M\beta_M(M,y) & M\beta_y(M,y) \\ -y\beta_M(M,y) & -\sigma - \beta(M,y) - y\beta_y(M,y) \end{pmatrix},$$

where $M\beta_M(M,y) := 0$ when $M = 0$.

At a boundary equilibrium $(M^\diamond, y^\diamond) = (0, y^\diamond)$, the Jacobian matrix is of lower triangular form

$$DF(0, y^\diamond) = \begin{pmatrix} \beta(0, y^\diamond) - \mu & 0 \\ -y^\diamond \beta_M(0, y^\diamond) & -\sigma - \beta(0, y^\diamond) - y^\diamond \beta_y(0, y^\diamond) \end{pmatrix},$$

and so has the two eigenvalues

$$\lambda_1 = \beta(0, y^\diamond) - \mu = \frac{a}{y^\diamond} - \sigma - \mu = a\left(\frac{1}{y^\diamond} - \frac{1}{y^*}\right),$$
$$\lambda_2 = -\sigma - \beta(0, y^\diamond) - y^\diamond \beta_y(0, y^\diamond) = -\phi'(y^\diamond).$$

Notice that the sign of λ_1 depends on how the boundary equilibrium is positioned in comparison to y^*, which is the y-coordinate of the interior equilibrium in the case where the latter exist. Proposition 10.7 can be rephrased as stating that no solution which starts with $M(0) > 0$, $y(0) \geqslant 0$, converges towards the boundary equilibrium $(0, y^\diamond)$ if the eigenvalue λ_1 associated with the Jacobian matrix is strictly positive. The eigenvalue λ_2 is related to the stability of $(0, y^\diamond)$ as an equilibrium of the solution flow restricted to the invariant set $M = 0$. Recall that a solution which satisfies $M(t) = 0$ for at least one t does so for all t. If $M \equiv 0$, $y' = a - \phi(y)$.

At the interior equilibrium (M^*, y^*) (if it exists), the Jacobian matrix has the form (remember $\beta(M^*, y^*) = \mu$)

$$DF(M^*, y^*) = \begin{pmatrix} M^* \beta_M(M^*, y^*) & M^* \beta_y(M^*, y^*) \\ -y^* \beta_M(M^*, y^*) & -\sigma - \beta(M^*, y^*) - y^* \beta_y(M^*, y^*) \end{pmatrix}.$$

The trace of the Jacobian matrix at an interior equilibrium is

$$\begin{aligned} \operatorname{trace} DF(M^*, y^*) &= M^* \beta_M(M^*, y^*) - \sigma - \beta(M^*, y^*) - y^* \beta_y(M^*, y^*) \\ &= M^* \beta_M(M^*, y^*) - \sigma - \mu - y^* \beta_y(M^*, y^*), \end{aligned}$$

while the determinant is

$$\begin{aligned} \det DF(M^*, y^*) &= -M^* \beta_M(M^*, y^*)(\sigma + \beta(M^*, y^*)) \\ &= -(\sigma + \mu) M^* \beta_M(M^*, y^*) > 0. \end{aligned}$$

This implies that the two eigenvalues have real parts of the same sign and that 0 is never an eigenvalue. So, both eigenvalues have strictly negative real parts if the trace of $DF(M^*, y^*) < 0$. In this case, the equilibrium is locally asymptotically stable, and is called a *sink*. If the trace is positive, both eigenvalues have strictly positive real parts, and the equilibrium is not only unstable, but a *source*, i.e., it is a sink if time is reversed. If both eigenvalues are strictly positive, the equilibrium is called a *stable node*; it is called an *unstable node*, if they are both strictly negative.

Theorem 10.8. *Assume that the interior equilibrium* (M^*, y^*) *exists. If*

$$\sigma + \mu + y^*\beta_y(M^*, y^*) - M^*\beta_M(M^*, y^*) > 0,$$

then the interior equilibrium is locally asymptotically stable. If this expression is strictly negative, the interior equilibrium is unstable, a source.

Corollary 10.9.

(a) *If*

$$\sigma + \beta(M, y) + y\beta_y(M, y)) - M\beta_M(M, y) > 0 \quad \forall M, y > 0,$$

then the endemic equilibrium is locally asymptotically stable for all $\mu, a > 0$ *for which it exists.*

(b) *If* $M^*, y^* > 0, 0 < \beta(M^*, y^*) =: \mu$ *and*

$$\sigma + \beta(M^*, y^*) + y^*\beta_y(M^*, y^*)) - M^*\beta_M(M^*, y^*) < 0,$$

then $a > 0$ *can be chosen such that* (M^*, y^*) *is an unstable equilibrium.*

Another important observation links the stability of the interior equilibrium to the stability of a sufficiently close boundary equilibrium.

Proposition 10.10. *Let* $(0, y^\diamond)$ *be a boundary equilibrium with* $\phi'(y^\diamond) \neq 0$. *If the interior equilibrium is sufficiently close to* $(0, y^\diamond)$, *then it has the same kind of stability as the boundary equilibrium with respect to the boundary flow on the set* $M = 0$, *with both eigenvalues being real.*

Proof. Notice that the determinant of $DF(M^*, y^*)$ approaches 0, while the trace of $DF(M^*, y^*)$ approaches $-\phi'(y^\diamond)$ as $(M^*, y^*) \to (0, y^\diamond)$, with $-\phi'(y^\diamond)$ being the eigenvalue describing the stability of $(0, y^\diamond)$ for the boundary flow on the set $M = 0$. Let τ denote the trace and d the determinant of $DF(M^*, y^*)$. Then $\lambda_1 + \lambda_2 = \tau$ and $\lambda_1\lambda_2 = d$, which implies that

$$2\lambda_{1,2} = \tau \pm \sqrt{\tau^2 - 4d}.$$

So, λ_1 and λ_2 are real for small $d > 0$ and have the same sign as τ and, consequently, as $-\phi'(y^\diamond)$. □

If we combine the information that the eigenvalues are real with our knowledge that their real parts have the same sign, we learn that the eigenvalues are either both negative or both positive, i.e., the interior equilibrium turns out to be a *node*, if it is sufficiently close to a boundary equilibrium.

Proposition 10.11. *Assume that $\beta(0, y^*) \neq \mu > 0$ is satisfied.*

(a) *Then $(0, y^*)$ is not a boundary equilibrium.*

If $\beta(0, y^*) > \mu$, i.e., if there exists an interior equilibrium, then there exists a boundary equilibrium $(0, y^\diamond)$ with $y^\diamond < y^*$ which is unstable and attracts no solution with initial values $M(0) > 0$, $y(0) \geqslant 0$.

If $\beta(0, y^*) < \mu$ such that there is no interior equilibrium, then there exists at least one boundary equilibrium $(0, y^\diamond)$ with $y^\diamond > y^*$.

Assume in addition that $a > 0$ is a regular value of ϕ, i.e., $\phi'(y)$ exists and $\phi'(y) \neq 0$ for any $y \in (0, \infty)$ with $\phi(y) = a$. Then the following holds.

(b) *The number of boundary equilibria is odd (Figure 10.1). If $\beta(0, y^*) > \mu$, then the number of boundary equilibria $(0, y^\diamond)$ with $y^\diamond < y^*$ is odd, and the number of boundary equilibria with $y^\diamond > y^*$ is zero or even. If $\beta(0, y^*) < \mu$, then it is the other way round.*

(c) *Label the boundary equilibria as $(0, y_j^\diamond)$ with*

$$0 < y_1^\diamond < \cdots < y_{2m+1}^\diamond, \quad m \geqslant 0,$$

such that equilibria with different indices are different. Then every equilibrium $(0, y_j^\diamond)$, with $y_j^\diamond < y^*$ and j odd, is a saddle with the stable manifold being the segment of the $M = 0$ axis between $y_{j-1}^\diamond$ and $y_{j+1}^\diamond$. Every equilibrium $(0, y_j^\diamond)$, with $y_j^\diamond < y^*$ and j even, is an unstable node. Every equilibrium $(0, y_j^\diamond)$, with $y_j^\diamond > y^*$ and j even, is a saddle with the unstable manifold being the segment on the $M = 0$ axis between $y_{j-1}^\diamond$ and $y_{j+1}^\diamond$. Every equilibrium $(0, y_j^\diamond)$, with $y_j^\diamond > y^*$ and j odd, is a stable node.

(d) *If, in the labeling of (c), $y^* \leqslant y_{2m+1}^\diamond$, then all solutions starting in the region $\{M \geqslant 0,\ y \geqslant y_{2m}^\diamond\}$, except at $(0, y_{2m}^\diamond)$, converge to $(0, y_{2m+1}^\diamond)$.*

Proof. (a) We have

$$a - y^*[\sigma + \beta(0, y^*)] = a - \frac{a}{\sigma + \mu}[\sigma + \beta(0, y^*)] = \frac{a}{\sigma + \mu}[\mu - \beta(0, y^*)] \neq 0.$$

So, $(0, y^*)$ is not an equilibrium. If $\beta(0, y^*) > \mu$, then this expression is negative. By the intermediate value theorem there exists some $y^\diamond \in (0, y^*)$ such that

$$a - y^\diamond[\sigma + \beta(0, y^\diamond)] = 0,$$

i.e., $(0, y^\diamond)$ is a boundary equilibrium. Since $y^\diamond < y^*$, it follows from our previous considerations that at least one eigenvalue, λ_1, is strictly positive. By Proposition 10.7, there exists no solution with $M(0) > 0$ that converges towards this boundary equilibrium as $t \to \infty$.

A similar consideration shows that there exists a boundary equilibrium with $y^\diamond > y^*$ if $\beta(0, y^*) < \mu$.

(b) (See Figure 10.1.) Since $\phi(0) = 0 < a$, we have $\phi'(y_1^\diamond) > 0$. This implies $\phi(y) > a$ for $y \in (y_1^\diamond, y_2^\diamond)$, so $\phi'(y_2^\diamond) < 0$. This implies $\phi(y) < a$ for $y \in (y_2^\diamond, y_3^\diamond)$. Continuing this argument, we have $\phi'(y_j^\diamond) > 0$ at all equilibria with odd numbers j and $\phi'(y_j^\diamond) < 0$ with even numbers j. Since $\phi(y) > a$ for $y > a/\sigma$, the total number of boundary equilibria is odd.

If $\beta(0, y^*) > \mu$, then $\phi(y^*) > a$, and the number of equilibria with $y^\diamond < y^*$ is odd as well.

(c) The nature of the equilibria follows from the eigenvalues satisfying

$$\lambda_1 = a\left(\frac{1}{y^\diamond} - \frac{1}{y^*}\right), \qquad \lambda_2 = -\phi'(y^\diamond).$$

So, $\lambda_1 > 0$ for all boundary equilibria with $y^\diamond < y^*$ and $\lambda_2 < 0$ for all boundary equilibria with odd numbers, etc. The statement concerning the stable and unstable manifolds of the saddles follows from the fact that the $M = 0$ axis is invariant. More precisely, if $M(0) = 0$, then $M(t) = 0$ for all t and $y' = a - \phi(y)$. If $y(0) \in (y_j^\diamond, y_{j+1}^\diamond)$ and j is odd, then, as we have seen in the proof of (b), $\phi(y(0)) > a$, so $y'(0) < 0$ and $y(t)$ decreases to $y_j^\diamond$ as $t \to \infty$ (cf. Theorem 6.1). If j is even, then all inequalities are reversed, and $y(t)$ increases to $y_{j+1}^\diamond$.

(d) By Proposition 10.7 (a), the region $\{M \geqslant 0,\ y \geqslant y_{2m}^\diamond\}$ is forward invariant. It does not contain an interior equilibrium because there is no interior equilibrium if $y^* \in [y_{2m}^\diamond, y_{2m+1}^\diamond]$. Since there are only the two boundary equilibria $(0, y_{2m}^\diamond)$ and $(0, y_{2m+1}^\diamond)$, with the second being a sink, all solutions starting in that region must converge to one of these two boundary equilibria as $t \to \infty$ by the Poincaré–Bendixson trichotomy (Theorem A.10 in the toolbox). We must exclude solutions that do not start at $(0, y_{2m}^\diamond)$ converge to the saddle $(0, y_{2m}^\diamond)$.

We already know that solutions starting at $M(0) = 0$, $y(0) > y_{2m}^\diamond$ converge to $(0, y_{2m+1}^\diamond)$. So, we consider initial data $M(0) > 0$, $y(0) \geqslant y_{2m}^\diamond$. Then $M(t) > 0$ for all $t \geqslant 0$. Notice that $y_{2m}^\diamond < (a/\sigma)$, i.e., $\beta(0, y_{2m}^\diamond) > 0$. Then, by assumption (2), $\beta(\cdot, y_{2m}^\diamond)$ is strictly decreasing. So, $y = y_{2m}^\diamond$, $M > 0$ implies $y' > 0$. This means that $y(t) > y_{2m}^\diamond$ for $t > 0$. Assume that $M(t) \to 0$, $y(t) \to y_{2m}^\diamond$. Then there exists a sequence $t_n \to \infty, n \to \infty$, such that

$$0 \geqslant y'(t_n) = a - y(t_n)[\sigma + \beta(M(t_n), y(t_n)] > a - \phi(y(t_n)).$$

This implies $\phi(y(t_n)) > a$, a contradiction because $\phi(y) \leqslant a$ for $y \in [y_{2m}^\diamond, y_{2m+1}^\diamond]$. See the proof of (b). □

The concept of regular values of a piecewise continuously differentiable function is an important one, because they form a set which is both topologically and measure-theoretically fat: topologically in so far as the set of regular values contains a set which is dense and is the countable intersection of open sets; and measure theoretically as its complement has Lebesgue measure 0 (see Toolbox B.7 for details).

The Poincaré–Bendixson theory (see Theorems A.7 and A.10) limits the complexity of the behavior of solutions of planar ordinary differential equation systems. The behavior of solutions of system (10.9) is further restricted if there exists exactly one boundary equilibrium.

Theorem 10.12. *Assume that the interior equilibrium (M^*, y^*) exists, i.e., the threshold condition $\beta(0, y^*) > \mu$ is satisfied.*

(a) *If there is a unique boundary equilibrium or, more generally, there is no boundary equilibrium $(0, y^\diamond)$ with $y^\diamond > y^*$, then all solutions that start at initial data $M(0) > 0$, $y(0) \geqslant 0$ tend to the interior equilibrium or to a surrounding period orbit as $t \to \infty$.*

(b) *Let a be a regular value of ϕ, i.e., a belong to a very fat set. Then, as $t \to \infty$, all solutions that start at initial data $M(0) > 0$, $y(0) \geqslant 0$ tend to the interior equilibrium, to a locally asymptotically stable boundary equilibrium $(0, y^\diamond)$ with $y^\diamond > y^*$, to a period orbit, or to a heteroclinic orbit. The heteroclinic orbit, in the labeling of Proposition 10.11, would connect adjacent boundary equilibria $(0, y^\diamond_{2k-1})$ and $(0, y^\diamond_{2k})$ with $y^\diamond_{2k-1} < y^* < y^\diamond_{2k}$. One part of the heteroclinic orbit is formed by the segment on the $M = 0$ axis between the two adjacent boundary equilibria, the other part goes through the interior of the first quadrant surrounding the interior equilibrium.*

Proof. We want to apply the Poincaré–Bendixson trichotomy, Theorem A.10. We extend β to $\mathbb{R} \times [0, \infty)$ by setting $\beta(M, y) = \beta(M_+, y)$. Notice that the overall assumptions of Toolbox A.2, uniqueness of solutions of (10.9), are satisfied for $X = \mathbb{R} \times (0, \infty)$ and that every solution with $M(0), y(0) \geqslant 0$ has its ω-limit set in X by Theorem 10.5.

(a) By Proposition 10.7, boundary equilibria $(0, y^\diamond)$ with $y^\diamond < y^*$ do not attract solutions with $M(0) > 0$. In particular, there is no orbit that leads from the interior equilibrium or a boundary equilibrium to a boundary equilibrium, so there is no heteroclinic cycle involving a boundary and the interior equilibrium or two boundary equilibria. Finally, there is no homoclinic orbit connecting the interior equilibrium to itself, because it is either a source or a sink. Recall that the two eigenvalues have real parts of the same sign. The only possibilities left by the Poincaré–Bendixson trichotomy (Theorem A.10) are that solutions with $M(0) > 0$ tend towards the interior equilibrium or a periodic orbit.

(b) In this constellation, there is also the possibility that the ω-limit set is formed by boundary equilibria that are saddles and orbits that connect them, including a cyclic chain (see Theorem A.10 and Remark A.11). One part of the chain is a segment on the $M = 0$ axis connecting two adjacent boundary equilibria. The sources and sinks on the one hand and the saddles on the other alternate in such a way (Proposition 10.11) that the boundary equilibria $(0, y^{\diamond}_{2k-1})$ and $(0, y^{\diamond}_{2k})$ with $y^{\diamond}_{2k-1} < y^* < y^{\diamond}_{2k}$ are the only adjacent boundary equilibria on the $M = 0$ axis which are both saddles. Moreover, the stable manifold of the first matches the unstable manifold of the second. □

Under which conditions are periodic orbits excluded? First we notice that every periodic orbit necessarily lies in the interior of the first quadrant. We try $\chi \equiv 1$ as a *Dulac function* (Theorem A.12). We have

$$F_{1M}(M, y) = \begin{cases} \beta(M, y) - \mu + M\beta_M(M, y), & M > 0, \\ \beta(0, y) - \mu, & M \leqslant 0, \end{cases}$$

$$F_{2y}(M, y) = \begin{cases} -\sigma - \beta(M, y) - y\beta_y(M, y), & M > 0, \\ -\sigma - \beta(0, y) - y\beta_y(0, y), & M \leqslant 0. \end{cases}$$

So,

$$\operatorname{div} F(M, y) = \begin{cases} -\mu - \sigma + M\beta_M(M, y) - y\beta_y(M, y), & M > 0 \\ -\mu - \sigma - y\beta_y(0, y), & M \leqslant 0, \end{cases}$$

and we have the following result from Corollary A.13.

Theorem 10.13. *Let $\beta(0, y^*) > \mu$, i.e., there exists an interior equilibrium. Assume that β is continuously differentiable on $[0, \infty)^2$ and*

$$\sigma + \mu + y\beta_y(M, y) - M\beta_M(M, y) > 0 \quad \forall M \geqslant 0, \quad y > 0.$$

If there are finitely many boundary equilibria on the $M = 0$ axis, then every solution of (10.9) with $M(0), y(0) \geqslant 0$ converges towards an equilibrium. If there is exactly one boundary equilibrium, then every solution of (10.9) with $M(0) > 0, y(0) \geqslant 0$ converges towards the interior equilibrium.

An alternative approach often works well if the interior equilibrium is known to be locally asymptotically stable. If we can show in addition that all periodic orbits are necessarily locally asymptotically orbitally stable (i.e., locally asymptotically stable sets, see Toolbox A.6), we are able to rule out periodic orbits entirely. The argument uses contradiction: suppose that there exists a periodic orbit, which we denote by ξ. ξ necessarily surrounds the unique interior equilibrium and is locally asymptotically orbitally stable. There is a point in the region R surrounded by ξ which does not lie on a periodic orbit. Let us consider the solution starting at such a point. It is defined for all times because it stays in the region surrounded by ξ. Since the interior equilibrium is locally asymptotically stable, the solution cannot get arbitrarily close to the interior equilibrium in backward time. By the Poincaré–Bendixson dichotomy (Theorem A.7), it converges to a periodic orbit in backward time, because there is no other equilibrium in the region R. But this is impossible because this periodic orbit is also orbitally stable.

Fortunately there is a condition for periodic orbits x in the plane to be locally asymptotically orbitally stable which is not so difficult to check, though its derivation via Floquet theory is not so easy, namely

$$\int_0^\tau \operatorname{div} F(x(t))\, dt < 0,$$

with $\tau > 0$ being the period (see Proposition A.38). Using these arguments we are able to establish the following result. To keep the formulas short, we suppress the arguments of β.

Lemma 10.14. *There are no periodic solutions provided that* $\sigma + \beta + y\beta_y - M\beta_M > 0$ *for all* $M, y > 0$.

Proof. By Corollary 10.9 (a), the interior equilibrium is locally asymptotically stable (if it exists). We only need to show that all periodic orbits are locally asymptotically orbitally stable, i.e., that the integral above is strictly negative.

Let $x(t) = (M(t), y(t))$ be a periodic solution of (10.9). Then

$$\begin{aligned}\operatorname{div} F(M(t), y(t)) &= \beta(M(t), y(t)) - \mu + M(t)\beta_M(M(t), y(t)) - \sigma \\ &\quad - \beta(M(t), y(t)) - \beta_y(M(t), y(t))y(t) \\ &= \frac{y'(t)}{y(t)} + M(t)\beta_M(M(t), y(t)) - \sigma \\ &\quad - \beta(M(t), y(t)) - \beta_y(M(t), y(t))y(t).\end{aligned}$$

Let $\tau > 0$ be the period of x. Noticing that

$$\int_0^\tau \frac{y'(t)}{y(t)}\, dt = \ln y(\tau) - \ln y(0) = 0,$$

we see that $\int_0^\tau \operatorname{div} F(x(t))\, dt < 0$ if $\sigma + \beta(M, y) + y\beta_y(M, y) - M\beta_M(M, y) > 0$ for all $M, y > 0$. So, under this condition, all periodic orbits are locally asymptotically orbitally stable. This finishes the proof. □

If we also require the condition in Lemma 10.14 for $M = 0$, uniqueness of boundary equilibria follows. So, we have the following result from Theorem 10.12 and Lemma 10.14.

Theorem 10.15. *Let* β *be continuously differentiable on* $(0, \infty)^2$. *Assume that the interior equilibrium exists and that*

$$\sigma + \beta + y\beta_y - M\beta_M > 0 \quad \forall M \geqslant 0, \quad y > 0.$$

Then all solutions starting at $M(0) > 0$, $y(0) \geqslant 0$ *converge towards the interior equilibrium.*

We now return to the case where β is of Beverton–Holt form, (10.6),

$$\beta(M, y) = \frac{\rho(y)}{1 + Mg(y)}, \qquad g(y) = \frac{\rho(y)}{\Lambda\gamma(y)}.$$

Recall that $\rho(y)$ is the basic biomass production rate, $\gamma(y)$ is the yield function, and $g(y)$ is proportional to the food uptake rate, all taken at body burden y. We have

$$\beta_M(M, y) = -\rho(y)\frac{g(y)}{(1 + Mg(y))^2},$$
$$\beta_y(M, y) = \frac{\rho'(y)(1 + Mg(y)) - \rho(y)Mg'(y)}{(1 + Mg(y))^2}.$$

We use the symbol $\bowtie$ to indicate that two expressions have the same sign. Then

$$\begin{aligned}
&\sigma + \beta(M, y) + y\beta_y(M, y)) - M\beta_M(M, y) \\
&\quad\bowtie \sigma(1 + Mg(y))^2 + \rho(y)(1 + Mg(y)) \\
&\qquad + y\rho'(y)(1 + Mg(y)) - y\rho(y)Mg'(y) + M\rho(y)g(y) \\
&\quad = \sigma + \rho(y) + y\rho'(y) + \sigma M^2(g(y))^2 \\
&\qquad + M[2\sigma g(y) + 2\rho(y)g(y) + y\rho'(y)g(y) - y\rho(y)g'(y)].
\end{aligned}$$

This can be written in two forms:

$$\begin{aligned}
&\sigma + \beta(M, y) + y\beta_y(M, y)) - M\beta_M(M, y) \\
&\quad\bowtie \sigma[1 + (Mg(y))^2] + \rho(y) + y\rho'(y) \\
&\qquad + M[\sigma g(y) + \rho(y)(g(y) - yg'(y)) + (\sigma + \rho(y) + y\rho'(y))g(y)] \\
&\quad = \sigma(Mg(y))^2 + \phi'(y) + M[\sigma g(y) + \rho(y)(g(y) - yg'(y)) + \phi'(y)g(y)],
\end{aligned}$$

and

$$\begin{aligned}
&\sigma + \beta(M, y) + y\beta_y(M, y) - M\beta_M(M, y) \\
&\quad\bowtie \sigma + \rho(y) + y\rho'(y) + M(g(y))^2\left[\sigma M + \frac{2\sigma}{g(y)} + 2\frac{\rho(y)}{g(y)} + y\frac{\mathrm{d}}{\mathrm{d}y}\frac{\rho(y)}{g(y)}\right].
\end{aligned}$$

Using the formula for g,

$$\begin{aligned}
&\sigma + \beta(M, y) + y\beta_y(M, y)) - M\beta_M(M, y) \\
&\qquad\bowtie \phi'(y) + M(g(y))^2\Lambda\left[\frac{\sigma M}{\Lambda} + 2\gamma(y) + y\gamma'(y) + \frac{2\sigma}{\Lambda g(y)}\right].
\end{aligned}$$

Theorem 10.16. *Assume that the interior equilibrium exists. All solutions starting at initial data $M(0) > 0$, $y(0) \geqslant 0$ converge towards the interior equilibrium* *if $y[\sigma + \rho(y)]$ is a strictly increasing function of $y \geqslant 0$ and one of the following additional conditions is satisfied:*

(i) *$g(y)/y$ is decreasing; or*

(ii) *$y^2\gamma(y)$ is increasing.*

Condition (ii) and the increase of $y[\sigma + \rho(y)]$ as a function of y can be interpreted as the body burden not affecting food conversion and basic biomass production too drastically. If the body burden reduces the food uptake rate, then (i) is satisfied automatically. There is also the possibility that individuals try to make up for a reduced food conversion rate by increasing the food uptake rate. Then condition (i) states that this effect is not too strong.

If $(d/dy)(y^2\gamma(y)) < 0$ for some $y = y^* > 0$ with $\rho(y^*) > 0$, by choosing a sufficiently large $M^* > 0$ and a sufficiently small $\sigma > 0$, one can achieve that

$$\sigma + \beta(M^*, y^*) + y^*\beta_y(M^*, y^*)) - M^*\beta_M(M^*, y^*) < 0.$$

By Corollary 10.9 (b), we can choose $a > 0$ such that (M^*, y^*) is an unstable equilibrium, actually a source (i.e., all eigenvalues have strictly positive real parts). Then the only solution converging to the interior equilibrium is the interior equilibrium itself. All other solutions that start with initial data $M(0) > 0$, $y(0) \geqslant 0$ converge towards a periodic orbit.

Exercises

10.4. Show that the following function,

$$\chi(M, y) = \frac{1}{yM(\sigma + \beta(M, y))},$$

is a Dulac function in the region $M, y > 0$, provided that

$$\sigma + \beta(M, y) + y\beta_y(M, y) > 0 \quad \forall M > 0, \quad y > 0.$$

(Recall that $\beta(M, y)$ decreases in both variables.) What can you conclude from this result for the global behavior of solutions? Compare your conclusion with Theorem 10.15.

10.5. Consider β in Beverton–Holt form,

$$\beta(M, y) = \frac{\rho(y)}{1 + Mg(y)}.$$

Show that the interior equilibrium is locally asymptotically stable if

$$(\sigma + \mu)\rho + \mu y^*\left[\rho' - \frac{g'}{g}(\rho - \mu)\right] + \mu(\rho - \mu) > 0,$$

where ρ, ρ', g, g' are evaluated at $y^* = a/(\sigma + \mu)$.

10.6 Multiple Extinction Equilibria, Bistability and Periodic Oscillations

We restrict ourselves to the case in which the system (10.9),

$$M' = M[\beta(M, y) - \mu], \qquad y' = a - [\sigma + \beta(M, y)]y,$$

has three different boundary equilibria (extinction equilibria), associated with roots $\phi(y_j^\diamond) = a$, $j = 1, 2, 3$, $\phi(y) = y[\sigma + \beta(0, y)]$, because it is representative for, but less confusing than, the general case of an odd number of multiple boundary equilibria. We assume that ϕ is continuously differentiable in neighborhoods of the subsequent sets and points and satisfies

$$\phi'(y) > 0 \quad \text{for } y \in [0, y_1^\diamond] \cup [y_3^\diamond, \infty), \qquad \phi'(y_2^\diamond) < 0. \qquad (\diamondsuit)$$

See Figure 10.1 for an example. Let us describe another situation where a and σ can be chosen such that we have the following scenario.

($\diamond_1$) $\beta(0, \cdot)$ is twice continuously differentiable.

($\diamond_2$) $y\beta(0, y)$, as a function of $y \geqslant 0$, has one maximum at $y_1 > 0$ and is strictly increasing on $[0, y_1]$ and strictly decreasing on $[y_1, \infty)$.

($\diamond_3$) $y\beta(0, y)$ has exactly one inflection point at $y_2 > y_1$ with $y\beta(0, y)$ being strictly concave on $[0, y_2]$ and strictly convex on $[y_2, \infty)$. y_2 is the only point at which the second derivative of $y\beta(0, y)$ is zero.

($\diamond_4$) $y\beta(0, y) \to 0$ as $y \to \infty$.

Examples are $\beta(0, y) = e^{-y}$ and $\beta(0, y) = \xi^{-1}(1 + y)^{-\xi}$, $\xi > 1$.

Let $\sigma^\sharp = -(\beta(0, y_2) + y_2\beta_y(0, y_2))$, which is positive by ($\diamond_2$) and ($\diamond_3$). For $\sigma \in (0, \sigma^\sharp)$, $\phi(y) = y[\sigma + \beta(0, y)]$ has at least one local maximum and one local minimum, and for all values a strictly between the minimum and the maximum value there are at least three boundary equilibria. Since ϕ has only one argument at which the second derivative is 0, there are exactly three boundary equilibria. ϕ has the same convexity and concavity behavior as $y\beta(0, y)$, which implies that ϕ' is strictly decreasing on $[0, y_2]$ and strictly increasing on $[y_2, \infty)$. So, there are exactly one local maximum and one local minimum, let us say at $\tilde{y}_1$ and $\tilde{y}_2$, $\tilde{y}_1 < \tilde{y}_2$. Then $y_1^\diamond < \tilde{y}_1 < y_2^\diamond < \tilde{y}_2 < y_3^\diamond$ and $y' > 0$ on $[0, \tilde{y}_1) \cup (\tilde{y}_2, \infty)$ and $y' < 0$ on $(\tilde{y}_1, \tilde{y}_2)$. In particular, ($\diamondsuit$) holds.

We return to the scenario described at the beginning of this section. There are two different cases. If $\beta(0, (a/\sigma)) > 0$, any boundary equilibrium $(0, y^\diamond)$ satisfies $y^\diamond < (a/\sigma)$. This means that $y^* = a/(\sigma + \mu)$ can be positioned anywhere relative to the boundary equilibria by choosing $\mu \in (0, \infty)$ appropriately.

If $\beta(0, (a/\sigma)) = 0$, then $y_3^\diamond = a/\sigma$ and it cannot occur that $y^* > y_3^\diamond$. Otherwise y^* can be positioned anywhere by adjusting μ. We go through the various cases discussing those where there is no interior equilibrium first because they give us clues as to what happens in the case with interior equilibrium. We will speak about the *domain of attraction* (also called stable set) of a locally asymptotically stable equilibrium $x = (M, y)$, which is the set of initial data for which the solutions of (10.9) converge to x as $t \to \infty$.

Case 1. $y^* < y_1^\diamond < y_2^\diamond < y_3^\diamond$ (Figure 10.2A).

By contraposition of Proposition 10.11 (a), there exists no interior equilibrium and, by Theorem 10.6, all solutions converge towards a boundary equilibrium. By Proposition 10.11 (c), $(0, y_1^\diamond)$ and $(0, y_3^\diamond)$ are stable nodes and $(0, y_2^\diamond)$ is a saddle with the unstable manifold being part of the $M = 0$ axis. So, there are no cyclic orbit connections of equilibria. By the Poincaré–Bendixson trichotomy (see Theorem A.10 and Remark A.11), all solutions converge towards a boundary equilibrium. The stable manifold of $(0, y_2^\diamond)$ separates the first quadrant into the domains of attraction of $(0, y_1^\diamond)$ and $(0, y_3^\diamond)$. By Proposition 10.11 (d), the region $\{M \geqslant 0,\ y \geqslant y_2^\diamond\} \setminus \{(0, y_2^\diamond)\}$ is contained in the domain of attraction of $(0, y_3^\diamond)$.

Summarizing, the population dies out inevitably, but with different internal toxicant concentrations, $y_1^\diamond$ and $y_2^\diamond$, depending on the initial conditions.

Case 2. $y_1^\diamond < y_2^\diamond < y^* < y_3^\diamond$ (Figure 10.2H).

By Proposition 10.11 (b), there exists no interior equilibrium and therefore no periodic orbit. By Proposition 10.11 (c), $(0, y_1^\diamond)$ is a saddle with the stable manifold being part of the $M = 0$ axis, $(0, y_2^\diamond)$ is a source, and $(0, y_3^\diamond)$ is a stable node. By Theorem 10.6, all solutions converge to a boundary equilibrium. By Proposition 10.7, all solutions with $M(0) > 0$ converge to $(0, y_3^\diamond)$. The unstable manifold of $(0, y_1^\diamond)$ which connects $(0, y_1^\diamond)$ to $(0, y_3^\diamond)$ divides the first quadrant into two regions: one bounded invariant region and an unbounded forward invariant region. The unbounded invariant region contains solutions with $y(0) < y_1^\diamond$ and $M(0) > 0$ very small. Such solutions cannot penetrate the bounded invariant region, so they have to go around it, before they converge to the boundary equilibrium $(0, y_3^\diamond)$. This means that $M(t)$, though very small initially, becomes quite large in between before it goes to 0 again.

Some important ecological lessons can be learned from this scenario. Though the population eventually dies out, there are transient dynamics during which there is a substantial build up of population biomass from small initial biomass. This may mislead an observer to make the assumption that the population is doing fine and to conclude later that environmental conditions must have changed to cause extinction. The fact is that the population was doomed from the very beginning and it only appeared as if it would flourish.

Case 3. $y_1^\diamond < y^* < y_2^\diamond < y_3^\diamond$.

By Proposition 10.11 (b), there is an interior equilibrium (M^*, y^*). By Proposition 10.11 (c), $(0, y_1^\diamond)$ is a saddle with the stable manifold being part of the

$M = 0$ axis, $(0, y_2^\diamond)$ is a saddle with the unstable manifold being part of the $M = 0$ axis, and $(0, y_3^\diamond)$ is a stable node. By Proposition 10.11 (d), the region $\{M \geqslant 0,\ y \geqslant y_2^\diamond\}\setminus\{(0, y_2^\diamond)\}$ is contained in the domain of attraction of $(0, y_3^\diamond)$.

Since $y[\sigma + \beta(M, y)] = a$ for $(M, y) = (M^*, y^*)$ and $(M, y) = (0, y_j^\diamond)$ and $\beta(M, y_j^\diamond) < \beta(0, y_j^\diamond)$ for $M > 0$, $j = 1, 2$, we have $M^* \to 0$ as $y^* \to y_1^\diamond$ or $y^* \to y_2^\diamond$, i.e., the interior equilibrium approaches the respective boundary equilibrium as $y^* \to y_j^\diamond$. By Proposition 10.10, the interior equilibrium is a stable node if y^* is close to $y_1^\diamond$, and an unstable node if y^* is close to $y_2^\diamond$.

Case 3.1. $y^* \in (y_1^\diamond, y_2^\diamond)$ is close to $y_1^\diamond$ (Figure 10.2B).

As y^* moves from $y^* < y_1^\diamond$ to $y^* > y_1^\diamond$, the boundary equilibrium $(0, y_2^\diamond)$ turns from a stable node into a saddle and a unique interior equilibrium arises which is a stable node.

Through this transition (called a *bifurcation*), the boundary equilibrium $(0, y_2^\diamond)$ remains a saddle with the unstable manifold being part of the $M = 0$ axis and the stable manifold undergoing only slight changes locally.

Recall that in case 1, $y^* < y_1^\diamond$, the stable manifold of $(0, y_2^\diamond)$ forms the boundary of the domain of attraction of $(0, y_3^\diamond)$. Since solutions depend continuously on parameters on finite time intervals, it is suggestive that this feature does not change if $y^* > y_1^\diamond$ with the two being close to each other. In the following we will argue that there are no periodic orbits surrounding the interior equilibrium if y^* is close enough to $y_1^\diamond$. This will imply that, for y^* close to $y_1^\diamond$, all solutions with $M(0) > 0$ converge to the interior equilibrium or to the boundary equilibrium $(0, y_3^\diamond)$ with the stable manifold of $(0, y_2^\diamond)$ separating the domains of attraction.

To rule out periodic orbits, we construct a forward invariant rectangle around the interior equilibrium as follows. One corner is $(0, y_1^\diamond)$. Recall that the region $M \geqslant 0$, $y \geqslant y_1^\diamond$ is forward invariant (Proposition 10.7 (a)). Since $\mu = \beta(M^*, y^*) < \beta(M^*, y_1^\diamond)$ and $\beta(\cdot, y_1^\diamond)$ is strictly decreasing, there is a unique number M_1, $M_1 > M^*$, such that $\beta(M_1, y_1^\diamond) = \mu$. Since $\beta(M_1, y) \leqslant \mu$ for $y \geqslant y^\diamond$, the region $0 \leqslant M \leqslant M_1$, $y \geqslant y_1^\diamond$ is forward invariant. Since $\beta(\cdot, y_1^\diamond)$ is strictly decreasing, $M^* - M_1 \to 0$ as $y^* \to y_1^\diamond$. We already notice that $M^* \to 0$ as $y^* \to y_1^\diamond$. Combining the two statements, we have that M_1 is close to 0 if y^* is close to $y_1^\diamond$. Now choose the smallest $\tilde{y} > y^*$ such that $\tilde{y}[\sigma + \beta(M_1, \tilde{y})] = a$. This is possible by ($\diamond$), because $y[\sigma + \beta(M_1, y)$ is a strictly increasing function of y in a neighborhood of $[0, y_1^\diamond]$ if M_1 is close enough to 0, which is the case if y^* is close enough to $y_1^\diamond$. Again one can show that $\tilde{y} \to y_1^\diamond$ if $y^* \to y_1^\diamond$. Furthermore, $y' \leqslant 0$ if $y = \tilde{y}$, $M \in [0, M_1]$. So, the rectangle with corners $(0, y_1^\diamond)$, $(M_1, y_1^\diamond)$, $(M_1, \tilde{y})$, $(0, \tilde{y})$ is forward invariant and shrinks to the point $(0, y_1^\diamond)$ if $y^* \to y_1^\diamond$. Any periodic orbit that surrounds the interior equilibrium (M^*, y^*) must pass through that rectangle and so lie within it. In particular, such a periodic orbit is arbitrarily close to $(0, y_1^\diamond)$ if y^* is chosen sufficiently close to $y_1^\diamond$. The divergence of the vector field, evaluated at $(0, y_1^\diamond)$, is $a(1/y_1^\diamond) - (1/y^*)) - \phi'(y_1^\diamond)$, which is strictly negative if y^* is close to $y_1^\diamond$. So, the divergence of the vector field is strictly negative in a neighborhood of $(0, y_1^\diamond)$, which excludes periodic orbits in this neighborhood by the (Dulac–)Bendixson criterion (Theorem A.12).

There possibly exists a heteroclinic cycle formed by the segment of the $M = 0$ axis between the boundary equilibria $(0, y_1^\diamond)$ and $(0, y_2^\diamond)$ and a connecting orbit leading through the interior of the first quadrant. This is the only heteroclinic cycle possible under the circumstances. This heteroclinic orbit would surround the interior equilibrium, which is a sink. We want to rule out the possibility that the heteroclinic orbit is the ω-limit set of a solution. Obviously this is not the case if the solution starts outside the heteroclinic orbit. If it starts inside, then the solution cannot approach the heteroclinic orbit in both forward and backward time without crossing itself, which is impossible by uniqueness of solutions. So, if a solution starts inside the heteroclinic orbit and approaches it in forward time, its α-limit set does not contain the heteroclinic cycle. The interior equilibrium is a sink, so it is not contained in the α-limit set either. Consequently the α-limit set is a periodic orbit by the Poincaré–Bendixson trichotomy (see Theorem A.10 and Remark A.11). But periodic orbits do not exist, as we have seen before.

Summarizing, we have a *bistable* situation. It depends on the initial conditions whether the population dies out (typically when the initial body burden is high) or whether it survives converging to a survival equilibrium.

Case 3.2. $y^* \in (y_1^\diamond, y_2^\diamond)$ is close to $y_2^\diamond$ (Figure 10.3G).

As y^* moves from $y^* > y_2^\diamond$ (case 2) to $y^* < y_2^\diamond$, the boundary equilibrium $(0, y_2^\diamond)$ changes from an unstable node to a saddle (with the unstable manifold being part of the $M = 0$ axis), and an interior equilibrium arises which is an unstable node.

This bifurcation does not affect the boundary equilibrium $(0, y_1^\diamond)$, which remains a saddle with the stable manifold being part of the $M = 0$ axis and the unstable manifold undergoing only slight local changes. Recall that in case 2, $y^* > y_2^\diamond$, the unstable manifold of $(0, y_1^\diamond)$ connects to the stable node $(0, y_3^\diamond)$. If $y^* < y_2^\diamond$ but is close to $y_2^\diamond$, by continuous dependence of solutions on parameters this feature is preserved, and the unstable manifold of $(0, y_1^\diamond)$ encloses a bounded invariant region which contains the interior equilibrium. The stable manifold of $(0, y_2^\diamond)$ connects to the interior equilibrium.

Summarizing, though a survival equilibrium exists, it is a source, and the population inevitably dies out, except in the extraordinary case in which it is exactly at the survival equilibrium initially.

Case 3.3. Bifurcation of periodic solutions at some $y^* \in (y_1^\diamond, y_2^\diamond)$ (Figures 10.2D and 10.3E).

As we have seen, the interior equilibrium changes its nature from being a sink to being a source as y^* is moved from being close to $y_1^\diamond$ to being close to $y_2^\diamond$.

Since the Jacobian is always positive, the eigenvalues of the Jacobian matrix at the interior equilibrium cannot cross the imaginary axis through the origin, but must cross it as a complex conjugate pair. This suggests a Hopf bifurcation of periodic solutions. The classical Hopf bifurcation theorem requires transversal crossing of the eigenvalues over the imaginary axis and nonresonance, with the second being automatically satisfied in the plane. Since the real part of the complex conjugate eigenvalues is half the divergence of the vector field, it is sufficient to determine y^* where the divergence is zero and show that the derivative of the divergence with respect to y^* is different from 0 at this point. One can also employ global Hopf bifurcation theorems, which show existence of periodic orbits, but not local uniqueness. Here one does not need to check the derivative of the divergence (e.g., Fiedler, 1986). In any case, there will be periodic orbits for certain values of $y^* \in (y_1^\diamond, y_2^\diamond)$.

Case 4. $y_1^\diamond < y_2^\diamond < y_3^\diamond < y^*$.

Recall that this case can only occur if $\beta(0, (a/\sigma)) > 0$.

By Proposition 10.11 (b), there is a unique interior equilibrium (M^*, y^*). By Proposition 10.11 (c), $(0, y_1^\diamond)$ and $(0, y_3^\diamond)$ are saddles with the stable manifold being part of the $M = 0$ axis and $(0, y_2^\diamond)$ is a source. By Theorem 10.12 (a), all solutions starting with $M(0) > 0$ converge towards (M^*, y^*) or a periodic orbit surrounding (M^*, y^*).

10.7 Linear Dose Response

We assume for simplicity that β is of the form (10.7), i.e., the body burden does not influence the food uptake rate, but lowers the food conversion rate. Switching to a dimensionless time and redefining the parameters, we can assume that $\beta(0, 0) = \rho(0) = 1$. Scaling y we can assume that $\beta_y(0, 0) = \rho'(0) = -1$ (provided that $\rho(0) \neq 0$). Replacing M by cM we can assume by (10.7) that

$$\beta(M, y) = \frac{\rho(y)}{1 + M}.$$

β has the following partial derivatives:

$$\beta_y(M, y) = \frac{\rho'(y)}{1 + M}, \qquad \beta_M(M, y) = -\frac{\rho(y)}{(1 + M)^2}.$$

The condition in Theorem 10.13 takes the form

$$0 < \sigma + \mu + y\beta_y(M, y) - M\beta_M(M, y) = \sigma + \mu + \frac{y\rho'(y)}{1 + M} + \frac{\rho(y)}{(1 + M)^2}.$$

Multiplying by $(1 + M)^2$,

$$0 < (\sigma + \mu)(1 + M)^2 + y\rho'(y)(1 + M) + \rho(y).$$

Theorem 10.17. *Let ρ be continuously differentiable on $[0, \infty)$. Assume that there are finitely many boundary equilibria and*

$$\sigma + \mu + y\rho'(y) + \rho(y) \geqslant 0 \quad \forall y > 0,$$
$$2(\sigma + \mu) + y\rho'(y) \geqslant 0 \quad \forall y > 0.$$

Then all solutions of (10.9) converge towards an equilibrium.

We specialize further and consider a linear dose response of the per unit biomass gain rate to the body burden (cf. Hallam, De Luna, 1984). Since, after scaling, we can assume that $\rho(0) = 1 = -\rho'(0)$, we choose

$$\beta(0, y) = \rho(y) = [1 - y]_+,$$

where $[r]_+$ is the positive part of the number r, i.e., $[r]_+ = r$ if $r \geqslant 0$ and $[r]_+ = 0$ if $r \leqslant 0$. So,

$$\beta(M, y) = \frac{[1 - y]_+}{1 + M}.$$

β is not differentiable on the line $y = 1$, but is continuously differentiable wherever it is strictly positive, as we have assumed before. We conclude the following from Theorem 10.17.

Corollary 10.18. *Let $\beta(M, y) = [1 - y]_+/(1 + M)$ and $\sigma + \mu \geqslant 1$, $a \neq \sigma$. Then all solutions converge towards an equilibrium.*

Proof. Actually, we cannot apply Theorem 10.17 immediately because β is not differentiable on the line $y = 1$. But we can still apply the Dulac–Bendixson condition.

Case 1. $a < \sigma$.

We then have $y' = a - \sigma y \leqslant a - \sigma < 0$ as long as $y \geqslant 1$. This means that all solutions enter the region $y < 1$ and stay in a compact subset of this region. This means that periodic orbits or homoclinic/heteroclinic cycles must lie in the region $y < 1$, where β is continuously differentiable.

Case 2. $a > \sigma$.

If $a > \sigma$, we have $y' = a - \sigma y$ for $y \geqslant 1$ and $y' = a - \sigma > 0$ for $y = 1$. This means that the region $y \geqslant 1$ is invariant. On that region, $M' = -\mu M$. So, every solution starting in the region $y \geqslant 1$ converges to the boundary equilibrium $(0, a/\sigma)$. Again all periodic or clinic cycles must lie in the region $y < 1$.

Case 3. $a = \sigma$.

Now $y \equiv 1$ and $M(t) = \mathrm{e}^{-\mu t}$ are solutions of (10.9). Since solutions of (10.9) are uniquely determined by their initial values, they cannot cross the line $y = 1$. So, both the regions $y < 1$, $y > 1$, and the line $y = 1$ are invariant. For $y \geqslant 1$, solutions of (10.9) satisfy $M' = -\mu M$, so no periodic orbits or homoclinic/heteroclinic cycles are contained in this region and they must lie in the region $y < 1$ if they exist.

So, in either of these three cases, periodic orbits and homoclinic/heteroclinic cycles must lie in the region $y < 1$, where ρ is continuously differentiable. This is enough for Theorem 10.17 to apply. □

The threshold condition for existence of the interior equilibrium,

$$\beta(0, y^*) = \rho\left(\frac{a}{\sigma + \mu}\right) > \mu,$$

becomes

$$\mu + \sigma - a > \mu(\sigma + \mu).$$

Rewriting this as a quadratic inequality in μ and solving this inequality, we find that the interior equilibrium exists if and only if

$$0 < a < (y^\sharp)^2 \quad \text{and} \quad \mu_- < \mu < \mu_+,$$

$$y^\sharp := \tfrac{1}{2}(1 + \sigma), \qquad \mu_\pm := \tfrac{1}{2}(1 - \sigma) \pm \sqrt{(y^\sharp)^2 - a}.$$

Notice that $\mu_+ > 0$ while μ_- has the same sign as $a - \sigma$. In order to determine the boundary equilibria $(0, y^\diamond)$, we notice that they satisfy $\phi(y^\diamond) = a$ with

$$\phi(y) = \begin{cases} \sigma y, & y \geqslant 1, \\ [\sigma + 1 - y]y, & 0 \leqslant y < 1. \end{cases}$$

So,

$$\phi'(y) = \begin{cases} \sigma, & y > 1, \\ \sigma + 1 - 2y, & 0 < y < 1. \end{cases}$$

In particular,

$$\phi'(0) = \sigma + 1, \qquad \phi'(1-) = \sigma - 1, \qquad \phi'(1+) = \sigma.$$

We find that ϕ is increasing on $[0, y^\sharp]$ and decreasing on $[y^\sharp, 1]$. So, if $\sigma \geqslant 1$, the interval $[y^\sharp, 1]$ is empty and $\phi'(y) > 0$ for $y \neq 1$. However, if $0 \leqslant \sigma < 1$, then ϕ has a local maximum at $y^\sharp$ and

$$\phi(y^\sharp) = (y^\sharp)^2 = \tfrac{1}{4}(\sigma + 1)^2.$$

Furthermore, ϕ has a local minimum at 1 and $\phi(1) = \sigma$. More precisely, ϕ is strictly increasing on $[0, y^\sharp]$ and on $[1, \infty)$ and strictly decreasing on $[y^\sharp, 1]$. Recalling that we are looking for roots $\phi(y^\diamond) = a$, we obtain the following lemma.

Lemma 10.19.

(a) *If* $\sigma \geqslant 1$, *then there exists exactly one boundary equilibrium,* $y^\diamond$,

$$y^\diamond = \begin{cases} y^\sharp - \sqrt{(y^\sharp)^2 - a}, & 0 < a \leqslant \sigma, \\ a/\sigma, & a \geqslant \sigma. \end{cases}$$

(b) *Let* $0 < \sigma < 1$. *If* $0 < a < \sigma$, *then there exists exactly one boundary equilibrium:*

$$y^\diamond = y^\sharp - \sqrt{(y^\sharp)^2 - a}.$$

If $a = \sigma$, *then there exist exactly two boundary equilibria:*

$$y_1^\diamond = y^\sharp - \sqrt{(y^\sharp)^2 - a}, \qquad y_2^\diamond = \sigma = a.$$

If $\sigma < a < (y^\sharp)^2$, *then there exist exactly three boundary equilibria (Figure 10.1):*

$$y_1^\diamond = y^\sharp - \sqrt{(y^\sharp)^2 - a}, \qquad y_2^\diamond = y^\sharp + \sqrt{(y^\sharp)^2 - a}, \qquad y_3^\diamond = a/\sigma.$$

If $a = (y^\sharp)^2$, *then there exist exactly two boundary equilibria:*

$$y_1^\diamond = y^\sharp, \qquad y_2^\diamond = a/\sigma.$$

If $a > (y^\sharp)^2$, *then there exists exactly one boundary equilibrium:*

$$y^\diamond = a/\sigma.$$

Apparently, one can expect the most interesting dynamics in the case of three boundary equilibria, which we therefore want to study in detail (for other cases see the exercises below), so we assume that

$$\sigma < a < (y^\sharp)^2 < 1, \qquad y^\sharp = \tfrac{1}{2}(1 + \sigma).$$

This is equivalent to

$$0 < \mu_- < \mu_+ < 1, \qquad \mu_\pm := \tfrac{1}{2}(1 - \sigma) \pm \sqrt{(y^\sharp)^2 - a}.$$

Recall that the interior equilibrium exists if and only if $\mu_- < \mu < \mu_+$. Furthermore,

$$0 < y_1^\diamond < \tfrac{1}{2}(1 + \sigma) < y_2^\diamond < 1 < y_3^\diamond = a/\sigma.$$

Notice the following relations between μ and y^*, which can be checked by some elementary algebra:

$$\left.\begin{array}{rcl} 0 < y^* < y_1^\diamond & \Longleftrightarrow & \mu_+ < \mu < \infty, \\ y_1^\diamond < y^* < y_2^\diamond & \Longleftrightarrow & \mu_- < \mu < \mu_+, \\ y_2^\diamond < y^* < y_3^\diamond & \Longleftrightarrow & 0 < \mu < \mu_-. \end{array}\right\} \tag{10.10}$$

$y^* \geqslant y_3^\diamond$ cannot occur in this example because $y_3^\diamond = a/\sigma$ and $y^* = a/(\sigma + \mu)$.

We notice that we have the situation described at the beginning of Section 10.6, in particular ($\diamond$). So, we have the same dynamics as we discussed before, except that case 4, $y^* > y_3^\diamond$, cannot occur. So, the boundary equilibrium $(0, y_3^\diamond)$ is always a stable node, and its domain of attraction contains the quadrant $M \geqslant 0$, $y \geqslant y_2^\diamond$, $(M, y) \neq (0, y_2^\diamond)$. See cases 1–3 in Section 10.6.

For the linear dose response, we can describe the bifurcation of periodic solutions in case 3.3 more precisely. We use μ rather than y^* as bifurcation parameter.

Recall that we want to determine $\mu^\sharp$ where the divergence is zero and show that the derivative of the divergence with respect to μ is different from 0 at $\mu^\sharp$ (see the remark in Toolbox A.7). In our case this involves finding the zeros of a cubic polynomial and showing that the zeros are not double zeros. This is the case for a dense open set of parameters a and σ.

Recall that

$$\begin{aligned} -\operatorname{div} F(M^*, y^*) &= \mu + \sigma + y^* \beta_y(M^*, y^*) - M^* \beta_M(M^*, y^*) \\ &= \mu + \sigma - \frac{y^*}{1 + M^*} + M^* \frac{1 - y^*}{(1 + M^*)^2}. \end{aligned}$$

Using the fact that $\mu = \beta(M^*, y^*) = (1 - y^*)/(1 + M^*)$, we have

$$\begin{aligned} -\operatorname{div} F(M^*, y^*) &= \mu + \sigma - \frac{y^*}{1 + M^*} + M^* \frac{\mu}{1 + M^*} \\ &= \sigma + 2\mu - \frac{y^*}{1 + M^*} - \frac{\mu}{1 + M^*} \\ &= \sigma + 3\mu - \frac{1}{1 + M^*} - \frac{\mu}{1 + M^*} \\ &= \sigma + 3\mu - \mu \frac{\mu + 1}{1 - y^*}. \end{aligned}$$

Substituting $y^* = a/(\mu + \sigma)$,

$$\begin{aligned} -\operatorname{div} F(M^*, y^*) &= \sigma + 3\mu - \mu \frac{(1 + \mu)(\sigma + \mu)}{\sigma + \mu - a} \\ &= [(\sigma + 3\mu)(\sigma + \mu - a) - \mu(1 + \mu)(\sigma + \mu)] \frac{1}{\sigma + \mu - a}. \end{aligned}$$

From our bifurcation consideration we already know that the last expression has a zero $\mu = \mu^\sharp$ under the constraint $\sigma + \mu - a > \mu(\mu + \sigma)$ related to the existence of the interior equilibrium. To confirm this, recall that the interior equilibrium exists for $\mu_- < \mu < \mu_+$ and that for $\mu = \mu_\pm$ we have $\sigma + \mu - a = \mu(\mu + \sigma)$. So, for $\mu = \mu_\pm$,

$$-\operatorname{div} F(M^*, y^*) = \sigma - 1 + 2\mu_\pm = \pm 2\sqrt{(y^\sharp)^2 - a}.$$

By the intermediate value theorem, $-\operatorname{div} F(M^*, y^*)$, as a function of μ, has a zero between μ_- and μ_+. Transversal crossing of the eigenvalues is equivalent to the derivative of this expression with respect to μ being nonzero when evaluated at $\mu^\sharp$.

This derivative, evaluated at the zero $\mu = \mu^\sharp$, has the same sign as the derivative of

$$\chi(\mu) := (\sigma + 3\mu)(\sigma + \mu - a) - \mu(1 + \mu)(\sigma + \mu),$$

evaluated at the zero $\mu = \mu^\sharp$, which is

$$\begin{aligned}
\chi'(\mu) &= 3(\sigma + \mu - a) + \sigma + 3\mu - (1 + \mu)(\sigma + \mu) - \mu(\sigma + \mu) - \mu(1 + \mu) \\
&= 3(\sigma + \mu - a) + \sigma + 3\mu - (\sigma + \mu) - 2\mu(\sigma + \mu) - \mu(1 + \mu) \\
&= 3(\sigma + \mu - a) + \mu - 2\mu(\sigma + \mu) - \mu^2 \\
&= 3[(\sigma + \mu - a) - \mu(\sigma + \mu)] + \mu(\sigma + 1).
\end{aligned}$$

So, $\chi'(\mu) > 0$ under the constraint $\sigma + \mu - a > \mu(\sigma + \mu)$, and we have shown that the eigenvalues cross the imaginary axis transversally. It also follows that they only cross once, because $\chi'(\mu) > 0$ under the constraint.

We summarize the behavior of solutions of (10.9) for linear dose response. The various cases in Section 10.6 have been reformulated in terms of μ rather than y^* (see (10.10)).

Theorem 10.20. *Let us assume that there exist three boundary equilibria, i.e.,* $0 < \mu_- < \mu_+$ *where*

$$\mu_\pm := \tfrac{1}{2}(1 - \sigma) \pm \sqrt{(y^\sharp)^2 - a}, \quad a < (y^\sharp)^2.$$

The boundary equilibria $(0, y_j^\diamond)$ *can be ordered as* $0 < y_1^\diamond < y_2^\diamond < 1 < a/\sigma = y_3^\diamond$. *The boundary equilibrium* $y_3^\diamond$ *is always a stable node which attracts all solutions starting with* $M(0) \geqslant 0$, $y(0) \geqslant y_2^\diamond$, *except those starting at* $(0, y_2^\diamond)$. *Furthermore, the following hold:*

(a) *Let* $\mu > \mu_+$, *i.e.,* $y^* < y_1^\diamond$ *(Figure 10.2A).*

There is no interior equilibrium. The boundary equilibria $y_1^\diamond$ *and* $y_2^\diamond$ *are stable nodes and the boundary equilibrium* $y_2^\diamond$ *is a saddle whose stable* *manifold separates the domains of attraction of $y_1^\diamond$ and $y_3^\diamond$. The population inevitably becomes extinct with a body burden which depends on the initial values.*

(b) *Let $0 < \mu < \mu_-$, i.e., $y_2^\diamond < y^* < y_3^\diamond$ (Figure 10.3H).*

There is no interior equilibrium. The boundary equilibrium $(0, y_2^\diamond)$ is an unstable node, the boundary equilibrium $(0, y_3^\diamond)$ is a stable node, and the boundary equilibrium $(0, y_1^\diamond)$ is a saddle whose unstable manifold connects to the stable node $(0, y_3^\diamond)$ through the positive quadrant. All solutions starting with $M(0) > 0$, $y(0) \geqslant 0$ converge towards $(0, y_3^\diamond)$. The population inevitably dies out, but can grow substantially in between.

(c) *Let $\mu_- < \mu < \mu_+$ and μ be sufficiently close to μ_+, i.e., $y^* \in (y_1^\diamond, y_2^\diamond)$ is close to $y_1^\diamond$ (Figure 10.2B).*

There then exists a unique interior equilibrium which is close to the boundary equilibrium $(0, y_1^\diamond)$ and is a stable node. The boundary equilibrium $(0, y_1^\diamond)$ is a saddle whose unstable manifold connects to the interior equilibrium. The boundary equilibrium $(0, y_2^\diamond)$ is a saddle whose stable manifold separates the domains of attraction of the interior equilibrium and the boundary equilibrium $(0, y_3^\diamond)$.

This is a bistable situation where the population either dies out or converges to a (unique) survival equilibrium depending on the initial conditions, with high initial body burden typically leading to extinction.

(d) *Let $\mu_- < \mu < \mu_+$ and μ be sufficiently close to μ_-, i.e., $y^* \in (y_1^\diamond, y_2^\diamond)$ is close to $y_2^\diamond$ (Figure 10.3G).*

There then exists a unique interior equilibrium which is close to the boundary equilibrium $(0, y_2^\diamond)$. The interior equilibrium is a source and the boundary equilibrium $(0, y_2^\diamond)$ is a saddle whose stable manifold connects to the unstable interior equilibrium. The boundary equilibrium $(0, y_3^\diamond)$ is a stable node and the boundary equilibrium $(0, y_1^\diamond)$ is a saddle whose unstable manifold connects to the stable node $(0, y_3^\diamond)$ through the positive quadrant. All solutions starting with $M(0) > 0$, $y(0) \geqslant 0$ except those starting on the stable manifold of $(0, y_2^\diamond)$ converge towards $(0, y_3^\diamond)$.

Though there exists a survival equilibrium (which is unique), the population dies out unless the system is exactly at the survival equilibrium initially. As in (b), populations can temporarily grow substantially before they become extinct.

(e) *There exists a unique $\mu^\sharp \in (\mu_-, \mu_+)$ (corresponding to a unique $y^\odot \in (y_1^\diamond, y_2^\diamond)$) such that the interior equilibrium is stable for $\mu \in (\mu^\odot, \mu_+)$ and unstable for $\mu \in (\mu_-, \mu^\odot)$ and undergoes a Hopf bifurcation of periodic solutions at $\mu = \mu^\odot$ (Figures 10.2B–D and 10.3E–G).*

The stability change of the interior equilibrium in the Hopf bifurcation is not the only transition that occurs for y^* between $y_1^\diamond$ and $y_2^\diamond$. If y^* is close to $y_2^\diamond$, the unstable manifold of the boundary equilibrium $(0, y_1^\diamond)$ connects to the boundary equilibrium $(0, y_3^\diamond)$ (Figure 10.3G). If y^* is close to $y_1^\diamond$, the unstable manifold of the boundary equilibrium $(0, y_1^\diamond)$ connects to the interior equilibrium, which is a stable node (Figure 10.2B). For some value y^* between $y_1^\diamond$ and $y_2^\diamond$, there must be a transition between these two configurations. We conjecture that for such a $y^* = y^\triangleright$, there is a heteroclinic orbit which connects the boundary equilibria $(0, y_1^\diamond)$ and $(0, y_2^\diamond)$. Our conjecture is supported by the theory of global Hopf bifurcation (see, for example, Fiedler, 1986). According to this theory, the branch of periodic solutions either extends to another Hopf bifurcation point (which cannot happen because there is exactly one y^* at which Hopf bifurcation occurs), or the amplitudes of the periodic solutions go to infinity, or (in two space dimensions) the periods go to infinity. Because periodic orbits surround an interior equilibrium, they only occur for y^* between $y_1^\diamond$ and $y_2^\diamond$; so Theorem 10.5 rules out that the amplitudes tend to infinity. The only possibility left is the periods going to infinity, which suggests that the branch of period orbits ends in a cycle of saddle connections. According to the nature of the interior equilibrium and the boundary equilibria, the only possibility is a heteroclinic orbit connecting $(0, y_1^\diamond)$ and $(0, y_2^\diamond)$ and surrounding the interior equilibrium.

Using MAPLE, we have performed several phase-plane plots of solutions to

$$M' = M\left(\frac{[1-y]_+}{1+M} - \mu\right),$$
$$y' = a - y\left(\sigma + \frac{[1-y]_+}{1+M}\right),$$

for $\sigma = 0.2$, $a = 0.25$, and various values of μ. See Figures 10.1–10.3 and Toolbox C for a commented MAPLE worksheet for one of the phase-plane plots. The boundary equilibria are $(0, y_j^\diamond)$ with

$$y_1^\diamond = 0.268\,337\,521\,0, \qquad y_2^\diamond = 0.931\,662\,479\,0, \qquad y_3^\diamond = 1.250\,000\,000.$$

The interior equilibrium exists for $y^* \in (y_1^\diamond, y_2^\diamond)$, i.e., for $\mu_- < \mu < \mu_+$, where $\mu_- \approx 0.068\,337\,521\,0$ and $\mu_+ \approx 0.731\,662\,479$. Solving the cubic polynomial for a zero numerically shows that the Hopf bifurcation occurs at $\mu = \mu^\odot \approx 0.134\,663\,082$, i.e., $y^* = y^\odot \approx 0.747\,019\,953\,2$. The interior equilibrium is a sink for $y^* \in (y_1^\diamond, y^\odot)$ and a source for $y^* \in (y^\odot, y_2^\diamond)$. The numerical simulations suggest that the Hopf bifurcation is subcritical with unstable periodic solutions for y^* existing in some interval $(y^\triangleright, y^\odot)$ with $y^\triangleright$ being somewhere between 0.715 and 0.717 (Figure 10.2C,D). The stable manifold of the saddle $(0, y_2^\diamond)$ connects to the unstable periodic orbit (Figures 10.2D

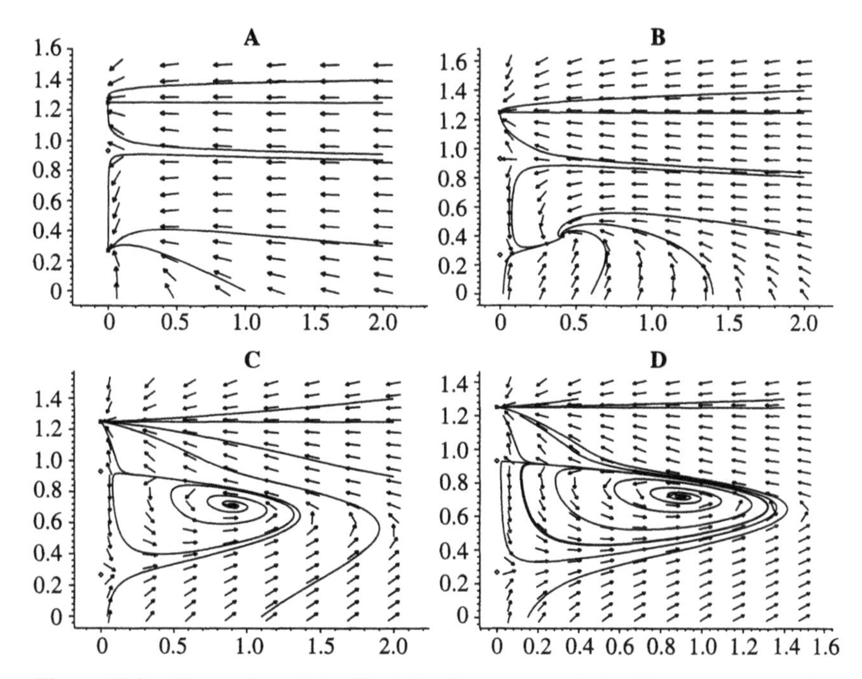

Figure 10.2. Phase-plane plots of body burden over population biomass concentration for various values of y^*. The units on both axes have no physical meaning. (**A**) $y^* = 0.2$. All solutions converge to the upper or lower boundary equilibrium (BE). (**B**) $y^* = 0.4$. All solutions converge to the upper BE or to the interior equilibrium (IE). (**C**) $y^* = 0.715$. As (**B**), but the IE is a spiral point. (**D**) $y^* = 0.72$. All solutions converge to the upper BE or to the IE. The domains of attraction are separated by a large unstable periodic orbit.

and 10.3E). For $y^* = y^{\triangleright}$, we suspect the existence of a heteroclinic orbit connecting the boundary equilibria $(0, y_1^{\diamond})$ and $(0, y_2^{\diamond})$ and surrounding the interior equilibrium.

The unstable periodic orbit separates the domains of attraction of the extinction equilibrium $(0, y_3^{\diamond})$ and the survival equilibrium (M^*, y^*) (Figures 10.2D and 10.3E). As in Theorem 10.20 (c) this is a bistable situation, but we have an Allee effect in addition: populations that start at low biomass concentrations die out even if the initial body burden is low.

It is interesting to follow the changes of the unstable manifold of the saddle $(0, y_2^{\diamond})$ as y^* varies from 0 to $y_2^{\diamond}$ (i.e., μ varies from ∞ to μ_-). First the unstable manifold looks like a rather straight line that first separates the domain of attractions of the extinction sinks $(0, y_1^{\diamond})$ and $(0, y_3^{\diamond})$ (Figure 10.2A) and then, for $y^* \in (y_1^{\diamond}, y^{\triangleright})$, of the stable survival equilibrium and the sink $(0, y_3^{\diamond})$ (Figure 10.2B). As y^* grows, the unstable manifold bends down, intersecting the positive M-axis with the intersection point moving towards the origin

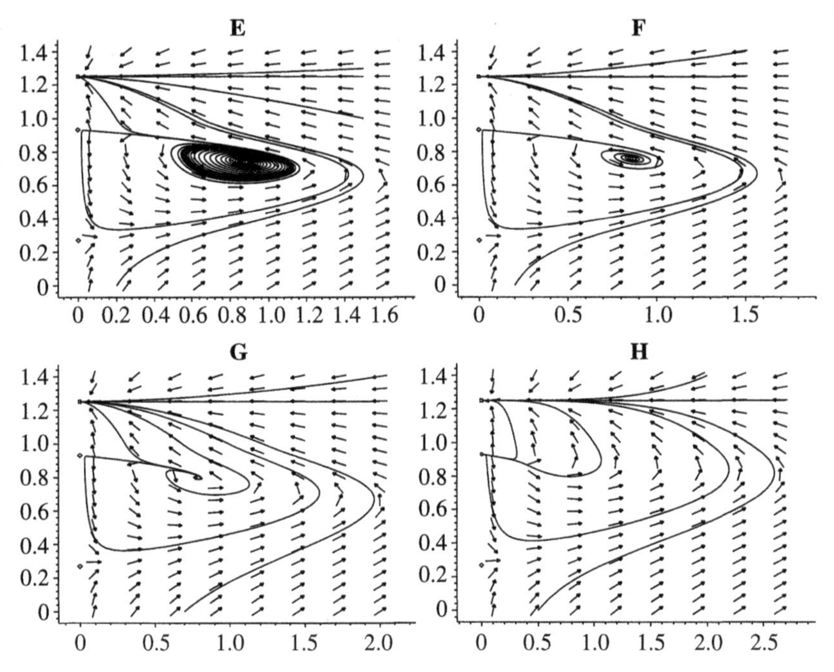

Figure 10.3. Continuation of Figure 10.2. (**E**) $y^* = 0.74$. As Figure 10.2D, but the periodic orbit has become smaller. (**F**) $y^* = 0.76$. All solutions except the IE converge to the upper BE. (**G**) $y^* = 0.8$. As (**F**), but the IE is no spiral point. (**H**) $y^* = 0.1$. All solutions converge to the upper BE.

(Figure 10.3C). At $y^* = y^{\triangleright}$ it jumps and becomes part of a heteroclinic orbit connecting $(0, y_1^{\diamond})$ and $(0, y_2^{\diamond})$. For $y \in (y^{\triangleright}, y^{\odot})$, it connects to the unstable period orbit (Figures 10.2D and 10.3E) and for $y \in (y^{\odot}, y_2^{\diamond})$ to the unstable survival equilibrium (Figure 10.3F,G). At $y^* = y_2^{\diamond}$ it stops existing because $(0, y_2^{\diamond})$ is a source for $y^* > y_2^{\diamond}$ (Figure 10.3H).

Summary

Even a rather simplistic model for an aquatic population in interaction with a toxicant displays quite a multifarious array of dynamics under a constant concentration of environmental toxin. While there is at most one survival equilibrium (interior equilibrium), there can be an odd number (greater than one) of extinction equilibria (boundary equilibria) where the population has died out under different body burdens of toxicant. Populations can survive at equilibrium or in periodic motion. In almost all scenarios, extinction is a definite possibility and occurs if the initial conditions are unfavorably chosen, and, even if a stable survival equilibrium exists, survival is only possible if the initial body burden is not too high. Even Allee effects are possible, where populations with low initial biomass concentrations and body burdens die out, while populations with intermediate initial biomass concentrations and body burdens survive. A disturbing scenario is the one with final extinction, where a considerable temporary build up of population biomass deceivingly distracts from the looming doom.

Exercises

10.6. Show that if $a \geqslant (y^\sharp)^2$, then, for all $\mu > 0$, there exists no interior equilibrium and all solutions of (10.9) converge towards the unique boundary equilibrium.

10.7. Assume that $\sigma > a > 0$.

(a) Show that $\mu_- < 0 < \mu_+$.

Assume in addition that $0 < \mu < \mu_+$.

(b) Show that the interior equilibrium exists and is locally asymptotically stable.

Hint: review the paragraphs on Hopf bifurcation before Theorem 10.20 and evaluate div $F(M^*, y^*)$ for $\mu = 0$.

(c) Show that all solutions of (10.9) starting with $M(0) > 0$, $y(0) \geqslant 0$ converge to the interior equilibrium.

10.8. Work through Section 10.7 again, this time with

$$\beta(M, y) = \frac{1}{(1+M)\xi(1+y)^\xi}, \quad \xi > 1.$$

In this case, it is possible that $y^* > y_3^\diamond$.

Bibliographic Remarks

A survey on the mathematical modeling of the effect of toxicants on aquatic populations has been given by Hallam *et al*. al. (1989). My interest in this topic was raised by the paper by Thomas et al. (1996) (see Navarova, Thieme 1998). A model that explicitly takes account of the dynamics of the food resource for the aquatic population has been analyzed by Nisbet et al. (1997). The effects of toxicant on individuals can be modeled in much more detail than we did here by dynamic energy budget models (Kooijman, 2000, Chapter 6) and can be lifted to the population level by structured population models (Metz, Diekmann, 1986; Tuljapurkar, Caswell, 1996; Cushing, 1998; Diekmann et al., 2001; see also Kooijman, 2000, Section 9.2.2, for further references).

Chapter Eleven

Population Growth Under Basic Stage Structure

Most population models consider two ways in which the population density couples back to population growth: by affecting the per capita birth rate or the per capita mortality rate. A third form of feedback lets high population density lower the rate of individual development, delaying the onset of reproduction. There are many ways to model this phenomenon (see the bibliographic remarks at the end of this chapter); the simplest consists of dividing the population into two parts, juveniles and adults, with the understanding that only adults can reproduce, and letting the per capita transition rate from the adult to the juvenile stage be density dependent. In this chapter we explore the dynamic consequences of dividing the population into two stages. As we have seen in Chapter 6, a one-species population with one stage only, which does not interact with the environment as it does in Chapter 10, tends towards an equilibrium. Here we will find that this picture does not change as long as the per capita stage transition rate does not depend on the number of juveniles. If it does, periodic solutions occur, meaning that the population stays in sustained oscillations. The existence of multiple steady states is also possible, and in a certain narrow parameter range quite a multitude of dynamics occur.

11.1 A Most Basic Stage-Structured Model

We split the population into juveniles and adults, with the understanding that juveniles do not reproduce,

$$N = L + A,$$

where L denotes the size of the juvenile population, A the size of the adult population, and N denotes the size of the total population. For the purposes of our model, these entities have to be understood as numbers or densities, not as biomass. We use the letter L because of its association with larvae as a special form of juveniles.

We assume that transition from the juvenile into the adult stage requires a certain minimum size. Such an assumption is valid, e.g., for the water flea *Daphnia magna*, where individuals typically have a length of 0.8 mm at birth and a length of 2.5 mm at the onset of reproduction (Kooijman, Metz, 1984). While the body size at metamorphosis is quite flexible for amphibians, a certain minimum size still seems to be required (Wilbur, Collins, 1973; Collins, 1979).

Anyway, it seems reasonable to assume that food scarcity will prolong the length of the juvenile period and that this effect is more pronounced than an increase of instantaneous juvenile mortality.

So, if there are more juveniles, then there is less food, which suggests that it will take longer for a single juvenile to transform into an adult. Let

$$f(L, A)$$

be the per capita transition rate from the juvenile into the adult stage if L is the size of the juvenile (e.g., larval) population and A the size of the adult population. In other words, if the juvenile and adult populations have constant sizes L and A, respectively, then the juvenile stage has an average length (death neglected) of $1/f(L, A)$. See formula (2.3) and the considerations thereafter. We have argued before that the length of the juvenile stage is an increasing function of the number of juveniles and of the number of adults; hence we assume that $f(L, A)$ is a nonincreasing nonnegative function of $L, A \geqslant 0$,

$$f(L, A) \geqslant f(\tilde{L}, \tilde{A}) \geqslant 0 \quad \text{whenever } \tilde{L} \geqslant L \geqslant 0, \ \tilde{A} \geqslant A \geqslant 0.$$

The model

Our simple two-stage model takes the following form:

$$\left.\begin{aligned} L' &= \beta(L, A)A - \mu(L, A)L - f(L, A)L, \\ A' &= f(L, A)L - \alpha(L, A)A. \end{aligned}\right\} \tag{11.1}$$

Here $\beta(A, L)$ is the per capita reproduction rate of an average adult individual if L and A are the sizes of the juvenile and adult populations, respectively, $\mu(L, A)$ the per capita mortality rate of a juvenile individual, and $\alpha(A, L)$ the per capita mortality rate of an adult individual.

Assumption 11.1. *$f, \alpha, \beta, \mu : [0, \infty)^2 \to (0, \infty)$ are continuous functions with the following additional properties:*

(1) *f, β are strictly positive, nonincreasing and continuously differentiable in both variables on $(0, \infty)^2$ with bounded derivatives.*

(2) *α, μ are strictly positive, nondecreasing and continuously differentiable in both variables on $(0, \infty)^2$ and the partial derivatives are bounded on every bounded subset of $(0, \infty)^2$.*

These assumptions express that there is competition within the population. If the rates f and μ only depend on L and the rates β and α only depend on A, we have pure *intrastage competition*, i.e., there is no competition between juveniles and adults. This may happen when the two stages have different resource requirements, like tadpoles and adult frogs. Theoretically, we can also consider *pure interstage competition*, where adults compete with juveniles and vice versa but adults and juveniles do not compete within their own stage; this is hardly a realistic scenario, however.

11.2 Well-Posedness and Dissipativity

It is a common belief in ecology that no given population can grow beyond a certain limit. Mathematically, that is reflected in the concept of dissipativity.

A system $x' = F(x)$ of ordinary differential equations is called *dissipative* and its solutions *uniformly eventually bounded*, if all solutions exist for all forward times and if there exists some $c > 0$ such that

$$\limsup_{t \to \infty} \|x(t)\| < c$$

for all solutions x. So, all solutions are bounded with the large-time bound $c > 0$ being independent of initial data. The time it takes until the norm of the solutions is below the bound c may depend on the initial data, however. The system $x' = F(x)$ is called *weakly dissipative* if all solutions exist for all forward times and if there exists some $c > 0$ such that

$$\liminf_{t \to \infty} \|x(t)\| < c$$

for all solutions x. Every finite system of ordinary differential equations that is weakly dissipative is also strongly dissipative (Corollary A.33).

Theorem 11.2. *Let Assumption 11.1 be satisfied, $L^\circ, A^\circ \geqslant 0$. Then there exists a unique solution to (11.1) on $[0, \infty)$ with initial data L°, A°. The solutions are nonnegative. The solutions are uniformly eventually bounded (i.e., system (11.1) is dissipative) if one of the following assumptions hold:*

(i) $\displaystyle \limsup_{A \to \infty} \frac{\beta(0, A)}{\alpha(0, A)} < 1.$

(ii) $\displaystyle \frac{f(L, 0)}{\mu(L, 0)} \to 0, \; L \to \infty.$

(iii) $\alpha(L, 0) \to \infty$ as $L \to \infty$ *and* $\displaystyle \limsup_{L \to \infty} \sup_{A \geqslant 0} \frac{\mu(L, A)}{\alpha(L, A)} < 1.$

Assumption (i) means that, at large adult population sizes, the average number of offspring produced by one adult throughout all its adult life drops below one. Recall that $1/\alpha(0, A)$ is the expected length of adult life if there are no juveniles and the size of the adult population is constantly at A. (See formula (2.3) and the considerations thereafter.) Assumption (ii) expresses that the quotient of per capita transition rate and the per capita juvenile mortality goes to 0 as the size of the juvenile population goes to infinity. Assumption (iii) has the interpretation that there is strong interstage competition at large juvenile population sizes with the effect that the per capita adult mortality rate outweighs the per capita juvenile mortality rate.

Proof. We apply Theorem A.4 in Toolbox A.1,

$$F_1(L, A) = \beta(L, A)A - \mu(L, A)L - f(L, A)L,$$
$$F_2(L, A) = f(L, A)L - \alpha(L, A)A.$$

If $A \geqslant 0$, $L = 0$, $F_1(A, L) = \beta(0, A)A \geqslant 0$, and, if $L \geqslant 0$, $A = 0$, $F_2(L, A) = f(L, 0)L \geqslant 0$.

By Theorem A.4 in Toolbox A.1, there exists some $b > 0$ and a unique nonnegative solution of (11.1) on $[0, b)$ such that, if $b < \infty$,

$$\limsup_{t\to\infty}(|L(t)| + |A(t)|) = \infty.$$

Adding all equations, we find for $N = L + A$ that

$$N' = \beta(L, A)A - \mu(L, A)L - \alpha(L, A)A \leqslant \beta(0, 0)A \leqslant \beta(0, 0)N.$$

Hence $N(t) \leqslant N(0)e^{-\beta(0,0)t}$ and the solutions are bounded on every finite interval. Thus $b = \infty$.

Let us now make the extra assumptions.

(i) There then exists some $\epsilon \in (0, 1)$ and some $N^\sharp > 0$ such that

$$\frac{\beta(L, A)}{\alpha(L, A)} \leqslant 1 - \epsilon \quad \text{whenever } A \geqslant N^\sharp.$$

This implies that

$$(\beta(L, A) - (1 - \epsilon)\alpha(L, A))A \leqslant 0 \quad \text{whenever } A \geqslant N^\sharp.$$

So, $(\beta(L, A) - (1 - \epsilon)\alpha(L, A))A \leqslant c$ for all $A \geqslant 0$, with some constant $c > 0$. Returning to our differential inequality,

$$N' \leqslant c - \mu(0, 0)L - \alpha(0, 0)\epsilon A \leqslant c - \delta N$$

with some $\delta > 0$. This implies that $N(t) \leqslant \max\{N(0), (c/\delta)\}$ and also that $\limsup_{t\to\infty} N(t) \leqslant (c/\delta)$. Recall that c and δ were independent of the initial data.

For the other assumptions, we will use the fact that a finite system of ordinary differential equations is dissipative whenever it is weakly dissipative (Corollary A.33).

(ii) Define $V = L + \xi A$, where ξ will be chosen later. Then

$$V' = (\beta(L, A) - \xi\alpha(L, A))A - \mu(L, A)L + f(L, A)(\xi - 1)L$$
$$\leqslant (\beta(0, 0) - \xi\alpha(0, 0))A + [f(L, A)(\xi - 1) - \mu(L, A)]L.$$

Choose $\xi = (\beta(0,0)/\alpha(0,0)) + 1$. Then

$$V' \leqslant -\alpha(0,0)A + \mu(L,A)\left((\xi - 1)\frac{f(L,A)}{\mu(L,A)} - 1\right)L.$$

Notice that $N \leqslant V \leqslant \xi N$. By assumption, there exists some $L^\sharp > 2/\mu(0,0)$ such that $(\xi - 1)(f(L,A)/\mu(L,A)) \leqslant \frac{1}{2}$ whenever $L \geqslant L^\sharp$. So, $V' \leqslant -\frac{1}{2}\mu(0,0)L^\sharp \leqslant -1$ if $L \geqslant L^\sharp$. Let $V > L^\sharp$, but $L \leqslant L^\sharp$. Then $A \geqslant (1/\xi)(V - L^\sharp)$ and

$$V' \leqslant \left[\frac{\alpha(0,0)}{\xi} + f(0,0)(\xi - 1)\right]L^\sharp - \frac{1}{\xi}V.$$

We find some $V^\sharp > L^\sharp$, such that $V' \leqslant -1$ whenever $V \geqslant V^\sharp$, $L \leqslant L^\sharp$.

Combining the two cases, we have $V' \leqslant -1$ whenever $V \geqslant V^\sharp$. This implies that $\liminf_{t\to\infty} V(t) \leqslant V^\sharp$; otherwise $V(t)$ would become negative in finite time, a contradiction. Since $\xi \geqslant 1$, $N(t) \leqslant V(t)$, and we have shown that the system is weakly dissipative.

(iii) is left as an exercise (Exercise 11.1). □

Proposition 11.3. *If the initial values L°, A° are nonnegative and at least one of them is strictly positive, then L, A are strictly positive on $(0, \infty)$.*

11.3 Equilibria and Reproduction Ratios

Recall that an equilibrium is a time-independent solution $L(t) \equiv L^*$, $A(t) \equiv A^*$ of system (11.1). The origin $(0, 0)$ is always an equilibrium of (11.1), called the trivial or extinction equilibrium. Any nontrivial equilibrium satisfies $L^* > 0$ and $A^* > 0$, and the determinant of the matrix

$$\begin{pmatrix} -(\mu(L^*,A^*) + f(L^*,A^*)) & \beta(L^*,A^*) \\ f(L^*,A^*) & -\alpha(L^*,A^*) \end{pmatrix}$$

is zero, i.e.,

$$\alpha(L^*,A^*)(\mu(L^*,A^*) + f(L^*,A^*)) = f(L^*,A^*)\beta(L^*,A^*).$$

There are various insightful ways of rewriting this equation. The following is mathematically useful,

$$\frac{\beta(L^*,A^*)}{\alpha(L^*,A^*)} = 1 + \frac{\mu(L^*,A^*)}{f(L^*,A^*)}, \tag{11.2}$$

while

$$1 = \frac{\beta(L^*,A^*)}{\alpha(L^*,A^*)}\frac{f(L^*,A^*)}{\mu(L^*,A^*) + f(L^*,A^*)} =: \mathcal{R}(L^*,A^*) \tag{11.3}$$

has the biological interpretation that the reproduction ratio is one at equilibrium. The *reproduction ratio* is the average number of offspring produced by one typical individual during its life. Notice that the first factor in the last formula is the number of offspring produced by one individual during adulthood, while the second gives the probability of surviving the juvenile stage (in analogy to (2.4)). $\mathcal{R}$ is a decreasing function of both variables, the number of juveniles and the number of adults.

A second equilibrium relation is obtained from the second equation in (11.1),

$$\alpha(L^*, A^*)A^* = L^* f(L^*, A^*). \tag{11.4}$$

11.4 Basic Reproduction Ratios and Threshold Conditions for Extinction versus Persistence

$\mathcal{R}_0 = \mathcal{R}(0, 0)$, with $\mathcal{R}$ from (11.3), is called the *basic reproduction ratio* and is the average number of offspring produced by one typical individual during its life if there is no intraspecies competition.

Theorem 11.4.

(a) *If $\mathcal{R}(L, A) < 1$ whenever $L, A > 0$ (in particular if $\mathcal{R}_0 < 1$), the population dies out.*

(b) *If $\mathcal{R}_0 > 1$, the total population size is uniformly persistent, i.e., there exists some $\epsilon > 0$ such that*

$$\liminf_{t\to\infty}[L(t) + A(t)] \geqslant \epsilon$$

for all solutions of (11.1) with $L(0) + A(0) > 0$.

Proof. (a) Let $V = L+\xi A$ with $\xi = \beta(0, 0)/\alpha(0, 0)$. It follows from Assumption 11.1 that V is a Lyapunov function; furthermore, $\dot{V} = 0$ if and only if $\mathcal{R}(L, A) = 1$. Since V is nonincreasing, it follows that every orbit is bounded in forward time. By the Lyapunov–LaSalle theorem, every ω-limit set is a connected invariant set contained in $\{L = 0\} \cup \{A = 0\}$. As the solutions flow, every ω-limit set is even contained in $\{L = 0\} \cap \{A = 0\}$ and so is the singleton containing the origin.

(b) Guided by persistence theory (Toolbox A.5), we let $\rho(L, A) = L + A$.

Step 1. The system (11.1) is uniformly weakly ρ-persistent.

Since $\mathcal{R}_0 > 1$, we find some $\epsilon > 0$ such that $\mathcal{R}(\epsilon, \epsilon) > 1$. Assume that $L(t) + A(t) \leqslant \epsilon$ for all $t \geqslant 0$. Set $V = L + \xi A$ with $\xi > 1$ to be determined later. Then

$$\dot{V} \geqslant [\beta(\epsilon, \epsilon) - \xi\alpha(\epsilon, \epsilon)]A + [f(\epsilon, \epsilon)(\xi - 1) - \mu(\epsilon, \epsilon)]L.$$

Since $\mathcal{R}(\epsilon, \epsilon) > 1$, we have

$$[f(\epsilon, \epsilon)(\xi_0 - 1) - \mu(\epsilon, \epsilon)] > 0$$

for $\xi_0 = \beta(\epsilon, \epsilon)/\alpha(\epsilon, \epsilon)$. In particular, $\xi_0 > 1$. So, we can choose $\xi \in (1, \xi_0)$ such that, in the last inequality for $\dot{V}$, both expressions in square brackets are positive. Hence there exists some $\delta > 0$ such that

$$\dot{V} \geqslant \delta V$$

and V increases exponentially, a contradiction.

Step 2. It follows from persistence theory (Theorem A.32) that system (11.1) is uniformly strongly ρ-persistent, i.e., the assertion holds. □

Corollary 11.5. *Let $\mathcal{R}_0 > 1$ and let one of the assumptions (i)–(iii) in Theorem 11.2 hold. Then the population is permanent, i.e., there exist constants $c > \epsilon > 0$ such that*

$$\epsilon \leqslant \liminf_{t\to\infty} L(t) \leqslant \limsup_{t\to\infty} L(t) \leqslant c,$$
$$\epsilon \leqslant \liminf_{t\to\infty} A(t) \leqslant \limsup_{t\to\infty} A(t) \leqslant c$$

for all nontrivial solutions L, A of (11.1). Furthermore, there exists an interior equilibrium.

Proof. By Theorems 11.2 and 11.4, we can assume that

$$\epsilon_0 \leqslant N(t) \leqslant c_0 \quad \forall t \geqslant 0.$$

Hence

$$A' \geqslant (N - A) f(N, N) - \alpha(N, N)A \geqslant (\epsilon_0 - A)\epsilon_1 - c_1 A$$

for appropriate constants $\epsilon_1 > 0$ and $c_1 > 0$. By the "fluctuation method" (Proposition A.22),

$$A_\infty \geqslant \frac{\epsilon_0 \epsilon_1}{c_1 + \epsilon_1}.$$

A similar statement of L_∞ is now obtained easily. Existence of a nontrivial equilibrium follows from the Poincaré–Bendixson theorem. □

11.5 Weakly Density-Dependent Stage-Transition Rates and Global Stability of Nontrivial Equilibria

We will show in this section that the dynamics of our simple stage-structured system converge to a uniquely determined nontrivial equilibrium, if there is only a weak negative feedback from the number of juveniles to the stage transition rate, i.e., if $Lf(L, A)$ is a nondecreasing function of L.

The stability of an equilibrium of an ordinary differential equation is related to the Jacobian matrix of the associated vector field (Toolbox A.4). The Jacobian matrix of the vector field associated with system (11.1) has the form

$$J(L, A) = \begin{pmatrix} -(\mu + \mu_L L + f + f_L L) + \beta_L A & \beta + \beta_A A - \mu_A L - f_A L \\ f + f_L L - \alpha_L A & f_A L - (\alpha + \alpha_A A) \end{pmatrix},$$

where f_L, f_A, etc., denote the partial derivatives with respect to L and A and the arguments of β, μ, f, α have been omitted. The stability of an equilibrium is related to the eigenvalues λ_1 and λ_2 of the Jacobian matrix evaluated at the equilibrium, and the eigenvalues in turn satisfy the relations

$$\lambda_1 + \lambda_2 = \text{trace}\, J(L^*, A^*), \qquad \lambda_1 \lambda_2 = \det J(L^*, A^*).$$

Recall that the trace of the matrix is the sum of the entries in the main diagonal. For our model,

$$\text{trace}\, J(L, A) = -(\mu + \mu_L L + f + f_L L) + \beta_L A + f_A L - (\alpha + \alpha_A A), \tag{11.5}$$

$$\begin{aligned} \det J(L, A) = {} & (\mu + \mu_L L + f + f_L L - \beta_L A)(\alpha + \alpha_A A - f_A L) \\ & - (\beta + \beta_A A - \mu_A L - f_A L)(f + f_L L - \alpha_L A). \end{aligned} \tag{11.6}$$

At the origin (the extinction equilibrium)

$$\begin{aligned} \text{trace}\, J(0, 0) &= -(\mu + f + \alpha)(0, 0) < 0, \\ \det J(0, 0) &= [(\mu + f)\alpha - f\beta](0, 0). \end{aligned}$$

Notice that $\det J(0, 0)$ has the same sign as $1 - \mathcal{R}_0$. Multiplying the terms in the determinant out,

$$\begin{aligned} \det J(L, A) = {} & (\mu + f)\alpha - \beta f + (\mu + f + f_L L)\alpha_A A \\ & - (\mu + \mu_L L - \beta_L A) f_A L + (\mu_L L - \beta_L A)\alpha \\ & + f_L L(\alpha - \beta) + (\mu_A L - \beta_A A)(f + f_L L) + \beta \alpha_L A \\ & + (\mu_L \alpha_A - \mu_A \alpha_L) LA + (\beta_A \alpha_L - \beta_L \alpha_A) A^2 - f_A \alpha_L AL. \end{aligned} \tag{11.7}$$

Theorem 11.6. *Let* $\mu(L, A) + \alpha(L, A) + f(L, A) + L f_L(A, L) > 0$ *for all* $L, A > 0$ *and let* $\alpha_A + \alpha_L + \mu_A + \mu_L - \beta_A - \beta_L - f_A - f_L$ *be strictly positive on* $(0, \infty)^2$.

(a) *There then exist no periodic orbits and no cyclic chains of equilibria.*

(b) *If, in addition,*

$$\beta_L \alpha_A \leqslant \beta_A \alpha_L, \qquad \mu_A \alpha_L \leqslant \mu_L \alpha_A,$$

then all interior equilibria are locally asymptotically stable.

(c) *Let, for every A,*

$$\frac{Lf(L, A)}{\alpha(L, A)} \quad \textit{be strictly increasing in } L > 0.$$

Then there exists at most one interior equilibrium.

(d) *If we finally add to the assumptions in (b) or to the assumptions in (c) one of assumption (i)–(iii) of Theorem 11.2, all bounded forward orbits converge towards the origin (the extinction equilibrium) or all nontrivial bounded forward orbits converge towards a uniquely determined interior equilibrium (the persistence equilibrium). Extinction occurs if*

$$\mathcal{R}_0 = \frac{\beta(0, 0)}{\alpha(0, 0} \frac{f(0, 0)}{\mu(0, 0) + f(0, 0)} \leqslant 1,$$

and persistence occurs if $\mathcal{R}_0 > 1$.

Notice that $\alpha_A + \alpha_L + \mu_A + \mu_L - \beta_A - \beta_L - f_A - f_L$ is always nonnegative by Assumption 11.1. The second assumption in part (b) is satisfied if α does not depend on L, i.e., the per capita mortality of adults is not affected by the density of juveniles. The result in this theorem shows that in our model the per capita stage transition rate is the only factor the density dependence of which can be made so strong that it would possibly lead to undamped oscillations. This message depends on our choice of model; in other models (e.g., models with a fixed rather than an exponentially distributed duration of the juvenile period (Theorem 16.4) or models with strong seasonality (Chapter 9)), density-dependent birth rates can cause undamped oscillations. But one can cautiously conclude that density dependence in the stage transition rate seems to have a higher potential to lead to undamped oscillations than density dependence in the birth rate or in the mortality rates. From the trace of the Jacobian matrix, equation (11.5), one can make the observation that interstage competition seems to have a dampening effect because it makes the trace more negative.

We also observe that if there is only intrastage competition and only weak density dependence in the stage transition rate, then there is no bistability, i.e., there are no multiple equilibria; in fact, all dynamics lead to a globally stable interior equilibrium.

Proof of Theorem 11.6. (a) Nonexistence of periodic orbits and cyclic chains of equilibrium follows from Bendixson's criterion (Theorem A.12). Notice that the divergence of the vector field (the trace of its Jacobian matrix) is strictly negative in $(0, \infty)^2$.

(b) At a nontrivial equilibrium, $(\mu + f)\alpha - \beta f = 0$ and $\beta > \alpha$, so our extra assumptions imply that all nontrivial equilibria are locally asymptotically stable because the trace of the Jacobian matrix is negative and the determinant positive.

(c) Under this assumption there exists a strictly increasing continuous function $\phi : [0, \infty) \to [0, \infty)$ such that $A^* = \phi(L^*)$ satisfies (11.4), i.e.,

$$\phi(L)\alpha(L, \phi(L)) = Lf(L, \phi(L)) \quad \forall L \geqslant 0.$$

Substituting this into (11.2) we see that the L-component of every interior equilibrium satisfies

$$1 + \frac{\mu(L, \phi(L))}{f(L, \phi(L))} - \frac{\beta(L, \phi(L))}{\alpha(L, \phi(L))} = 0.$$

Since the left-hand side of this equation is strictly decreasing, there is at most one solution.

(d) Let us assume that the assumption in (b) and one of assumptions (i)–(iii) in Theorem 11.2 hold such that all solutions are bounded in forward time. By the Poincaré–Bendixson theorem, every ω-limit set contains an equilibrium or is a periodic orbit. Since all nontrivial equilibria are locally asymptotically stable and periodic orbits do not exist, every orbit converges towards an interior equilibrium or its ω-limit set contains the origin. In the second case, the ω-limit set actually consists of the origin; otherwise a homoclinic orbit connects the origin to itself, which is again ruled out by Bendixson's criterion. If $\mathcal{R}_0 \leqslant 1$, our extra assumptions imply that $\mathcal{R}(L, A) < 1$ whenever $L, A > 0$ and extinction follows from Theorem 11.4 (a). If $\mathcal{R}_0 > 1$, we have permanence in the sense of Corollary 11.5, i.e., there exists a compact set in $X = (0, \infty)^2$ which attracts all orbits starting in X (see Theorems 11.2 and 11.4). To show the uniqueness of nontrivial equilibria, we apply Theorem A.37. The minimal sets of planar ordinary differential equation systems are equilibria and periodic orbits: equilibria in our case. Since all interior equilibria are locally asymptotically stable, so are all minimal sets in X. By Theorem A.37, there exists a globally asymptotically stable minimal set for X. Since every nontrivial orbit enters X, there is a locally asymptotically stable equilibrium that attracts all nontrivial orbits.

The proof is similar, actually easier, if the assumption in (c) and one of assumptions (i)–(iii) in Theorem 11.2 hold (Exercise 11.2). □

11.6 The Number and Nature of Possible Multiple Nontrivial Equilibria

If we allow strong negative feedback from the number of juveniles to the stage transition rate (i.e., $Lf(L)$ is not monotone increasing in $L \geqslant 0$), several nontrivial equilibria may exist. The monotonicity structure of equation (11.4) (see

Assumption 11.1) shows that different equilibria have different L-components. So, we can number the nontrivial equilibria according to the size of the L-component as $L_1 < L_2 < \cdots$. Strengthening our Assumption 11.1 only slightly, this implies that the A components satisfy $A_1 > A_2 > \cdots$. This reverse order of the juvenile and adult components of nontrivial equilibria also holds if there is no competition between juveniles and adults. Furthermore, it turns out that generically we have an alternate occurrence of sources and sinks on the one hand and of saddles on the other hand. Finally, under natural conditions, the number of nontrivial equilibria is odd, provided there are any.

Theorem 11.7. *Let $\mathcal{R}_0 > 1$ and let the nontrivial equilibria $(L_1, A_1), \ldots,$ be numbered as just indicated.*

(a) *If the function $(\mu/f) - (\beta/\alpha)$, which is nondecreasing in both arguments by assumption, is actually strictly increasing in both arguments, then $A_1 > A_2 > \cdots$.*

Assume that every interior equilibrium has nonzero Jacobian.

(b) *The odd-numbered equilibria are then either sources or sinks, while the even-numbered equilibria are saddles.*

(c) *Furthermore, the number of interior equilibria is odd, provided that at least one of the following two assumptions holds.*

 (i) $f(L, A)/\mu(L, A) \to 0$ *as* $L, A \to \infty$.

 (ii) $\beta(L, A)/\alpha(L, A) \to 0$ *as* $L, A \to \infty$.

Proof. (a) This follows from (11.2).

(b) We rewrite our system as

$$L' = G(L, A), \qquad A' = H(L, A).$$

It follows from Assumption 11.1 that the partial derivatives exist and are continuous on $(0, \infty)^2$. Furthermore, $H(0, 0) = 0$, $H(L, 0) > 0$ for $L > 0$, $H(L, A) \to -\infty$ as $A \to \infty$ and $H_A(L, A) < 0$. Hence there exists a continuous function $\phi : [0, \infty) \to [0, \infty)$ which is continuously differentiable on $(0, \infty)$ such that $\phi(0) = 0$, $H(L, \phi(L)) = 0$, and the equilibria of our system correspond to the zeros of the function ψ:

$$\psi(L) = G(L, \phi(L)).$$

Differentiating, we obtain

$$\begin{aligned}\psi'(L) &= G_L(L, \phi(L)) + G_A(L, \phi(L))\phi'(L),\\ 0 &= H_L(L, \phi(L)) + H_A(L, \phi(L))\phi'(L).\end{aligned}$$

Solving the second equation for $\phi'(L)$ and substituting the result in the first yields

$$\psi'(L) = \frac{\det J(L, A)}{H_A(L, A)}, \qquad A = \phi(L).$$

This means that $\psi'(L)$ has the opposite sign of $\det J(L, A)$, $A = \phi(L)$. Since $\mathcal{R}_0 > 1$,

$$\psi(0) = 0, \qquad \psi'(0) = \frac{\det J(0,0)}{H_A(0,0)} > 0.$$

By assumption, $\det J(L, A) \neq 0$ at every interior equilibrium, so $\psi'(L) \neq 0$ for every zero L of ψ. Since ψ is positive between 0 and the first positive zero, L_1, we have $\psi'(L_1) < 0$, i.e., $\det J(L_1, A_1) > 0$ at the first interior equilibrium, which is a sink or a source because both eigenvalues either have the same sign or are complex conjugates. Now ψ is strictly negative between L_1 and its second positive zero, L_2, so $\psi'(L_2) > 0$, i.e., $\det J(L_1, A_1) < 0$ at the second equilibrium, which is a saddle because the two eigenvalues have opposite signs. Continuing this reasoning, we find that all odd-numbered equilibria are sinks or sources and all even-numbered equilibria are saddles.

(c) Finally, we find that the number of nontrivial equilibria is odd if $\psi(L) < 0$ for all sufficiently large $L > 0$. Suppose that the latter is not true. Then there exists a sequence $L_n \to \infty$, $n \to \infty$, such that $\psi(L_n) \geqslant 0$ for all $n \in \mathbb{N}$. Notice that

$$\left.\begin{aligned} \psi(L_n) &= \left[f(L_n, A_n)\left(\frac{\beta(L_n, A_n)}{\alpha(L_n, A_n)} - 1\right) - \mu(L_n, A_n)\right] L_n, \\ A_n\alpha(L_n, A_n) &= f(L_n, A_n) L_n. \end{aligned}\right\} \tag{11.8}$$

Since $\psi(L_n) \geqslant 0$, we see from the first equation in (11.8) that

$$\frac{\beta(L_n, A_n)}{\alpha(L_n, A_n)} \geqslant 1,$$

in particular $\alpha(L_n, A_n)$ is bounded; furthermore, $\mu(L_n, A_n)$ is bounded and $f(L, A_n)$ is bounded away from 0. The second equation in (11.8) now implies that $A_n \to \infty$ as $n \to \infty$. So, neither (i) nor (ii) holds, a contradiction. □

11.7 Strongly Density-Dependent Stage-Transition Rates and Periodic Oscillations

We now turn to the case that the stage transition rate $Lf(L)$ is not nondecreasing in L, i.e., there is a strong negative feedback from the number of juveniles. To make our point that density-dependent development rates may cause oscillations as clear as possible, we assume that the stage transition rate does not depend on the number of adults and that all other per capita rates are positive constants.

This means that only the juvenile stage is subject to (intrastage) competition. The equations specialize to the following form:

$$L' = \beta A - \mu L - f(L)L,$$
$$A' = f(L)L - \alpha A.$$

We also assume that $f(L) \to 0$ as $L \to \infty$ such that all solutions are bounded in forward time (Theorem 11.2 (ii)). By assuming that $f'(L) < 0$ for all $L > 0$, a nontrivial equilibrium is uniquely determined by equation (11.2):

$$\frac{\beta}{\alpha} = 1 + \frac{\mu}{f(L^*)}. \tag{11.9}$$

The trace and determinant of the Jacobian matrix of the vector field at the nontrivial equilibrium are given by (11.5) and (11.7) as

$$\text{trace } J(L^*, A^*) = -(\mu + \alpha + f(L^*) + f'(L^*)L^*),$$
$$\det J(L^*, A^*) = f'(L^*)L^*(\alpha - \beta).$$

Remember that $(\mu + f)\alpha - \beta f = 0$ at a nontrivial equilibrium. Since $f'(L^*) < 0$, $\det J(L^*, A^*) > 0$ at every interior equilibrium. An unstable nontrivial equilibrium can now easily be constructed by fixing some $L^* > 0$ with $f(L^*) + f'(L^*)L^* < 0$ and choosing μ, α small enough and then determining β from (11.9) to actually make L^* the first component of the nontrivial equilibrium.

Theorem 11.8. *Assume that $f(L) \to 0$ as $L \to \infty$. Let $L^* > 0$ such that $f(L^*) + L^* f'(L^*) < 0$. Then the positive parameters α, β, μ can be chosen such that L^* becomes the first coordinate of a (unique) nontrivial equilibrium that is an unstable spiral point. Every solution that does not start at the nontrivial equilibrium or the origin converges towards a periodic orbit. In particular, there exists an orbitally stable periodic orbit (which is asymptotically orbitally stable if f is real analytic).*

Proof. The first part follows from the Poincaré–Bendixson theorem, the second from Theorem A in Zhu, Smith (1994) (see Theorem A.15). □

Bifurcation and Multiplicity of Periodic Orbits

In Theorem 11.8 we have learned that, with proper adjustment of the parameters, $L^* > 0$ can become the first coordinate of an unstable spiral point if

$$f(L^*) + L^* f'(L^*) < 0.$$

In particular, there exists a periodic orbit. Though there are global conditions for a periodic orbit to be (orbitally) stable (see Proposition A.38), they seem to be difficult to apply here in general. A local approach is offered by the theory of

Hopf bifurcation, in particular by the concepts of *supercritical* and *subcritical* bifurcation with the first leading to locally asymptotically stable periodic orbits and the second to unstable ones. Applying the criteria in the literature requires transformation of the equation to a certain normal form in which the Jacobian matrix at the bifurcation point has zeros in the main diagonal. In general this can be quite tedious and requires calculation of the eigenvectors of the Jacobian matrix to introduce a new coordinate system. Luckily, we do not have to do it in our case; again the function V, which served as a Lyapunov function under other circumstances, will be of great help. The point is that $\dot{V}$ only depends on L.

So, we rewrite the system in terms of the dependent variables $V = A + (\alpha/\beta)L$ and L:

$$V' = [(\beta - \alpha)f(L) - \mu\alpha]\frac{L}{\beta},$$
$$L' = \beta V - (\mu + \alpha)L - Lf(L).$$

We introduce

$$u = \frac{\beta}{\gamma}V,$$

where γ will be chosen appropriately. Then

$$\left.\begin{aligned} u' &= [(\beta - \alpha)f(L) - \mu\alpha]\frac{L}{\gamma}, \\ L' &= \gamma u - (\mu + \alpha)L - Lf(L). \end{aligned}\right\} \tag{11.10}$$

Let L^* be the first coordinate of the nontrivial steady state. Then the Jacobian matrix of the vector fields in (11.10) takes the form

$$\begin{pmatrix} 0 & (\beta - \alpha)L^* f'(L^*)(1/\gamma) \\ \gamma & -(\mu + \alpha) - f(L^*) - L^* f'(L^*) \end{pmatrix}.$$

We keep L^* fixed and use either μ or α as a bifurcation parameter and adjust β via (11.9) such that L^* is the first coordinate of the nontrivial equilibrium. By Theorem A.40, we have a Hopf bifurcation when, at the bifurcation point, the Jacobian matrix has the form

$$\begin{pmatrix} 0 & -\gamma \\ \gamma & 0 \end{pmatrix}.$$

Therefore we choose

$$\gamma^2 = -(\beta - \alpha)L^* f'(L^*), \quad \gamma > 0. \tag{11.11}$$

With this choice, we have a Hopf bifurcation for $\mu + \alpha = -f(L^*) - L^* f'(L^*)$, and the bifurcating periodic solutions have a period of approximately $2\pi/\gamma$ as long as they are sufficiently close to the bifurcation point.

Let $F = (F^1, F^2)$ be the vector field associated with (11.10). Then the stability of the bifurcating periodic orbit is determined by the following number (see Toolbox A.7, (A.14)):

$$a = \gamma[F^1_{uuu} + F^1_{uLL} + F^2_{uuL} + F^2_{LLL}] \\ + [F^1_{uL}(F^1_{uu} + F^1_{LL}) - F^2_{uL}(F^2_{uu} + F^2_{LL}) - F^1_{uu}F^2_{uu} + F^1_{LL}F^2_{LL}]. \tag{11.12}$$

Here the subscripts indicate partial derivatives. The partial derivatives need to be evaluated at the bifurcation point. If $a < 0$, the bifurcating periodic orbits are locally asymptotically stable (*supercritical bifurcation*); if $a > 0$, they are unstable (*subcritical bifurcation*). While the evaluation of a can be quite excruciating in general, most of the partial derivatives are 0 in our case and (11.12) becomes

$$a = -\gamma[f(L)L]_{LLL} + \frac{\beta - \alpha}{\gamma}([Lf(L)]_{LL})^2.$$

Theorem 11.9. *Assume that $f(L) \to 0$ as $L \to \infty$. Let $L^* > 0$ be such that*

$$f(L^*) + L^* f'(L^*) < 0.$$

Then we have a Hopf bifurcation at

$$\mu + \alpha = -(f(L^*) + L^* f'(L^*)), \qquad \beta = \alpha + \frac{\mu\alpha}{f(L^*)}.$$

The Hopf bifurcation is supercritical if the following number ξ is negative, and subcritical if $\xi > 0$:

$$\xi = Lf_L[f(L)L]_{LLL} + ([f(L)L]_{LL})^2,$$

evaluated at $L = L^*$. The bifurcating periodic solutions have a period of approximately $2\pi/\gamma$ with

$$\gamma = \sqrt{\alpha\mu\frac{\alpha + \mu + f(L^*)}{f(L^*)}},$$

as long as they are sufficiently close to the bifurcation point.

The formulas for ξ and γ follow from the formula for a (equation (11.12)) and from (11.11) and the relations at the Hopf point. Recalling that

$$\frac{\mu + f(L^*)}{f(L^*)} = \frac{\beta}{\alpha} := \mathcal{R}_A$$

is the average total number of offspring produced by one adult,

$$\gamma = \sqrt{\alpha^2\frac{\mu}{f(L^*)} + \alpha\mu\mathcal{R}_A}.$$

Noticing that

$$\frac{\mu}{f(L^*)} = \frac{\mu + f(L^*)}{f(L^*)} - 1 = \mathcal{R}_A - 1,$$

we have

$$\gamma = \alpha\sqrt{\mathcal{R}_A - 1 + r\mathcal{R}_A}, \qquad r = \frac{\mu}{\alpha},$$

where r is the ratio of the per capita mortality of juveniles over the per capita mortality of adults. Since $D_A := 1/\alpha$ is the average duration of adulthood (expectation of adult life), we obtain that the period of bifurcating solutions that are still close to the equilibrium is approximately

$$\text{period} \approx D_A \frac{2\pi}{\sqrt{\mathcal{R}_A - 1 + r\mathcal{R}_A}}.$$

Unfortunately, r is a parameter that is difficult to estimate. Often the mean sojourn time in the juvenile stage at equilibrium (including juvenile mortality), $S_L^* := 1/(\mu + f(L^*))$, will be much shorter than the average duration of adulthood, $D_A = 1/\alpha$. Then

$$\gamma \approx \sqrt{\alpha\mu\frac{\mu + f(L^*)}{f(L^*)}} = \sqrt{\alpha[\mu + f(L^*) - f(L^*)]\mathcal{R}_A}$$
$$= \sqrt{\alpha[\mu + f(L^*)]\left(1 - \frac{1}{\mathcal{R}_A}\right)\mathcal{R}_A}.$$

So, the period of the bifurcating periodic solutions will be close to

$$2\pi\sqrt{S_L^* D_A}\frac{1}{\sqrt{\mathcal{R}_A - 1}}.$$

We learn from Theorem 11.9 that we obtain a supercritical Hopf bifurcation, for example, if, when evaluated at L^*, we have

$$[Lf(L)]_L < 0, \qquad [Lf(L)]_{LL} = 0, \qquad [Lf(L)]_{LLL} > 0.$$

11.8 An Example for Multiple Periodic Orbits and Both Supercritical and Subcritical Hopf Bifurcation

In our model, a subcritical Hopf bifurcation will lead to multiple periodic orbits. Since the unique interior equilibrium is surrounded by an unstable periodic orbit, every trajectory that starts outside the unstable periodic orbit converges towards a limit cycle. This follows from the Poincaré–Bendixson theorem and the dissipativity of our system. The limit cycle must be different from the unstable periodic orbit, and therefore there exist at least two periodic orbits. Let the stage transition rate be of Ricker type, i.e., after scaling L and A,

$$f(L) = \eta e^{-L}.$$

Then

$$(Lf(L))_L = \eta e^{-L}(1 - L),$$
$$(Lf(L))_{LL} = -\eta e^{-L}(2 - L),$$
$$(Lf(L))_{LLL} = \eta e^{-L}(3 - L).$$

The number ξ in Theorem 11.9 has the same sign as

$$-L(3 - L) + (L - 2)^2 = 2L^2 - 7L + 4.$$

So, the Hopf bifurcation is supercritical when $L^* \in (1, L^\sharp)$, and subcritical when $L^* > L^\sharp$, where $L^\sharp = \frac{1}{4}(7 + \sqrt{17}) \approx 2.781$. Let us determine the periods at the Hopf bifurcation point. By Theorem 11.9, $\mu + \alpha = (L^* - 1)f(L^*)$. Substituting this into the formula for γ, we obtain $\gamma = \sqrt{\alpha\mu L^*}$. The per capita mortality of juveniles is a parameter that is hard to estimate because it will depend on what kind of and how many predators are around. So, we use the other relation at the bifurcation point to replace μ, $\mu = ((\beta/\alpha) - 1)\eta e^{-L}$. $D_L = 1/\eta$ is the average length of the juvenile period if there is no competition for resources (death neglected). Recall that $\mathcal{R}_A = \beta/\alpha$ is the average total number of offspring produced by one adult. This gives us

$$\text{period} \approx 2\pi\sqrt{D_A D_L}\frac{e^{L^*/2}}{\sqrt{(\mathcal{R}_A - 1)L^*}}.$$

Before we perform some simulations for supercritical Hopf bifurcation, let us scale our equation somewhat more. Scaling time, we can assume that $\alpha = 1$. This means that the expectation of adult life (average duration of adulthood) is our reference time unit. Furthermore, we set $A = \eta y$ and rename $L = x$ in order to be consistent with some of the labeling in the figures. Furthermore, we set $q = \mu/\eta$. Our system then takes the form

$$x' = \eta(\beta y - qx - e^{-x}x), \qquad y' = e^{-x}x - y.$$

In order to get a supercritical Hopf bifurcation, we choose $x^* = 2$. Then $y^* = 2e^{-2}$ and, by Theorem 11.9, $\beta = 1 + qe^2$. The trace of the Jacobian matrix at the equilibrium point becomes

$$\text{trace} = \eta(e^{-2} - q) - 1.$$

So, we can expect a periodic orbit if and only if $q < e^{-2}$ and $\eta > 1/(e^{-2} - q)$. In particular, there will be a periodic orbit only if $\beta \leqslant 2$. According to our scaling, β is the average total number of offspring by one adult. So, $\beta \leqslant 2$ is unrealistically low for an amphibian population, for example, but this is due to our choice of the nonlinearity f. For most of our simulations we choose $q = \frac{1}{2}\mathrm{e}^{-2}$; therefore, we will have periodic solutions for $\eta > 2\mathrm{e}^2$, i.e., for juvenile stages that are quite short compared to the adult stage. Furthermore, $\mu = q\eta > 1$; thus juveniles face at least the same per capita mortality as adults. In Figure 11.1, we present simulations for $\eta = 10, 20, 50, 400$. The last two values are chosen to present the phenomenon of a so-called *relaxation oscillation*. For large η, the solution tries to follow the *x-isocline*, i.e., the curve given by $x' = 0$, which is

$$y = \frac{1}{\beta}(q + \mathrm{e}^{-x})x.$$

For our choice of q, this curve first increases, then decreases, and eventually increases again. See the various parts of Figure 11.1 which show solutions of our planar system together with the x-isocline. The *y-isocline*, the curve given by $y' = 0$, is

$$y = \mathrm{e}^{-x}x,$$

which is above the x-isocline for small $x > 0$, but below it for large $x > 0$. The y-isocline is not displayed in Figure 11.1 to avoid confusion.

If η is large, at small x values, the trajectories of the solutions try to follow the first increasing segment of the x-isocline; they do so in an upward direction until they reach its local maximum, because here the y-isocline is above the x-isocline. After reaching the local maximum of the x-isocline, the trajectories move very fast to the second increasing segment of the x-isocline. They follow this segment in a downward direction until they reach its local minimum, because here the y-isocline lies below the x-isocline. From the local minimum of the x-isocline they rapidly move back to the first increasing segment, and the cycle closes. By Theorem 11.2 (i) or (ii) and the Poincaré–Bendixson theorem, the trajectories approach a periodic orbit, a limit cycle, which surrounds the local maximum and the local minimum of the x-isocline.

The form of the periodic solutions can depend on the parameters, as illustrated in Figure 11.2, which presents a comparison for $\eta = 20$ and $q = \frac{1}{2}\mathrm{e}^{-2}$ and $q = \frac{1}{10}\mathrm{e}^{-2}$.

11.9 Multiple Interior Equilibria, Bistability, and Many Bifurcations for Pure Intrastage Competition

At the end of Section 11.6 we mentioned the possibility of multiple interior equilibria if there is strong negative feedback from the numbers of juveniles to the stage transition rate. This alone is not sufficient, as can be seen from Section 11.7, where we still have a unique interior equilibrium. In this section we will also allow a negative feedback of the number of adults to the per capita birth rate β. Again we assume there is no interstage competition. This can happen if the juveniles (as larvae, for example) are quite different from the

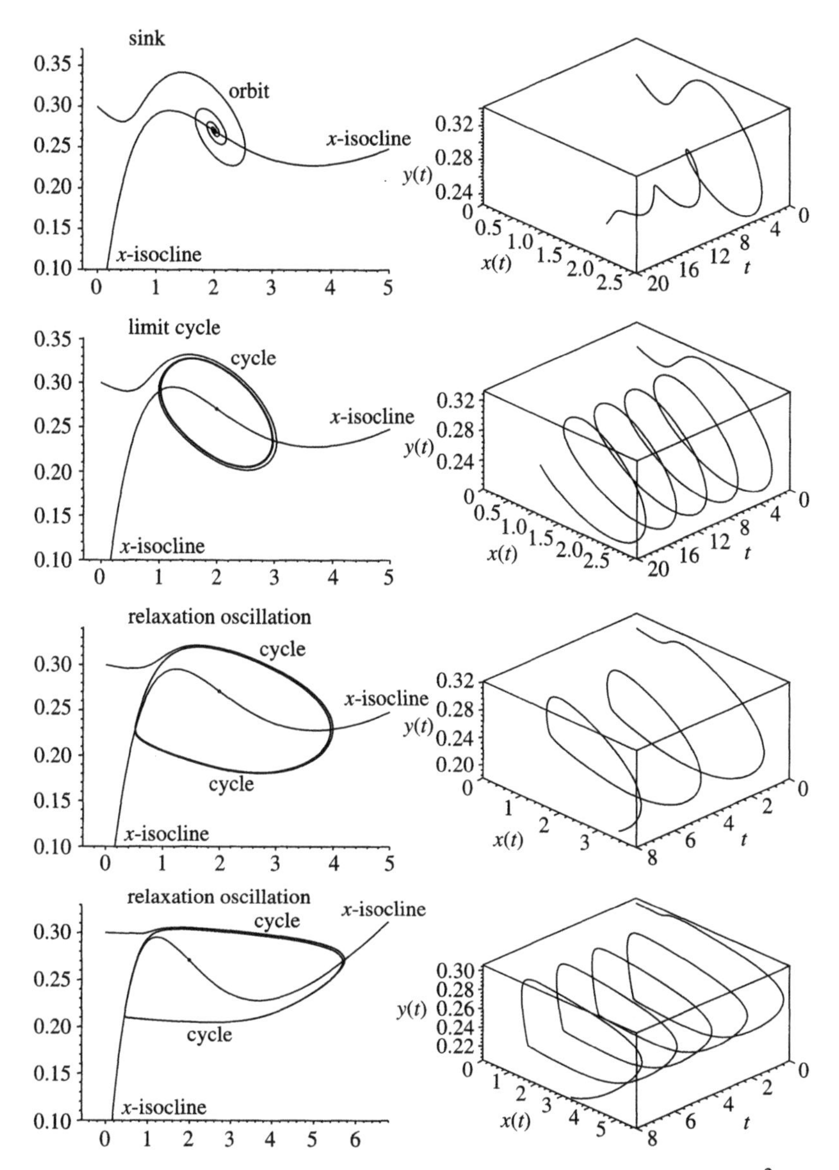

Figure 11.1. Illustrations for the example in Section 11.8 with $q = 0.5e^{-2}$. The first column shows phase-plane plots of adults over juveniles for various lengths of the juvenile period; the second column displays the associated three-dimensional plots with t denoting time, x juveniles, y adults. The x and y variables have been scaled, and their units have no physical meaning. The time unit is the expectation of adult life (average length of adulthood). Notice that the time windows are different. First row: $\eta = 10$. Second row: $\eta = 20$. Third row: $\eta = 50$. Fourth row: $\eta = 400$.

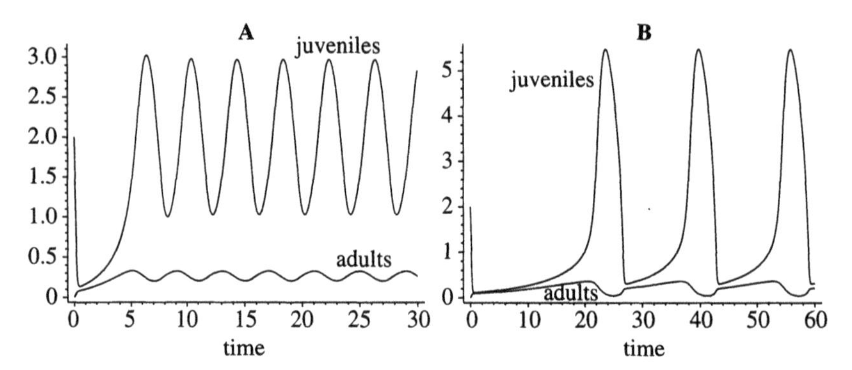

Figure 11.2. Illustration of how the form of periodic solutions may depend on parameters. (**A**) $q = 0.5e^{-2}$. (**B**) $q = 0.1e^{-2}$. In both (**A**) and (**B**), $\eta = 20$ such that (**A**) has the same parameters as the third row in Figure 11.1. The time unit is the expectation of adult life. Notice that the time windows are different.

adults and have different food requirements and/or if they have different habitats (e.g., the juveniles live in water and the adults live on land). We further assume that the per capita mortality rates are constant. This assumption is motivated by the idea that many species cover their maintenance costs first and only put the surplus into growth and/or reproduction such that competition for food would have an earlier and more profound impact on the per capita maturation and birth rates than on the per capita mortality rates.

Scaling time, we can assume that $\alpha = 1$, i.e., our time unit is the expectation of adult life (average length of the adult stage, see the derivation of (2.3)). Then the system takes the form

$$\begin{aligned} L' &= \beta(A)A - \mu L - f(L)L, \\ A' &= f(L)L - A. \end{aligned}$$

The solutions are uniformly eventually bounded by Theorem 11.2, because (i) or (ii) applies. Theorem 11.7 and Corollary 11.5 specialize to the following result.

Theorem 11.10. *Let* $\alpha = 1$ *and*

$$\zeta := \beta(0) - 1 - \frac{\mu}{f(0)}.$$

Assume that β *and* f *are strictly decreasing and* $f(L) \to 0$ *as* $L \to \infty$.

(a) *If* $\zeta \leqslant 0$, *the population dies out.*

(b) *If* $\zeta > 0$, *the population is permanent and there exists an interior equilibrium.*

If all interior equilibria have nonzero Jacobian determinants, their number is odd. If the equilibria (L_j^*, A_j^*) *are numbered* $L_1^* \leqslant L_2^* \leqslant \cdots$, *then* $L_1^* < L_2^* < \cdots$ *and* $A_1^* \geqslant A_2^* \geqslant \cdots$ *and the odd-numbered equilibria are sources or sinks, while the even-numbered ones are saddles.*

Remark. There is a fat set of parameters μ such that all interior equilibria have nonzero Jacobian determinants. This set is *fat* in a topological sense, namely that it contains a dense set which is the countable intersection of open sets, and in a measure theoretical sense as its complement has Lebesgue measure 0 (see Toolbox B.7).

Proof. (L^*, A^*) is an interior equilibrium if and only if

$$\mu = [\beta(L^* f(L^*)) - 1] f(L^*) := \tilde{\psi}(L^*),$$

and the Jacobian determinant at (L^*, A^*) is 0 if and only if $\tilde{\psi}'(L^*) = 0$. This means that all equilibria (L^*, A^*) have nonzero Jacobian determinants if and only if μ is a regular value of $\tilde{\psi}$, i.e., it is not a critical value of $\tilde{\psi}$. Apply Theorem B.42. □

The trace of the Jacobian matrix of the equilibrium (equation (11.5)) is

$$\operatorname{trace} J(L^*, A^*) = -(\mu + 1) - f(L^*) - L^* f'(L^*), \tag{11.13}$$

and the determinant (equation (11.7)) is

$$\begin{aligned} \det J(L^*, A^*) &= f_L L^*(1 - \beta) - \beta_A A^*(f + f_L L^*) \\ &= -f_L L^* \frac{\mu}{f} - \beta_A A^*(f + f_L L^*), \end{aligned} \tag{11.14}$$

where the functions and their partial derivatives need to be evaluated at the equilibrium point. Assume that $Lf(L)$ is unimodal (a function with one maximum at $L = L^\diamond$). Any equilibrium $L^* \leqslant L^\diamond$ will be locally asymptotically stable.

As an example, we choose a Ricker-type function for the stage transition rate,

$$f(L) = \frac{\mu}{m} e^{-L} \quad \text{and} \quad \beta(A) = g\left(\frac{m}{\mu} A\right),$$

with $m > 0$. The form e^{-L} as compared to, let us say, $e^{-\gamma L}$ presents no restriction of generality, as it is achieved by scaling, more precisely by introducing the new variables γL and γA and calling them L and A again. To interpret the parameter m, we recall that

$$p_0 = \frac{f(0)}{\mu + f(0)} = \frac{1}{1 + m}$$

is the probability of surviving the juvenile stage in the absence of any competition. The number ζ in Theorem 11.10 becomes

$$\zeta = g(0) - 1 - m.$$

In particular, there exists an interior equilibrium if and only if $m < g(0) - 1$. We introduce the new dependent variable

$$y = \frac{m}{\mu} A.$$

Then our system takes the form

$$\left.\begin{aligned} L' &= \frac{\mu}{m}[g(y)y - mL - Le^{-L}], \\ y' &= Le^{-L} - y. \end{aligned}\right\} \tag{11.15}$$

The divergence of the vector field and the Dulac criterion (Theorem A.12 in the toolbox) lead us to the following observation (Exercise 11.3).

Remark 11.11. Periodic orbits and heteroclinic/homoclinic orbits can only exist for

$$m < e^{-2} \approx 0.1353, \qquad \mu > \frac{m}{e^{-2} - m}.$$

To find the interior equilibrium we solve the second equation, with $y' = 0 = L'$, for y and substitute the result into the first:

$$m = [g(y^*) - 1]e^{-L^*}, \qquad y^* = e^{-L^*} L^*. \tag{11.16}$$

This suggests that we parametrize the equilibria over $L^* > 0$ and determine m as a function of L^*:

$$m = M(L^*) := [g(e^{-L^*} L^*) - 1]e^{-L^*}.$$

The trace of the Jacobian matrix at L^* is

$$\operatorname{trace} J(L^*, A^*) = \frac{\mu}{m}(-m + e^{-L^*}[L^* - 1]) - 1.$$

Depending on whether we choose μ or L^* as bifurcation parameter, there are two useful ways of expressing the trace. For μ,

$$\operatorname{trace} J(L^*, A^*) = \frac{\mu}{m} e^{-L^*}(L^* - 1 - me^{L^*}) - 1 = \frac{\mu}{m} e^{-L^*}(L^* - g(y^*)) - 1.$$

For L^* as bifurcation parameter,

$$\operatorname{trace} J(L^*, A^*) = \mu\left(-1 + \frac{1}{me^{L^*}}[L^* - 1]\right) - 1 = \mu\left(\frac{L^* - 1}{g(y^*) - 1} - 1\right) - 1.$$

The determinant of the Jacobian matrix at L^* is given by

$$\frac{m}{\mu} \det J(L^*, A^*) = m + e^{-L^*}[1 - L^*] - [g(y^*) + g'(y^*)y^*]e^{-L^*}[1 - L^*].$$

Substituting y^* and m,

$$\begin{aligned} e^{L^*}\frac{m}{\mu} \det J(L^*, A^*) &= g(y^*) - L^* - [g(y^*) + g'(y^*)y^*][1 - L^*] \\ &= [g(y^*) - 1]L^* + g'(y^*)y^*[L^* - 1], \end{aligned}$$

where $y^* = e^{-L^*}L^*$. Differentiating M shows that

$$\text{sign of } \det J(L^*, A^*) = -\text{sign of } M'(L^*).$$

Depending on the monotonicity properties of the function M, the following scenarios of Hopf bifurcations occur. It is not difficult to see that there exist numbers $L_2 \geqslant L_1 > 1$ such that M is strictly monotone decreasing on the intervals $[0, L_1]$ and $[L_2, \infty)$. The monotonicity of M fails for intermediate values if $g(e^{-1})$ is negative or close to 0. Notice that e^{-1} is the maximum of xe^{-x} on $[0, \infty)$.

Case 1. Let M be strictly monotone decreasing.

Then the Jacobian determinant is always positive. Considering L^* as a bifurcation parameter and keeping in mind that $y^* = e^{-L^*}L^* \to 0$ as $L^* \to \infty$, we see that the trace of the Jacobian matrix is negative for small L^* and positive for large L^*, so for fixed but arbitrary $\mu > 0$, there is at least one Hopf bifurcation in L^*.

Alternatively, we can choose μ as the bifurcation parameter. If $L^* > g(y^*) > 1$, which holds for large enough L^* (i.e., sufficiently small $m > 0$), we see that the trace of the Jacobian matrix changes from negative to positive as μ increases, again giving rise to a Hopf bifurcation.

Case 2. M is not monotone decreasing.

If μ is chosen as a bifurcation parameter, we make the same observation as in case 1: that we have a Hopf bifurcation if L^* is chosen sufficiently large. In terms of m, if $m > 0$ is chosen small enough, we have a solution with $L^* > g(y^*) > 1$ and $M'(L^*) < 0$. Again, for sufficiently small $m > 0$, there is a Hopf bifurcation with μ as bifurcation parameter.

If we choose L^* as bifurcation parameter, we notice as in case 1 that the trace of the Jacobian matrix changes its sign, but not necessarily at an L^* value where the determinant is positive. If μ is chosen sufficiently large, the trace changes sign for some $L^* \in (1, L_1)$ where the determinant is positive such that we have a Hopf bifurcation in this part of the (L^*, m) curve. If $\mu > 0$ is chosen sufficiently small, the trace changes sign at some $L^* > L_2$ where the determinant is again positive, leading to a Hopf bifurcation. For intermediate μ, the trace may switch sign for L^* in an interval on which the determinant is negative and no bifurcation occurs because the associated equilibrium remains a saddle.

For more specific considerations, we also let the birth rate be of Ricker type,

$$g(y) = \mathrm{e}^{(1/b)[a-y]},$$

with positive parameters a and b, and the system takes the form

$$\left.\begin{aligned} L' &= \frac{\mu}{m}[\mathrm{e}^{(1/b)(a-y)}y - mL - L\mathrm{e}^{-L}], \\ y' &= L\mathrm{e}^{-L} - y. \end{aligned}\right\} \tag{11.17}$$

The equilibrium equation can now be written as

$$a - b\ln(1 + me^{L^*}) = y^* = \mathrm{e}^{-L^*}L^*, \quad L^* > 0, \tag{11.18}$$

and its L-component, L^*, is the point of intersection of the graphs given by the left- and right-hand side expressions of this equation. Using a graphing device (like a computer with MAPLE), it is not difficult to come up with a choice of parameters for which there are three interior equilibria, henceforth called the left equilibrium, middle equilibrium, and right equilibrium according to the position of the L-coordinate in the LA-plane. Notice that the left-hand side of the equation represents a strictly decreasing, strictly concave function of L^* the graph of which can be easily shifted up and down by adjusting a. Let us show that there cannot be more than three equilibria.

Proposition 11.12. *For any parameter choice, there are at most three nontrivial equilibria of system (11.17). If there are three nontrivial equilibria, numbered such that $L_1^* < L_2^* < L_3^*$, then $L_1^* < 2$ and $L_2^* > 1$.*

Proof. If there are more than three solutions to (11.18), there must be at least five (counting multiplicities) by Theorem 11.7 and its proof. By the mean value theorem, there are four points $L > 0$ such that

$$\mathrm{e}^{-L}(1 - L) = -\frac{bm\mathrm{e}^L}{1 + m\mathrm{e}^L} = -\frac{bm}{\mathrm{e}^{-L} + m}.$$

These four L must lie in $(1, \infty)$. Again, by the mean value theorem, there are at least three $L > 1$ such that

$$\mathrm{e}^{-L}(L - 2) = -bm\frac{\mathrm{e}^{-L}}{(\mathrm{e}^{-L} + m)^2}.$$

These three L necessarily lie in $(1, 2)$. Simplifying, we have three $L \in (1, 2)$ such that

$$-bm = (L - 2)(\mathrm{e}^{-L} + m)^2.$$

So, there are at least two $L \in (1, 2)$ such that

$$0 = e^{-L} + m - (L-2)2e^{-L} = m - 2Le^{-L} + 5e^{-L},$$

implying that there is at least one $L \in (1, 2)$ such that

$$0 = 2Le^{-L} - 7e^{-L} = e^{-L}(2L - 7).$$

We obtain the contradiction $\frac{7}{2} = L \in (1, 2)$, so there are at most three nontrivial equilibria. The second statement follows from similar steps and Theorem 11.10. □

By Theorem 11.10, if there are three equilibria, the middle one is always a saddle while the other two are sources or sinks. It is possible to choose the parameters in such a way that the left equilibrium and the right equilibrium are both sources or both sinks or that the right equilibrium is a source and the left equilibrium a sink. To be more specific, we choose

$$a = 0.43 \quad \text{and} \quad b = 2.2.$$

Using MAPLE to analyze m as a function of L^* defined by (11.16), we find numbers $m_3 > m_2 > m_1 > 0$ with the following properties:

- There exists an interior equilibrium if and only if $m \in (0, m_3)$.
- There exists a unique interior equilibrium if and only if $m \in (m_2, m_3)$ or $m \in (0, m_1)$.
- There exist three interior equilibria if and only if $m \in (m_1, m_2)$.

The numerical values of these numbers are

$$m_3 \approx 0.215\,863\,526, \qquad m_2 \approx 0.102\,640\,576\,6, \qquad m_1 \approx 0.009\,513\,628\,742.$$

Notice that the m-interval for which three interior equilibria exist is very narrow. Furthermore, for m-values from this interval, the juvenile stage is so short that the probability of surviving it, $p_0 = 1/(1+m)$, is almost 1. Let us choose a specific value,

$$m = 0.01.$$

The L-coordinates of the associated equilibria are

$$L_1^* = 1.063\,855\,501, \qquad L_2^* = 1.565\,163\,982, \qquad L_3^* = 2.104\,560\,051,$$

the y-coordinates are

$$y_1^* = 0.367\,160\,599\,2, \quad y_2^* = 0.327\,203\,367\,4, \quad y_3^* = 0.256\,544\,380\,0.$$

We can use the traces of the Jacobian matrices and MAPLE to follow how the stability properties of the left equilibrium and the right equilibrium change as we vary μ. For small $\mu > 0$, both equilibria are locally asymptotically stable. The right equilibrium switches from stable to unstable at $\mu \approx 0.080\,227\,790\,38$, while the left equilibrium switches at $\mu \approx 0.830\,704\,300\,3$.

Let us follow the changes of the dynamics as μ varies from 0 to ∞ as revealed by a combination of AUTO bifurcation analysis (Figure 11.5), phase-plane arguments, and MAPLE simulations (Figures 11.3 and 11.4). The other parameters are still $m = 0.01, a = 0.43, b = 2.2$. For μ between 0 and 0.080 23 (approximately), both the left equilibrium and the right equilibrium are locally asymptotically stable, and the stable manifold of the middle equilibrium (which is a saddle for all $\mu > 0$) separates the domain of attraction. The two parts of the unstable manifold connect the saddle to the left equilibrium and the right equilibrium (Figure 11.3A,B).

At $\mu \approx 0.080\,23$, a supercritical Hopf bifurcation occurs from the right equilibrium, giving rise to a stable periodic orbit around the right equilibrium for μ between 0.080 23 and 0.081 33 (Figure 11.4). Again the domains of attraction of the left equilibrium and the periodic orbit are separated by the stable manifold of the middle equilibrium (Figure 11.3C).

At $\mu \approx 0.081\,33$, the periodic orbit is replaced by a homoclinic orbit connecting the middle equilibrium to itself and surrounding the right equilibrium (Figure 11.5). The homoclinic orbit attracts all solutions starting inside, while all solutions starting outside tend to the left equilibrium, except those starting at the origin and the other part of the stable manifold, which connect infinity to the middle equilibrium.

For μ between 0.081 33 and approximately 0.7900, the left equilibrium attracts all solutions starting in the first quadrant except the other equilibria and the stable manifold of the middle equilibrium (Figures 11.3D,E and 11.4A). One part of the stable manifold connects the right equilibrium to the middle one, while the other connects infinity to the middle equilibrium.

At $\mu \approx 0.7900$, there is a homoclinic orbit connecting the middle equilibrium to itself and surrounding the left equilibrium. As μ approaches this number from the right, the homoclinic orbit is the limit (from inside) of a branch of periodic orbits surrounding the left equilibrium (Figure 11.5). For μ larger than the approximate value of 0.7900, there are also periodic orbits surrounding all three equilibria. It is not quite clear at this point whether the two different types of periodic orbits appear at exactly the same value for μ. The branch of periodic orbits surrounding all three equilibria will persist for all greater μ and be locally orbitally stable with the possible exception of μ close to 0.7900 (Figure 11.3F). The branch of periodic orbits which just surround the first equilibrium will persist until approximately $\mu = 0.839\,07$ and be unstable, with the possible exception of μ close to 0.7900, and separate the domains of attraction between the locally stable first equilibrium and the orbitally stable periodic orbit surrounding all three equilibria (Figure 11.4B).

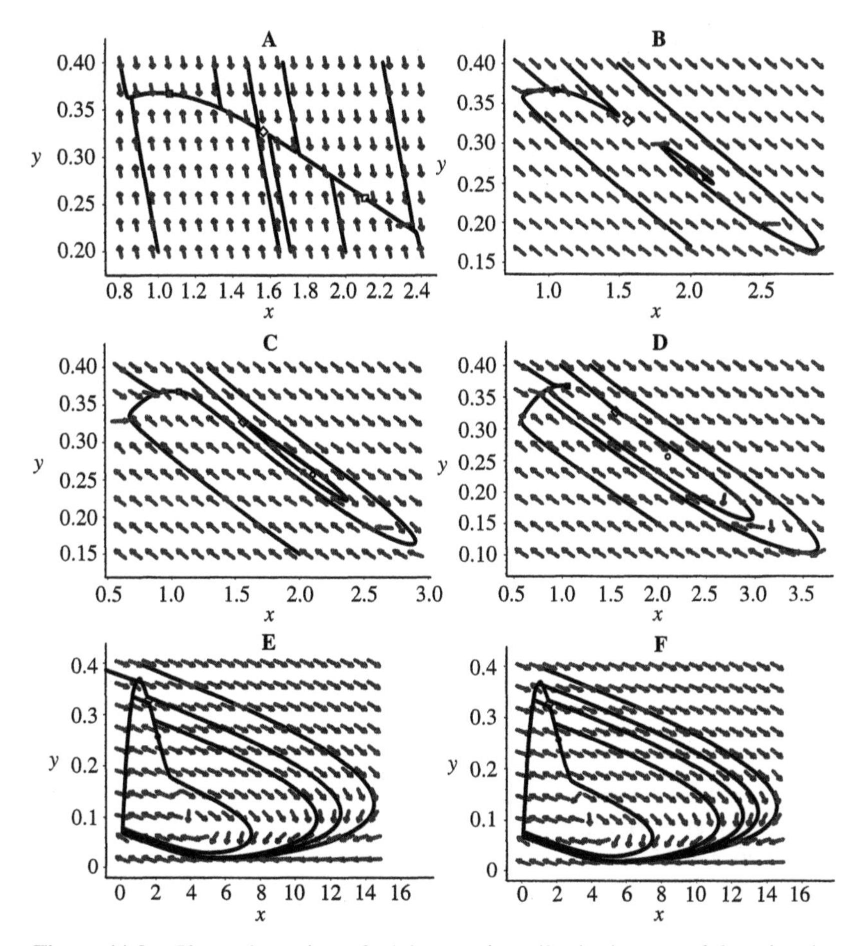

Figure 11.3. Phase-plane plots of adults over juveniles in the case of three interior equilibria for various values of μ. The middle equilibrium is always a saddle. Sinks are symbolized by a square, sources by a circle, and saddles by a diamond. **(A)** $\mu = 0.01$, all trajectories are attracted to the left or the right equilibrium with the stable manifold of the middle equilibrium separating the domains of attraction. **(B)** $\mu = 0.07$, as **(A)**, but the right equilibrium is now a spiral point. **(C)** $\mu = 0.081$, the right equilibrium has lost its stability to a surrounding periodic orbit. **(D)** $\mu = 0.1$, all trajectories are attracted to the left equilibrium except the other equilibria and the stable manifold of the middle equilibrium. **(E)** $\mu = 0.79$, as **(D)**, but the trajectories look quite different. **(F)** $\mu = 0.8$, a stable periodic orbit surrounds all three equilibria. The left equilibrium is still a sink, the domains of attraction are separated by an unstable periodic orbit which is too small to be seen in this picture. At this scale, the picture does not change for $\mu = 0.9$, where the left equilibrium is a source. See the next figure.

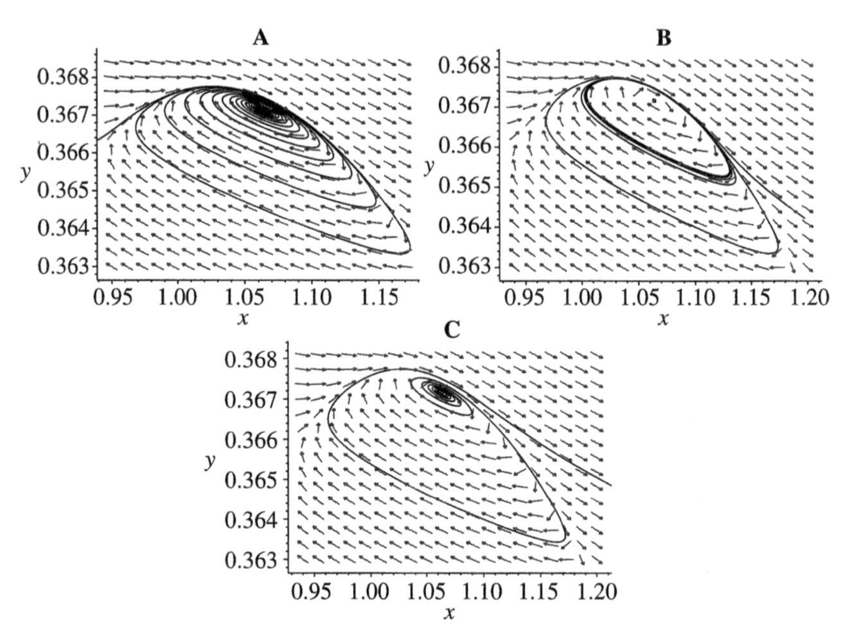

Figure 11.4. Microscopic views of the phase-plane in the neighborhood of the left equilibrium. (**A**) $\mu = 0.79$, the left equilibrium attracts almost all trajectories. (**B**) $\mu = 0.8$, the left equilibrium is still a sink, but is surrounded by an unstable periodic orbit which forms the boundary of its domain of attraction. (**C**) $\mu = 0.9$, the left equilibrium is a source.

At $\mu \approx 0.8307$, there is a subcritical Hopf bifurcation from the left equilibrium which gives rise to the unstable periodic orbits which we have observed for μ between 0.7900 and 0.8307 (Figure 11.5).

For $\mu > 0.8307$, all equilibria are unstable (Figure 11.4C), and the periodic orbit which surrounds them all attracts all trajectories starting in the first quadrant except the other equilibria and the stable manifold of the middle equilibrium which connects the left equilibrium and the right equilibrium to the middle one. Since μ/m is large, the periodic orbit is a relaxation oscillation, which can similarly be explained as in Section 11.8.

We can also fix $\mu = 0.81$ and make a bifurcation analysis as we vary m from 1 to 0 (Figure 11.6). For m between 1 and approximately 0.2146, the origin is locally asymptotically stable. At $m \approx 0.2146$, a saddle-node bifurcation occurs and the origin is a saddle for m between 0 and 0.2146, and one or more interior equilibria exist. They form a global branch with two turning points, one at approximately $m \approx 0.095\,13$ and the other at $m \approx 0.0102$. This is associated with the existence of three interior equilibria for m between m_1 and m_2, which we mentioned before. The bending of the branch is very weak, however. At $m \approx 0.009\,994$ there is a subcritical Hopf bifurcation of unstable periodic

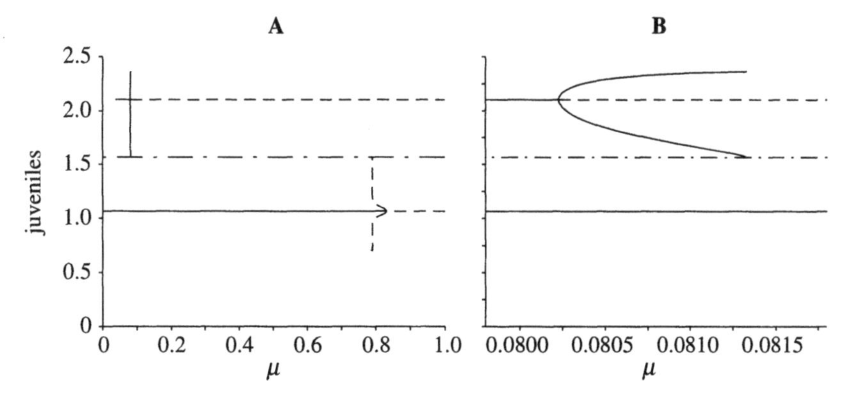

Figure 11.5. Bifurcation diagram of equilibria and periodic solutions drawn by AUTO (Doedel, 1981); the bifurcation parameter is the per capita mortality rate of juveniles, μ. (**B**) A magnification of (**A**) around $\mu = 0.08$. The three horizontal lines represent the juvenile components of the three equilibria. The dash-dotted (middle) line represents a saddle. Dashed lines represent sources or unstable periodic orbits. Solid lines represent sinks or stable periodic orbits. Around $\mu = 0.080\,23$, there is a supercritical Hopf bifurcation of stable periodic solutions from the equilibrium which we call the "right" equilibrium in the text, and around $\mu = 0.8307$ a subcritical bifurcation of unstable periodic solutions from the "left" equilibrium. The upper branches show the maximum amplitudes and the lower branches the minimum amplitudes of the periodic orbits. Homoclinic orbits connecting the middle equilibrium to itself are at those μ values, where the branches of periodic solutions touch the middle horizontal line representing this saddle.

orbits. It occurs before the branch of equilibria turns for the first time, but for an m value for which three equilibria exist. With our labeling, the periodic branch bifurcates from the first, or left, equilibrium. The middle branch of equilibria, between the two turning points, consists of saddles, the upper branch of sources. As m decreases, the branch of periodic orbits seems to blow up to a homoclinic orbit around the left equilibrium. For m approximately between 0 and 0.0101, there exists a large stable periodic orbit which surrounds all interior equilibria.

Summary

The powerful mathematical tools available for planar systems of ordinary differential equations (the Poincaré–Bendixson theorem, the Bendixson–Dulac criterion, isoclines, phase-plane) allow a rather complete discussion of the elementary one-species two-stage model from which the following can be learned.

Density dependence has a stronger potential to lead to undamped oscillation or bistability when occurring in the per capita stage transition rate than when occurring in the per capita birth rate or mortality rate. In fact, in this model,

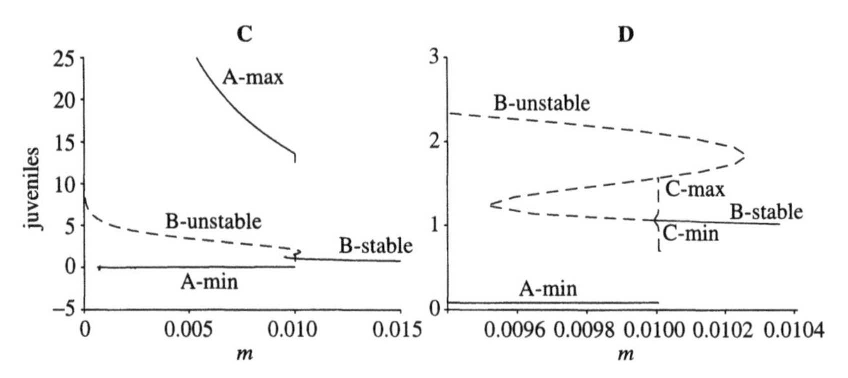

Figure 11.6. Bifurcation diagram of equilibria and periodic orbits drawn by AUTO (Doedel, 1981). The bifurcation parameter is $m = \mu/\eta$, where $\mu = 0.81$ is the per capita mortality rate of the juveniles and η is the per capita transition rate from the juvenile to the adult stage if there is no competition. (**D**) A magnification of (**C**) around $m = 0.01$. Solid lines represent stable equilibria and periodic orbits, dashed lines represent unstable equilibria and periodic orbits. For $m > 0.01$, the equilibria on the lower branch are locally asymptotically stable, indicated as B-stable. At $m \approx 0.001$ there is a subcritical Hopf bifurcation of unstable periodic orbits. C-min and C-max denote their maximal and minimal amplitudes. The middle branch consists of saddles and the upper branch of sources, indicated as B-unstable. For m between 0 and approximately 0.0101, there exists a branch of large stable periodic orbits which surround all interior equilibria. A-max and A-min denote their maximal and minimal amplitudes.

undamped oscillations are only possible if the per capita stage transition rate is density dependent.

Density-dependent per capita transition rates alone cannot lead to bistability, but they can do so in cooperation with intrastage competition if they are strong enough. If they are both weak, there is no bistability. Pure intrastage competition cannot lead to bistability as long as the per capita stage transition rate is constant.

These statements have to be taken with a grain of salt, because they do not necessarily hold when another model framework is chosen. In that case they have to be read rather as tendencies than as absolute facts.

If pure intrastage competition is combined with a per capita stage transition rate that strongly depends on juvenile density, bistability can occur involving two locally asymptotically stable equilibria or a locally asymptotically stable equilibrium and a locally asymptotically orbitally stable periodic orbit. Bistability involving two locally asymptotically stable orbits has not been observed.

If the per capita stage transition rate depends sufficiently strongly on juvenile density, for extreme parameter combinations, there exist three unstable equilibria. They are surrounded by one periodic orbit which appears to attract all points except those on the stable manifold of the middle equilibrium, which is a saddle.

Exercises

11.1. Prove Theorem 11.2 (iii). Hint: use the total population size N, as in (i), and consider the differential equation for $u := A/L$.

11.2. Prove Theorem 11.6 (d) under the assumption in (c).

11.3. Prove Remark 11.11.

11.4. Prove the second statement in Proposition 11.12.

Bibliographic Remarks

Our model considers density-dependent individual development in the most elementary way, assuming an exponentially distributed length of the juvenile period. Alternatively, one can assume that there is no variation of this length among individuals (Theorem 16.4; Aiello et al., 1992; Gurney, Nisbet, 1998, Section 8.5); differently from our model, models with discrete delay between birth and maturity have the feature that density-dependent per capita reproduction or mortality can lead to undamped oscillations (Nisbet, 1997). Even more realistically, maturation can be modeled by transport equations with density-dependent speed (Kooijman, Metz, 1984; Metz, Diekmann, 1986; Thieme, 1988b; Kooijman, 1993, 2000; de Roos, 1997; Diekmann et al., 1998, 2001). Here the density dependence is mediated by the amount of available food, and individual maturity is explicitly tracked. Surveys on and many references to stage-structured or physiologically structured population models are available in the volume edited by Tuljapurkar, Caswell (1997) and the monograph by Cushing (1998), which also include chapters on discrete-time models.

The concept of dissipativity has been extended to more general situations (Hale, 1988). Differently from this book, permanence is sometimes used instead of strong uniform persistence (Hofbauer, Sigmund, 1998).

The phenomenon that stage structure can lead to multiple interior equilibria has also been observed by Nisbet, Onyiah (1994) in discrete-time models. Stage structure in prey can cause Allee effects in predators (de Roos et al., 2003).

More material on relaxation oscillations can be found in Edelstein-Keshet (1988, Chapter 8).

Other planar differential equation models in biology are discussed in Edelstein-Keshet (1988, Chapters 5–7), Murray (1989, Chapters 3, 5–7), Busenberg, Cooke (1993, Chapter 2), Hofbauer, Sigmund (1998, Chapters 2–4), Bazykin (1998), Gurney, Nisbet (1998, Chapters 6, 7), Brauer, Castillo-Chavez (2001, Part II), and Kot (2001, Part I.B).

PART 2

Stage Transitions and Demographics

Chapter Twelve

The Transition Through a Stage

Biology is the science of life (bios). Human (earthly) life is the stage between conception and death, clearly divided into two separate stages by birth. There are more stages that are less clearly separated. Invertebrate animals may have quite a few clearly separated stages, like egg, larva (or instar), pupa, imago. There may even be several larval stages. Infectious diseases take their course through various stages: a latent period, an infectious period without symptoms (the first two together usually form the incubation period), an infectious period with symptoms, and often an immunity period. With some diseases, HIV/AIDS for example, the infectious period is further subdivided.

The modeling of stage transition is therefore important. We already touched on it in Chapter 11, where we divided the population into a juvenile and an adult stage and noticed that this can drastically change its dynamics if the per capita stage transition rate is density dependent. In the following we will not assume that stage transition is density dependent, but we will let it depend on the time an individual has already spent in the stage, a realistic assumption. Per capita mortality, for example, depends on age. Typical questions are: how long does it take to go through a certain stage? How is the number of individuals in a stage and the number leaving the stage (stage output) related to the number of individuals entering the stage (stage input)? How does the average age in a population depend on the mortality?

12.1 The Sojourn Function

Assume that you have entered a certain stage. You would like to know how long you are going to remain in this stage. In other words, you would like to know your sojourn time or the duration of the stage. If the stage under consideration is life, you would like to know for how long you are going to live.

All relevant information about the duration of the stage is contained in a function

$$\mathcal{F} : [0, \infty) \to [0, \infty),$$

where $\mathcal{F}(a)$ gives the probability of still being in the stage a time units after you have entered it. If life is the stage under consideration, $\mathcal{F}(a)$ gives the probability of being still alive at age a. In this case, $\mathcal{F}$ is sometimes called the *survival function*, or the *survivor function*, or the *survivorship*. If we consider an arbitrary stage, we call $\mathcal{F}$ the *sojourn function* of the stage. Illustrations of $\mathcal{F}$ can be seen in Figures 12.1B and 12.3. If the stage is not life, but a certain part of life, for example, then we call the variable a the *stage age* or *class age*, which means the time that has passed since entering the stage. With this understanding, we can call the sojourn time (or the stage duration) the *stage age at exit*.

Since the sojourn function will play a prominent role for a few sections, some words may be in order on the choice of the symbol $\mathcal{F}$. $1 - \mathcal{F}(\cdot)$ can be considered to be the probability distribution function of a random variable describing the duration of the stage (see Section 12.9), and there is a tradition in probability theory (e.g., Cox, 1962, Section 1.2) to let F denote *(cumulative) probability distribution functions* and $\mathcal{F} = 1 - F$. We do not completely follow this tradition here because we have some other use for the letter F and will instead let $1 - \mathcal{F} = \mathcal{G}$, but we will stick with $\mathcal{F}$ to remind us of this connection. Furthermore, if $\mathcal{F}$ is a survival function, then $\mathcal{F}(a)$ is the *fraction* of a cohort (collection of individuals born at the same time) that is still alive at age a.

$\mathcal{F}$ has two obvious properties:

- $\mathcal{F}(0) = 1$.
- $\mathcal{F}$ is a monotone nonincreasing function.

We can introduce the function

$$\mathcal{G}(a) = 1 - \mathcal{F}(a), \quad a \geqslant 0.$$

$\mathcal{G}(a)$ is the probability of already having left the stage (e.g., being dead) at stage age a. $\mathcal{G}$ has the following properties:

- $\mathcal{G}(0) = 0$.
- $\mathcal{G}$ is a monotone nondecreasing function.

If $0 \leqslant a < b$, then

$$\mathcal{F}(a) - \mathcal{F}(b) = \mathcal{G}(b) - \mathcal{G}(a)$$

is the probability of leaving the stage in the stage-age-interval from a to b.

$\mathcal{F}$ and $\mathcal{G}$ are not necessarily continuous functions. Since they are monotone, their limits from the right and left exist,

$$\mathcal{F}(a+) = \lim_{a<b\to a} \mathcal{F}(b), \qquad \mathcal{F}(a-) = \lim_{a>r\to a} \mathcal{F}(r).$$

For instance, if there is a nonzero chance of dying immediately after birth, one has $1 = \mathcal{F}(0) > \mathcal{F}(0+)$ and $\mathcal{F}$ is not continuous at 0. Similarly, if a stage is harvested at a particular stage age $a_0 > 0$, this leads to a discontinuity at a_0. $\mathcal{G}(a+)$ is often called the *probability distribution function of the stage sojourn time*.

The monotonicity of $\mathcal{F}$ also implies that the limit

$$\mathcal{F}(\infty) = \lim_{a\to\infty} \mathcal{F}(a)$$

exists. If $\mathcal{F}(\infty) = 0$, everybody has to leave the stage eventually; if $\mathcal{F}(\infty) > 0$, it is possible to stay in the stage forever.

Let us introduce the *maximum stage age* c (or *maximum stage duration, maximum sojourn time*) as follows. If $\mathcal{F}(a) > 0$ for all $a \geqslant 0$, then $c = \infty$; otherwise c is the uniquely determined number such that $\mathcal{F}(a) > 0$ for all $a \in [0, c)$ and $\mathcal{F}(a) = 0$ for all $a > c$.

12.2 Mean Sojourn Time, Expected Exit Age, and Expectation of Life

We want to determine how long a stage lasts on average, i.e., the average duration or mean sojourn time, which is identical to the expected age at stage exit. If the stage is life, this is the life expectancy or the expected age at death. At this point, we use the terms *duration* and *sojourn time* synonymously, because we assume that the stage can be left in only one way. If we consider a stage that can be left by either death or, let us say, maturation, then the average duration will be the average time needed to mature out of the stage (neglecting death), while mean sojourn time will be the average time spent in the stage including removal by death (cf. Section 13.6). We also do not distinguish between the terms 'mean' and 'average', while we will later distinguish between 'expected age' and 'average age' at stage exit (death, for example).

Let us first assume that we have a finite maximum sojourn time c. We choose some number $b > c$ and a partition $0 = t_0 < \cdots < t_{n+1} = b$ such that $t_{j+1} - t_j$ is very small for all j. As we have seen before,

$$p_j = \mathcal{F}(t_j) - \mathcal{F}(t_{j+1})$$

gives the probability of leaving the stage in the time interval from t_j to t_{j+1}. In this case, the sojourn time is some number s_j between t_j and t_{j+1}. Notice that

$$\sum_{j=0}^{n} p_j = \sum_{j=0}^{n} \mathcal{F}(t_j) - \sum_{j=0}^{n} \mathcal{F}(t_{j+1}) = \mathcal{F}(0) - \mathcal{F}(b) = 1.$$

Since the differences $t_{j+1} - t_j$ are very small, the mean (or average) sojourn time (expectation of life), which is denoted by D, is close to

$$\sum_{j=0}^{n} s_j p_j = \sum_{j=0}^{n} s_j(\mathcal{F}(t_j) - \mathcal{F}(t_{j+1})). \tag{12.1}$$

Taking the sums apart and changing the index of summation in the first sum and recombining them (recall $\mathcal{F}(t_{n+1}) = 0$), we obtain

$$\sum_{j=0}^{n} s_j p_j = s_0 \mathcal{F}(0) + \sum_{j=0}^{n-1} \mathcal{F}(t_{j+1})(s_{j+1} - s_j).$$

Notice that $t_{j+1} \in [s_j, s_{j+1}]$. If $s_n = b$ and $s_0 = 0$, the points s_j form a partition of the interval from 0 and b; otherwise we introduce $s_{-1} = 0$ and/or $s_{n+1} = b$, $t_{n+1} = b$ and can write

$$\sum_{j=0}^{n} s_j p_j = \sum_{j=-1}^{n} \mathcal{F}(t_{j+1})(s_{j+1} - s_j).$$

If the differences $t_{j+1} - t_j$ tend to 0 for every j, so do the differences $s_{j+1} - s_j$, and the sums converge towards the Riemann integral

$$\int_0^b \mathcal{F}(t)\, \mathrm{d}t = \int_0^\infty \mathcal{F}(a)\, \mathrm{d}a =: D \tag{12.2}$$

and D is the *mean sojourn time* in the stage or the *average duration* of the stage.

This integral exists as an improper Riemann integral because $\mathcal{F}$ is monotone and $\mathcal{F}(a) = 0$ for all $a \geqslant b > c$.

Alternatively, as the partitions get finer, the sums (12.1) converge to the Stieltjes integral (see Theorem B.8)

$$-\int_0^b t\, \mathrm{d}\mathcal{F}(t) = D,$$

and we obtain formula (12.2) by integrating by parts (Corollary B.11).

Let us now assume that $c = \infty$ and assume that $\mathcal{F}(\infty) = 0$. We then define a family of sojourn functions $\mathcal{F}_\alpha$, $\alpha > 0$, by setting

$$\mathcal{F}_\alpha(a) = \begin{cases} \mathcal{F}(a), & 0 \leqslant a < \alpha, \\ 0, & a \geqslant \alpha. \end{cases}$$

For the sojourn function $\mathcal{F}_\alpha$, the corresponding mean sojourn time is given by

$$D_\alpha = \int_0^\infty \mathcal{F}_\alpha(a)\, \mathrm{d}a = \int_0^\alpha \mathcal{F}(a)\, \mathrm{d}a.$$

Since $\mathcal{F}_\alpha(a) \to \mathcal{F}(a)$ as $\alpha \to \infty$, uniformly in $a \geqslant 0$, it is suggestive to define

$$D = \lim_{\alpha \to \infty} D_\alpha = \lim_{\alpha \to \infty} \int_0^\alpha \mathcal{F}(a)\, \mathrm{d}a = \int_0^\infty \mathcal{F}(a)\, \mathrm{d}a.$$

We need to assume, however, that this limit exists, which involves that $\mathcal{F}(\infty) = 0$ as we have assumed above (though this is not sufficient).

More generally, if f is a continuous function on $[0, \infty)$, the sums

$$\sum_{j=0}^{n} f(s_j)p_j \to -\int_0^\infty f(t)\,\mathrm{d}\mathcal{F}(t),$$

provided that this improper Riemann–Stieltjes integral exists. If $\mathcal{F}$ is continuously differentiable,

$$-\int_0^\infty f(t)\,\mathrm{d}\mathcal{F}(t) = -\int_0^\infty f(t)\mathcal{F}'(a)\,\mathrm{d}a.$$

If f is continuously differentiable, we have the integration-by-parts formula

$$-\int_0^\infty f(t)\,\mathrm{d}\mathcal{F}(t) = f(0) + \int_0^\infty f'(t)\mathcal{F}(t)\,\mathrm{d}t.$$

See Toolbox B.1 for details.

The average duration D must be discriminated from the *median duration* $\tilde{D}$. The intuitive meaning of the median stage duration is the following: if a cohort of individuals has entered the stage at time 0, half of the cohort has left and the other half is still in the stage at time $\tilde{D}$. Mathematically, this means that $\mathcal{F}(a) \geqslant \frac{1}{2}$ for $a < \tilde{D}$ and $\mathcal{F}(a) \leqslant \frac{1}{2}$ for $a > \tilde{D}$, equivalently $\mathcal{F}(\tilde{D}+) \leqslant \frac{1}{2} \leqslant \mathcal{F}(\tilde{D}-)$. If the sojourn function $\mathcal{F}$ is strictly decreasing, $\tilde{D}$ is uniquely determined by this property.

12.3 The Variance of the Sojourn Time, Moments and Central Moments

In order to determine the variance of the sojourn time, V, we first assume again that the maximum stage age is finite and choose some $b > 0$ such that $\mathcal{F}(a) \geqslant 0$ for all $a \geqslant b$. Then V is approximated by sums

$$\sum_{j=0}^{n}(s_j - D)^2 p_j = \sum_{j=0}^{n}(s_j - D)^2(\mathcal{F}(t_j) - \mathcal{F}(t_{j+1}))$$

with partitions $0 = t_0 < \cdots < t_{n+1} = b$, where, as before, p_j is the probability of leaving the stage between t_j and t_{j+1} and s_j is some number in $[t_j, t_{j+1}]$. This means that the variance of the sojourn time is given by the Stieltjes integral

$$V = -\int_0^b (a - D)^2\,\mathrm{d}\mathcal{F}(a)$$

(see Theorem B.8). Integrating by parts (Corollary B.11),

$$V = D^2 + 2\int_0^b (a - D)\mathcal{F}(a)\,\mathrm{d}a = 2\int_0^b a\mathcal{F}(a)\,\mathrm{d}a - D^2.$$

The approximation above also tells us that $V \geqslant 0$.

A similar consideration as for the mean sojourn time provides the following formula that also holds for infinite maximum stage age provided that the improper integral exists,

$$0 \leqslant V = 2 \int_0^\infty a\mathcal{F}(a)\,\mathrm{d}a - D^2. \tag{12.3}$$

For a stochastic interpretation, see Section 12.9. The nth moment of the sojourn time is defined by

$$-\int_0^\infty t^n \,\mathrm{d}\mathcal{F}(t) = n \int_0^\infty t^{n-1} \mathcal{F}(t)\,\mathrm{d}t,$$

and the nth central moment by

$$C_n = -\int_0^\infty (t - D)^n \,\mathrm{d}\mathcal{F}(t).$$

Notice $V = C_2$. The *skewness*

$$C_3/C_2^{3/2}$$

is a measure of the symmetry of the distribution of sojourn times, and the *kurtosis* (coefficient of excess)

$$C_4/C_2^2$$

is related to its flatness.

12.4 Remaining Sojourn Time and Its Expectation

Assume that you have already stayed in a stage until stage age a. What is the time you can expect to remain in the stage? We call this time the *expected remaining sojourn (duration)* at stage age a and denote it by $D(a)$. If we consider life, we can call this the *expected remaining life* at age a.

The probability of still being in the stage at time $a+t$ provided the individual is still in the stage at time a is given by

$$\mathcal{F}(t \mid a) = \frac{\mathcal{F}(t+a)}{\mathcal{F}(a)}, \quad 0 \leqslant a < c, \quad t \geqslant 0. \tag{12.4}$$

We set

$$\begin{aligned} \mathcal{F}(t \mid a) &= 0, \quad \text{whenever } a \geqslant c,\ t > 0, \\ \mathcal{F}(0 \mid a) &= 1, \quad a \geqslant c. \end{aligned}$$

Notice that, for fixed but arbitrary $a \geqslant 0$, $\mathcal{F}(\cdot \mid a)$ has the properties of a sojourn function: $\mathcal{F}(0 \mid a) = 1$ and $\mathcal{F}(\cdot \mid a)$ is a nonincreasing function. Apparently,

$\mathcal{F}(\cdot \mid a)$ is the probability of still being in the substage of the original state that starts at age a.

Analogously to the consideration in Section 12.2, the *expected remaining sojourn* at stage age a can be determined by

$$D(a) = \int_0^\infty \mathcal{F}(t \mid a)\,\mathrm{d}t.$$

Using the definition of $\mathcal{F}(t \mid a)$ in (12.4),

$$D(a) = \int_0^\infty \frac{\mathcal{F}(t+a)}{\mathcal{F}(a)}\,\mathrm{d}t, \qquad 0 \leqslant a < c. \tag{12.5 a}$$

By substitution,

$$D(a) = \frac{1}{\mathcal{F}(a)} \int_a^\infty \mathcal{F}(t)\,\mathrm{d}t, \quad 0 \leqslant a < c. \tag{12.5 b}$$

Obviously,

$$D(a) = 0, \quad a \geqslant c.$$

Apparently, $D(0) = D$. At first glance, one might think that $D(a)$ should be a monotone nonincreasing function on a. But a moment's reflection will tell us that, in a country where infant mortality is considerably higher than the mortality in later age classes, the expected remaining life at age a, $D(a)$, may actually increase for small a. See the bibliographic remarks at the end of this section for references.

We use the letter D both for the mean sojourn time and for the function whose values provide the expected remaining sojourn. To avoid confusion, we use the following convention. An isolated D usually denotes the mean sojourn time, and sometimes we instead write $D(0)$. If we refer to the function D, we usually write $D(\cdot)$.

We can rewrite (12.5 b) in the form

$$D(a) = \frac{1}{\mathcal{F}(a)}\left(D - \int_0^a \mathcal{F}(t)\,\mathrm{d}t\right).$$

Using the fact that $\mathcal{F}$ is nonincreasing, we have the estimates

$$\frac{1}{\mathcal{F}(a)}(D-a) \leqslant D(a) \leqslant \frac{D}{\mathcal{F}(a)} - a. \tag{12.6}$$

Without some restrictions on $\mathcal{F}$, $D(a)$ may be an unbounded function of a. See Exercise 12.5 for an example.

Proposition 12.1. *Let $c = \infty$. Then the following three statements are equivalent.*

(a) *$D(a)$ is a bounded function of $a \in [0, \infty)$.*

(b) *$(\mathcal{F}(a+t)/\mathcal{F}(a)) \to 0$ for $t \to \infty$, uniformly in $a \in [0, \infty)$.*

(c) *There exist $\epsilon > 0$, $M \geqslant 1$ such that*

$$\mathcal{F}(t \mid a) = \frac{\mathcal{F}(a+t)}{\mathcal{F}(a)} \leqslant M e^{-\epsilon t} \quad \forall t, a \geqslant 0.$$

Proof. (a) $\Rightarrow$ (b).

Let K be an upper bound of $D(a)$, $a \geqslant 0$. Then, since $\mathcal{F}$ is nonincreasing,

$$K \geqslant \frac{\int_a^{a+t} \mathcal{F}(r)\, dr}{\mathcal{F}(a)} \geqslant t \frac{\mathcal{F}(a+t)}{\mathcal{F}(a)}.$$

Regrouping the terms,

$$\frac{\mathcal{F}(a+t)}{\mathcal{F}(a)} \leqslant \frac{K}{t} \to 0, \quad t \to \infty, \quad a \geqslant 0.$$

(b) $\Rightarrow$ (c).

For $t > 0$ define

$$\alpha(t) := \sup_{a \geqslant 0} \frac{\mathcal{F}(a+t)}{\mathcal{F}(a)}.$$

If (b) holds, then there exists some $t_0 > 0$ such that

$$\alpha(t_0) < 1.$$

Let $n \in \mathbb{N}$. Then

$$\frac{\mathcal{F}(a+[n+1]t_0)}{\mathcal{F}(a)} = \frac{\mathcal{F}(a+nt_0+t_0)}{\mathcal{F}(a+nt_0)} \frac{\mathcal{F}(a+nt_0)}{\mathcal{F}(a)} \leqslant \alpha(t_0)\alpha(nt_0) \quad \forall a \geqslant 0.$$

Using the definition of α,

$$\alpha([n+1]t_0) \leqslant \alpha(t_0)\alpha(nt_0) \quad \forall n \in \mathbb{N}.$$

By induction,

$$\alpha(nt_0) \leqslant (\alpha(t_0))^n, \quad n \in \mathbb{N}.$$

Set $\epsilon = -(1/t_0) \ln \alpha(t_0)$. Let $t > 0$. Then there exists some $n \in \mathbb{N}$ such that

$$t = nt_0 + v, \quad \text{where } v \in [0, t_0).$$

Then, since $\mathcal{F}$ is nonincreasing and $t \geqslant nt_0$,

$$\frac{\mathcal{F}(a+t)}{\mathcal{F}(a)} \leqslant \frac{\mathcal{F}(a+nt_0)}{\mathcal{F}(a)} \leqslant \alpha(nt_0) \leqslant (\alpha(t_0))^n$$
$$= \mathrm{e}^{-\epsilon n t_0} = \mathrm{e}^{-\epsilon t}\mathrm{e}^{\epsilon v} \leqslant \mathrm{e}^{-\epsilon t}\mathrm{e}^{\epsilon t_0}.$$

(c) $\Rightarrow$ (a).

$$D(a) = \frac{\int_0^\infty \mathcal{F}(a+t)\,\mathrm{d}t}{\mathcal{F}(a)} \leqslant \int_0^\infty M\mathrm{e}^{-\epsilon t}\,\mathrm{d}t = M/\epsilon.$$

□

If life is the stage under consideration, the function $D(\cdot)$ is of interest to retirement and pension plans and life insurance. In this book, it will later play a role in optimal vaccination schedules (see the end of Section 22.5).

Data for $D(\cdot)$ seem to go farther back than for the survival function $\mathcal{F}$ itself, even to the Roman Empire (Smith, Keyfitz, 1977, p. 5; Hutchinson, 1978, figures 27 and 31). So, it is of some interest to recover $\mathcal{F}$ from $D(\cdot)$. Set $u(a) = \mathcal{F}(a)D(a)$. From (12.5 b), by the fundamental theorem of calculus,

$$u'(a) = -\mathcal{F}(a) = -\frac{u(a)}{D(a)}.$$

Integrating,

$$u(a) = u(0)\exp\left(-\int_0^a \frac{\mathrm{d}s}{D(s)}\right).$$

By definition of u, we have the following result.

Proposition 12.2. *The following relation holds between the sojourn function and the expected remaining sojourn:*

$$\mathcal{F}(a) = \frac{D(0)}{D(a)}\exp\left(-\int_0^a \frac{\mathrm{d}s}{D(s)}\right), \quad 0 \leqslant a < c.$$

Proposition 12.2 offers an alternative approach starting from expected remaining sojourn rather than from the sojourn function. But not every function $D : [0,\infty) \to [0,\infty)$ qualifies.

Remark 12.3. Let $D : [0,c) \to (0,\infty)$ and $\mathcal{F}$ be related by the formula in Proposition 12.2.

(a) Let D be differentiable. Then $\mathcal{F}$ is monotone nonincreasing if and only if $D'(a) + 1 \geqslant 0$ for all $a \in (0,c)$.

(b) Let D be continuous on $[0,c)$. Then $\mathcal{F}$ is monotone nonincreasing if $D(a) + a$ is a monotone nondecreasing function of $a \in [0,c)$.

Proof. Part (a) of the remark follows from differentiating

$$\ln \mathcal{F}(a) = \ln D(0) - \ln D(a) - \int_0^a \frac{\mathrm{d}s}{D(s)}.$$

For (b) define

$$D_k(a) = k \int_a^{a+(1/k)} D(s)\,\mathrm{d}s = \int_0^1 D(a + (s/k))\,\mathrm{d}s.$$

Then $D_k(a) \to D(a)$ as $k \to \infty$ uniformly for a in every closed interval in $[0, c)$. Furthermore,

$$\begin{aligned} D_k'(a) + 1 &= k[D(a + (1/k)) - D(a) + (1/k)] \\ &= k[(D(a + (1/k)) + a + (1/k)) - (D(a) + a)] \geqslant 0. \end{aligned}$$

By part (a) the function $\mathcal{F}_k$ related to D_k via the formula in Proposition 12.2 is monotone nonincreasing. Since $\mathcal{F}_k$ converges towards $\mathcal{F}$, this property is inherited by $\mathcal{F}$. □

An application of Proposition 12.2 and Remark 12.3 is given in Exercises 12.9 and 12.10. If $D(a) = \alpha e^{-\gamma a}$, with appropriate constants $\alpha, \gamma > 0$, we obtain a modification of Gompertz's law which goes into a direction opposite to Makeham's improvement of Gompertz's law. See Impagliazzo (1985) for a discussion and the original references concerning Gompertz's and Makeham's papers, and Smith, Keyfitz (1977) for excerpts of the original papers.

Reconstruction of a Survival Function for Ulpian's Table

One of the oldest tables for expectations of remaining life is attributed to Ulpian, who lived in the third century A.D. See the piecewise-linear curve in Figure 12.1A, which has been adapted from Trenerry (1926) (see also Hutchinson (1978), the figure on p. 5 in Smith, Keyfitz (1977), and the excerpt from Trenerry's book on p. 7 of the same reference). In spite of its obvious shortcomings, in particular in the age bracket from 0 to 20, Ulpian's table was used as the official annuity table in northern Italy until the end of the 18th century. Since Ulpian's table spells out for how many years provisions should be allowed after surviving to a given age, a step function like the one in figure 27 in Hutchinson (1978) better reflects the original purpose than a piecewise-linear function, but the latter may be more appropriate if one thinks in terms of expected remaining life. We try to find $D(a)$ in the form

$$D(a) = \frac{c_0 + a}{c_1 + c_2 a^n},$$

with positive constants c_0, c_1, c_2 and $n \in \mathbb{N}$. This choice is somewhat arbitrary; we want $D(\cdot)$ first to increase (to take account of infant mortality) and then

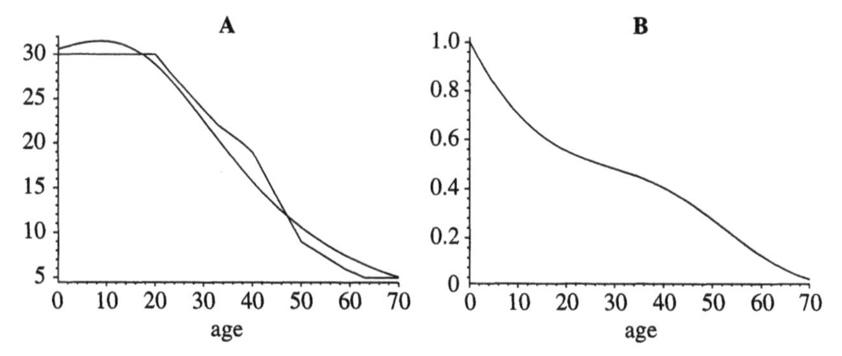

Figure 12.1. (**A**) The piecewise-linear curve represents expectations of remaining life from Ulpian's table. The smooth curve is our fit. (**B**) The survival function $\mathcal{F}$ has been reconstructed from the fit illustrated in (**A**) using the formula in Proposition 12.2. Realistically, the decrease of $\mathcal{F}$ should be steeper at the early ages.

to decrease. We also want to obtain a closed form for $\mathcal{F}$. Figure 12.1A shows $D(a)$ for $n = 3$, $c_0 = 200$, $c_1 = 6.534$, and $c_2 = 0.000\,136$. The values for n and c_0 were obtained by trial and error, those for c_1 and c_2 by a linear fit of $0 = c_0 + a - D(a)(c_1 + c_2 c^n)$ using MAPLE. I tried to fit c_0, c_1, and c_2 together, but this provided negative values. Proposition 12.2 yields the following formula for $\mathcal{F}$:

$$\mathcal{F}(a) = \frac{D(0)}{D(a)} \exp\left[-\tfrac{1}{3}c_2 a^3 + \tfrac{1}{2}c_0 c_2 a^2 - c_0^2 c_2 a\right]\left(1 + \frac{a}{c_0}\right)^{c_0^3 c_2 - c_1}.$$

$\mathcal{F}$ is plotted in Figure 12.1B.

Average Expectation of Remaining Sojourn

In order to find the average expectation of remaining sojourn (duration), E, we have to average the expected remaining sojourns $D(a)$ over all stage ages a. Since the probability of still being in the stage at stage age a is $\mathcal{F}(a)$, we need to form the average weighed by $\mathcal{F}(a)$:

$$E = \frac{\int_0^\infty D(a)\mathcal{F}(a)\,\mathrm{d}a}{\int_0^\infty \mathcal{F}(a)\,\mathrm{d}a} = \int_0^\infty \frac{D(a)}{D(0)}\mathcal{F}(a)\,\mathrm{d}a. \tag{12.7}$$

Substituting $D(a)$ from (12.5 b), we obtain

$$E = \frac{\int_0^\infty (\int_a^\infty \mathcal{F}(t)\,\mathrm{d}t)\,\mathrm{d}a}{D(0)}.$$

Interchanging the order of integration yields

$$E = \frac{\int_0^\infty (\int_0^t 1\,\mathrm{d}a)\mathcal{F}(t)\,\mathrm{d}t}{D(0)}.$$

Hence we obtain the following formula for the average expectation of remaining sojourn,

$$E = \frac{\int_0^\infty a\mathcal{F}(a)\,\mathrm{d}a}{D(0)} = \frac{\int_0^\infty a\mathcal{F}(a)\,\mathrm{d}a}{\int_0^\infty \mathcal{F}(a)\,\mathrm{d}a}. \tag{12.8}$$

A simple calculation shows that E is uniquely determined by the relation

$$\int_0^\infty (a - E)\mathcal{F}(a)\,\mathrm{d}a = 0.$$

From (12.7) and Proposition 12.2 we obtain the following relation between the average expectation of remaining sojourn and expected remaining sojourn:

$$E = \int_0^c \exp\left(-\int_0^a \frac{1}{D(s)}\,\mathrm{d}s\right)\mathrm{d}a.$$

From formula (12.3) we derive the following relation between the mean and the variance of the stage duration and the expected remaining sojourn:

$$V = D(2E - D). \tag{12.9}$$

This can be rewritten in various useful ways,

$$2\frac{E}{D} = 1 + \frac{V}{D^2}, \qquad V - D^2 = 2D(E - D).$$

From the second equation we learn that $V - D^2$ has the same sign as $E - D$.

Proposition 12.4.

(a) *Let $c < \infty$. Then $D(a) \leqslant c - a$ for $0 \leqslant a < c$.*

(b) *$D/2 \leqslant E \leqslant c/2$.*

Proof. Let $0 < a < c$. Since $\mathcal{F}$ is monotone nonincreasing,

$$D(a) = \frac{\int_a^c \mathcal{F}(r)\,\mathrm{d}r}{\mathcal{F}(a)} \leqslant (c - a)\frac{\mathcal{F}(a)}{\mathcal{F}(a)}.$$

By (12.7) and (12.8),

$$E \leqslant \frac{\int_0^c (c - a)\mathcal{F}(a)\,\mathrm{d}a}{\int_0^c \mathcal{F}(a)\,\mathrm{d}a} = c - E.$$

Hence $E \leqslant c/2$. This inequality is trivially true if $c = \infty$. The other inequality follows from (12.9) and $V \geqslant 0$. □

We will see in the next section that $E = \frac{1}{2}D$ if and only if the variance of the stage duration is 0, i.e., if the sojourn function is the step function equal to 1 for $a < c$ and 0 for $a > c$. In this case the maximum stage duration c is finite and coincides with the mean stage duration. In the exercises (Exercises 12.4–12.6) and in Section 12.8, we will see examples in which E/D is arbitrarily large; actually it can be any number greater than $\frac{1}{2}$.

12.5 Fixed Stage Durations

There are two extreme but widely used assumptions about the duration of a stage, the first being that it is fixed and the second that it is exponentially distributed. Here we consider a slight generalization of the first assumption, namely that

$$\mathcal{F}(a) = \begin{cases} 1, & a = 0, \\ p, & 0 < a < c, \\ 0, & a > c, \end{cases}$$

where $p \in [0, 1]$. At $a = c$, $\mathcal{F}(a)$ can be any number between 0 and p. If $p = 1$, everybody leaves the stage at stage age c. In this case, $\mathcal{F}$ is called the *step sojourn function*, or, if the stage is life, the *step function survivorship*. If $p = 0$, everybody leaves the stage immediately after having entered it. If $p \in (0, 1)$, one either leaves the stage immediately after having entered it (this happens with probability $1 - p$) or one leaves the stage exactly at the maximum possible stage age c (with probability p). If the stage is life, the jump at age $a = 0$ is the parameter-sparse approximation of a very short period after birth with very high infant mortality, and $\mathcal{F}$ is called the *step survival function with infant mortality*. Though $\mathcal{F}$ is an extreme idealization of a realistic sojourn function, it is widely used in mathematical models which would become too complicated otherwise. Here we want to reflect upon what such an idealization means in terms of mean sojourn time, expected remaining sojourn, and average expectation of remaining sojourn.

From (12.6) we readily see that

$$D(a) = \begin{cases} pc, & a = 0, \\ c - a, & 0 < a < c, \\ 0, & a \geqslant c. \end{cases}$$

We notice that the expected remaining sojourn jumps upward immediately after birth, but then declines steadily until it becomes 0 when reaching the maximum sojourn time. From formula (12.8) we obtain that

$$E = \frac{\int_0^c a \, \mathrm{d}a}{c} = \tfrac{1}{2}c.$$

So, we have the relations

$$D = pc, \qquad E = \tfrac{1}{2}c = \frac{1}{2p}D.$$

Interestingly enough, some converses hold.

Remarks.

(a) If the maximum stage duration, c, is finite and the average expectation of remaining sojourn satisfies $E = \frac{1}{2}c$, then $\mathcal{F}$ is constant on $(0, c)$.

(b) If $E = \frac{1}{2}D$ and $a^2\mathcal{F}(a) \to 0$ as $a \to \infty$, then the maximum stage duration is finite and equals D, and $\mathcal{F}(a) = 1$ for all $a \in [0, D)$ and $\mathcal{F}(a) = 0$ for all $a > D$.

Proof. (a) If $E = \frac{1}{2}c$, the formula subsequent to (12.8) implies that

$$0 = \int_0^c (a - \tfrac{1}{2}c)\mathcal{F}(a)\,\mathrm{d}a = \int_0^{(c/2)} (a - \tfrac{1}{2}c)\mathcal{F}(a)\,\mathrm{d}a + \int_{(c/2)}^c (a - \tfrac{1}{2}c)\mathcal{F}(a)\,\mathrm{d}a.$$

By appropriate substitution in both integrals we obtain

$$0 = \int_0^{(c/2)} a(\mathcal{F}(\tfrac{1}{2}c + a) - \mathcal{F}(\tfrac{1}{2}c - a))\,\mathrm{d}a.$$

Now

$$\mathcal{F}(\tfrac{1}{2}c + a) - \mathcal{F}(\tfrac{1}{2}c - a)$$

is nonpositive and monotone nonincreasing. Since the integral is 0, this implies that

$$\mathcal{F}(\tfrac{1}{2}c + a) - \mathcal{F}(\tfrac{1}{2}c - a) = 0 \quad \text{for all } a \in (0, \tfrac{1}{2}c).$$

Using the monotonicity of $\mathcal{F}$ again, $\mathcal{F}$ is constant on $(0, c)$.

(b) If $D = \frac{1}{2}E$, then the variance of the stage duration, V, is 0 by formula (12.9). Intuitively this tells us that the stage duration is already fixed. Let

$$\phi(x) = -\int_x^\infty (a - D)^2\,\mathrm{d}\mathcal{F}(a).$$

Then $\phi(0) = V = 0$. Since the integrand is nonnegative and $\mathcal{F}$ is monotone nonincreasing, we obtain that $\phi(x) = 0$ for all $x \geqslant 0$ (Proposition B.5 (e) and Theorem B.12). Integrating by parts (Corollary B.11) implies that

$$0 = (x - D)^2\mathcal{F}(x) + 2\int_x^\infty (a - D)\mathcal{F}(a)\,\mathrm{d}a.$$

For $x \geqslant D$, this implies $(x - D)^2\mathcal{F}(x) \leqslant 0$, so $\mathcal{F}(x) = 0$ for $x > D$. By (12.2),

$$D = \int_0^D \mathcal{F}(a)\,\mathrm{d}a,$$

i.e.,

$$0 = \int_0^D (1 - \mathcal{F}(a))\,\mathrm{d}a.$$

Since the integrand is nonnegative and monotone nondecreasing, this implies $1 = \mathcal{F}(a)$ for all $a \in (0, D)$. □

12.6 Per Capita Exit Rates (Mortality Rates)

The conditional probability of exiting the stage in the interval between stage age a and stage age $a + h$ under the condition to still be in the stage at stage age a is

$$\frac{\mathcal{F}(a) - \mathcal{F}(a+h)}{\mathcal{F}(a)} = 1 - \mathcal{F}(h \mid a).$$

If $\mathcal{F}$ is differentiable at a, then

$$\mu(a) = \lim_{h \searrow 0} \frac{1}{h} \frac{\mathcal{F}(a) - \mathcal{F}(a+h)}{\mathcal{F}(a)} = -\frac{\mathcal{F}'(a)}{\mathcal{F}(a)} = -\frac{\mathrm{d}}{\mathrm{d}a} \ln \mathcal{F}(a) \qquad (12.10)$$

can be interpreted as the (per capita) *exit rate* at stage age a. We can rewrite (12.10) as

$$\mathcal{F}'(a) = -\mu(a)\mathcal{F}(a).$$

If $\mathcal{F}$ is differentiable on $(0, c)$, we can recover $\mathcal{F}$ from μ as

$$\mathcal{F}(a) = \exp\left(-\int_0^a \mu(s)\,\mathrm{d}s\right), \quad 0 < a < c. \qquad (12.11)$$

If $0 \leqslant b < c$ and $\mathcal{F}$ is differentiable on (b, c), we obtain from (12.4) that

$$\mathcal{F}(t \mid a) = \exp\left(-\int_a^{a+t} \mu(s)\,\mathrm{d}s\right) = \exp\left(-\int_0^t \mu(a+s)\,\mathrm{d}s\right),$$
$$b < a < c, \quad t \geqslant 0, \quad a + t < c. \qquad (12.12)$$

$\mathcal{F}$ is differentiable on $(0, c)$ if and only if the expected remaining sojourn $D(\cdot)$ is differentiable, as can be seen from (12.5 b) and Proposition 12.2. From Remark 12.3 and its proof we discover the relation

$$\mu(a) = \frac{D'(a) + 1}{D(a)}. \qquad (12.13)$$

It can be used to reconstruct the per capita mortality rate for Ulpian's table (Figure 12.2). Figures 12.1 and 12.2 and the figures in the literature suggest that there is some opposite relationship between the graphs of $\mu(\cdot)$ and $D(\cdot)$ (see, for example, figure 27 on p. 45 and figure 32 on p. 52 of Hutchinson, 1978). We show that this is not accidental.

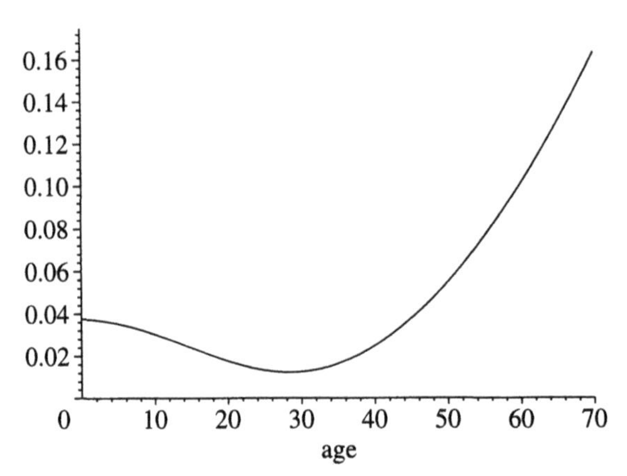

Figure 12.2. Reconstruction of the per capita mortality rate μ from the fit of $D(\cdot)$, illustrated in Figure 12.1A, via formula (12.13). Realistically, it should be higher and decrease more steeply at the early ages.

Proposition 12.5. *Let $b \in [0, c)$.*

(a) *Assume that $\mathcal{F}$ is differentiable on $[0, b)$ and $D(\cdot)$ is nondecreasing and concave on $[0, b)$. Then μ is nonincreasing on $[0, b)$.*

(b) *Assume that $\mathcal{F}$ is differentiable on (b, c) and that μ is nondecreasing on (b, c). Then $D(a)$ is nonincreasing in $a \in (b, c)$.*

(c) *Assume that $c = \infty$ and μ is nonincreasing on (b, ∞). Then D is nondecreasing on (b, ∞).*

Proof. Part (a) of the proposition follows from (12.13). Recall that the derivative of a concave function is nonincreasing.

(b) We have from (12.5) and (12.12) that

$$D(a) = \int_0^{c-a} \mathcal{F}(t \mid a)\, \mathrm{d}t = \int_0^{c-a} \exp\left(-\int_0^t \mu(a+s)\, \mathrm{d}s\right) \mathrm{d}t.$$

Now let $b \leqslant a \leqslant v < c$. Then

$$D(a) \geqslant \int_0^{c-v} \exp\left(-\int_0^t \mu(a+s)\, \mathrm{d}s\right) \mathrm{d}t.$$

Now $\mu(a+s) \leqslant \mu(v+s)$ whenever $0 < s \leqslant t < c - v$. Hence

$$D(a) \geqslant \int_0^{c-v} \exp\left(-\int_0^t \mu(v+s)\, \mathrm{d}s\right) \mathrm{d}t = D(v).$$

The proof of part (c) is similar. □

Proposition 12.6.

(a) *Assume that* $\mathcal{F}$ *is differentiable on* $(0, c)$ *and that* μ *is nondecreasing on the interval* $(0, c)$. *Then* $E \leqslant D$.

(b) *Assume that* $c = \infty$ *and* μ *is nonincreasing on* $(0, \infty)$. *Then* $E \geqslant D$.

Proof. (a) By Proposition 12.5 (b), $D(a) \leqslant D(0) = D$. The statement now follows from (12.7). The proof of part (b) is similar. □

If $\mu(a)$ is the age-dependent per capita mortality rate, it typically decreases first and then increases as a function of age (see Hutchinson (1978, figure 32, p. 52), Impagliazzo (1985, figure 1.12, p. 32), and our Figure 12.2).

12.7 Exponentially Distributed Stage Durations

An important special case is the (per capita) exit rate μ being a constant. Then

$$\mathcal{F}(a) = \mathrm{e}^{-\mu a}$$

and we say that the duration of the stage is *exponentially distributed*. Obviously,

$$\mathcal{F}(a+s) = \mathcal{F}(a)\mathcal{F}(s),$$

which can be written more suggestively as

$$\mathcal{F}(s \mid a) = \mathcal{F}(s) \quad \forall a, s \geqslant 0,$$

i.e., the probability of staying longer in the stage is independent of the present stage age.

Actually, the converse is also true: if $\mathcal{F}(a+s) = \mathcal{F}(a)\mathcal{F}(s)$ for all $a, s \geqslant 0$, then either $\mathcal{F}(a) = 0$ for all $a > 0$ or there exists some $\mu \in \mathbb{R}$ such that $\mathcal{F}(a) = \mathrm{e}^{-\mu a}$ (see Exercise 12.1).

If the exit rate is a constant μ, then the average duration of the stage (mean sojourn time in the stage) is given by

$$D = \int_0^\infty \mathrm{e}^{-\mu a}\, \mathrm{d}a = \frac{1}{\mu} = D(a) \quad \text{for all } a \geqslant 0.$$

The last equality is a special case of Proposition 12.5, but also follows easily from formula (12.5). Actually, the following characterization follows from relation (12.13).

Proposition 12.7. *The sojourn time is exponentially distributed if and only if the expected remaining sojourn* $D(a)$ *at age* a *does not depend on* a.

For the average expectation of remaining sojourn we get

$$E = \frac{\int_0^\infty a\mathcal{F}(a)\,\mathrm{d}a}{D} = \frac{\mu}{\mu^2} = \frac{1}{\mu}$$

as well. Finally, from (12.9),

$$V = D(2E - D) = (1/\mu)^2.$$

Proposition 12.8. *If the sojourn time in a stage is exponentially distributed, the mean sojourn time, the expected remaining sojourn at a certain age, the average expectation of remaining sojourn, and the standard deviation of the sojourn time are all identical and equal to the reciprocal of the per capita exit rate.*

By definition, the standard deviation is the square root of the variance. In particular, if the life duration is exponentially distributed, life expectancy and average expectation of remaining life coincide and are the reciprocal of the constant mortality rate.

In later models we will make the idealizing assumption that the life duration is exponentially distributed, though this is not the case in reality. This section teaches us that this assumption makes the life expectancy, the average expectation of remaining life, and the standard deviation of life length all the same, though they typically do not coincide in reality. It also puts us in a dilemma. Should we choose the constant per capita mortality rate as the reciprocal of the life expectancy or of the average expectation of remaining life? This will depend on what features we would like to be preserved (see Chapter 13).

12.8 Log-Normally Distributed Stage Durations

Presumably, the most commonly used probability distribution is the normal distribution. From a theoretical point of view, it is not very satisfying to use the normal distribution for stage durations (sojourn times), because its domain is the whole real line, while stage durations are nonnegative by their very nature. Furthermore, it is symmetric and so does not fit asymmetric distributions very well. Nevertheless, it has been used, for example, to determine the length and variance of the latency in measles (Bailey, 1975, Chapter 15), mainly because many techniques are available to estimate its mean and standard deviation. It is more satisfying theoretically to assume that it is not the sojourn time itself, but its natural logarithm that is normally distributed. Sartwell (1950, 1966) finds that the incubation periods of various infectious diseases are often nicely fitted by log-normal distributions and that the fits are better than those by normal distributions. The duration of a stage is *log-normally distributed* if its probability distribution function satisfies

$$\mathcal{G}(a) = \mathcal{N}(\ln a),$$

where $\mathcal{N}$ is the probability distribution function of a normal distribution; in other words,

$$\mathcal{F}(a) = \mathcal{F}_{m,\sigma}(a) = \int_{\ln a}^{\infty} \exp\left(-\frac{1}{2}\left[\frac{x-\ln m}{\sigma}\right]^2\right)\frac{1}{\sigma\sqrt{2\pi}}\,\mathrm{d}x$$

with $\ln m$ being the mean and σ the standard deviation of the normal distribution. Following Sartwell we call m the *median* and e^{σ} the *dispersion factor* of the distribution. After a couple of substitutions we obtain

$$\mathcal{F}(a) = \int_{\ln(a/m)/\sigma}^{\infty} \exp(-\tfrac{1}{2}y^2)\frac{1}{\sqrt{2\pi}}\,\mathrm{d}y = 1 - \mathcal{N}_0\left(\frac{\ln(a/m)}{\sigma}\right), \tag{12.14}$$

where $\mathcal{N}_0$ is the probability distribution function of a normal distribution with mean 0 and standard deviation 1. Notice that m has the properties of the median; indeed,

$$\mathcal{F}_{m,\sigma}(m) = \tfrac{1}{2},$$

and m is the only argument for which this value is taken. Furthermore,

$$\mathcal{F}_{m,\sigma}(a) \to \begin{cases} 0, & a > m, \\ 1, & 0 \leqslant a < m, \end{cases} \qquad \sigma \to 0,$$

i.e., $\mathcal{F}_{m,\sigma}$ pointwise approaches the sojourn function for stages with fixed duration m as $\sigma \to 0$ (cf. Figure 12.3).

We notice that $\mathcal{F}$ is differentiable and that the probability density of stage duration is given by

$$-\mathcal{F}'(a) = \exp\left(-\frac{1}{2}\left[\frac{\ln(a/m)}{\sigma}\right]^2\right)\frac{1}{a\sigma\sqrt{2\pi}}.$$

One of the nice features of the log-normal distribution is the easy calculation of moments:

$$\begin{aligned} n\int_0^{\infty} a^{n-1}\mathcal{F}(a)\,\mathrm{d}a &= -\int_0^{\infty} a^n\mathcal{F}'(a)\,\mathrm{d}a \\ &= \int_0^{\infty} a^{n-1}\exp\left(-\frac{1}{2}\left[\frac{\ln(a/m)}{\sigma}\right]^2\right)\frac{1}{\sigma\sqrt{2\pi}}\,\mathrm{d}a \\ &= m^n\int_0^{\infty} t^{n-1}\exp\left(-\frac{1}{2}\left[\frac{\ln(t)}{\sigma}\right]^2\right)\frac{1}{\sigma\sqrt{2\pi}}\,\mathrm{d}t \\ &= m^n\int_{-\infty}^{\infty} \mathrm{e}^{ns}\exp\left(-\frac{1}{2}\left[\frac{s}{\sigma}\right]^2\right)\frac{1}{\sigma\sqrt{2\pi}}\,\mathrm{d}s \end{aligned}$$

$$= m^n \int_{-\infty}^{\infty} \mathrm{e}^{nt\sigma} \exp(-\tfrac{1}{2}t^2) \frac{1}{\sqrt{2\pi}}\, \mathrm{d}t$$
$$= m^n \int_{-\infty}^{\infty} \exp(-\tfrac{1}{2}(t^2 - 2tn\sigma)) \frac{1}{\sqrt{2\pi}}\, \mathrm{d}t$$
$$= m^n \int_{-\infty}^{\infty} \exp(-\tfrac{1}{2}(t^2 - n\sigma)^2 + \tfrac{1}{2}(n\sigma)^2) \frac{1}{\sqrt{2\pi}}\, \mathrm{d}t$$
$$= m^n \mathrm{e}^{(n\sigma)^2/2}. \tag{12.15}$$

So, the mean sojourn time in the stage is given by

$$D = m\mathrm{e}^{\sigma^2/2}, \tag{12.16}$$

and the average expectation of remaining sojourn by

$$E = \tfrac{1}{2} m \mathrm{e}^{3\sigma^2/2}.$$

Notice that

$$\frac{E}{D} = \tfrac{1}{2}\mathrm{e}^{\sigma^2},$$

so log-normal distributions are flexible enough to fit all D and E values in the feasible range $E/D \geqslant \frac{1}{2}$ except fixed-stage durations, which are the limiting case for $\sigma \to 0$. The variance is

$$V = m^2(\mathrm{e}^{2\sigma^2} - \mathrm{e}^{\sigma^2}) = m^2 \mathrm{e}^{\sigma^2}(\mathrm{e}^{\sigma^2} - 1).$$

Hence

$$\frac{\sqrt{V}}{D} = \sqrt{\mathrm{e}^{\sigma^2} - 1}.$$

In order to recognize the skewness of the distribution (cf. Sectoin 12.3), we take the third central moment,

$$-\int_0^{\infty} (a - D)^3 \mathcal{F}'(a)\, \mathrm{d}a$$
$$= m^3 \mathrm{e}^{(3\sigma)^2/2} - 3m^3 \mathrm{e}^{(2\sigma)^2/2} \mathrm{e}^{\sigma^2/2} + 2m^3 \mathrm{e}^{3\sigma^2/2}$$
$$= m^3(\mathrm{e}^{9\sigma^2/2} - 3\mathrm{e}^{5\sigma^2/2} + 2\mathrm{e}^{3\sigma^2/2})$$
$$= m^3 \mathrm{e}^{3\sigma^2/2}(\mathrm{e}^{3\sigma^2} - 3\mathrm{e}^{\sigma^2} + 2)$$
$$= D^3(\mathrm{e}^{\sigma^2} + 2)(\mathrm{e}^{\sigma^2} - 1)^2 > 0.$$

So, with $\omega := \mathrm{e}^{\sigma^2} = 2E/D$,

$$\text{skewness} = (\omega + 2)(\omega - 1)^{1/2}.$$

This shows that log-normally distributed stage durations are always positively skewed. Taking derivatives we see that the skewness increases from 0 for $\sigma = 0$

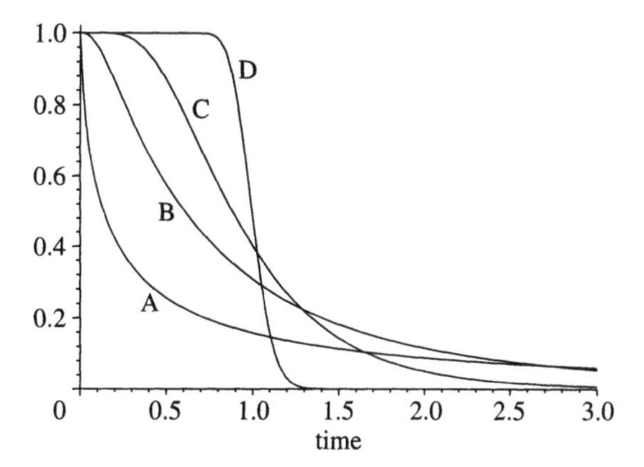

Figure 12.3. Log-normally distributed sojourn functions for various values of σ. The average durations have been normalized to 1 by choosing the medians as $m = \mathrm{e}^{-\sigma^2/2}$. (A) $\sigma = 2$. (B) $\sigma = 1$. (C) $\sigma = 0.5$. (D) $\sigma = 0.1$. As $\sigma \to 0$ and so $m \to 1$, the sojourn functions pointwise converge to the sojourn function for fixed duration 1.

to 1 for $\sigma \to \infty$. The kurtosis is $\omega^4 + 2\omega^3 + 3\omega^2 - 3$ and increases with σ. See Section 12.3, Crow, Shimizu (1988, p. 10), or Evans et al. (2000, p. 130).

Since the normal distribution is symmetric, its mean and its median coincide, which in our case is $\ln m$. Since medians are preserved under transformation, m is the median of our log-normally distributed stage sojourn, but not the mean, which is D. The fact that $D > m$ is not surprising, because the log-normal distribution is positively skewed. It is known from the properties of the normal distribution that about 16% of the logarithmic sojourn times lie above $\ln m + \sigma$ and again about 16% below $\ln m - \sigma$. This means that about 16% of the sojourn times will exceed $m\mathrm{e}^{\sigma}$ and about 16% will fall below $m\mathrm{e}^{-\sigma}$. This nice relation has motivated Sartwell (1950, 1966) to list the dispersion factor e^{σ} in addition to the median for the incubation times of many infectious diseases. From the data of Stillerman and Thalhimer (1944) on 199 cases of measles, Sartwell (1950) estimates a median of 12.2 days and a dispersion factor of 1.18 for the incubation period (actually the time interval between the onsets of rash in the first and second cases in households). This gives $\mathrm{e}^{\sigma^2} \approx 1.028$ and $\mathrm{e}^{\sigma^2/2} \approx 1.014$. So, the average duration of the incubation period, D, would be about 12.4 days, while the average expectation of remaining sojourn, E, would be about 6.4 days. Furthermore, $\sqrt{V}/D \approx 0.17$.

For Sartwell's estimate of another set of measles data (Goodall, 1931), one obtains $\sqrt{V}/D \approx 0.28$. Remember that for an exponential distribution this number would be 1. Very few of the many diseases reported by Sartwell have dispersion factors above 1.6. If the dispersion factor is below 1.6, then $\sqrt{V}/D \leqslant 0.5$ and $2E/D \leqslant 1.25$. Recall that $2E/D \geqslant 1$ always, with equality for fixed durations, and $E = D$ for exponential distributions. For comparison, Bailey (1975, Chapter 15) estimates the latent and infectious periods for three sets of measles data, assuming that the latent period is normally distributed while the infectious period has fixed length. He comes up with different values for the mean length and the variance, but in all three cases $\sqrt{V}/D \approx 0.21$ for the latent period. By formula (12.9), $2E/D \approx 1.044$.

12.9 A Stochastic Interpretation of Stage Transition

The previous sections have been written from the point of view of an analyst who intends to use these concepts, in spite of their stochastic flavor, in deterministic models. At least very briefly, we would like to make the stochastic connection. The transition through a stage can also be described by a random variable T on an appropriate probability space (Ω, Σ, P) with a set Ω, a σ-algebra Σ, and a probability measure P. For $x \in \Omega$, $T(x)$ would be a particular sojourn time in the stage, e.g., as observed in an experiment. We can then consider the induced probability measure

$$P^T(S) = P(T^{-1}(S))$$

for each Borel set S in $[0, \infty)$. The relation to the sojourn function $\mathcal{F}$ is then

$$\begin{aligned}
\mathcal{F}(0) &= P(\{T \geqslant 0\}), \\
\mathcal{F}(a-) &= P^T([a, \infty)) = P(\{T \geqslant a\}), \quad a > 0, \\
\mathcal{F}(a+) &= P^T((a, \infty)) = P(\{T > a\}), \quad a \geqslant 0.
\end{aligned}$$

For every continuous function f on $[0, \infty)$, the measure-theoretic change-of-variable formula implies that

$$\int_\Omega f(T(x))P(\mathrm{d}x) = \int_{[0,\infty)} f(a)P^T(\mathrm{d}a),$$

provided that at least one of the integrals exists absolutely. If f is continuously differentiable and nonnegative, we have

$$\int_\Omega f(T(x))P(\mathrm{d}x) = -\int_0^\infty f(a)\mathcal{F}(\mathrm{d}a) = f(0) + \int_0^\infty f'(a)\mathcal{F}(a)\,\mathrm{d}a.$$

The last equality holds if $f(a)\mathcal{F}(a) \to 0$ as $a \to \infty$.

Choosing $f(a) = a$, we obtain that the average duration of the stage, D, is given by

$$D = \int_\Omega T(x)P(\mathrm{d}x) = -\int_0^\infty a\mathcal{F}(\mathrm{d}a) = \int_0^\infty \mathcal{F}(a)\,\mathrm{d}a.$$

Choosing $f(a) = (a - D)^2$, we find the variance of the stage duration, V, as

$$\begin{aligned} V &= \int_\Omega (T(x) - D)^2 P(\mathrm{d}x) = -\int_0^\infty (a - D)^2 \mathcal{F}(\mathrm{d}a) \\ &= D^2 + 2\int_0^\infty (a - D)\mathcal{F}(a)\,\mathrm{d}a = D^2 + 2ED - 2D^2 \\ &= 2ED - D^2 = D(2E - D), \end{aligned}$$

where E is the average expectation of remaining sojourn. Here we needed to assume that $a^2\mathcal{F}(a) \to 0$ as $a \to \infty$. Recall that this is formula (12.9).

Exercises

12.1. Let $\mathcal{F}$ be a survival function such that $\mathcal{F}(a+b) = \mathcal{F}(a)\mathcal{F}(b)$ for all $a, b \geqslant 0$. Show the following alternative: either $\mathcal{F}(a) = 0$ for all $a > 0$ or there exists some $\mu \geqslant 0$ such that $\mathcal{F}(a) = \mathrm{e}^{-\mu a}$.

Hint: you may like to consider $f(a) = \ln \mathcal{F}(a)$.

12.2. Let $\mathcal{F}$ be a survival function such that $\mathcal{F}(a+b) \leqslant \mathcal{F}(a)\mathcal{F}(b)$ for all $a, b \geqslant 0$. Then $D(a) \leqslant D(0)$ for all $a \geqslant 0$ and the average expectation of remaining life does not exceed the expectation of life.

12.3. Let $\mathcal{F}$ be a sojourn function, $\gamma > 0$. Define

$$\mathcal{F}_\gamma(a) = \mathcal{F}(\gamma a), \quad a \geqslant 0.$$

(a) Show that $\mathcal{F}_\gamma$ has the properties of a sojourn function. Also show that $\mathcal{F}_\gamma$ has a finite maximum sojourn time c_γ if and only if this is the case for $\mathcal{F}$ and that they are related by $c_\gamma = c/\gamma$.

(b) Let $D_\gamma(a)$ be the expected remaining sojourn at stage age a associated with $\mathcal{F}_\gamma$. Show that

$$D_\gamma(a) = \frac{1}{\gamma} D(\gamma a).$$

(c) Let D_γ, E_γ be the mean sojourn time and the average expectation of remaining sojourn associated with $\mathcal{F}_\gamma$. Show that E_γ/D_γ is independent of γ.

(d) Let V_γ be the variance of the sojourn time associated with $\mathcal{F}_\gamma$. Relate V_γ to V, the variance associated with $\mathcal{F}$.

Remark. This exercise shows that the quotients of the average expectations of remaining sojourn and of the mean sojourn times do not change in the subsequent exercises when one more parameter is introduced (if this is done in the same way as here).

12.4. Let $\kappa > 0$ and consider a generalized de Moivre sojourn function

$$\mathcal{F}_\kappa(a) = \begin{cases} (1-a)^\kappa, & 0 \leqslant a < 1, \\ 0, & a \geqslant 1. \end{cases}$$

(a) Compute D_κ and E_κ, where D_κ, E_κ are the mean sojourn time and the average expectation of remaining sojourn associated with $\mathcal{F}_\kappa$. Discuss the behavior of E_κ / D_κ as κ varies between 0 and ∞.

(b) Compute the associated exit rate $\mu_\kappa(a)$ on $[0, 1)$ and the expected remaining sojourn $D_\kappa(a)$.

12.5. Let $\kappa > 2$ and consider a sojourn function of Pareto type

$$\mathcal{F}_\kappa(a) = [1+a]^{-\kappa}, \quad a \geqslant 0.$$

(a) Compute D_κ and E_κ. Discuss the behavior of E_κ / D_κ as κ varies between 2 and ∞.

(b) Compute the associated exit rates $\mu_\kappa(a)$ and the expected remaining sojourn $D_\kappa(a)$. Notice that $D_\kappa(\cdot)$ is an unbounded function.

12.6. Let $\kappa > 0$ and consider the sojourn function

$$\mathcal{F}_\kappa(a) = \begin{cases} 1 - a^\kappa, & 0 \leqslant a < 1, \\ 0, & a \geqslant 1. \end{cases}$$

(a) Compute D_κ and E_κ. Discuss the behavior of E_κ / D_κ as κ varies between 0 and ∞.

(b) Compute the associated exit rates $\mu_\kappa(a)$ and the expected remaining sojourn $D_\kappa(a)$.

12.7. Let $\mathcal{F}(a) = 1/(1+a^2)^2$. Calculate the life expectancy and the average expectation of remaining life.

12.8. Let $\mathcal{F}(a) = \mathrm{e}^{-a^2}$. Calculate the life expectancy and the average expectation of remaining life.

12.9. (Modified Gompertz law.) Let $D(a) = \kappa \mathrm{e}^{-a}$.

(a) Derive restrictions for κ such that $D(\cdot)$ qualifies as the expected remaining sojourn (Proposition 12.2 and Remark 12.3).

(b) Determine the sojourn function $\mathcal{F}$ and the exit rates $\mu(a)$ that are associated with $D(\cdot)$.

12.10. Let $\kappa > 0$ and $D(a) = \kappa(1 - a)$ for $0 \leqslant a \leqslant 1$, $D(a) = 0$ for $a > 1$.

(a) Derive restrictions for κ such that $D(\cdot)$ qualifies as the expected remaining sojourn (Proposition 12.2 and Remark 12.3).

(b) Determine the sojourn function $\mathcal{F}$ and the exit rates $\mu(a)$ that are associated with $D(\cdot)$.

12.11. Consider a generalized de Moivre survival function

$$\mathcal{F}(a) = \left[1 - \frac{a}{c}\right]_+^{\kappa}$$

with finite maximum age c. Show the following relation between maximum age, life expectancy, and average expectation of remaining life:

$$\frac{c}{D} = \frac{E}{D - E}.$$

Bibliographic Remarks

Hutchinson (1978, Chapter 2) displays many figures of survivorships. Most of them are displayed on semilogarithmic paper so that it is difficult to see the concavity and convexity properties of $\mathcal{F}$. An exception are the survivorships for the water flea *Daphnia pulex* at various population densities (figure 36, p. 57). There $\mathcal{F}(a)$ switches back and forth between concavity and convexity several times. Impagliazzo (1985) shows the survivorship of Danish females in 1967 (figure 1.5, p. 23) and the associated per capita mortality rates (figure 1.12, p. 32). Keyfitz, Flieger (1971, figure 10, p. 43) compare male survivorships and the associated per capita mortality rates from the United States, 1967, and from England and Wales, 1966–1968. These survival functions are strictly concave except for very small and large ages. Wills (1996) displays several survivorships, among them one from Cata Huyuk (Anatolia), determined from burials that took place between 6500 and 5300 B.C. (figure 3.3, p. 41), and two from London, 1728 (figure 3.4, p. 42) and 1830 (figure 3.5, p. 43).

Sartwell (1950, figure 1) shows discrete versions of $-\mathcal{F}'$ for the sojourn functions of the incubation periods of food-borne streptococcal sore throat and serum hepatitis.

Examples for life expectancies, D, can be found in Table D in Keyfitz, Flieger (1971) for the United States and Europe, and in Table E in that reference for five European countries and their principal cities, separately for men and women. Figures displaying expected remaining lives at age a, $D(a)$, for various countries at various times, can be found in Hutchinson (1978, figure 27, p. 45), Impagliazzo (1985, figure 1.3, p. 11 and figure 1.8, p. 24), and Smith, Keyfitz (1977, p. 5). In some cases, $D(a)$ is a decreasing function of a, but in other cases $D(a)$ first increases and then decreases. Figure 9 in Keyfitz, Flieger (1971) shows how $D(a)$ has changed in the United States from 1920 to 1967 for $a = 0, 50, 70$, separately for men and women. Hutchinson (1978, figure 32, p. 52), compares mortality rates from different times or seasons and presents interesting background information concerning Ulpian's table.

The paper by Kim, Aron (1989) motivated us to consider the average expectation of remaining sojourn. I did not find any data for the average expectation of remaining life in the literature.

Commented excerpts of the original papers by Gompertz (1825) and by Makeham (1867) can be found in Smith, Keyfitz (1977).

The log-normal distribution is also used in describing how populations partition resources (Roughgarden, 1979, Chapter 24) and to model insurance claim severity or investment returns (Hassett, Stewart, 1999, Chapter 4). The book edited by Crow and Shimura (1988) contains applications to time event distributions in medicine, business and economics, and to tissue and population growth as well as basic and background information. Hassett, Stewart (1999) also give additional information on the sojourn functions used in some of the exercises.

Condensed presentations of the material in this section and applications to epidemic models of infectious diseases can be found in Feng, Thieme (2000b) and Thieme (2002).

Chapter Thirteen

Stage Dynamics with Given Input

In order to describe the dynamics in a particular stage, we stratify, by stage age, the number $N(t)$ of individuals that are in that stage at time t,

$$N(t) = \int_0^\infty u(t,a)\,\mathrm{d}a. \tag{13.1}$$

Here $u(t,a)$ is the density of individuals at time t that have stage age a, i.e., the integral

$$\int_a^b u(t,r)\,\mathrm{d}r$$

gives the number of individuals that are in the stage at time t and have a stage age between a and b.

We recall that the *stage age* of an individual in a certain stage is the time the individual has already spent in the stage. We call $u(t,\cdot)$ the *stage-age-density* and $N(t)$ the *stage content* at time t.

13.1 Input and Stage-Age Density

Assume that $B(t)$ is the input into the stage at time t. In other words, $B(t)$ is the rate at which individuals enter the stage at time t, i.e., the number of individuals entering the stage at time t per unit of time. If the stage is life, B is the birth rate of the population under consideration.

Let 0 be the time at which we start to consider the development of the stage, and $u_0(a)$ the age-density of individuals that are in the stage at time 0.

As before let $\mathcal{F}(a)$ be the probability of still being in the stage at stage age a.

Recall the definition of the maximum stage age c. If $\mathcal{F}$ is strictly positive on $[0,\infty)$, then $c=\infty$. Otherwise $0<c<\infty$ is determined by the property that $\mathcal{F}(a)>0$ for $a<c$ and $\mathcal{F}(a)=0$ for $a>c$. Consistently with the interpretation of maximum age we assume that $u_0(a)=0$ for $a>c$.

The stage-age-density u can now be expressed as follows:

$$\left.\begin{aligned} u(t,a) &= B(t-a)\mathcal{F}(a), && t>a\geqslant 0,\\ u(t,a) &= u_0(a-t)\frac{\mathcal{F}(a)}{\mathcal{F}(a-t)}, && 0\leqslant t<a<c,\\ u(t,a) &= 0, && t\geqslant 0,\quad a>c. \end{aligned}\right\} \tag{13.2}$$

The first equation comes from the following consideration. An individual that has stage age a at time $t > a$ has entered the stage at time $t-a$. The probability that it is still in the stage is $\mathcal{F}(a)$.

As for the second equation, the reasoning goes as follows. An individual that has stage age a at time $t < a$ was already in the stage at time 0 and, at time 0, had the stage age $a-t$. The probability of still being in the stage at time t, with stage age a, is then the conditional probability $\mathcal{F}(a)/\mathcal{F}(a-t)$, the probability of still being in the stage at stage age a under the condition one was in the stage at stage age $a-t$.

One typically leaves $u(t,a)$ undefined for $t = a$, because it is not clear whether one should choose $B(0)\mathcal{F}(a)$ or $u_0(0)\mathcal{F}(a)$ unless $B(0) = u_0(0)$. But there is no biological reason why such an equality should hold.

Equation (13.2) is basically the same as the equation suggested by Lotka (1907), with the difference that Lotka uses the first equation exclusively and assumes the input to also be defined for negative times.

The following convention will make our formalism much easier:

$$\frac{u_0(a)}{\mathcal{F}(a)} := 0, \quad a > c.$$

We have from (13.1) that

$$N(t) = \int_0^t B(t-a)\mathcal{F}(a)\,\mathrm{d}a + \int_t^\infty \mathcal{F}(a)\frac{u_0(a-t)}{\mathcal{F}(a-t)}\,\mathrm{d}a.$$

This means that we can split up the stage content into two parts, one consisting of individuals that were already in the stage at time $t = 0$ and another consisting of individuals that have entered later. The size of the first part is denoted by $N_0(t)$ and the size of the second by $N_1(t)$; so, after substitution,

$$\left.\begin{aligned} N(t) &= N_0(t) + N_1(t), \\ N_0(t) &= \int_0^c u_0(a)\frac{\mathcal{F}(a+t)}{\mathcal{F}(a)}\,\mathrm{d}a, \\ N_1(t) &= \int_0^t B(t-a)\mathcal{F}(a)\,\mathrm{d}a. \end{aligned}\right\} \tag{13.3}$$

13.2 The Partial Differential Equation Formulation

Assume that B and $u_0/\mathcal{F}$ are differentiable. Then, by (13.2),

$$\frac{u(t,a)}{\mathcal{F}(a)} = \begin{cases} B(t-a), & t > a \geqslant 0, \\ \dfrac{u_0(a-t)}{\mathcal{F}(a-t)}, & 0 \leqslant t < a < c, \end{cases}$$

is differentiable for $t \neq a$ and satisfies the partial differential equation

$$\frac{\partial}{\partial t}\frac{u(t,a)}{\mathcal{F}(a)} + \frac{\partial}{\partial a}\frac{u(t,a)}{\mathcal{F}(a)} = 0, \quad t \neq a.$$

From (13.2) and $\mathcal{F}(0) = 1$, we also have the boundary condition

$$u(t,0) = B(t),$$

and the initial condition

$$u(0,a) = u_0(a).$$

Now assume that B, u_0 and $\mathcal{F}$ are differentiable. Then u is differentiable for $t \neq a$ and

$$0 = \frac{(\partial/\partial t)u(t,a)}{\mathcal{F}(a)} + \frac{(\partial/\partial a)u(t,a)}{\mathcal{F}(a)} - u(t,a)\frac{\mathcal{F}'(a)}{(\mathcal{F}(a))^2}.$$

Multiplying this equation by $\mathcal{F}$ and using the per capita exit rate (Section 12.6)

$$\mu(a) = -\frac{\mathcal{F}'(a)}{\mathcal{F}(a)},$$

we obtain the following partial differential equation for u with boundary and initial conditions:

$$\frac{\partial}{\partial t}u(t,a) + \frac{\partial}{\partial a}u(t,a) + \mu(a)u(t,a) = 0, \quad t \neq a,$$
$$u(t,0) = B(t),$$
$$u(0,a) = u_0(a).$$

This equation is nowadays called McKendrick's equation (McKendrick, 1926), after it was associated with VonFoerster (1959) for a while.

One can recover (13.2) from this equation using the method of characteristics, which is a general method for solving first-order partial differential equations. Actually, we will look at a stage-transition problem in which the per capita exit rate μ depends on both stage age a and time t:

$$\left.\begin{aligned} \frac{\partial}{\partial t}u(t,a) + \frac{\partial}{\partial a}u(t,a) + \mu(t,a)u(t,a) &= 0, \quad t \neq a, \\ u(t,0) &= B(t), \\ u(0,a) &= u_0(a). \end{aligned}\right\} \tag{13.4}$$

Rather than explaining the method of characteristics, we proceed in the following way, which is more suggestive from a modeling point of view. Introduce $v(a,s)$ as the number of individuals in the stage that have stage age a and have entered the stage at time s. These coincide with the individuals in the stage at time $t = s + a$ and stage age a, hence

$$v(a, s) = u(a + s, a).$$

Then

$$\begin{aligned}\frac{\partial}{\partial a} v(a, s) &= \frac{\partial}{\partial t} u(t, a)\bigg|_{t=a+s} + \frac{\partial}{\partial a} u(t, a)\bigg|_{t=a+s}\\ &= -\mu(t, a) u(t, a)|_{t=a+s}\\ &= -\mu(a + s, a) v(a, s).\end{aligned}$$

Furthermore,

$$v(0, s) = u(s, 0) = B(s).$$

Then

$$v(a, s) = B(s) \exp\left(-\int_0^a \mu(r + s, r)\, \mathrm{d}r\right).$$

Recovering u from v,

$$u(t, a) = v(a, t - a) = B(t - a) \exp\left(-\int_0^a \mu(r + t - a, r)\, \mathrm{d}r\right), \quad t > a. \tag{13.5}$$

Now let

$$w(t, s) = u(t, t + s).$$

$w(t, s)$ can be interpreted as the number of individuals that are in the stage at time t and have been in the stage already at time 0, and had the stage age s at time 0. We have

$$\begin{aligned}\frac{\partial}{\partial t} w(t, s) &= \frac{\partial}{\partial t} u(t, a)\bigg|_{a=t+s} + \frac{\partial}{\partial a} u(t, a)\bigg|_{a=t+s}\\ &= -\mu(t, t + s) u(t, t + s)\\ &= -\mu(t, t + s) w(t, s).\end{aligned}$$

Furthermore,

$$w(0, s) = u(0, s) = u_0(s).$$

Hence

$$w(t, s) = u_0(s) \exp\left(-\int_0^t \mu(r, r + s)\, \mathrm{d}r\right).$$

Recovering u from w,

$$u(t, a) = w(t, a - t) = u_0(a - t) \exp\left(-\int_0^t \mu(r, r + a - t)\, \mathrm{d}r\right), \quad a > t. \tag{13.6}$$

Three Ways to Write u

By appropriate substitutions in (13.5) and (13.6), we can write u in three different ways.

First Way

$$u(t,a) = \begin{cases} B(t-a)\exp\left(-\int_{t-a}^{t} \mu(r, r+a-t)\,\mathrm{d}r\right), & t > a, \\ u_0(a-t)\exp\left(-\int_{0}^{t} \mu(r, r+a-t)\,\mathrm{d}r\right), & a > t. \end{cases}$$

We introduce

$$P(t,s,b) = \exp\left(-\int_{s}^{t} \mu(r, b+r-s)\,\mathrm{d}r\right)$$

and can express u as

$$u(t,a) = \begin{cases} B(t-a)P(t,t-a,0), & t > a, \\ u_0(a-t)P(t,0,a-t), & 0 \leqslant a < t. \end{cases} \tag{13.7}$$

$P(t,s,b)$ can be interpreted as the probability of still being in the stage at time t provided one was in the stage at time s and, at time s, had the stage age b. P is connected to the time-autonomous case (with μ not depending on t) by

$$P(t,s,b) = \frac{\mathcal{F}(b+t-s)}{\mathcal{F}(b)}.$$

Second Way

Another way of expressing u is

$$u(t,a) = \begin{cases} B(t-a)\exp\left(-\int_{0}^{a} \mu(t-r, a-r)\,\mathrm{d}r\right), & t > a \geqslant 0, \\ u_0(a-t)\exp\left(-\int_{0}^{t} \mu(t-r, a-r)\,\mathrm{d}r\right), & 0 \leqslant t < a. \end{cases}$$

Introducing

$$\Pi(a,t,x) = \exp\left(-\int_{0}^{x} \mu(t-r, a-r)\,\mathrm{d}r\right),$$

we can write

$$u(t,a) = \begin{cases} B(t-a)\Pi(a,t,a), & t > a \geqslant 0, \\ u_0(a-t)\Pi(a,t,t), & t < a. \end{cases} \tag{13.8}$$

$\Pi(a,t,x)$ can be interpreted as the probability of still being in the stage at time t with stage age a provided one was in the stage at time $t-x$ with age $a-x$.

Π is related to the time-autonomous case by

$$\Pi(a,t,x) = \frac{\mathcal{F}(a)}{\mathcal{F}(a-x)}.$$

Third Way

Finally, we can write

$$u(t,a) = \begin{cases} B(t-a)\exp\left(-\int_0^a \mu(r+t-a,r)\,\mathrm{d}r\right), & t > a \geqslant 0, \\ u_0(a-t)\exp\left(-\int_{a-t}^a \mu(t+r-a,r)\,\mathrm{d}r\right), & 0 \leqslant t < a. \end{cases}$$

We introduce

$$\mathcal{F}(a,\sigma) = \exp\left(-\int_0^a \mu(r+\sigma,r)\,\mathrm{d}r\right), \quad a \geqslant 0, \quad \sigma \in \mathbb{R},$$

and can write

$$u(t,a) = \begin{cases} B(t-a)\mathcal{F}(a,t-a), & t > a \geqslant 0, \\ u_0(a-t)\dfrac{\mathcal{F}(a,t-a)}{\mathcal{F}(a-t,t-a)}, & 0 \leqslant t < a. \end{cases} \tag{13.9}$$

$\mathcal{F}(a,\sigma)$ can be interpreted as the probability of still being in the stage at stage age a having entered the stage at time σ. Actually, the definition of $\mathcal{F}$ requires $\mu(t,a)$ to also be given for $t \leqslant 0$. This can easily be achieved by extending μ by $\mu(t,a) = \mu(0,a)$ for $t < 0$. Note that it does not affect the formula for u how μ is extended.

Formula (13.9) is closest to the time-autonomous case:

$$\mathcal{F}(a,\sigma) = \mathcal{F}(a).$$

Exercises

13.1. Consider the following stage-transition model with immigration:

$$\begin{gathered} \frac{\partial}{\partial t}u(t,a) + \frac{\partial}{\partial a}u(t,a) + \mu(a)u(t,a) = f(t,a), \quad t \neq a, \\ u(t,0) = 0, \\ u(0,a) = 0. \end{gathered}$$

Here f is a continuous function from $[0, \infty)^2$ to $\mathbb{R}$ representing the immigration rate. Solve for u. Try to find a solution in terms of $\mathcal{F}$ and f, rather than μ and f.

13.2. Assume that the stage duration is exponentially distributed, i.e., $\mathcal{F}(a) = e^{-\mu a}$ with $D = 1/\mu$ being the mean stage duration. Show that

$$N'(t) = B(t) - \mu N(t).$$

13.3. Assume that $\mathcal{F}(a) = 1$ for $0 \leqslant a < c$ and $\mathcal{F}(a) = 0$ for $a > c$. Let B and u_0 be continuous. Show that N is differentiable for $t \neq c$ and

$$N'(t) = \begin{cases} B(t) - u_0(c-t), & 0 < t < c, \\ B(t) - B(t-c), & t > c. \end{cases}$$

13.3 Stage Content and Average Stage Duration

In this section we want to relate stage content, stage input, and average stage duration under the assumption that the stage input stabilizes. We start with the observation that the individuals that were in the state already at time 0 vanish as time tends to infinity. Recall formula (13.3).

We assume that the function u_0 on $[0, c)$ is nonnegative and that

$$N_0(0) = \int_0^c u_0(a)\,\mathrm{d}a$$

exists and is finite. This means that the initial stage content has finite size.

Proposition 13.1. *Let $N_0(0) < \infty$ and $\mathcal{F}(\infty) = 0$. Then $N_0(t) \leqslant N_0(0) < \infty$ for $t > 0$ and*

$$N_0(t) \longrightarrow 0, \quad t \to \infty.$$

Proof. The first statement follows from $\mathcal{F}$ being nonincreasing. The second statement is obvious if the maximum age c is finite, because then $N_0(t) = 0$ for $t > c$. Hence we assume $c = \infty$. Let $r > 0$. Then

$$\begin{aligned} N_0(t) &\leqslant \int_0^r u_0(a) \frac{\mathcal{F}(t)}{\mathcal{F}(r)}\,\mathrm{d}a + \int_r^\infty u_0(a)\,\mathrm{d}a \\ &\leqslant \frac{\mathcal{F}(t)}{\mathcal{F}(r)} N_0(0) + \int_r^\infty u_0(a)\,\mathrm{d}a. \end{aligned}$$

Since $\mathcal{F}(t) \to 0$ as $t \to \infty$,

$$\limsup_{t\to\infty} N_0(t) \leqslant \int_r^\infty u_0(a)\,\mathrm{d}a.$$

Since this holds for every $r > 0$, we have

$$\limsup_{t\to\infty} N_0(t) = 0.$$

□

If the stage input approaches a finite limit, so does the stage content, and the two limits are related to each other by the average stage duration (or mean sojourn time), D.

Theorem 13.2. *Assume $N_0(0) < \infty$ and $D < \infty$ and that $B(t)$ tends to a finite strictly positive limit $B(\infty)$ as $t \to \infty$. Then*

$$N(t) \to DB(\infty),$$

where D is the mean sojourn time in the stage.

Proof. By assumption,

$$\infty > D = \int_0^\infty \mathcal{F}(a)\,\mathrm{da} \geqslant \int_0^t \mathcal{F}(a)\,\mathrm{da} \geqslant t\mathcal{F}(t).$$

Hence

$$\mathcal{F}(t) \leqslant D/t \to 0, \quad t \to \infty.$$

By Proposition 13.1, $N_0(t) \to 0$ as $t \to \infty$. So, it is sufficient to prove that $N_1(t) \to DB(\infty)$ as $t \to \infty$. Let $t > r > 0$. Then

$$\begin{aligned}
|N_1(t) - DB(\infty)| \\
&= \left|\int_0^t B(t-a)\mathcal{F}(a)\,\mathrm{da} - B(\infty)\int_0^\infty \mathcal{F}(a)\,\mathrm{da}\right| \\
&= \left|\int_0^r B(t-a)\mathcal{F}(a)\,\mathrm{da} - B(\infty)\int_0^r \mathcal{F}(a)\,\mathrm{da}\right. \\
&\qquad \left. + \int_r^t B(t-a)\mathcal{F}(a)\,\mathrm{da} - B(\infty)\int_r^\infty \mathcal{F}(a)\,\mathrm{da}\right| \\
&\leqslant \int_0^r |B(t-a) - B(\infty)|\mathcal{F}(a)\,\mathrm{da} + 2\sup|B|\int_r^\infty \mathcal{F}(a)\,\mathrm{da}.
\end{aligned}$$

Let $\epsilon > 0$. Since $D < \infty$,

$$\int_r^\infty \mathcal{F}(a)\,\mathrm{da} \to 0, \quad r \to \infty.$$

Hence there exists some $r > 0$ such that

$$2\sup B \int_r^\infty \mathcal{F}(a)\,\mathrm{da} < \tfrac{1}{2}\epsilon.$$

Since $B(s) \to B(\infty)$ as $s \to \infty$, there exists some $t_\epsilon > 0$ such that

$$|B(s) - B(\infty)| < \frac{\epsilon}{2D} \quad \forall s \geqslant t_\epsilon.$$

Let $t > t_\epsilon + r$. Then

$$|B(t-a) - B(\infty)| < \frac{\epsilon}{2D} \quad \forall a \in [0, r]$$

and

$$\int_0^r |B(t-a) - B(\infty)| \mathcal{F}(a)\, da \leqslant \tfrac{1}{2}\epsilon.$$

Hence

$$\left| \int_0^t B(t-a)\mathcal{F}(a)\, da - B(\infty)D \right| < \tfrac{1}{2}\epsilon + \tfrac{1}{2}\epsilon = \epsilon.$$

□

If we consider the dynamics of a population, this theorem means the following. If the population birth rate stabilizes to some limit $B(\infty)$, the population size stabilizes to this limit times the life expectancy.

Exercises

13.4. Assume that the maximum stage duration, c, is finite and that there exist $B(\infty), r \geqslant 0$ such that

$$B(t) = B(\infty) \quad \text{for all } t \geqslant r.$$

Show that $N(t) = B(\infty)D$ for all $t \geqslant r + c$.

13.4 Average Stage Age

The mean sojourn time in a stage has to be discriminated from the average stage (or class) age of the individuals in the stage. If the stage is life, the average stage age coincides with the average age of a population. Unfortunately, the average stage age depends very much on the input into the stage, i.e., the way in which individuals enter the stage.

The *average stage age* is the average of the individual stage ages taken over all individuals in the stage. At time t, it is given by

$$A(t) = \frac{1}{N(t)} \int_0^\infty a u(t,a)\, da.$$

By (13.2),

$$A(t) = \frac{1}{N(t)} \left(\int_0^t a B(t-a)\mathcal{F}(a)\, da + \int_0^\infty (a+t) u_0(a) \frac{\mathcal{F}(a+t)}{\mathcal{F}(a)}\, da \right).$$

Assumption 13.3. *We assume that the expected remaining sojourn at stage age a, $D(a)$, is a bounded function of a.*

We further assume that the function u_0 on $[0, c)$ is nonnegative and that

$$\int_0^c u_0(a)\,\mathrm{d}a$$

is finite. This means that the initial stage content has finite size. Since we also want to speak about the initial mean age, we assume that

$$\int_0^c a u_0(a)\,\mathrm{d}a$$

is finite.

Proposition 13.4. *Under these assumptions,*

$$\int_0^c (a+t)u_0(a)\frac{\mathcal{F}(a+t)}{\mathcal{F}(a)}\,\mathrm{d}a \to 0, \quad t \to \infty.$$

Proof. The statements are obvious if the maximum age c is finite, because this expression is 0 for $t > c$. Hence we assume $c = \infty$. We recall from Proposition 12.1 that the boundedness of $D(a)$ implies that

$$\frac{\mathcal{F}(a+t)}{\mathcal{F}(a)} \leqslant M\mathrm{e}^{-\epsilon t} \quad \forall t \geqslant 0.$$

Using this estimate,

$$\begin{aligned}
&\int_0^c (a+t)u_0(a)\frac{\mathcal{F}(a+t)}{\mathcal{F}(a)}\,\mathrm{d}a \\
&\qquad \leqslant \int_0^c (a+t)u_0(a)M\mathrm{e}^{-\mu t}\,\mathrm{d}a \\
&\qquad \leqslant M\mathrm{e}^{-\mu t}\int_0^c a u_0(a)\,\mathrm{d}a + Mt\mathrm{e}^{-\mu t}\int_0^c u_0(a)\,\mathrm{d}a \to 0, \quad t \to \infty.
\end{aligned}$$

□

Theorem 13.5. *Assume that Assumption 13.3 is satisfied. If $B(t)$ tends to a finite strictly positive limit $B(\infty)$ as $t \to \infty$, then*

$$A(t) \to E, \quad t \to \infty,$$

where E is the average expectation of remaining sojourn.

Proof. By Theorem 13.2, $N(t) \to DB(\infty)$ as $t \to \infty$. By Proposition 13.4 it is enough to show that

$$\int_0^t aB(t-a)\mathcal{F}(a)\,\mathrm{d}a \to B(\infty)\int_0^\infty a\mathcal{F}(a)\,\mathrm{d}a = B(\infty)DE.$$

Since $D(\cdot)$ is bounded,

$$\int_0^\infty a\mathcal{F}(a)\,\mathrm{d}a < \infty$$

by Proposition 12.1. The proof now is very similar to the proof of Theorem 13.2 and is omitted. □

If we consider the dynamics of a population, this theorem means the following. If the population birth rate stabilizes to some limit $B(\infty)$, the average age in the population converges to the expectation of remaining life (cf. Kim, Aron, 1989).

Let us return to the dilemma we face when we want to replace a realistic survival distribution by an exponentially distributed one (Proposition 12.6). This dilemma originates from $D = E = 1/\mu$ for an exponentially distributed survivorship, while the equality $D = E$ does not hold in reality, creating the question of whether one should choose $\mu = 1/D$ or $\mu = 1/E$ as the per capita mortality rate. From Theorems 13.2 and 13.5 we have the relations

$$N(t) \to DB(\infty), \quad A(t) \to E, \quad t \to \infty.$$

Most of the time, it is the first relation one wants to be preserved and so one typically chooses μ as the reciprocal of the life expectancy, and not as the reciprocal of the expectation of remaining life (or of the standard variation of life length).

Exercises

13.5. Assume that the maximum stage duration, c, is finite and that there exist $B(\infty) > 0$, $r \geqslant 0$ such that

$$B(t) = B(\infty) \quad \text{for all } t \geqslant r.$$

Show that $A(t) = E$ for all $t \geqslant r + c$.

13.6. Complete the proof of Theorem 13.5, $A(t) \to E$ as $t \to \infty$.

13.5 Stage Exit Rates

The exit rate of the stage, C, is the rate at which individuals leave the stage, i.e., the number of individuals leaving the stage per unit of time. Let us assume that $\mathcal{F}$ is differentiable on $[0, \infty)$, i.e., the per capita exit rate $\mu(\cdot)$ is defined for all ages,

$$\mu(a) = -\frac{\mathcal{F}'(a)}{\mathcal{F}(a)}, \quad a \in [0, \infty).$$

Then the stage exit rate is given by

$$\left.\begin{aligned}
C(t) &= \int_0^\infty \mu(a)u(t,a)\,\mathrm{d}a = C_1(t) + C_0(t),\\
C_1(t) &= \int_0^t \mu(a)B(t-a)\mathcal{F}(a)\,\mathrm{d}a = -\int_0^t B(t-a)\mathcal{F}'(a)\,\mathrm{d}a,\\
C_0(t) &= \int_t^\infty \mu(a)\mathcal{F}(a)\frac{u_0(a-t)}{\mathcal{F}(a-t)}\,\mathrm{d}a = -\int_0^\infty \mathcal{F}'(a+t)\frac{u_0(a)}{\mathcal{F}(a)}\,\mathrm{d}a.
\end{aligned}\right\} \tag{13.10}$$

Here we have used formula (13.2) for the stage-age-density u. C_0 is the part of the exit rate associated with the individuals that were in the state initially, while C_1 is the part associated with those that have entered later.

13.5.1 The Fundamental Balance Equation of Stage Dynamics

Next we present conditions such that the state content $N(t)$ is differentiable in t and

$$N'(t) = B(t) - C(t), \tag{13.11}$$

i.e., the rate of change of the population size is the difference between the stage input rate and the stage exit rate.

If the stage is life, N can be interpreted as the population size, B as the population birth rate, and C as the population mortality rate.

With this interpretation (13.11) is called the *fundamental balance equation of population dynamics* for closed populations, i.e., without immigration and emigration. For a general stage, we call it the *fundamental balance equation of stage dynamics* (cf. Section 2.1).

In the following, we assume that the input B is continuous and that the function $\mathcal{F}$ is continuously differentiable on $[0,\infty)$ with a bounded derivative $\mathcal{F}'$. More restrictively, we assume that there exists a constant $K > 0$ such that

$$\mathcal{F}'(a+t) \leqslant K\mathcal{F}(a) \quad \forall t, a \geqslant 0.$$

If $c < \infty$, this implies that $\mathcal{F}'(a) = 0$ for $a \geqslant c$.

Proposition 13.6. *Under these assumptions, the stage content $N(t)$ is continuously differentiable and*

$$N'(t) = B(t) - \int_0^c \mu(a)u(t,a)\,\mathrm{d}a = B(t) - C(t).$$

Proof. By formula (13.3),

$$N(t) = N_1(t) + N_0(t),$$

$$N_1(t) = \int_0^t B(t-a)\mathcal{F}(a)\,\mathrm{d}a,$$

$$N_0(t) = \int_0^c u_0(r)\frac{\mathcal{F}(t+r)}{\mathcal{F}(r)}\,\mathrm{d}r.$$

In order to be able to use the differentiability of $\mathcal{F}$, we substitute $t-a=r$ in the integral for N_1:

$$N_1(t) = \int_0^t B(r)\mathcal{F}(t-r)\,\mathrm{d}r.$$

Differentiating N_1 amounts to differentiating with respect to t in the upper integration limit and differentiating with respect to t within the integral. The latter is done by interchanging differentiation and integration. This yields

$$N_1'(t) = B(t)\mathcal{F}(0) + \int_0^t B(r)\frac{\mathrm{d}}{\mathrm{d}t}\mathcal{F}(t-r)\,\mathrm{d}r$$
$$= B(t) + \int_0^t B(r)\mathcal{F}'(t-r)\,\mathrm{d}r.$$

This formal calculation can be made rigorous and extended to a more general setting (see Toolbox B.4). Reversing the substitution and using (13.10) we obtain

$$N_1'(t) = B(t) - C_1(t).$$

As for N_0, we again interchange integration and differentiation and obtain from (13.10) that

$$N_0'(t) = \int_0^c u_0(r)\frac{\mathrm{d}}{\mathrm{d}t}\frac{\mathcal{F}(r+t)}{\mathcal{F}(r)}\,\mathrm{d}r = \int_0^c u_0(r)\frac{\mathcal{F}'(r+t)}{\mathcal{F}(r)}\,\mathrm{d}r$$
$$= -C_0(t).$$

For a rigorous proof we refer to Toolbox B.4. Adding the differential equations for N_1 and N_0 yields the result (see (13.10) again). □

If $\mathcal{F}$ is not differentiable, we can formally express the stage exit rate using Stieltjes integrals

$$C(t) = -\int_0^t B(t-a)\,\mathrm{d}\mathcal{F}(a) - \int_t^\infty \frac{u_0(a-t)}{\mathcal{F}(a-t)}\,\mathrm{d}\mathcal{F}(a)$$

with the usual understanding that

$$\frac{u_0(a)}{\mathcal{F}(a)} = 0, \quad a \geqslant c,$$

if $c < \infty$. Since $\mathcal{F}$ is monotone nonincreasing, $\mathcal{F}$ is of bounded variation and the integral exists, provided that B and u_0 are continuous. We still obtain the fundamental balance equation of stage dynamics (equation (13.11)), though it only holds for almost all $t > 0$ (see Toolbox B.4).

It is intuitively clear and also perceptible from the formula $N_0' = -C_0$ in the proof of Proposition 13.6 that the rate of change of the number of individuals that were already in the stage initially only contributes to the stage exit rate, but not to the stage input. One might conjecture that $N_0'(t) \to 0$ as $t \to \infty$, in particular because we have shown that $N_0(t) \to 0$ as $t \to \infty$ (Proposition 13.1). This is true, of course, if the maximum stage duration, c, is finite. For $c = \infty$, however, one can construct $\mathcal{F}$ and u_0 with the following properties (see Exercise 13.8).

- $D(\cdot)$ is bounded on $[0, \infty)$.
- u_0 is bounded and continuously differentiable,

$$\int_0^\infty u_0(a)\,\mathrm{d}a < \infty, \qquad \sum_{j=1}^\infty u_0(j) = \infty.$$

- $N_0(t)$ is continuously differentiable on every interval $(m, m+1)$, $m \in \mathbb{N}$, and

$$\lim_{t \to m-} N_0'(t) = \infty \quad \forall m \in \mathbb{N}.$$

13.5.2 Average Age at Stage Exit

Since, at time t, the stage exit rate at (stage) age a is given by $\mu(a)u(t,a)$, the average age at stage exit, $A^\sharp(t)$, is defined as

$$A^\sharp(t) = \frac{\int_0^c a\mu(a)u(t,a)\,\mathrm{d}a}{\int_0^c \mu(a)u(t,a)\,\mathrm{d}a} = \frac{1}{C(t)}\int_0^c a\mu(a)u(t,a)\,\mathrm{d}a.$$

We assume that $c < \infty$ or that u_0 is bounded and zero for large ages and that $B(t) \to B(\infty) > 0$ as $t \to \infty$. Furthermore, we assume that $t\mathcal{F}(t) \to 0$ as $t \to \infty$. We claim that the average age at stage exit converges to the stage duration (expected age at stage exit):

$$A^\sharp(t) \to D, \quad t \to \infty.$$

Set

$$M(t) = \int_0^c a\mu(a)u(t,a)\,\mathrm{d}a.$$

By (13.2),

$$M(t) = M_1(t) + M_0(t),$$

$$M_1(t) = \int_0^t a\mu(a)B(t-a)\mathcal{F}(a)\,\mathrm{d}a$$

$$= -\int_0^t B(t-a)a\mathcal{F}'(a)\,\mathrm{d}a,$$

$$M_0(t) = \int_t^\infty a\mu(a)u_0(a-t)\frac{\mathcal{F}(a)}{\mathcal{F}(a-t)}\,\mathrm{d}a$$

$$= -\int_0^\infty (a+t)\mathcal{F}'(a+t)\frac{u_0(a)}{\mathcal{F}(a)}\,\mathrm{d}a.$$

Here we have again used the convention that $u_0(a)/\mathcal{F}(a) = 0$ whenever $\mathcal{F}(a) = 0$. Taking a formal limit for $t \to \infty$ in the formula for M_1, we obtain

$$M_1(t) \to -\int_0^\infty B(\infty)a\mathcal{F}'(a)\,\mathrm{d}a = B(\infty)\int_0^\infty \mathcal{F}(a)\,\mathrm{d}a = B(\infty)D.$$

A precise proof can be given along the lines of the proof of Theorem 13.2. Obviously, $M_0(t) \to 0$ as $t \to \infty$ if $c < \infty$. Otherwise let $|u_0(a)| \leqslant K$ for all $a \geqslant 0$ and $u_0(a) = 0$ for $a \geqslant b$ with some appropriate $b > 0$. Since F is nonincreasing,

$$|M_0(t)| \leqslant -\frac{K}{\mathcal{F}(b)}\int_0^b (a+t)\mathcal{F}'(a+t)\,\mathrm{d}a$$

$$\leqslant -\frac{K}{\mathcal{F}(b)}\int_0^b (b+t)\mathcal{F}'(a+t)\,\mathrm{d}a$$

$$= \frac{K}{\mathcal{F}(b)}(b+t)[\mathcal{F}(t) - \mathcal{F}(t+b)].$$

Since $t\mathcal{F}(t) \to 0$ as $t \to \infty$ by assumption, $|M_0(t)| \to 0$ as $t \to \infty$. Similarly, one shows that $C_1(t) \to B(\infty)$, $C_0(t) \to 0$ as $t \to \infty$, and so

$$A^\sharp(t) \to \frac{B(\infty)D}{B(\infty)} = D.$$

Exercises

13.7. Let $c < \infty$, $\mu > 0$, $u_0, B : [0, \infty) \to [0, \infty)$, and let

$$\mathcal{F}(a) = \begin{cases} \mathrm{e}^{-\mu a}, & 0 \leqslant a < c, \\ 0, & a > c. \end{cases}$$

Show that N is differentiable in $t \neq c$ and

$$N'(t) = B(t) - \mathrm{e}^{-\mu t} u_0(c - t) - \mu N(t), \quad 0 < t < c,$$
$$N'(t) = B(t) - \mathrm{e}^{-\mu c} B(t - c) - \mu N(t), \quad t > c.$$

13.8. Define

$$\mathcal{F}(a) = 2^{-n}, \quad a \in [n, n+1), \quad n = 0, 1, 2, \ldots,$$
$$u_0(a) = \begin{cases} \phi(2^n(x - n)), & |x - n| < 2^{-n}, \quad n \in \mathbb{N}, \quad n \geqslant 1, \\ 0, & \text{otherwise}, \end{cases}$$

where $\phi(x) = (1 - x^2)^2$. Let

$$N_0(t) = \int_0^\infty \frac{u_0(a)}{\mathcal{F}(a)} \mathcal{F}(a + t)\, \mathrm{d}a.$$

Show that $\mathcal{F}$, u_0, and N_0 have the properties described at the end of Section 13.5.1.

13.6 Stage Outputs

If a stage does not coincide with the whole life, but only includes a part of it, the stage exit rate is made of individuals leaving by death and of individuals leaving alive. We call the rate at which individuals leave the stage alive the output of the stage. If one looks at various stages of an insect species, the input of the pupal stage, for example, is the living part of the exit rate of the larval stage, i.e., the output of the larval stage.

In order to address the output, we factor the sojourn function $\mathcal{F}$ as

$$\mathcal{F}(a) = \mathcal{F}_\mathrm{s}(a)\mathcal{F}_\mathrm{d}(a),$$

where $\mathcal{F}_\mathrm{s}(a)$ is the probability of still being alive at stage age a and $\mathcal{F}_\mathrm{d}(a)$ is the probability of not having left the stage before stage age a in other ways than dying. Both $\mathcal{F}_\mathrm{s}$ and $\mathcal{F}_\mathrm{d}$ are nonnegative, nonincreasing functions on $[0, \infty)$, $\mathcal{F}_\mathrm{s}(0) = \mathcal{F}_\mathrm{d}(0) = 1$.

While we have not done so before, we now distinguish between stage duration and stage sojourn if a stage is only a part of life. *Stage duration* is the length of the stage discounting death; *stage sojourn* may also be terminated by death. We call $\mathcal{F}_\mathrm{s}$ the *stage survival function*, $\mathcal{F}_\mathrm{d}$ the *stage duration function*, and the product $\mathcal{F}$ the *stage sojourn function*.

Assume that $\mathcal{F}_\mathrm{s}$ and $\mathcal{F}_\mathrm{d}$ are differentiable. Then

$$\mu_\mathrm{s}(a) = -\frac{\mathcal{F}_\mathrm{s}'(a)}{\mathcal{F}_\mathrm{s}(a)}$$

is the per capita mortality rate, while

$$\mu_{\mathrm{d}}(a) = -\frac{\mathcal{F}_{\mathrm{d}}'(a)}{\mathcal{F}_{\mathrm{d}}(a)}$$

is the per capita rate of leaving the stage alive. The (living) output of the stage is then given by

$$\begin{aligned} C(t) &= \int_0^\infty \mu_{\mathrm{d}}(a) u(t,a)\,\mathrm{d}a \\ &= \int_0^t B(t-a)\mu_{\mathrm{d}}(a)\mathcal{F}(a)\,\mathrm{d}a + \int_t^\infty \frac{u_0(a-t)}{\mathcal{F}(a-t)}\mu_{\mathrm{d}}(a)\mathcal{F}(a)\,\mathrm{d}a. \end{aligned}$$

Using the formulas for μ_{s} and μ_{d},

$$C(t) = -\int_0^t B(t-a)\mathcal{F}_{\mathrm{s}}(a)\mathcal{F}_d'(a)\,\mathrm{d}a - \int_t^\infty \frac{u_0(a-t)}{\mathcal{F}(a-t)}\mathcal{F}_{\mathrm{s}}(a)\mathcal{F}_{\mathrm{d}}'(a)\,\mathrm{d}a.$$

If $\mathcal{F}_{\mathrm{s}}$ or $\mathcal{F}_{\mathrm{d}}$ are not differentiable, then we still have the Stieltjes formulation

$$\left.\begin{aligned} C(t) &= C_1(t) + C_0(t), \\ C_1(t) &= -\int_0^t B(t-a)\mathcal{F}_{\mathrm{s}}(a)\,\mathrm{d}\mathcal{F}_{\mathrm{d}}(a), \\ C_0(t) &= -\int_t^\infty \frac{u_0(a-t)}{\mathcal{F}(a-t)}\mathcal{F}_{\mathrm{s}}(a)\,\mathrm{d}\mathcal{F}_{\mathrm{d}}(a), \end{aligned}\right\} \tag{13.12}$$

with C_1 being the part originating from the stage input and C_0 being the part originating from the initial stage content. Recall that

$$\frac{u_0(a)}{\mathcal{F}(a)} = 0, \quad a \geqslant c.$$

A similar formula holds for the stage mortality rate. You just need to interchange the indices "s" and "d."

Theorem 13.7. *Let $\mathcal{F}_{\mathrm{s}}$ or $\mathcal{F}_{\mathrm{d}}$ be continuous and*

$$p = -\int_0^\infty \mathcal{F}_{\mathrm{s}}(a)\,\mathrm{d}\mathcal{F}_{\mathrm{d}}(a) < \infty.$$

(a) *If $B(t) \to B(\infty)$ as $t \to \infty$, then $C_1(t) \to pB(\infty)$.*

(b) *If $u_0/\mathcal{F}$ is bounded on $[0, c)$, then $C_0(t) \to 0$ as $t \to \infty$.*

Remark. Notice that

$$p = -\int_0^\infty \mathcal{F}_{\mathrm{s}}(a)\,\mathrm{d}\mathcal{F}_{\mathrm{d}}(a) \tag{13.13}$$

is a number between 0 and 1. It can be interpreted as the probability of getting through the stage alive. This can be seen by approximating the right-hand side of (13.13) by Stieltjes sums

$$\sum_{j=0}^{\infty} \mathcal{F}_s(t_j)(\mathcal{F}_d(a_j) - \mathcal{F}_d(a_{j+1}))$$

with strictly increasing sequences a_j and t_j such that $a_0 = 0$, $a_j \to \infty$ as $j \to \infty$, and $a_j \leqslant t_j \leqslant a_{j+1}$. Recall that $\mathcal{F}_d(a_j) - \mathcal{F}_d(a_{j+1})$ is the probability of leaving the stage during the stage-age-interval from a_j and a_{j+1} (without dying) and that $\mathcal{F}_s(t_j)$ is the probability of still being alive at some time t_j in that age-interval. Furthermore, notice via integration by parts (Theorem B.10) that

$$1 - p = -\int_0^{\infty} \mathcal{F}_d(a)\, d\mathcal{F}_s(a) \tag{13.14}$$

is the probability of dying while in the stage.

Proof of Theorem 13.7. (a) We use the Stieltjes integral formulation in the following. If this looks too alien, replace $d\mathcal{F}_d(a)$ by $\mathcal{F}_d'(a)\, da$.

Formally, replacing t by ∞ in the formula for C_1 in (13.12), we could argue

$$C_1(t) \to -\int_0^{\infty} B(\infty - a)\mathcal{F}_s(a)\, d\mathcal{F}_d(a) = -B(\infty)\int_0^{\infty} \mathcal{F}_s(a)\, d\mathcal{F}_d(a).$$

In order to make the formal argument rigorous, let $t > r > 0$. Then

$$\begin{aligned}
&|C_1(t) - pB(\infty)| \\
&\quad \leqslant -\int_0^{r} |B(t-a) - B(\infty)|\mathcal{F}_s(a)\, d\mathcal{F}_d(a) \\
&\qquad - \int_r^{t} |B(t-a)|\mathcal{F}_s(a)\, d\mathcal{F}_d(a) - B(\infty)\int_r^{\infty} \mathcal{F}_s(a)\, d\mathcal{F}_d(a) \\
&\quad \leqslant -\int_0^{r} |B(t-a) - B(\infty)|\, d\mathcal{F}_d(a) \\
&\qquad - \int_r^{t} |B(t-a)|\, d\mathcal{F}_d(a) - B(\infty)\int_r^{\infty} d\mathcal{F}_d(a) \\
&\quad \leqslant -\int_0^{r} |B(t-a) - B(\infty)|\, d\mathcal{F}_d(a) \\
&\qquad - \sup_{v\geqslant 0} |B(v)| \int_r^{t} d\mathcal{F}_d(a) - B(\infty)\int_r^{\infty} d\mathcal{F}_d(a) \\
&\quad \leqslant -\int_0^{r} |B(t-a) - B(\infty)|\, d\mathcal{F}_d(a) \\
&\qquad + 2\sup_{v\geqslant 0} |B(v)|(\mathcal{F}_d(r) - \mathcal{F}_d(\infty)).
\end{aligned}$$

Let $\epsilon > 0$. Then there exists some t_ϵ such that $|B(t) - B(\infty)| < \epsilon$ for all $t > t_\epsilon$. Let $t - r > t_\epsilon$. Then

$$\begin{aligned}|C_1(t) - pB(\infty)| &\leqslant -\int_0^r \epsilon \, \mathrm{d}\mathcal{F}_\mathrm{d}(a) + 2\sup_{v \geqslant 0} |B(v)|(\mathcal{F}_\mathrm{d}(r) - \mathcal{F}_\mathrm{d}(\infty)) \\ &\leqslant \epsilon + 2\sup_{v \geqslant 0} |B(v)|(\mathcal{F}_\mathrm{d}(r) - \mathcal{F}_\mathrm{d}(\infty)).\end{aligned}$$

Hence

$$\limsup_{t \to \infty} |C_1(t) - pB(\infty)| \leqslant \epsilon + 2\sup_{v \geqslant 0} |B(v)|(\mathcal{F}_\mathrm{d}(r) - \mathcal{F}_\mathrm{d}(\infty)).$$

Since this holds for all $\epsilon > 0$ and all $r > 0$, we have

$$\limsup_{t \to \infty} |C_1(t) - pB(\infty)| = 0.$$

(b) Let $u_0(a)/\mathcal{F}(a) \leqslant M$ for all $a \in [a, c)$ with some constant $M > 0$. Then

$$0 \leqslant C_0(t) \leqslant -\int_t^\infty M\mathcal{F}_\mathrm{s}(a) \, \mathrm{d}\mathcal{F}_\mathrm{d}(a) \to 0, \quad t \to \infty.$$

□

Example 13.8. Let the stage have a fixed duration $c_2 \in (0, \infty)$, i.e., $\mathcal{F}_\mathrm{d}(a) = 1$ for $a \in [0, c_2)$ and $\mathcal{F}_\mathrm{d}(a) = 0$ for $a > c_2$, and let the stage survival function $\mathcal{F}_\mathrm{s}$ be continuous on $[0, \infty)$. Then $p = \mathcal{F}_\mathrm{s}(c_2)$.

Proof. Since $\mathcal{F}_\mathrm{d}(a) = 0$ for $a \geqslant c_2 + 1$,

$$p = -\int_0^{c_2+1} \mathcal{F}_\mathrm{s}(a) \, \mathrm{d}\mathcal{F}_\mathrm{d}(a).$$

Let us first assume that $\mathcal{F}_\mathrm{s}$ is continuously differentiable on $[0, c_2 + 1]$. Integrating by parts (Corollary B.11),

$$\begin{aligned}&-\int_0^{c_2+1} \mathcal{F}_\mathrm{s}(a) \, \mathrm{d}\mathcal{F}_\mathrm{d}(a) \\ &\qquad = 1 - \mathcal{F}_\mathrm{s}(c_2+1)\mathcal{F}_\mathrm{d}(c_2+1) + \int_0^{c_2+1} \mathcal{F}_\mathrm{s}'(a)\mathcal{F}_\mathrm{d}(a) \, \mathrm{d}a \\ &\qquad = 1 + \int_0^{c_2} \mathcal{F}_\mathrm{s}'(a) \, \mathrm{d}a \\ &\qquad = \mathcal{F}_\mathrm{s}(c_2).\end{aligned}$$

If $\mathcal{F}_\mathrm{s}$ is continuous, we approximate $\mathcal{F}_\mathrm{s}$ by continuously differentiable functions uniformly on $[0, c_2 + 1]$ and our assertion follows from Theorem B.7. □

Exercises

13.9. (a) Consider a stage with a constant per capita exit rate (death disregarded) γ. Let D be the mean sojourn time in the stage (mortality included). Show that the probability to get through the stage alive is γD.

(b) Consider a stage with a constant per capita mortality rate μ and a constant per capita exit rate (death disregarded) γ. Determine the probability to get through the stage alive.

13.10. Consider a stage with a constant per capita mortality rate μ and let D be the mean sojourn time in the stage (mortality included). Show that the probability of dying during the stage is given by μD.

13.7 Which Recruitment Curves Can Be Explained by Cannibalism of Newborns by Adults?

Cannibalism (or intraspecific predation) is a significant and widespread trait in animal species. In Chapter 5 we derived both the Beverton–Holt and the Ricker function from cannibalism of newborns by adults. We assumed that the stage during which the juveniles are prone to cannibalism is short and has an exponentially distributed duration in the first case and a fixed duration in the second. We want to show that a much wider class of recruitment functions can be derived from cannibalism if other length distributions of the (passive) cannibalism stage are considered. A recruitment function relates the number of adults, N, at a certain time or in a certain season to the number of their offspring, R, born at that time or season, that reach adulthood.

We assume that adults cannibalize their young in an early distinct stage of their life, e.g., an egg or larval stage, which starts immediately after birth (or oviposition). Let $u(t,\cdot)$ denote the age-density of juvenile individuals in this stage that is prone to cannibalism. While we will later assume that the duration of the stage has an arbitrary distribution, let us start by assuming that the duration function $\mathcal{F}_{\mathrm{d}}$ is differentiable and that $\gamma(a) = -\mathcal{F}_{\mathrm{d}}'(a)/\mathcal{F}_{\mathrm{d}}(a)$ is the per capita rate of leaving the stage at age a. Similarly, we assume that there is a natural per capita death rate $\mu(a) = -\mathcal{F}_{\mathrm{s}}'(a)/\mathcal{F}_{\mathrm{s}}(a)$, where $\mathcal{F}_{\mathrm{s}}$ is the survival function associated with natural (i.e., other than cannibalism) causes of death. Then the transition through the stage has the following McKendrick formulation (cf. equation (13.4) with $\mu(t,a) = \gamma(a) + \mu(a) + \kappa(a)N(t)$):

$$\begin{aligned}\frac{\partial}{\partial t}u(t,a) + \frac{\partial}{\partial a}u(t,a) &= -(\gamma(a)+\mu(a))u(t,a) - \kappa(a)N(t)u(t,a),\\ u(t,0) &= \beta N(t),\\ u(0,a) &= u_0(a).\end{aligned}$$

In the partial differential equation, cannibalism has been modeled according to the law of mass action with $\kappa(a)$ being the rate at which one average juvenile of age a is cannibalized by one average adult, provided that it is still in the cannibalism-prone stage. β denotes the per capita birth rate. The *recruitment* from the cannibalism-prone stage is described by

$$R(t) = \int_0^\infty u(t,a)\gamma(a)\,\mathrm{d}a.$$

The function $\mathcal{F}(a) = \mathcal{F}_\mathrm{s}(a)\mathcal{F}_\mathrm{d}(a)$ is the sojourn function of the cannibalism-prone stage if there are no cannibals around (cf. Section 13.6). The recruitment from the cannibalism stage corresponds to the stage output in the previous section, though now we distinguish between natural death and a time-dependent death by cannibalism. Using the quotient rule of differentiation, we see that this system is equivalent to

$$\begin{aligned}
\frac{\partial}{\partial t}\frac{u(t,a)}{\mathcal{F}(a)} + \frac{\partial}{\partial a}\frac{u(t,a)}{\mathcal{F}(a)} &= -\kappa(a)N(t)\frac{u(t,a)}{\mathcal{F}(a)},\\
u(t,0) &= \beta N(t),\\
u(0,a) &= u_0(a),\\
R(t) &= -\int_0^\infty \frac{u(t,a)}{\mathcal{F}_\mathrm{d}(a)}\,\mathrm{d}\mathcal{F}_\mathrm{d}(a),
\end{aligned}$$

with the last integral being a Stieltjes integral. This shows that the system can be formulated without assuming that the duration and survival function are differentiable. Setting $v(t,a) = u(t,a)/\mathcal{F}(a)$, these equations transform into

$$\begin{aligned}
\frac{\partial}{\partial t}v(t,a) + \frac{\partial}{\partial a}v(t,a) &= -\kappa(a)N(t)v(t,a),\\
v(t,0) &= \beta N(t),\\
v(0,a) &= \frac{u_0(a)}{\mathcal{F}(a)} =: v_0(a),\\
R(t) &= -\int_0^\infty v(t,a)\mathcal{F}_\mathrm{s}(a)\,\mathrm{d}\mathcal{F}_\mathrm{d}(a).
\end{aligned}$$

The partial differential equation can be solved by integration along characteristics (Section 13.2), and we obtain the Lotka formulation:

$$\begin{aligned}
v(t,a) &= \beta N(t-a)\exp\left(-\int_0^a \kappa(a-s)N(t-s)\,\mathrm{d}s\right), \qquad 0 \leqslant a < t,\\
v(t,a) &= v_0(a-t)\exp\left(-\int_0^t \kappa(a-s)N(t-s)\,\mathrm{d}s\right), \qquad 0 \leqslant t < a.
\end{aligned}$$

We assume that the period during which the newborn are cannibalized by the adults is very short, but that the cannibalism is very severe. To express this we assume that

$$\mathcal{F}_d(a) = \mathcal{F}_0(a/D),$$

with D being the average duration of the cannibalism-prone period (without taking account of natural or cannibalism deaths) and $\mathcal{F}_0$ having the normalized average duration:

$$\int_0^\infty \mathcal{F}_0(a)\,\mathrm{d}a = 1. \tag{13.15}$$

Then $\mathcal{F}$ has the average duration D. Furthermore, we assume that the cannibalism rate scales like

$$\kappa(a) = \frac{1}{D}\kappa_0\left(\frac{a}{D}\right).$$

We assume that natural death has two components, a severe one and a mild one, which is expressed by

$$\mathcal{F}_s(a) = \mathcal{F}_1(a/D)\mathcal{F}_2(a),$$

with $\mathcal{F}_1$ describing the severe component, which has the same time scale as the length of the cannibalism stage, and $\mathcal{F}_2$ describing the mild one, which has the same scale as ordinary time.

By substituting the two formulas for v into the formula for R, the recruitment rate splits into two parts:

$$\begin{aligned}
R(t) &= R_1(t) + R_0(t), \\
R_1(t) &= \int_0^t \beta N(t-a)\exp\left(-\int_0^a \kappa(a-s)N(t-s)\,\mathrm{d}s\right)\mathcal{F}_s(a)\,\mathrm{d}\mathcal{F}_d(a), \\
R_0(t) &= \int_t^\infty v_0(a-t)\exp\left(-\int_0^t \kappa(a-s)N(t-s)\,\mathrm{d}s\right)\mathcal{F}_s(a)\,\mathrm{d}\mathcal{F}_d(a).
\end{aligned}$$

Using the substitution rule for Stieltjes integrals (Exercise B.4),

$$\begin{aligned}
R_1(t) &= \int_0^{t/D} \beta N(t-Da) \\
&\qquad \times \exp\left(-\int_0^{Da} \kappa(Da-s)N(t-s)\,\mathrm{d}s\right)\mathcal{F}_1(a)\mathcal{F}_2(Da)\,\mathrm{d}\mathcal{F}_0(a) \\
&= \int_0^{t/D} \beta N(t-Da) \\
&\qquad \times \exp\left(-\int_0^{a} \kappa_0(a-s)N(t-Ds)\,\mathrm{d}s\right)\mathcal{F}_1(a)\mathcal{F}_2(Da)\,\mathrm{d}\mathcal{F}_0(a).
\end{aligned}$$

Taking the limit for $D \to 0$ and assuming that N is continuous:

$$R_1(t) = \int_0^\infty \beta N(t)\exp\left(-\int_0^a \kappa_0(s)N(t)\,\mathrm{d}s\right)\mathcal{F}_1(a)\mathcal{F}_2(0+)\,\mathrm{d}\mathcal{F}_0(a).$$

In order to show that $R_0(t) \to 0$ as $D \to 0$, we assume that v_0 is bounded. This is a restrictive assumption in so far as the most general natural assumption is that u_0 is integrable. But it is a property that is preserved, because if v_0 is bounded, $v(t, \cdot)$ will be bounded as long as $N(t)$ remains bounded. We have

$$R_0(t) \leqslant (\sup v_0) \int_{t/D}^{\infty} \mathcal{F}_1(a)\mathcal{F}_2(Da)\,\mathrm{d}\mathcal{F}_0(a)$$
$$\leqslant (\sup v_0)\mathcal{F}_0(t/D) \to 0, \quad D \to 0.$$

Actually, this convergence is uniform on every interval $[\epsilon, \infty)$, $\epsilon > 0$. So, in the limit $D \to 0$,

$$R(t) = \alpha N(t)g(N(t)), \qquad \alpha = \beta\mathcal{F}_2(0+),$$
$$g(N) = -\int_0^{\infty} \exp\left(-N\int_0^a \kappa_0(s)\,\mathrm{d}s\right)\mathcal{F}_1(a)\,\mathrm{d}\mathcal{F}_0(a).$$

The recruitment function can be used as a building block in various contexts. If not only the cannibalism-prone stage but the entire juvenile stage is very short, we can use it in a continuous-time model as in Chapter 6,

$$N' = \alpha N g(N) - \mu N.$$

$f(N) = \alpha N g(N) - \mu N$ has all the properties required in Chapter 6, and so this equation gives rise to sigmoid growth.

If reproduction only occurs once a year during a very short reproductive season which is followed by a very short season during which the newborns are cannibalized by the adults and if all surviving newborns are adults after one year (cf. Chapter 9), then we can employ the recruitment function in a discrete-time model

$$N_{n+1} = pN_n + q\alpha N_n g(N_n),$$

where p is the probability that an adult survives for one year and q the probability that a juvenile, once it has survived cannibalism, survives for the rest of the year.

If the cannibalism rate κ_0 is constant and if there is no severe natural death, then

$$g(N) = -\check{\mathcal{F}}_0(\kappa_0 N)$$

with

$$-\check{\mathcal{F}}_0(\lambda) = -\int_0^{\infty} \mathrm{e}^{-\lambda a}\,\mathrm{d}\mathcal{F}_0(a) = 1 - \lambda\int_0^{\infty} \mathrm{e}^{-\lambda a}\mathcal{F}_0(a)\,\mathrm{d}a$$

being the Laplace–Stieltjes transform of $\mathcal{F}_0$ (see the end of Toolbox B.1). Dropping the normalization (13.15), κ_0 can be absorbed into the duration function by a change of variables. Even if the cannibalism rate is age dependent and/or there is severe natural death, g can still be written as a Laplace–Stieltjes transform of an increasing function $\mathcal{G}$ with $\mathcal{G}(\infty) \leqslant 1$ after an appropriate change of variables (Exercise B.4 with $g(x) = \int_0^x \kappa_0(s)\,\mathrm{d}s$),

$$g(N) = \check{\mathcal{G}}(N).$$

This shows that the recruitment rate, if derived in this way, necessarily involves a function g that is completely monotonic, i.e., g is infinitely often differentiable and $(-1)^n g^{(n)}$ is nonnegative for all $n \in \mathbb{N}$ (see Theorem B.13). In particular, g is decreasing and convex, with these properties actually being strict except in the case where all individuals die from natural causes during the stage. So, a recruitment function

$$R = Ng(N) = \frac{\alpha N}{1 + \xi N^{\eta}}$$

(Maynard Smith, 1974) cannot be derived from cannibalism, at least not in this framework, as soon as $\eta > 1$, because $1/(1 + x^{\eta})$ is not a convex function for $\eta > 1$.

Conversely, by Bernstein's theorem (Theorem B.13) that any completely monotonic function can be written as the Laplace–Stieltjes transform of an increasing function, every recruitment function $R = Ng(N)$ with a completely monotonic g can be, theoretically, explained by cannibalism, choosing an appropriate length distribution of the passive cannibalism stage.

In the following we present the recruitment functions one obtains by assuming some standard distributions of the length of the cannibalism stage, assuming that there is no severe natural death during the cannibalism stage, $\mathcal{F}_1 \equiv 1$, and that the per capita rate of passive cannibalism is constant. The parameter

$$\alpha = \beta \mathcal{F}_2(0+)$$

is the effective per capita birth rate, $\mathcal{F}_2(0+)$ is the probability of dying immediately after birth from noncannibalistic causes. When comparing with the literature, note that stochasticians and statisticians prefer the concept of a *moment-generating function*, which is the Laplace–Stieltjes transform of the probability distribution function evaluated at the negative argument (Patel et al., 1976; Johnson, Kotz, 1970); the name comes from the fact that the moments are related to the derivatives of the Laplace–Stieltjes transform evaluated at 0. Furthermore, when stage duration can be described by a probability density,

$$-\mathcal{F}_0(a) = \int_a^{\infty} f_0(a)\,\mathrm{d}a,$$

the Laplace–Stieltjes transform of $\mathcal{F}_0$ is the negative Laplace transform of the associated density:

$$\check{\mathcal{F}}_0(\lambda) = \int_0^{\infty} \mathrm{e}^{-\lambda a} f(a)\,\mathrm{d}a.$$

Finally, notice that the normalization $1 = \int_0^\infty \mathcal{F}_0(a)\,\mathrm{d}a = -\check{\mathcal{F}}_0{}'(0)$ often enforces a relation between the parameters of the distribution.

If the cannibalism stage has a *fixed duration*,

$$\mathcal{F}_0(a) = \begin{cases} 1, & 0 < a < 1, \\ 0, & 1 < a < \infty, \end{cases}$$

we obtain the Ricker function

$$R = \alpha N \mathrm{e}^{-\kappa_0 N}.$$

If the duration of the cannibalism stage is a *Gamma distribution*, then $\mathcal{F}_0'(a) = -\eta(\eta a)^{\eta-1}(\mathrm{e}^{-\eta a}/\Gamma(\eta))$, with $\eta > 0$, and

$$-\check{\mathcal{F}}_0(\lambda) = \int_0^\infty \mathrm{e}^{-((\lambda/\eta)+1)a} \frac{a^{\eta-1}}{\Gamma(\eta)}\,\mathrm{d}a = \left(\frac{\eta}{\eta+\lambda}\right)^\eta.$$

So, the recruitment has the form

$$R = \frac{\alpha \eta^\eta N}{(\eta + \kappa_0 N)^\eta}$$

(Hassell, 1975). For $\eta = 1$, the *exponential distribution*, we obtain the Beverton–Holt recruitment rate.

If the cannibalism stage is *uniformly distributed*, $\mathcal{F}_0(a) = 1 - \frac{1}{2}a$, $0 \leqslant a \leqslant 2$, and 0 otherwise, then

$$\check{\mathcal{F}}_0(\lambda) = \frac{1}{2}\int_0^2 \mathrm{e}^{-\lambda a}\,\mathrm{d}a = \frac{1}{2\lambda}(1 - \mathrm{e}^{-2\lambda}),$$

and

$$R = \frac{\alpha}{2\kappa_0}(1 - \mathrm{e}^{-2\kappa_0 N}).$$

An *inverse Gaussian* distribution gives

$$R = \alpha N \exp\left(\eta - \eta\sqrt{1 + \frac{2\kappa_0}{\eta}N}\right).$$

A *noncentral* χ^2 distribution leads to

$$R = \alpha N(1 + 2\kappa_0 N)^{(\eta-1)/2} \exp\left(-\frac{\eta\kappa_0 N}{1 + 2\kappa_0 N}\right),$$

where $\eta \in (0, 1]$ is the noncentrality parameter and $1 - \eta$ the degree of freedom.

The *truncated exponential* distribution is not one of the standard statistical distributions, but stage-structure modelers use it as a combination of the two extreme cases of fixed duration and exponential distribution:

$$\mathcal{F}_\mathrm{d}(a) = \begin{cases} \mathrm{e}^{-\gamma a}, & 0 \leqslant a < c, \\ 0, & a > c. \end{cases}$$

The Laplace–Stieltjes transform can be calculated by partial integration:

$$\begin{aligned} -\check{\mathcal{F}}_\mathrm{d}(\lambda) &= 1 - \lambda \int_0^\infty \mathrm{e}^{-\lambda a} \mathcal{F}_\mathrm{d}(a)\, \mathrm{d}a = 1 - \lambda \int_0^c \mathrm{e}^{-(\lambda+\gamma)a}\, \mathrm{d}a \\ &= 1 + \frac{\lambda}{\lambda+\gamma}(\mathrm{e}^{-(\lambda+\gamma)c} - 1) = \frac{\gamma}{\lambda+\gamma} + \frac{\lambda}{\lambda+\gamma}\mathrm{e}^{-(\lambda+\gamma)c}. \end{aligned}$$

For the normalization $-\check{\mathcal{F}}_\mathrm{d}{}'(0) = 1$, we notice

$$\check{\mathcal{F}}_\mathrm{d}'(0) = \frac{1}{\gamma} - \frac{1}{\gamma}\mathrm{e}^{-\gamma c}.$$

We realize that we need to require $\gamma < 1$ and choose $c = -\ln(1-\gamma)/\gamma$ in order to get the normalization $\mathcal{F}_0'(0) = 1$. So,

$$\begin{aligned} -\check{\mathcal{F}}_0(\lambda) &= \frac{\gamma}{\lambda+\gamma} + \frac{\lambda}{\lambda+\gamma}(1-\gamma)^{1+(\lambda/\gamma)} \\ &= \frac{1}{1+(\lambda/\gamma)} + \frac{(\lambda/\gamma)}{1+(\lambda/\gamma)}(1-\gamma)^{1+(\lambda/\gamma)}. \end{aligned}$$

This results in the recruitment

$$R = \alpha \frac{N}{1+\xi N}(1 + \xi N(1-\gamma)^{1+\xi N}), \quad \xi = \frac{\kappa_0}{\gamma}.$$

In a similar way, the interaction between an egg-eating predator and its prey can be modeled, where a predator eats the prey only during a very short stage (van den Bosch, Diekmann, 1986). If $B(t)$ is the prey birth rate and $N(t)$ the number of predators, then the age-density of the prey in the vulnerable stage is modeled by

$$\begin{aligned} \frac{\partial}{\partial t}u(t,a) + \frac{\partial}{\partial a}u(t,a) &= -(\gamma(a) + \mu(a))u(t,a) - \kappa(a)N(t)u(t,a), \\ u(t,0) &= B(t), \\ u(0,a) &= u_0(a). \end{aligned}$$

The same derivation as above provides the output from this stage as

$$C(t) = B(t)g(N(t))$$

with the same g as above. In particular,

$$g(N) = e^{-\kappa_0 N} \qquad \text{(fixed stage duration)},$$

$$g(N) = \frac{1}{1+\kappa_0 N} \qquad \text{(exponential distribution)},$$

$$g(N) = \frac{\eta^\eta}{(\eta+\kappa_0 N)^\eta} \qquad (\Gamma\text{-distribution}),$$

$$g(N) = \frac{1-e^{-2\kappa_0 N}}{2\kappa_0 N} \qquad \text{(uniform distribution)},$$

$$g(N) = \exp\left(\eta - \eta\sqrt{1+\frac{2\kappa_0}{\eta}N}\right) \qquad \text{(inverse Gaussian distribution)},$$

$$g(N) = (1+2\kappa_0 N)^{(\eta-1)/2}\exp\left(-\frac{\eta\kappa_0 N}{1+2\kappa_0 N}\right) \qquad (\text{noncentral } \chi^2 \text{ distribution}).$$

Bibliographic Remarks

In describing the stage dynamics we have preferred the older Lotka (1907) approach (formula (13.2)) over the somewhat more recent McKendrick (1926) approach (the partial differential equation (13.4)) because it appears to be more elementary and is more general, at least for the time-autonomous case. While the partial differential equation is readily extended to the nonautonomous case, this extension is somewhat more clumsy for the Lotka approach (see the three possibilities (13.7)–(13.9)). Equation (13.4) may in particular be more communicable when one wants to incorporate nonlinear effects. Still, the Lotka approach has definite advantages even here, in particular if populations are not structured with respect to age, but to body size or other characteristics. See Diekmann et al. (1998).

The widespread occurrence of cannibalism in animals has been described by Fox (1975) and Polis (1981). Our treatise does not present a full model for the dynamics of a population, but only the modeling of the recruitment rate. A full population model with short stages of severe passive cannibalism with fixed duration has been considered by Diekmann et al. (1986). It has been demonstrated both theoretically and experimentally that cannibalism can lead to complicated population dynamics (aperiodic motions including chaos) in flour beetles (Costantino et al., 1995; Dennis et al., 1995; Cushing, 1998). The bibliography in Cushing (1998) contains many more references concerning the role of cannibalism in population dynamics. Cannibalism can be a very complex phenomenon, as shown by the work of Collins and his group at ASU on tiger salamanders (see Collins, Cheek, 1983; Collins et al., 1993; Loeb, Collins, Maret, 1994; Pfennig et al., 1994; Maret, Collins, 1996; and the references therein). Cannibalism among salamander larvae is affected by food quantity and quality on the one hand and population density, body size distribution, and even kinship of larvae on the other.

Feng and Thieme (2000a,b) apply the stage-dynamics theory developed in the last two sections to a model for an infectious disease which takes its course through arbitrarily many infection stages. A condensed presentation of the combined material can be found in Thieme (2002).

Chapter Fourteen

Demographics in an Unlimiting Constant Environment

Though human life can be divided into stages like childhood, youth, adulthood, senescence, these are not so distinct as the stages egg, larva, pupa, imago which occur in some insect species. Moreover, they can be easily identified with certain age brackets. Though there is some individual variation in the ages at which these stages are entered and left, age gives a reasonably exact characterization of a person's stage. More importantly, with humans, age is a characteristic that can easily be checked, while it requires a thorough medical examination to reveal whether somebody physiologically belongs to adulthood or senescence. For these reasons, human life is typically modeled as one single stage with a continuous age-structure, or it is divided into artificial stages which comprise age groups each extending, let us say, over a decade. This book only considers the first type of model; for the second see the bibliographic remarks.

In the following, $\mathcal{F}$ is the survival function, i.e., $\mathcal{F}(a)$ denotes the probability of still being alive at age a. Recall the definition of the maximum age c. If $\mathcal{F}$ is strictly positive on $[0, \infty)$, then $c = \infty$. Otherwise $0 < c < \infty$ is determined by the property that $\mathcal{F}(a) > 0$ for $a < c$ and $\mathcal{F}(a) = 0$ for $a > c$.

In this demographic context, $B(t)$ denotes the population birth rate (i.e., the population input), the number of newborn individuals at time t per unit of time. Let $u(t, \cdot)$ be the age-density of the population at time t, while u_0 is the age-density at time 0. As in Section 13.1 we have the following relation:

$$\left.\begin{aligned} u(t,a) &= B(t-a)\mathcal{F}(a), && t > a \geqslant 0, \\ u(t,a) &= \frac{u_0(a-t)}{\mathcal{F}(a-t)}\mathcal{F}(a), && 0 \leqslant t < a < \infty, \end{aligned}\right\} \tag{14.1}$$

where

$$\frac{u_0(a)}{\mathcal{F}(a)} = 0, \quad a > c, \quad \text{if } c \in (0, \infty).$$

We now introduce a linear feedback from the population density to the population birth rate:

$$B(t) = \int_0^\infty \beta(a) u(t,a)\, da. \tag{14.2}$$

Here $\beta(a)$ is the per capita birth (or reproduction) rate at age a. For simplicity, we assume that β is a piecewise continuous nonnegative bounded function.

For references concerning the illustration of survival functions, we refer to the bibliographic remarks at the end of Chapter 12, and for figures displaying per capita reproduction rates (or natalities) to those at the end of Chapter 15.

14.1 The Renewal Equation

The easiest way to solve the system (14.1) and (14.2) consists of reducing it to a Volterra integral equation for the population birth rate B, the so-called *renewal equation*:

$$\left.\begin{aligned} B(t) &= \int_0^t \beta(a)u(t,a)\,\mathrm{d}a + \int_t^\infty \beta(a)u(t,a)\,\mathrm{d}a \\ &= \int_0^t \beta(a)B(t-a)\mathcal{F}(a)\,\mathrm{d}a + B_0(t), \\ B_0(t) &= \int_t^\infty \beta(a)\mathcal{F}(a)\frac{u_0(a-t)}{\mathcal{F}(a-t)}\,\mathrm{d}a \\ &= \int_0^c \beta(a+t)\mathcal{F}(a+t)\frac{u_0(a)}{\mathcal{F}(a)}\,\mathrm{d}a. \end{aligned}\right\} \tag{14.3}$$

This equation can be shown to have a unique solution by using the contraction mapping theorem in a similar way as in the existence and uniqueness proof for ordinary differential equations. We will not go into this here, but do this as a homework problem (see Exercise 14.1 and the solution in Toolbox B.6).

In order to find concrete information about this linear problem, we can use Laplace transforms. Recall that the Laplace transform of a function B on $[0,\infty)$ is defined by

$$\hat{B}(\lambda) = \mathcal{L}\{B\} = \int_0^\infty \mathrm{e}^{-\lambda t}B(t)\,\mathrm{d}t = \lim_{r\to\infty}\int_0^r \mathrm{e}^{-\lambda t}B(t)\,\mathrm{d}t \tag{14.4}$$

provided that the last expression exists. The inverse Laplace transform of a function F is denoted by

$$\mathcal{L}^{-1}\{F\}.$$

Furthermore, recall that the Laplace transform converts a convolution into a product. Hence

$$\hat{B}(\lambda) = \hat{B}(\lambda)\hat{G}(\lambda) + \hat{B}_0(\lambda),$$

where

$$\left.\begin{aligned} G(a) &= \beta(a)\mathcal{F}(a), \\ B_0(t) &= \int_t^c \beta(a)u_0(a-t)\frac{\mathcal{F}(a)}{\mathcal{F}(a-t)}\,\mathrm{d}a. \end{aligned}\right\} \tag{14.5}$$

Since $\hat{\mathcal{F}}(\lambda) \leqslant (1/\lambda)$, we have $\hat{\mathcal{F}}(\lambda) < 1$ for $\lambda > 1$, and so we can write

$$\hat{B}(\lambda) = \frac{\hat{B}_0(\lambda)}{1 - G(\lambda)}, \quad \lambda > \lambda_0, \tag{14.6}$$

where λ_0 is the supremum of β. Hence B is the inverse Laplace transform of the right-hand side of this equation.

14.2 Balanced Exponential Growth

We say that the population exhibits balanced (or asynchronous) exponential growth if the population density is defined for all times t and takes the form

$$u(t, a) = \mathrm{e}^{st} v(a), \quad t \in \mathbb{R}, \quad a \geqslant 0.$$

Since the total population size should be finite,

$$\int_0^\infty |v(a)| \,\mathrm{d}a < \infty.$$

Then

$$B(t) = \int_0^\infty \beta(a) \mathrm{e}^{st} v(a) \,\mathrm{d}a = \alpha \mathrm{e}^{st},$$

for some α. By (14.1),

$$u(t, a) = B(t - a)\mathcal{F}(a) = \alpha \mathrm{e}^{st} \mathrm{e}^{-sa} \mathcal{F}(a), \quad t > a.$$

Since, on the other hand, we have $u(t, a) = \mathrm{e}^{st} v(a)$ for all $t \in \mathbb{R}$, $a \geqslant 0$, we can conclude that $v(a) = \alpha \mathrm{e}^{-sa} \mathcal{F}(a)$ and so we have the relation

$$u(t, a) = \alpha \mathrm{e}^{st} \mathrm{e}^{-sa} \mathcal{F}(a), \quad t \in \mathbb{R}, \quad a \geqslant 0. \tag{14.7}$$

Substituting this formula into the equation for B,

$$\alpha \mathrm{e}^{st} = B(t) = \int_0^\infty \beta(a) u(t, a) \,\mathrm{d}a = \int_0^\infty \alpha \beta(a) \mathrm{e}^{st} \mathrm{e}^{-sa} \mathcal{F}(a) \,\mathrm{d}a.$$

Dividing by $\alpha \mathrm{e}^{st}$ we obtain the following relation for s:

$$\left.\begin{aligned} 1 = \int_0^\infty \beta(a) \mathrm{e}^{-sa} \mathcal{F}(a) \,\mathrm{d}a = \hat{G}(s), \\ G(a) = \beta(a)\mathcal{F}(a). \end{aligned}\right\} \tag{14.8}$$

If $\beta(a)\mathcal{F}(a)$ is not 0 almost everywhere, the number s is uniquely determined by this relation.

In the literature, s has several names: intrinsic exponential growth rate, intrinsic rate of natural increase, *Malthusian parameter*. The latter is in honor of

Thomas Malthus, who, in 1798, predicted that the human population would grow exponentially in time with all the catastrophic consequences one could imagine. Equation (14.8) is called the *characteristic equation* for the Malthusian parameter. Balanced exponential growth is of interest not because natural populations satisfy the underlying conditions (time-independent per capita mortality and reproductivity rates), which they generally do not, but because it provides basic insight which helps to understand more complex types of age-structured phenomena. Furthermore, it provides a theoretic laboratory for demographic experiments like the one in Section 15.4: Frauenthal's (1986) abrupt shift in maternity.

The Malthusian parameter does not always exist. Let

$$\mathcal{F}(a) = \frac{1}{1+a^2}, \qquad \beta(a) = \text{const.} \in \left(0, \frac{2}{\pi}\right).$$

Then, for $s \geqslant 0$,

$$\int_0^\infty \mathrm{e}^{-sa} \beta(a) \mathcal{F}(a)\,\mathrm{d}a < 1,$$

while, for $s < 0$, this integral is infinity. A sufficient condition for the existence of the Malthusian parameter is the following one.

Proposition 14.1. *Assume that there exists some $\lambda \in \mathbb{R}$ such that*

$$1 \leqslant \int_0^\infty \mathrm{e}^{-\lambda a} \beta(a) \mathcal{F}(a)\,\mathrm{d}a < \infty.$$

Then the Malthusian parameter s exists and $s \geqslant \lambda$.

Proof. Set

$$G(a) = \beta(a)\mathcal{F}(a).$$

By assumption, the Laplace transform

$$\hat{G}(\nu) = \int_0^\infty \mathrm{e}^{-\nu a} \beta(a) \mathcal{F}(a)\,\mathrm{d}a$$

exists on $[\lambda, \infty)$ and is a continuous function on that interval. We have $\hat{G}(\lambda) \geqslant 1$ and $\hat{G}(\nu) \to 0$ as $\nu \to \infty$. By the intermediate value theorem there exists some $s \geqslant \lambda$ such that $\hat{G}(s) = 1$. □

Corollary 14.2. *Let $c < \infty$ or let β have compact support, i.e., there exists some $a_0 \in (0, \infty)$ such that $\beta(a) = 0\, \forall a \geqslant a_0$. Furthermore, assume that $\beta \cdot \mathcal{F}$ is not 0 almost everywhere. Then the Malthusian parameter exists.*

Proof. Under these assumptions the Laplace transform of $G = \beta \cdot \mathcal{F}$ exists for all s and $\hat{G}(\nu) \to \infty$ as $\nu \to -\infty$. In particular, there exists some $\lambda \in \mathbb{R}$ such that $\hat{G}(\lambda) \geqslant 1$. □

The sign of s determines whether the population increases or decreases exponentially and is of special interest. It is closely related to the *basic reproduction ratio* of the population,

$$\mathcal{R}_0 = \int_0^\infty \beta(a)\mathcal{F}(a)\,da. \tag{14.9}$$

$\mathcal{R}_0$ gives the average number of offspring an individual produces during its lifetime.

Proposition 14.3.

(i) *If* $\mathcal{R}_0 > 1$, *the Malthusian parameter* s *exists and* $s > 0$.

(ii) *If* $\mathcal{R}_0 = 1$, *then the Malthusian parameter* s *exists and* $s = 0$.

(iii) *If* $\mathcal{R}_0 < 1$, *then* $s < 0$ *if it exists.*

Let us consider the very special case in which the duration of life is exponentially distributed with an age-independent per capita mortality μ and where the per capita birth rate β is also constant.

Then

$$\mathcal{R}_0 = \beta/\mu \tag{14.10}$$

and $1 = \beta/(s+\mu)$, i.e.,

$$s = \beta - \mu. \tag{14.11}$$

Recall that $\mu = 1/D$, where D is the life expectancy(Section 12.7). We can rewrite the equality (14.10) as

$$\beta = \frac{\mathcal{R}_0}{D}. \tag{14.12}$$

Let us check this special case for balanced exponential growth. By Proposition 13.6, for the total population size N,

$$N'(t) = B(t) - \int_0^\infty \mu u(t,a)\,da = (\beta - \mu)N(t).$$

Hence

$$N(t) = N_0 e^{st}, \quad t \geqslant 0,$$

with $s = \beta - \mu$. By (14.2), $B(t) = \beta N_0 e^{st}$. By (14.1),

$$u(t,a) = B(t-a)\mathcal{F}(a) = \beta N_0 e^{s(t-a)} e^{-\mu a}, \quad t > a \geqslant 0,$$
$$u(t,a) = u_0(a-t)e^{-\mu t}, \quad 0 \leqslant t < a < c.$$

Hence

$$e^{-st}u(t,a) = \begin{cases} \beta N_0 e^{-\beta a}, & t > a \geqslant 0, \\ u_0(a-t)e^{-\beta t}, & 0 \leqslant t < a < c. \end{cases}$$

Hence the population is not completely in balanced exponential growth, but it approaches balanced exponential growth as

$$e^{-st}u(t,a) \to \beta N_0 e^{-\beta a}, \quad t \to \infty \text{ pointwise in } a \geqslant 0.$$

We also have

$$\begin{aligned}\int_0^\infty &|e^{-st}u(t,a) - \beta N_0 e^{-\beta a}|\,da \\ &\leqslant \int_t^\infty |u_0(a-t)|e^{-\beta t}\,da + \int_t^\infty \beta N_0 e^{-\beta a}\,da \\ &\leqslant 2e^{-\beta t}N_0 \longrightarrow 0, \quad t \to \infty.\end{aligned}$$

In the next section we learn that this relation holds in a far more general situation.

14.3 The Renewal Theorem: Approach to Balanced Exponential Growth

The celebrated renewal theorem states that an age-structured population that develops without constraints in a constant environment approaches balanced exponential growth,

$$u(t,a) \sim e^{st}v(a) \quad \text{as } t \to \infty.$$

Theorem 14.4. *Assume that there exists some $\lambda \in \mathbb{R}$, $M > 0$, such that*

$$\mathcal{F}(a+t) \leqslant Me^{\lambda t}\mathcal{F}(a) \quad \forall t, a \geqslant 0,$$

and that

$$1 < \int_0^\infty e^{-\lambda a}\beta(a)\mathcal{F}(a)\,da < \infty.$$

Then the Malthusian parameter s exists, $s > \lambda$, and

$$e^{-st}u(t,a) \to v(a), \quad t \to \infty \text{ pointwise in } a \geqslant 0,$$

and

$$\int_0^\infty |e^{-st}u(t,a) - v(a)|\,da \to 0, \quad t \to \infty,$$

where

$$\begin{aligned} v(a) &= \alpha e^{-sa}\mathcal{F}(a), \\ \alpha &= \frac{\int_0^\infty e^{-sr}B_0(r)\,dr}{\int_0^\infty ae^{-sa}\beta(a)\mathcal{F}(a)\,da}, \\ B_0(r) &= \int_0^c \frac{u_0(a)}{\mathcal{F}(a)}\beta(a+r)\mathcal{F}(a+r)\,da. \end{aligned}$$

Remark. The assumptions of Theorem 14.4 are satisfied if the maximum age c is finite and $\beta \cdot \mathcal{F}$ is not zero almost everywhere. They are also satisfied if $\mathcal{R}_0 > 1$. In that case choose $\lambda = 0$. Finally, they are satisfied if $\mathcal{R}_0 \geqslant 1$ and the expectations of remaining life at age a, $D(a)$, form a bounded function of a. See Section 12.4, in particular Proposition 12.1. In the latter case, we have $\mathcal{F}(a+t) \leqslant \mathrm{e}^{-\epsilon t}\mathcal{F}(a)$ for all $t, a \geqslant 0$, with some $\epsilon > 0$, by Proposition 12.1, and we choose $\lambda = -\epsilon/2$.

The proof of Theorem 14.4 can be reduced to showing the convergence of $\mathrm{e}^{-st}B(t)$.

Proposition 14.5. *Let the assumptions of Theorem 14.4 be satisfied. Then the assertions in Theorem 14.4 hold if and only if* $\mathrm{e}^{-st}B(t) \to \alpha$ *as* $t \to \infty$.

Proof. As in Proposition 14.1, we see that the Malthusian parameter, s, exists and $s > \lambda$.

Assume that the assertions of Theorem 14.4 hold. Then, by (14.2) and (14.8),

$$\mathrm{e}^{-st}B(t) = \int_0^c \beta(a)\mathrm{e}^{-st}u(t,a)\,\mathrm{d}a \longrightarrow \int_0^\infty \beta(a)\alpha\mathrm{e}^{-sa}\mathcal{F}(a)\,\mathrm{d}a = \alpha, \quad t \to \infty.$$

Now assume that $\mathrm{e}^{-st}B(t) \to \alpha$ as $t \to \infty$. To get the idea let us first look at pointwise convergence. If $t > a$,

$$\mathrm{e}^{-st}u(t,s) = \mathrm{e}^{-s(t-a)}B(t-a)\mathrm{e}^{-sa}\mathcal{F}(a) \to \alpha\mathrm{e}^{-sa}\mathcal{F}(a).$$

Then, by (14.1),

$$\begin{aligned}
&\int_0^\infty |\mathrm{e}^{-st}u(t,a) - \alpha\mathrm{e}^{-sa}\mathcal{F}(a)|\,\mathrm{d}a \\
&\quad\leqslant \int_0^t |\mathrm{e}^{-st}B(t-s)\mathcal{F}(a) - \alpha\mathrm{e}^{-sa}\mathcal{F}(a)|\,\mathrm{d}a \\
&\qquad + \int_t^\infty \left|\mathrm{e}^{-st}u_0(a-t)\frac{\mathcal{F}(a)}{\mathcal{F}(a+t)} - \alpha\mathrm{e}^{-sa}\mathcal{F}(a)\right|\mathrm{d}a \\
&\quad\leqslant \int_0^t |\mathrm{e}^{-s(t-a)}B(t-a) - \alpha|\mathrm{e}^{-sa}\mathcal{F}(a)\,\mathrm{d}a \\
&\qquad + \int_0^\infty \mathrm{e}^{-st}|u_0(a)|\frac{\mathcal{F}(a+t)}{\mathcal{F}(a)}\,\mathrm{d}a + \int_t^\infty |\alpha|\mathrm{e}^{-sa}\mathcal{F}(a)\,\mathrm{d}a.
\end{aligned}$$

The first integral tends to 0 as $t \to \infty$, because $\mathrm{e}^{-st}B(t) \to \alpha$ as $t \to \infty$. The second integral can be estimated by

$$\int_0^\infty \mathrm{e}^{-st}|u_0(a)|M\mathrm{e}^{\lambda t}\,\mathrm{d}a = \mathrm{e}^{(\lambda-s)t}\int_0^\infty |u_0(a)|\,\mathrm{d}a \to 0, \quad t \to \infty,$$

because $s > \lambda$. The third integral can be estimated by

$$|\alpha| \int_t^\infty \mathrm{e}^{-sa} M \mathrm{e}^{\lambda a}\, \mathrm{d}a = |\alpha| \frac{M}{s-\lambda} \mathrm{e}^{(\lambda - s)t} \to 0, \quad t \to \infty.$$

□

We will give no proof of the renewal theorem here. The most difficult part consists of showing the convergence of $\mathrm{e}^{-st} u(t, \cdot)$ or, equivalently, of $\mathrm{e}^{-st} B(t)$. The proof of the latter starts from the formula for the Laplace transform of B and uses techniques of evaluating inverse Laplace transforms. We will derive the formula for $\lim_{t\to\infty} \mathrm{e}^{-st} B(t)$ under the provision that this limit exists. We recall the following fact.

Lemma 14.6. *Let ϕ be a piecewise continuous real-valued function on* $[0, \infty)$ *such that* $\phi(\infty) = \lim_{t\to\infty} \phi(t)$ *exists. Then*

$$\epsilon \hat{\phi}(\epsilon) \to \phi(\infty), \quad \epsilon \to 0+.$$

The last type of convergence is called *convergence in Abel average.*

Proof. Let us first give an intuitive argument.

$$\epsilon \hat{\phi}(\epsilon) = \epsilon \int_0^\infty \mathrm{e}^{-\epsilon t} \phi(t)\, \mathrm{d}t.$$

Substituting $r = \epsilon t$,

$$\epsilon \hat{\phi}(\epsilon) = \int_0^\infty \mathrm{e}^{-r} \phi(r/\epsilon)\, \mathrm{d}r.$$

Now, for every $r > 0$, $\phi(r/\epsilon) \to \phi(\infty)$ as $\epsilon \to 0+$. So,

$$\epsilon \hat{\phi}(\epsilon) \to \int_0^\infty \mathrm{e}^{-r} \phi(\infty)\, \mathrm{d}r = \phi(\infty), \quad \epsilon \to 0+.$$

This argument can be made rigorous by using the Lebesgue theorem of dominated convergence. Otherwise we can argue as follows:

$$\epsilon \hat{\phi}(\epsilon) - \phi(\infty) = \epsilon \int_0^\infty \mathrm{e}^{-\epsilon t} (\phi(t) - \phi(\infty))\, \mathrm{d}t.$$

Hence

$$|\epsilon \hat{\phi}(\epsilon) - \phi(\infty)| \leqslant \epsilon \int_0^\infty \mathrm{e}^{-\epsilon t} |\phi(t) - \phi(\infty)|\, \mathrm{d}t.$$

Let $\delta > 0$. Since $\phi(t) \to \phi(\infty)$ as $t \to \infty$, we find some $r > 0$ such that

$$|\phi(t) - \phi(\infty)| < \delta/2 \quad \forall t > r.$$

Hence

$$\begin{aligned}
|\epsilon\hat{\phi}(\epsilon) - \phi(\infty)| \\
&\leqslant \epsilon \int_0^r \mathrm{e}^{-\epsilon t} |\phi(t) - \phi(\infty)| \,\mathrm{d}t + \epsilon \int_r^\infty \mathrm{e}^{-\epsilon t} |\phi(t) - \phi(\infty)| \,\mathrm{d}t \\
&\leqslant 2 \sup |\phi| \epsilon \int_0^r \mathrm{e}^{-\epsilon t} \,\mathrm{d}t + \epsilon \int_r^\infty \mathrm{e}^{-\epsilon t} (\delta/2) \,\mathrm{d}t \\
&\leqslant 2 \sup |\phi| (1 - \mathrm{e}^{-\epsilon r}) + (\delta/2).
\end{aligned}$$

Now we find some ϵ_0 (which depends on δ and r) such that

$$2 \sup |\phi| (1 - \mathrm{e}^{-\epsilon r}) < \delta/2 \quad \forall \epsilon \in (0, \epsilon_0).$$

Hence $|\epsilon\hat{\phi}(\epsilon) - \phi(\infty)| < \delta$ for all $\epsilon \in (0, \epsilon_0)$ and the proof is finished. □

Proposition 14.7. *Let the assumptions of Theorem 14.4 be satisfied. Then* $\mathrm{e}^{-st} B(t)$, *as* $t \to \infty$, *converges in Abel average towards*

$$\alpha = \frac{\int_0^\infty \mathrm{e}^{-sr} B_0(r) \,\mathrm{d}r}{\int_0^\infty a\mathrm{e}^{-sa} \beta(a) \mathcal{F}(a) \,\mathrm{d}a}.$$

Proof. As before, let $G(a) = \beta(a)\mathcal{F}(a)$. As in Proposition 14.1, we see that the Malthusian parameter s exists and that $s > \lambda$.

Let us first give the gist of the proof and then go into the technical details. Set

$$C(t) = \mathrm{e}^{-st} B(t).$$

Then, for $\epsilon > 0$,

$$\hat{C}_s(\epsilon) = \int_0^\infty \mathrm{e}^{-\epsilon t} \mathrm{e}^{-st} B(t) \,\mathrm{d}t = \hat{B}(s + \epsilon).$$

By (14.6),

$$\epsilon \hat{C}(\epsilon) = \epsilon \frac{\hat{B}_0(s + \epsilon)}{1 - \hat{G}(s + \epsilon)} = \hat{B}_0(s + \epsilon) \frac{\epsilon}{\hat{G}(s) - \hat{G}(s + \epsilon)}.$$

So,

$$\lim_{\epsilon \to 0+} \epsilon \hat{C}_s(\epsilon) = -\frac{\hat{B}_0(s)}{\hat{G}'(s)} =: \alpha$$

with

$$\hat{G}'(s) = -\int_0^\infty a\mathrm{e}^{-sa} \beta(a) \mathcal{F}(a) \,\mathrm{d}a < 0.$$

To make this argument rigorous we establish below that $\hat{B}_0(\nu)$ exists for $\nu \geqslant s$ and $\hat{B}(\nu)$ exists for $\nu > s$. We know from the assumptions of Theorem 14.4 that $\hat{G}(\nu)$ exists for $\nu \geqslant \lambda$, where $\lambda < s$. This implies that $\hat{G}'(s) > -\infty$.

Step 1. There exists $K > 0$ such that $|B_0(t)| \leqslant K e^{\lambda t}$ for all $t \geqslant 0$.

Hence the Laplace transforms of B_0 exist for $\nu > \lambda$, in particular for $\nu \geqslant s$. Indeed, since $\mathcal{F}(a+t) \leqslant M e^{\lambda t} \mathcal{F}(a)$ by the assumptions of Theorem 14.4,

$$|B_0(t)| \leqslant \int_0^c \beta(a+t)|u_0(a)| \frac{\mathcal{F}(a+t)}{\mathcal{F}(a)} \, da \leqslant e^{\lambda t} M \sup \beta \int_0^\infty |u_0(a)| \, da.$$

Step 2. For every $\nu > s$, there exists some $M_\nu > 0$ such that $|B(t)| \leqslant M_\nu e^{\nu t}$ for all $t \geqslant 0$. In particular, the Laplace transforms $\hat{B}(\nu)$ exist for $\nu > s$.

For $\nu > s$,

$$\hat{G}(\nu) < \hat{G}(s) = 1.$$

Fix $r > 0$. For $\nu > s$, set

$$\gamma_r^\nu = \sup\{e^{-\nu t}|B(t)|;\ 0 \leqslant t \leqslant r\}.$$

Let

$$\xi^\nu = \sup\{e^{-\nu t}|B_0(t)|;\ 0 \leqslant t < \infty\}.$$

By (14.3), for $0 \leqslant t \leqslant r$,

$$\begin{aligned} |e^{-\nu t} B(t)| &\leqslant \int_0^t e^{-\nu(t-a)}|B(t-a)| e^{-\nu a} G(a) \, da + e^{-\nu t}|B_0(t)| \\ &\leqslant \gamma_r^\nu \hat{G}(\nu) + \xi^\nu. \end{aligned}$$

Hence

$$\gamma_r^\nu \leqslant \gamma_r^\nu \hat{G}(\nu) + \xi^\nu.$$

Since $\hat{G}(\nu) < 1$,

$$\gamma_r^\nu \leqslant (1 - \hat{G}(\nu))^{-1} \xi^\nu.$$

Since the right-hand side does not depend on r, we have

$$e^{-\nu t}|B(t)| \leqslant (1 - \hat{G}(\nu))^{-1} \xi^\nu \quad \forall t \geqslant 0.$$

□

Theorem 14.4 can be conveniently reformulated as follows. We introduce the *age-profile*

$$p(t,a) = u(t,a)/N(t), \qquad N(t) = \int_0^c u(t,a) \, da.$$

Notice that $\int_0^c p(t,a) \, da = 1$, so, at any time t, $p(t,\cdot)$ is a probability density (or normalized density) in age.

Corollary 14.8. *Let the assumptions of Theorem 14.4 be satisfied. Furthermore, let*

$$\int_0^c \beta(a)\left(\int_0^a u_0(r)\,\mathrm{d}r\right)\mathrm{d}a > 0.$$

Then

$$\int_0^c |p(t,a) - p_\infty(a)|\,\mathrm{d}a \to 0, \quad t \to \infty,$$

where

$$p_\infty(a) = \frac{\mathrm{e}^{-sa}\mathcal{F}(a)}{\hat{\mathcal{F}}(s)}, \quad \hat{\mathcal{F}}(s) = \int_0^c \mathrm{e}^{-sr}\mathcal{F}(r)\,\mathrm{d}r.$$

Furthermore,

$$\mathrm{e}^{-st}N(t) \to \alpha\hat{\mathcal{F}}(s), \quad t \to \infty,$$

and

$$\frac{p(t,a)}{p_\infty(a)} \to 1, \quad t \to \infty,$$

uniformly in a from finite intervals.

Here α is defined as in Theorem 14.4. The extra assumption guarantees that the initial population will have offspring, which means that $\alpha > 0$. If it is violated, everybody in the initial population is beyond child-bearing age. Indeed, suppose $\alpha = 0$. From the formula for α in Theorem 14.4 we see that

$$\int_0^c \frac{u_0(a)}{\mathcal{F}(a)}\beta(a+r)\mathcal{F}(a+r)\,\mathrm{d}a = 0 \quad \text{for all } r \geqslant 0.$$

Then

$$0 = \int_0^\infty \left(\int_0^c \frac{u_0(a)}{\mathcal{F}(a)}\beta(a+r)\mathcal{F}(a+r)\,\mathrm{d}a\right)\mathrm{d}r.$$

Changing the variables and the order of integration,

$$\begin{aligned} 0 &= \int_0^\infty \left(\int_r^c \frac{u_0(a-r)}{\mathcal{F}(a-r)}\beta(a)\mathcal{F}(a)\,\mathrm{d}a\right)\mathrm{d}r \\ &= \int_0^c \beta(a)\mathcal{F}(a)\left(\int_0^a \frac{u_0(a-r)}{\mathcal{F}(a-r)}\,\mathrm{d}r\right)\mathrm{d}a \\ &\geqslant \int_0^c \beta(a)\mathcal{F}(a)\left(\int_0^a u_0(r)\,\mathrm{d}r\right)\mathrm{d}a. \end{aligned}$$

Since $\mathcal{F}$ is strictly positive on $[0, c)$, this implies

$$\int_0^c \beta(a)\left(\int_0^a u_0(r)\,\mathrm{d}r\right)\mathrm{d}a = 0$$

for all $a \geqslant 0$, a contradiction.

Proof. By Theorem 14.4,

$$e^{-st}N(t) \to \int_0^c v(a)\,da = \alpha\hat{\mathcal{F}}(s).$$

The first assertion follows from Theorem 14.4, because $\alpha > 0$ and $\hat{\mathcal{F}}(s) > 0$. As for the third assertion, for $t > a$,

$$\frac{p(t,a)}{p_\infty(a)} = \frac{u(t,a)\hat{\mathcal{F}}(s)}{N(t)e^{-sa}\mathcal{F}(a)} = \frac{B(t-a)\hat{\mathcal{F}}(s)}{N(t)e^{-sa}} = \frac{e^{-s(t-a)}B(t-a)\hat{\mathcal{F}}(s)}{e^{-st}N(t)}.$$

Since $e^{-st}B(t) \to \alpha$ as $t \to \infty$ (Proposition 14.5), we have $e^{-s(t-a)}B(t-a) \to \alpha$ as $t \to \infty$, uniformly in a from finite intervals. □

The renewal theorem relies on the assumption that per capita mortality and natality are constant in time, while they are actually affected by time-dependent factors such as famine, wars, epidemics, health treatment, birth control, etc. Corollary 14.8, however, remains valid for time-dependent per capita mortality rates $\mu(t,a)$ that additively split up into an age-dependent and a time-dependent part:

$$\mu(t,a) = \mu_0(a) + \mu_1(t).$$

Adding the age-independent term, $\mu_1(t)$ is sometimes called a *neutral change in mortality* because it affects all age classes in the same way (Keyfitz, 1977, Section 7.4; Preston et al., 2001, Section 7.6.2).

While the model with time-dependent mortalities can be cast in the Lotka framework (see the end of Section 13.2), the McKendrick formulation with per capita mortality rates $\mu(t,a)$ (cf. system (13.4)) is more convenient:

$$\left.\begin{aligned} \frac{\partial}{\partial t}u(t,a) + \frac{\partial}{\partial a}u(t,a) + [\mu_1(t) + \mu_0(a)]u(t,a) = 0, \quad t \neq a,\\ u(t,0) = \int_0^c \beta(a)u(t,a)\,da,\\ u(0,a) = u_0(a). \end{aligned}\right\} \quad (14.13)$$

Define

$$\mathcal{F}_1(t) = \exp\left(-\int_0^t \mu_1(s)\,ds\right), \qquad \mathcal{F}(a) = \exp\left(-\int_0^a \mu_0(s)\,ds\right)$$

and

$$\breve{u}(t,a) = \frac{u(t,a)}{\mathcal{F}_1(t)}.$$

Then

$$\frac{\partial}{\partial t}\breve{u}(t,a) + \frac{\partial}{\partial a}\breve{u}(t,a) + \mu_0(a)\breve{u}(t,a) = 0, \quad t \neq a,$$
$$\breve{u}(t,0) = \int_0^c \beta(a)\breve{u}(t,a)\,\mathrm{d}a,$$
$$\breve{u}(0,a) = u_0(a).$$

Apparently, $\breve{u}$ describes the development of an age-structured population with time-independent birth and mortality rates and the renewal theorem applies. Set

$$N(t) = \int_0^c u(t,a)\,\mathrm{d}t, \qquad \breve{N}(t) = \int_0^c \breve{u}(t,a)\,\mathrm{d}a.$$

Then

$$\breve{N}(t) = \frac{N(t)}{\mathcal{F}_1(t)} \quad \text{and} \quad p(t,a) = \frac{u(t,a)}{N(t)} = \frac{\breve{u}(t,a)}{\breve{N}(t)}$$

such that the following result holds.

Theorem 14.9. *Corollary 14.8 also holds for time-dependent per capita mortality rates of the special form $\mu(t,a) = \mu_1(t) + \mu_0(a)$. In particular, the large-time age-profile $p_\infty(a)$ is independent of μ_1.*

This additive splitting of the per capita mortality rate means that all age groups are equally affected by the time-dependent mortality factors. While this is an improvement over a time-independent mortality rate, it is still unrealistic. Furthermore, time-dependent per capita birth rates are not accounted for. Therefore, we give a second reformulation of the approach to balanced exponential growth that can be generalized to the situation where per capita mortalities and natalities change with time. Let u, $\tilde{u}$ describe the age-densities of populations with (different) initial age-densities u_0 and $\tilde{u}_0$, i.e., let u satisfy (14.13) and $\tilde{u}$ satisfy (14.13) with u_0 being replaced by $\tilde{u}_0$.

Let p be as in Corollary 14.8 and, analogously,

$$\tilde{p}(t,a) = \frac{\tilde{u}(t,a)}{\tilde{N}(t)}, \qquad \tilde{N}(t) = \int_0^c \tilde{u}(t,a)\,\mathrm{d}a$$

be the associated age-profile for the population with initial age-density $\tilde{u}_0$. Since the initial data do not affect the final age-profile of the population, Theorem 14.9 implies $\tilde{p}(t,a)/p_\infty(a) \to 1$ as $t \to \infty$ uniformly for $a \geqslant 0$ in finite age-intervals and also the analogous statement for the age-profile of the population with initial density u_0. So, $\tilde{p}(t,a)/p(t,a) \to 1$, $t \to \infty$, uniformly for $a \geqslant 0$ in finite age-intervals, where p and $\tilde{p}$ are age-profiles associated with different initial age-densities. This property is preserved if both per capita mortalities and natalities are time dependent. In the demographic literature it is sometimes called *weak ergodicity*, while the approach to balanced exponential growth is sometimes called *strong ergodicity*. We prefer the notion of *asymptotic equality*, which we have already used in Chapter 8. For a more precise formulation, we again use the McKendrick formulation instead of the Lotka formulation,

$$\frac{\partial}{\partial t}u(t,a) + \frac{\partial}{\partial a}u(t,a) + \mu(t,a)u(t,a) = 0, \quad t \neq a,$$
$$u(t,0) = \int_0^c \beta(t,a)u(t,a)\,\mathrm{d}a,$$

together with appropriate initial conditions for $u(0,\cdot)$. The property of asymptotic equality can now be formulated as follows.

Theorem 14.10 (Norton 1928). *Let β, μ be nonnegative functions on $[0,\infty)\times[0,c)$ satisfying appropriate extra assumptions. Let u and $\tilde{u}$ be solutions of the last system associated with different initial age-densities $u(0,\cdot)$ and $\tilde{u}(0,\cdot)$, which are chosen such that not everybody is beyond child-bearing age. Let p and $\tilde{p}$ be the associated age-profiles. Then*

$$\frac{\tilde{p}(t,a)}{p(t,a)} \to 1, \quad t \to \infty,$$

uniformly for $a \geqslant 0$ in finite age-intervals.

The "appropriate extra assumptions" are somewhat technical, but reasonably general. They are omitted here, as is the lengthy and difficult proof. Asymptotic equality of age-profiles is a consequence of the linear nature of the equations, i.e., of the assumption that there is no feedback from the population to the per capita mortality and fecundity rates, μ and β, though nonlinear generalizations are possible to a certain degree. See the bibliographic remarks at the end of Chapter 15. Asymptotic equality is the opposite of some types of chaos characterized by extreme long-term sensitivity to tiny changes in the initial conditions.

Exercises

14.1. Let $\mathcal{F}$ be a survival function. Let β be a continuous, bounded, nonnegative function on $[0,\infty)$, u_0 a continuous, nonnegative function on $[0,c)$, $\int_0^\infty u_0(a)\,\mathrm{d}a < \infty$.

Then equation (14.3) has a continuous nonnegative solution.

Hint. Use the contraction mapping theorem in a way similar to its use in the Picard–Lindelöf existence and uniqueness proof for ordinary differential equations. If you do not succeed in getting the solutions on $[0,\infty)$, try to get them on a small interval $[0,\epsilon]$. See Toolbox B.6 for a possible proof.

14.2. Prove the convolution theorem for the Laplace transform.

Let B and G be piecewise continuous functions on $[0, \infty)$ such that their Laplace transforms exist at $\lambda \in \mathbb{R}$. Let $B * G$ be the convolution of B and G:

$$(B * G)(t) = \int_0^t B(t-a)G(a)\,\mathrm{d}a.$$

Then the Laplace transform of $B * G$ exists at λ and

$$\widehat{B * G}(\lambda) = \hat{B}(\lambda)\hat{G}(\lambda).$$

Chapter Fifteen

Some Demographic Lessons from Balanced Exponential Growth

As we learned in the previous section, an age-structured population, developing under unrestrained and stationary conditions, approaches balanced exponential growth, associated with a stable age-distribution (profile). The assumption of stationary conditions is crucial, but seldom satisfied in reality. For Honduras, in the period 1959–1961, the theoretically predicted stable age-distribution is very close to the observed (rather coarse) age-distribution, while for Japan, in the same years, there is drastic departure (see table below, taken from Keyfitz, 1977, Table 3.2), presumably caused by substantial changes in mortality and fertility (Keyfitz, Flieger, 1990, p. 372).

	Honduras				Japan			
age groups	0–14	15–29	30–44	45+	0–14	15–29	30–44	45+
% observed	46.3	25.8	15.2	12.6	29.0	27.2	20.5	23.3
% stable	44.7	26.3	15.2	13.8	18.8	19.3	19.7	42.2

Nevertheless, balanced exponential growth provides a convenient framework for theoretical experiments from which useful demographic lessons can be learned. However, these lessons should only be generalized with sufficient caution.

15.1 Inequalities and Estimates for the Malthusian Parameter

Apparently, it is important to determine the Malthusian parameter s.

Our starting point is its characteristic equation (14.8):

$$1 = \int_0^\infty \beta(a)e^{-sa}\mathcal{F}(a)\,da.$$

We introduce

$$g(a) = \frac{\beta(a)\mathcal{F}(a)}{\mathcal{R}_0}.$$

Then

$$\int_0^\infty g(a)\,\mathrm{d}a = 1.$$

We introduce the moments of the probability density g:

$$\begin{aligned} m_j &= \int_0^\infty a^j g(a)\,\mathrm{d}a \\ &= \frac{\mathcal{R}_j}{\mathcal{R}_0}, \quad j = 1, 2, \ldots, \\ \mathcal{R}_j &= \int_0^\infty a^j \beta(a)\mathcal{F}(a)\,\mathrm{d}a, \quad j = 0, 1, 2, \ldots. \end{aligned}$$

Since g is the probability density of giving birth at a certain age,

$$E_{\mathcal{R}} := m_1 = \mathcal{R}_1/\mathcal{R}_0$$

is the *expected reproductive age* or *expected age at reproduction (child bearing)*. The variance of the age at reproduction, $V_{\mathcal{R}} = \sigma^2(\mathcal{R})$, is given by

$$V_{\mathcal{R}} = \int_0^\infty (a - E_{\mathcal{R}})^2 g(a)\,\mathrm{d}a = m_2 - m_1^2 = \frac{\mathcal{R}_2}{\mathcal{R}_0} - \left(\frac{\mathcal{R}_1}{\mathcal{R}_0}\right)^2.$$

$\sigma(\mathcal{R})$ is the standard deviation of the age at reproduction.

With the introduction of g, the characteristic equation takes the form

$$\frac{1}{\mathcal{R}_0} = \int_0^\infty \mathrm{e}^{-sa} g(a)\,\mathrm{d}a =: \hat{g}(s). \tag{15.1}$$

Notice that $\hat{g}(s)$ is the Laplace transform of the probability density g evaluated at s. Since the function e^{-sa} is convex in a, we obtain from Jensen's inequality (Theorem B.34 in Toolbox B.5)

$$\frac{1}{\mathcal{R}_0} \geqslant \exp\left(-s \int_0^\infty a g(a)\,\mathrm{d}a\right).$$

Taking logarithms we obtain the following estimate of the Malthusian parameter from below.

Theorem 15.1. *The Malthusian parameter (intrinsic rate of exponential growth) satisfies the estimate*

$$s \geqslant \frac{\ln \mathcal{R}_0}{E_{\mathcal{R}}},$$

where $\mathcal{R}_0$ is the basic reproduction ratio and $E_{\mathcal{R}}$ is the expected reproductive age.

Remark. If the birth rate β is constant, $E_{\mathcal{R}} = E$ coincides with the expectation of remaining life. Table 5.5 in Keyfitz (1977) illustrates $\mathcal{R}_0$ and $E_{\mathcal{R}}$ for various countries at certain times.

For $\mathcal{R}_0$ close to one, the formula $s \approx \ln R_0/E_{\mathcal{R}}$ is actually a reasonable approximation, which can be improved by incorporating the variance of the age at reproduction.

Lemma 15.2. *Let g be a probability density on $[0, \infty)$. Let $\eta > 0$, $M > 0$ such that*

$$|g(a)| \leqslant M\mathrm{e}^{-\eta a} \quad \forall a \geqslant 0.$$

Then the Laplace transform $\hat{g}(s)$ exists for $s > -\eta$. Furthermore, there exist a sequence of numbers (a_n) and a sequence of functions (g_n) that are differentiable for $s > -\eta$ such that

$$\hat{g}(s) = \exp\biggl(\sum_{j=1}^{n} \frac{(-1)^j}{j!} a_j s^j + s^{n+1} g_n(s)\biggr), \quad s > -\eta.$$

For s in a neighborhood of 0,

$$\hat{g}(s) = \exp\biggl(\sum_{j=1}^{\infty} \frac{(-1)^j}{j!} a_j s^j\biggr).$$

Here a_1 is the mean and a_2 the variance of the probability density g and a_3 is the third central moment of g:

$$a_1 = \int_0^\infty a g(a)\,\mathrm{d}a, \qquad a_2 = \int_0^\infty (a - a_1)^2 g(a)\,\mathrm{d}a,$$

$$a_3 = \int_0^\infty (a - a_1)^3 g(a)\,\mathrm{d}a, \qquad a_4 = \int_0^\infty (a - a_1)^4 g(a)\,\mathrm{d}a - 3a_2^2.$$

The numbers a_j are called the *cumulants* of the probability density g. The third central moment a_3 gives an idea of how much the probability density g is skewed, while a_4 indicates whether g has a large tail or not (kurtosis). See Kendall, Stuart (1958, vol. I, pp. 67–74) for the general theory and Keyfitz (1977, Table 5.5) for demographic illustrations covering various countries at various times.

Proof. Set $G(s) = \hat{g}(s)$ and

$$h(s) = \ln \hat{g}(s) = \ln G(s).$$

Then h is infinitely often differentiable for $s > -\eta$. Using Taylor's formula in integral form (see Kirkwood, 1995, Theorem 8-13, for example), we have

$$\begin{aligned} h(s) &= \sum_{j=0}^{n} h^{(j)}(0) \frac{s^j}{j!} + \frac{1}{n!} \int_0^s r^n h^{(n+1)}(r)\,\mathrm{d}r \\ &= \sum_{j=0}^{n} h^{(j)}(0) \frac{s^j}{j!} + \frac{s^{n+1}}{n!} \int_0^1 r^n h^{(n+1)}(rs)\,\mathrm{d}r. \end{aligned}$$

Hence the assertion holds with $a_j = (-1)^n h^{(j)}(0)$ and

$$g_n(s) = \int_0^1 r^n h^{(n+1)}(rs)\,\mathrm{d}r.$$

Now $a_0 = h(0) = \ln G(0) = 0$. Furthermore,

$$h'(s) = \frac{G'(s)}{G(s)}$$

and

$$h'' = \frac{G''}{G} - \left(\frac{G'}{G}\right)^2.$$

Hence

$$a_1 = -h'(0) = -G'(0) = \int_0^\infty a g(a)\,\mathrm{d}a$$

and

$$a_2 = h''(0) = G''(0) - a_1^2 = \int_0^\infty a^2 g(a)\,\mathrm{d}a - a_1^2 = \int_0^\infty (a - a_1)^2 g(a)\,\mathrm{d}a.$$

Finally, G is a holomorphic function of $s \in \mathbb{C}$, $\Re s > -\eta$, and so real analytic in $s \geqslant -\eta$, while $\ln x$ is a real analytic function in a neighborhood of $1 = G(0)$. So, the composition $h = \ln G$ is real analytic in a neighborhood of 0. □

If $\mathcal{R}_0$ is close to one, s is close to 0 and we obtain

$$\ln \mathcal{R}_0 \approx E_{\mathcal{R}} s - \tfrac{1}{2} V_{\mathcal{R}} s^2.$$

Solving for s yields

$$s \approx \frac{E_{\mathcal{R}}}{V_{\mathcal{R}}} \pm \sqrt{\left(\frac{E_{\mathcal{R}}}{V_{\mathcal{R}}}\right)^2 - 2\frac{\ln \mathcal{R}_0}{V_{\mathcal{R}}}}.$$

Since $s = 0$ for $\mathcal{R}_0 = 1$, we need to choose

$$s \approx \frac{E_{\mathcal{R}}}{V_{\mathcal{R}}} - \sqrt{\left(\frac{E_{\mathcal{R}}}{V_{\mathcal{R}}}\right)^2 - 2\frac{\ln \mathcal{R}_0}{V_{\mathcal{R}}}}$$

(cf. Metz, Diekmann 1986, pp. 153, 154). After some elementary transformations,

$$\begin{aligned} s &\approx \frac{\ln \mathcal{R}_0}{E_{\mathcal{R}}} \frac{2}{1 + \sqrt{1 - 2(V_{\mathcal{R}}/E_{\mathcal{R}}^2)\ln \mathcal{R}_0}} \approx \frac{\ln \mathcal{R}_0}{E_{\mathcal{R}}} \frac{2}{2 - (V_{\mathcal{R}}/E_{\mathcal{R}}^2)\ln \mathcal{R}_0} \\ &\approx \frac{\ln \mathcal{R}_0}{E_{\mathcal{R}}}\left(1 + \frac{1}{2}\frac{V_{\mathcal{R}}}{E_{\mathcal{R}}^2}\ln \mathcal{R}_0\right). \end{aligned} \tag{15.2}$$

We conclude that s depends on the basic reproduction ratio in a monotone increasing way, on the expected age at reproduction in a monotone decreasing way, and on the variance of the age at reproduction in a monotone increasing way.

Let us look at the case of an exponentially increasing population. Since the concept of intrinsic rate of exponential increase is not very suggestive, one sometimes likes to look at the doubling time, the time T_2 it takes a population to double its size:

$$T_2 = \frac{\ln 2}{s}.$$

We obtain that

$$T_2 \approx \frac{\ln 2}{\ln \mathcal{R}_0} E_{\mathcal{R}} \frac{1 + \sqrt{1 - 2(V_{\mathcal{R}}/E_{\mathcal{R}}^2) \ln \mathcal{R}_0}}{2} \approx \frac{\ln 2}{\ln \mathcal{R}_0} E_{\mathcal{R}} \left(1 - \frac{\ln \mathcal{R}_0}{2} \frac{V_{\mathcal{R}}}{E_{\mathcal{R}}^2}\right). \tag{15.3}$$

Notice that the doubling time is almost directly proportional to the expected reproductive age.

Deriving an upper estimate for s is more involved than deriving a lower estimate.

Recalling the characteristic equation for s (equation (14.8)), we use Jensen's inequality (Theorem B.34 in Toolbox B.5):

$$\mathcal{R}_0 = \int_0^\infty \mathrm{e}^{sa} \mathrm{e}^{-sa} \beta(a) \mathcal{F}(a)\, \mathrm{d}a \geqslant \exp\left(s \int_0^\infty a \mathrm{e}^{-sa} \beta(a) \mathcal{F}(a)\, \mathrm{d}a\right).$$

In order to interpret this formula we recall that

$$p_\infty(a) = \text{const.} \times \mathrm{e}^{-sa} \mathcal{F}(a)$$

is the age-profile of the population, where the constant has been chosen such that the integral over p_∞ is 1. Then the *average age at reproduction (in the population)*, $A_{\mathcal{R}}$, is defined as

$$A_{\mathcal{R}} = \frac{\int_0^\infty a \beta(a) p_\infty(a)\, \mathrm{d}a}{\int_0^\infty \beta(a) p_\infty\, \mathrm{d}a}.$$

It follows from the characteristic equation for the Malthusian parameter that

$$A_{\mathcal{R}} = \frac{\int_0^\infty a \beta(a) \mathrm{e}^{-sa} \mathcal{F}(a)\, \mathrm{d}a}{\int_0^\infty \beta(a) \mathrm{e}^{-sa} \mathcal{F}(a)\, \mathrm{d}a} = \int_0^\infty a \beta(a) \mathrm{e}^{-sa} \mathcal{F}(a)\, \mathrm{d}a.$$

Notice that the expected age at reproduction and the average age at reproduction coincide if the Malthusian parameter is 0.

Taking logarithms in the above inequality, we obtain the following upper estimate.

Proposition 15.3. *The Malthusian parameter s satisfies the estimate*

$$s \leqslant \frac{\ln \mathcal{R}_0}{A_{\mathcal{R}}},$$

where $A_{\mathcal{R}}$ is the average age at reproduction in the population.

We can combine this estimate with the one in Theorem 15.1 to give

$$\frac{\ln \mathcal{R}_0}{E_{\mathcal{R}}} \leqslant s \leqslant \frac{\ln \mathcal{R}_0}{A_{\mathcal{R}}}.$$

From this formula we conclude that $E_{\mathcal{R}} - A_{\mathcal{R}}$ has the same sign as the Malthusian parameter s, i.e., in a growing population the average age at childbirth is smaller than the expected age at childbirth, while it is the other way around in a decreasing population.

Recalling our previous notation $g(a) = (1/\mathcal{R}_0)\beta(a)\mathcal{F}(a)$, the estimate in Proposition 15.3 can be rewritten as

$$\ln \mathcal{R}_0 \geqslant s\mathcal{R}_0 \int_0^{\infty} a\mathrm{e}^{-sa} g(a)\, da.$$

In the following we assume that $s > 0$, i.e., $\mathcal{R}_0 > 1$.

Using Jensen's inequality again,

$$\ln \mathcal{R}_0 \geqslant s m_1 \mathcal{R}_0 \exp\left(-s\frac{m_2}{m_1}\right).$$

We rewrite the inequality in dimensionless variables and parameters:

$$\left.\begin{aligned} s &= \frac{\ln \mathcal{R}_0}{m_1} x = \frac{\ln \mathcal{R}_0}{E_{\mathcal{R}}} x, \\ 1 &\geqslant x\mathcal{R}_0^{1-\gamma x} =: h(x), \\ \gamma &= \frac{m_2}{m_1^2} = 1 + \frac{V_{\mathcal{R}}}{E_{\mathcal{R}}^2}. \end{aligned}\right\} \tag{15.4}$$

The function $y\mathcal{R}_0^{1-\gamma y}$ has its maximum at $y_0 = 1/(\gamma \ln \mathcal{R}_0)$, where it takes the value $\mathcal{R}_0/(\mathrm{e}\gamma \ln \mathcal{R}_0)$. h is strictly increasing on $[0, y_0]$ and strictly decreasing on $[y_0, \infty)$. Furthermore, $h(0) = 0$ and $h(1) < 1$. The last follows from $\gamma > 1$.

We will only be able to extract useful information from this inequality if $h(y_0) > 1$, i.e.,

$$\gamma \mathrm{e} \ln \mathcal{R}_0 < \mathcal{R}_0.$$

This condition is satisfied if $\mathcal{R}_0$ is large or close to 1, but not if $\mathcal{R}_0 \approx \mathrm{e}$ because $\gamma > 1$. Furthermore, we are only interested in the case in which $x > 1$. Because of $h(x) \leqslant 1$ and $h(1) < 1$ this means that $1 < y_0$. Thus we assume

$$\gamma \ln \mathcal{R}_0 < 1.$$

Since $\gamma > 1$, this restricts our consideration to $\mathcal{R}_0 < \mathrm{e}$.

We rewrite the inequality in fixed point form.

$$x \leqslant \mathcal{R}_0^{\gamma x - 1} = f(x).$$

We are interested in the case where $s > 0$ is small, in particular $x \leqslant 1/(\gamma \ln \mathcal{R}_0)$. We consider sequences

$$x_n = f(x_{n-1}), \quad n \in \mathbb{N},$$

where x_0 has been chosen such that $x_1 < x_0 \leqslant y_0 = 1/(\gamma \ln \mathcal{R}_0)$. This means that $1 < x_0 \mathcal{R}_0^{1-\gamma x_0}$. Since $y\mathcal{R}^{1-\gamma y}$ is strictly increasing in $y \in [0, y_0]$, we have $x < x_0$.

By induction we see that

$$x < x_{n+1} < x_n < x_0, \quad n \in \mathbb{N},$$

i.e., we obtain an improving sequence of upper estimates for x.

For the doubling time we obtain the following estimates:

$$\mathcal{R}_1 \frac{\ln 2}{\ln \mathcal{R}_0} \frac{1}{x_n} \leqslant T_2 \leqslant \mathcal{R}_1 \frac{\ln 2}{\ln \mathcal{R}_0}.$$

Keyfitz (1977, figure 5.5) gives the following numbers for Puerto Rico in 1964:

$$\mathcal{R}_0 = 1.813, \qquad E_{\mathcal{R}} = 27.14, \qquad V_{\mathcal{R}} = 46.33.$$

After many iterations we find $x_n \approx 1.17$ and

$$27.02 \leqslant T_2 \leqslant 31.62.$$

The approximate formula (15.3) gives $T_2 \approx 30.85$.

Exercises

15.1. The following data have been taken from Table 5.5 in Keyfitz (1977).

Country (females)	$\mathcal{R}_0$	$E_{\mathcal{R}}$	$V_{\mathcal{R}}$
Canada 1965	1.496	27.75	37.09
Chile 1964	1.824	29.08	48.62
Honduras 1965	2.613	29.12	53.77

Calculate approximations and estimates from below and from above for the doubling times of these countries. (For Honduras it may not be possible to find a lower estimate, but try anyway.) Look for confirmation in the literature or on the web.

15.2 Let $\mathcal{F}$ and β be given survival and per capita birth rate functions. Introduce new survival and per capita birth rate functions by

$$\mathcal{F}_\gamma(a) = \mathcal{F}(\gamma a), \quad \beta_\gamma(a) = \gamma\beta(\gamma a), \quad a \geqslant 0,$$

where $\gamma > 0$. Determine how the Malthusian parameters s_γ associated with $\mathcal{F}_\gamma$ and β_γ are related with the Malthusian parameter s associated with the original functions. What about the doubling times, the basic reproduction ratios, and the average ages at reproduction?

15.2 Average Age and Average Age at Death in a Population at Balanced Exponential Growth. Average Per Capita Death Rate

We recall (Corollary 14.8) that, at balanced exponential growth, the age-profile (the probability density for having a certain age) is given by

$$p_\infty(a) = \frac{e^{-sa}\mathcal{F}(a)}{D_S}, \qquad D_S = \int_0^\infty e^{-sa}\mathcal{F}(a)\, da.$$

Hence the *average age of the population*, A, is

$$A = \int_0^\infty a p_\infty(a)\, da = \frac{1}{D_S}\int_0^\infty a e^{-sa}\mathcal{F}(a)\, da = \frac{\int_0^\infty a e^{-sa}\mathcal{F}(a)\, da}{\int_0^\infty e^{-sa}\mathcal{F}(a)\, da}. \tag{15.5}$$

If the Malthusian parameter s is 0, i.e., if the population is in a balanced time-independent state, the average age coincides with the expectation of remaining life, E (see formula (12.8)).

The age variance of the population, $\sigma^2(A)$, is given by

$$\sigma^2(A) = \int_0^\infty (a-A)^2 p_\infty(a)\, da = \frac{\int_0^\infty (a-A)^2 e^{-sa}\mathcal{F}(a)\, da}{\int_0^\infty e^{-sa}\mathcal{F}(a)\, da}. \tag{15.6}$$

For $s = 0$ we obtain the variance of remaining life, $\sigma^2(E)$:

$$\sigma^2(E) = \frac{\int_0^\infty (a-E)^2\mathcal{F}(a)\, da}{D}. \tag{15.7}$$

Changes in Fertility

It seems to be intuitive that the average age in a population with exponentially balanced growth should decrease if the Malthusian parameter increases. This can be verified for the case where the increase in s comes from a modification of the birth rate (cf. Impagliazzo, 1985, Corollary 7.23), but not from a modification of the survival function (cf. Keyfitz, 1977, Section 7.4). The same holds for the average age at death, $A^\sharp$:

$$A^\sharp = \frac{\int_0^\infty a\mu(a)p_\infty(a)\,\mathrm{d}a}{\int_0^\infty \mu(a)p_\infty(a)\,\mathrm{d}a}.$$

Recall that for a stationary population, i.e., $s = 0$, the average age at death equals the life expectancy, D (Section 13.5.2). So, the average age at death is smaller than the life expectancy in a growing population, while it is larger in a decreasing population.

Proposition 15.4. *For a fixed survival function $\mathcal{F}$, the average age A in a population with balanced exponential growth is a decreasing function of the Malthusian parameter s. The same holds for the average age $A^\sharp$ at death. In particular, $A < E$ if $s > 0$ and $A > E$ if $s < 0$, as well as $A^\sharp < D$ if $s > 0$ and $A^\sharp > D$ if $s < 0$.*

Proof. Observe that

$$\begin{aligned}\frac{\mathrm{d}}{\mathrm{d}s}\frac{\int_0^\infty a\mathrm{e}^{-sa}\mathcal{F}(a)\,\mathrm{d}a}{\int_0^\infty \mathrm{e}^{-sa}\mathcal{F}(a)\,\mathrm{d}a} &= -\frac{\int_0^\infty a^2\mathrm{e}^{-sa}\mathcal{F}(a)\,\mathrm{d}a}{\int_0^\infty \mathrm{e}^{-sa}\mathcal{F}(a)\,\mathrm{d}a} + A^2 \\ &= -\frac{\int_0^\infty (a-A)^2\mathrm{e}^{-sa}\mathcal{F}(a)\,\mathrm{d}a}{\int_0^\infty \mathrm{e}^{-sa}\mathcal{F}(a)\,\mathrm{d}a} \\ &= -\sigma^2(A) < 0.\end{aligned}$$

As for the average age at death,

$$A^\sharp = \frac{\int_0^\infty a\mathrm{e}^{-sa}\mathcal{F}'(a)\,\mathrm{d}a}{\int_0^\infty \mathrm{e}^{-sa}\mathcal{F}'(a)\,\mathrm{d}a}.$$

The same proof works with $\mathcal{F}$ being replaced by $-\mathcal{F}'$ and A being replaced by $A^\sharp$. □

From this proof we keep in mind that

$$E - A = s\sigma^2(A) + O(s^2) = s\sigma^2(E) + O(s^2). \tag{15.8}$$

Here we use the convention that $O(s^2)$ denotes an expression the absolute value of which can be estimated by a constant time s^2.

Proposition 15.4 can be illustrated by an example in Keyfitz (1977, Table 7.4), where the average age of a population has been calculated as a function of hypothetical values of the Malthusian parameter based on the life table for United States females, 1964.

Malthusian parameter	0.01	0.015	0.02	0.025	0.03	0.035	0.04
average age	33.3	30.8	28.4	26.2	24.1	22.3	20.6

Figure 7.1 in Keyfitz (1977) nicely shows the linear relationship for small s, which becomes parabolic for somewhat larger s.

The *average per capita mortality rate* of a population at balanced exponential growth is defined as

$$\bar{\mu} = \frac{\int_0^\infty \mu(a) p_\infty(a)\,\mathrm{d}a}{\int_0^\infty p_\infty(a)\,\mathrm{d}a}.$$

Using $p_\infty(a) = \mathrm{e}^{-sa}\mathcal{F}(a)$ and taking logarithms,

$$\ln\bar{\mu} = \ln\bar{\mu}(s) = \ln\int_0^\infty \mu(a)\mathrm{e}^{-sa}\mathcal{F}(a)\,\mathrm{d}a - \ln\int_0^\infty \mathrm{e}^{-sa}\mathcal{F}(a)\,\mathrm{d}a.$$

Using logarithmic differentiation with respect to s,

$$\frac{\mathrm{d}}{\mathrm{d}s}\ln\bar{\mu} = A - A^\sharp.$$

This implies that the average mortality rises with the Malthusian parameter if the average age exceeds the average age at death and falls otherwise (cf. Preston et al., 2001, Section 7.6.1).

Changes in Mortality

Let us assume that we have a basic survivorship described by a survival function $\mathcal{F}_0$ and an additional constant per capita mortality rate α. Then the survival function is

$$\mathcal{F}_\alpha(a) = \mathrm{e}^{-\alpha a}\mathcal{F}_0(a).$$

Let A_α be the average age in the population governed by the survivorship $\mathcal{F}_\alpha$. This *neutral change in mortality* does not affect the age-profile or the average age (Keyfitz, 1977, Section 7.4; Preston et al., 2001, Section 7.6.2). If $\mathcal{F}_0$ is associated with a per capita mortality rate $\mu_0(a)$ such that $\mathcal{F}_\alpha$ is associated with the overall per capita mortality rate $\mu(a) = \alpha + \mu_0(a)$, this already follows from Theorem 14.9; but it is also true in general.

Proposition 15.5. *The age-profile and the average age, A_α, do not depend on the extra mortality rate α.*

Proof. Let

$$h_\alpha(a) = \beta(a)\mathrm{e}^{-\alpha a}\mathcal{F}_0(a) = \mathrm{e}^{-\alpha a}h_0(a).$$

Then the Malthusian parameter s_α associated with the survivorship $\mathcal{F}_\alpha$ is determined by

$$\begin{aligned} 1 = \hat{h}_\alpha(s_\alpha) &= \int_0^\infty \mathrm{e}^{-s_\alpha a}\mathrm{e}^{-\alpha a}h_0(a)\,\mathrm{d}a \\ &= \int_0^\infty \mathrm{e}^{-(s_\alpha+\alpha)}h_0(a)\,\mathrm{d}a = \hat{h}_0(s_\alpha+\alpha). \end{aligned}$$

On the other hand we have

$$1 = \hat{h}_0(s_0),$$

and s_0 is uniquely determined by this relation. Hence

$$s_0 = s_\alpha + \alpha.$$

This implies that

$$e^{-s_0 a} \mathcal{F}(a) = e^{-s_\alpha a} \mathcal{F}_\alpha(a).$$

It follows from (15.5) that A_α is independent of α. □

Corollary 15.6. *If the length of life is exponentially distributed, then the average age of the population is independent of the age-independent mortality rate* μ.

If the length of life is exponentially distributed, i.e., we have a constant per capita mortality rate μ, then

$$A = \frac{1}{s + \mu}.$$

If the per capita birth rate β is also constant, then

$$1 = \frac{\beta}{s + \mu},$$

i.e.,

$$A = \frac{1}{\beta}.$$

This means that, in a population with balanced exponential growth and with constant per capita mortality and birth rates, the birth rate is the reciprocal of the average age of the population. Recall that $E = 1/\mu = D$.

We now can easily construct an example of two populations at balanced exponential growth, one increasing and the other decreasing, such that the average age in the decreasing population is smaller than in the increasing one: for the increasing population we set $\beta = 2$ and $\mu = 1$, while for the decreasing population we set $\beta = 3$ and $\mu = 4$.

Exercises

15.3. Use Keyfitz's (1977) hypothetical data for s and A to determine E and $\sigma^2(A)$. For instance, plot or draw them or use curve-fitting techniques.

15.3 Ratio of Population Size and Birth Rate

If the basic reproduction ratio $\mathcal{R}_0$ is 1, then the Malthusian parameter s is 0, and we can conclude from the renewal theorem (Theorem 14.4 and Proposition 14.5) that B converges to a finite limit as time tends to infinity. If this limit is positive, from Theorem 13.2, we have that the population size, N, also converges and

$$\frac{N(t)}{B(t)} \to D, \quad t \to \infty. \tag{15.9}$$

Here D is again the life expectancy (average duration of life, average age at death). We want to investigate what happens if $s \neq 0$, in which case the population eventually grows at a nonzero exponential rate.

Set

$$u_s(t,a) = \mathrm{e}^{-st} u(t,a), \qquad N_s(t) = \mathrm{e}^{-st} N(t),$$
$$C_s(t) = \mathrm{e}^{-st} B(t), \qquad \mathcal{F}_s(a) = \mathrm{e}^{-sa} \mathcal{F}(a).$$

Then

$$u_s(t,a) = C_s(t-a)\mathcal{F}_s(a), \qquad t > a \geqslant 0,$$
$$u_s(t,a) = u_0(a-t)\frac{\mathcal{F}_s(a)}{\mathcal{F}_s(a-t)}, \quad 0 \leqslant t < a < c.$$

Furthermore, by the renewal theorem, $C_s(t)$ converges, with the limit being strictly positive if the extra condition in Corollary 14.8 holds. From Theorem 13.2,

$$\frac{N(t)}{B(t)} = \frac{N_s(t)}{C_s(t)} \to D_S$$
$$:= \int_0^\infty \mathcal{F}_s(a)\,\mathrm{d}a = \int_0^\infty \mathrm{e}^{-sa}\mathcal{F}(a)\,\mathrm{d}a = \hat{\mathcal{F}}(s).$$

By Jensen's inequality (Toolbox B.5), applied to the convex function $a \mapsto \mathrm{e}^{-sa}$,

$$\frac{D_S}{D} = \frac{\int_0^\infty \mathrm{e}^{-sa}\mathcal{F}(a)\,\mathrm{d}a}{\int_0^\infty \mathcal{F}(a)\,\mathrm{d}a} \geqslant \exp\left(-s\frac{\int_0^\infty a\mathcal{F}(a)\,\mathrm{d}a}{\int_0^\infty \mathcal{F}(a)\,\mathrm{d}a}\right) = \mathrm{e}^{-sE}.$$

Again by Jensen's inequality, now applied to the convex function $a \mapsto \mathrm{e}^{sa}$,

$$\frac{D}{D_S} = \frac{\int_0^\infty \mathrm{e}^{sa}\mathrm{e}^{-sa}\mathcal{F}(a)\,\mathrm{d}a}{\int_0^\infty \mathrm{e}^{-sa}\mathcal{F}(a)\,\mathrm{d}a} \geqslant \exp\left(s\frac{\int_0^\infty a\mathrm{e}^{-sa}\mathcal{F}(a)\,\mathrm{d}a}{\int_0^\infty \mathrm{e}^{-sa}\mathcal{F}(a)\,\mathrm{d}a}\right).$$

Reorganizing and using (15.5),

$$\frac{D_S}{D} \leqslant \mathrm{e}^{-sA}.$$

If $\mathcal{R}_0$ is sufficiently close to 1, then $\mathrm{e}^{-sE_{\mathcal{R}}} \approx \mathcal{R}_0^{-1}$ by (15.2). So we come to the following conclusion.

Theorem 15.7. *As the population approaches balanced exponential growth, with the basic reproduction ratio $\mathcal{R}_0$ being sufficiently close to 1, we have*

$$\lim_{t\to\infty} \frac{N(t)}{B(t)} \approx D\mathcal{R}_0^{-(\tilde{A}/E_{\mathcal{R}})}$$

with $\tilde{A}$ being between E and A.

Let us repeat that D is the life expectancy, E the expectation of remaining life, A the average age of the population, and $E_{\mathcal{R}}$ an individual's expected age at child bearing.

Frauenthal (1986) formulated a rule of thumb that can be rephrased as $\tilde{A}/E_{\mathcal{R}}$ being approximately 1 for all human populations. If *Frauenthal's rule of thumb* holds, we have

$$\lim_{t\to\infty} \frac{N(t)}{B(t)} \approx \frac{D}{\mathcal{R}_0}. \tag{15.10}$$

Notice that (15.10) is precise when the birth rate β is constant because then $B(t) = \beta N(t)$ by (14.2) and $\mathcal{R}_0 = \beta D$.

15.4 Consequences of an Abrupt Shift in Maternity: Momentum of Population Growth

As an application of the theory of balanced exponential growth we consider the following example taken from Frauenthal (1986).

Imagine that a population has been growing under stationary conditions for a long time such that it has reached balanced exponential growth. At time $t = 0$, the maternity behavior abruptly changes such that the new basic reproduction ratio is 1. This change occurs by scaling the birth rate proportionally at all ages such that

$$\tilde{\beta}(a) = \beta(a)/\mathcal{R}_0,$$

where $\mathcal{R}_0 > 1$ is the old basic reproduction ratio,

$$\mathcal{R}_0 = \int_0^\infty \beta(a)\mathcal{F}(a)\,\mathrm{d}a.$$

Notice that such an adjustment of the birth rate does not affect the expected age at reproduction, $E_{\mathcal{R}}$ (Section 15.1).

Since the population is in balanced exponential growth before $t = 0$,

$$B(t) = B(0)\mathrm{e}^{st}, \quad t \leqslant 0.$$

This means that

$$u(t,a) = B(t-a)\mathcal{F}(a) = B(0)\mathrm{e}^{s(t-a)}\mathcal{F}(a), \quad t \leqslant 0, \quad a \geqslant 0,$$

in particular

$$u_0(a) = u(0, a) = B(0)\mathrm{e}^{-sa}\mathcal{F}(a), \quad a \geqslant 0.$$

After readjustment of the birth rate, the Malthusian parameter is 0, and we have from Theorem 14.4 and Proposition 14.5 (see also Proposition 14.7) that

$$B(t) \to B(\infty), \quad t \to \infty,$$

$$B(\infty) = \frac{\int_0^\infty (\int_0^c \tilde{\beta}(a+r)u_0(a)(\mathcal{F}(a+r)/\mathcal{F}(a))\,\mathrm{d}a)\,\mathrm{d}r}{\int_0^\infty a\tilde{\beta}(a)\mathcal{F}(a)\,\mathrm{d}a}.$$

Substituting the formula for u_0 and $\tilde{\beta}$,

$$B(\infty) = \frac{\int_0^\infty (\int_0^c \beta(a+r)B(0)\mathrm{e}^{-sa}\mathcal{F}(a+r)\,\mathrm{d}a)\,\mathrm{d}r}{\int_0^\infty a\beta(a)\mathcal{F}(a)\,\mathrm{d}a}.$$

Notice that the denominator is $E_{\mathcal{R}}\mathcal{R}_0$. Making the substitution $a+r=t$ in the inner integral we get

$$\frac{B(\infty)}{B(0)} = \frac{\int_0^\infty (\int_r^\infty \beta(t)\mathrm{e}^{-s(t-r)}\mathcal{F}(t)\,\mathrm{d}t)\,\mathrm{d}r}{\mathcal{R}_0 E_{\mathcal{R}}}.$$

Changing the order of integration

$$\frac{B(\infty)}{B(0)} = \frac{\int_0^\infty \beta(t)(\int_0^t \mathrm{e}^{-s(t-r)}\,\mathrm{d}r)\mathcal{F}(t)\,\mathrm{d}t}{\mathcal{R}_0 E_{\mathcal{R}}},$$

and

$$\frac{B(\infty)}{B(0)} = \frac{\int_0^\infty \beta(t)(1-\mathrm{e}^{-st})\mathcal{F}(t)\,\mathrm{d}t}{s\mathcal{R}_0 E_{\mathcal{R}}}.$$

Hence

$$\frac{B(\infty)}{B(0)} = \frac{\mathcal{R}_0 - 1}{s\mathcal{R}_0 E_{\mathcal{R}}}. \tag{15.11}$$

If the variance of the reproductive age is not too large compared to the square of the expected reproductive age and $\mathcal{R}_0$ is close to 1, then, by (15.2),

$$s \approx \frac{\ln \mathcal{R}_0}{E_{\mathcal{R}}}$$

is a reasonable approximation and we obtain from (15.11) that

$$\frac{B(\infty)}{B(0)} \approx \frac{\mathcal{R}_0 - 1}{\mathcal{R}_0 \ln \mathcal{R}_0}.$$

We obtain the approximation in Frauenthal (1986) by continuing:

$$\begin{aligned}\frac{B(\infty)}{B(0)} &\approx \frac{1}{\sqrt{\mathcal{R}_0}} \frac{\mathcal{R}_0^{1/2} - \mathcal{R}_0^{-1/2}}{\ln \mathcal{R}_0} \\ &= \frac{1}{\sqrt{\mathcal{R}_0}} \frac{\sinh[(1/2)\ln \mathcal{R}_0]}{(1/2)\ln \mathcal{R}_0} \\ &\approx \frac{1}{\sqrt{\mathcal{R}_0}}. \end{aligned} \tag{15.12}$$

With the per capita reproduction rate being reduced by the factor $1/\mathcal{R}_0$ in each age class, one might assume that the quotient of the new asymptotic birth rate over the birth rate at the time of change should also be $1/\mathcal{R}_0$ rather than $1/\sqrt{\mathcal{R}_0}$. But the population still has a lot of momentum which is stored in the age-distribution and causes this effect.

In order to relate the population size at $t = 0$, $N(0)$, and the final population size, N_∞, we remember from Theorem 13.2 that

$$N(\infty) = B(\infty)D.$$

Thus

$$\frac{N(\infty)}{N(0)} = \frac{N(\infty)}{B(\infty)} \frac{B(\infty)}{B(0)} \frac{B(0)}{N(0)} = D \frac{B(\infty)}{B(0)} \frac{B(0)}{N(0)}.$$

Now recall (15.12) and Theorem 15.7, which in this case reads

$$\frac{B(0)}{N(0)} \approx \frac{\mathcal{R}_0^{\tilde{A}/E_{\mathcal{R}}}}{D}, \qquad \text{for some } \tilde{A} \in [A, E].$$

Combining the two we obtain

$$\frac{N(\infty)}{N(0)} \approx \mathcal{R}_0^{(\tilde{A}/E_{\mathcal{R}})-(1/2)}. \tag{15.13}$$

To interpret this formula remember that E is the expectation of remaining life, $E_{\mathcal{R}}$ the expected age at childbearing, and A is the average age of the population at balanced exponential growth before the shift in maternity. Also remember that $E \geqslant A$ because the population size was increasing before the adjustment (Proposition 15.4). One can say with reasonable certainty that the exponent of $\mathcal{R}_0$ in (15.13) is strictly positive. Even though the basic reproduction ratio (average number of offspring per individual in a lifetime) is adjusted from being larger than 1 to being equal to 1, the final population size will be larger than the population size at the time of the adjustment. This is again due to the momentum stored in the age-distribution of the population.

One can also safely assume that the exponent of $\mathcal{R}_0$ in (15.13) is less than 1. If one accepts *Frauenthal's rule of thumb*, $E_{\mathcal{R}} \approx \tilde{A}$, then the exponent is $\frac{1}{2}$ and we obtain the formula by Frauenthal (1986):

$$\frac{N(\infty)}{N(0)} \approx \sqrt{\mathcal{R}_0}.$$

Exercises

15.4. Formula (15.13) implies

$$\frac{N(\infty)}{N(0)} \geqslant \mathcal{R}_0^{(A/E_{\mathcal{R}})-(1/2)} \text{ approximatively.}$$

The following data have been combined from Keyfitz (1977, Table 5.5) and Keyfitz, Flieger (1971, Table G, p. 34). Unfortunately, the two tables do not give data from the same year.

Country (females)	$\mathcal{R}_0$	$E_{\mathcal{R}}$	A
England	1.343 (1963)	27.36 (1963)	35.5 (1968)
Sweden	1.209 (1778–1782)	32.24 (1778–1782)	29.1 (1803–1807)
United States	1.395 (1965)	26.52 (1965)	35.4 (1967)

Using these data, give estimates from below for $N(\infty)/N(0)$ and compare them to the approximation you would obtain from Frauenthal's rule of thumb.

Bibliographic Remarks

Figures displaying the reproduction rates (or natalities) for various countries at various times can be found in Hutchinson (1978, figure 70, p. 92) for Taiwan (around 1906), Coale (1972, figure 1.3, p. 13) for the United States (1950 and 1959), and Coale, Trussell (1974, figure 3) for Hungary (1970), Japan (1964) and Sweden (1891–1900).

Approach to balanced exponential growth (or, equivalently, the renewal theorem, or strong ergodicity) was stated and heuristically proved by Sharpe, Lotka (1911) and by Lotka (1922). Two different rigorous proofs were given by Feller (1941, 1966). A heuristic explanation can be found in Hoppensteadt (1975) or Frauenthal (1986). Iannelli (1995) gives a rather complete proof of the case where the maximum age is finite. Proofs using operator semigroups are presented by Webb (1984, 1985). A more detailed historical exposition can be found in Smith, Keyfitz (1977) (see the introduction to the chapter entitled "Stable Population Theory"). The existence of exponential solutions can be extended to homogeneous two-sex populations (Martcheva, 1999).

Approach to balanced exponential growth is a special case of the weak ergodicity theory (see Theorem 14.10) developed in the paper by Norton (1928), which has been overlooked for a long time (see also Charlesworth (1980, 1994, Section 1.4.1), Caswell (1989, Section 8.3), and Thieme (1988a)). Nonlinear versions of weak ergodicity (or asymptotic equality) can be found in Nussbaum (1990), Nesemann (1999), or Thieme (1988a).

Keyfitz (1977) gives examples for the basic reproduction ratio, the mean age at reproduction, and the second to fourth cumulants for quite a few countries, mainly in the mid 1960s (Table 5.5). For a brief history of the basic reproduction ratio in demography, ecology, and epidemics see Heesterbeek (2002).

The equality of the average age and the average expectation of remaining life for a population in balanced exponential growth with Malthusian parameter 0 has been observed by Kim, Aron (1989). For the optimal control and harvesting of age-structured population we refer to Anita et al. (1998), Anita (2000), Barbu et al. (2001), and the references cite therein.

Chapter Sixteen

Some Nonlinear Demographics

In Part 1 we argued that linear models are only realistic under particular circumstances. This also applies to population models with age-structured population models. In this chapter, we will only scratch the surface of the large body of nonlinear age-dependent models, as we consider the following nonlinear version of the demographic model in Chapter 14. The age-density of the population is given in the usual way (Section 13.1):

$$\left.\begin{aligned} u(t,a) &= B(t-a)\mathcal{F}(a), && t > a \geqslant 0, \\ u(t,a) &= u_0(a-t)\frac{\mathcal{F}(a)}{\mathcal{F}(a-t)}, && 0 \leqslant t < a < c, \\ u(t,a) &= 0, && t \geqslant 0,\ a > c. \end{aligned}\right\} \tag{16.1}$$

As the stage under consideration is life as a whole, the stage input is identical to the population birth rate (cf. Chapter 14), which now relates to the age-density in a nonlinear way:

$$B(t) = g\left(\int_0^c \beta(a)u(t,a)\,\mathrm{d}a\right), \quad t > 0. \tag{16.2}$$

We set

$$x(t) = \int_0^c \beta(a)u(t,a)\,\mathrm{d}a.$$

The new variable x can be interpreted as the potential birth rate and B as the actual birth rate. The nonlinear relationship between the two may be due to a limited number of suitable birth sites (nests and dens, for example).

Since $B(t) = g(x(t))$, x satisfies the Volterra integral equation (VIE)

$$x(t) = \int_0^t k(a)g(x(t-a))\,\mathrm{d}a + x_\diamond(t), \tag{VIE}$$

where

$$\left.\begin{aligned} k(a) &= \beta(a)\mathcal{F}(a), \\ x_\diamond(t) &= \int_t^\infty \beta(a)u_0(a-t)\frac{\mathcal{F}(a)}{\mathcal{F}(a-t)}\,\mathrm{d}a, \end{aligned}\right\} \tag{16.3}$$

with the usual convention that $u_0(a)/\mathcal{F}(a) = 0$ if $\mathcal{F}(a) = 0$.

Lemma 16.1. *$x_\diamond(t) \to 0$ as $t \to \infty$ if one of the following assumptions holds.*

(i) $c < \infty$.

(ii) *u_0 is integrable on $[0, \infty)$ and $\beta(a) \to 0$ as $a \to \infty$.*

(iii) *$u_0/\mathcal{F}$ is integrable on $[0, \infty)$ and $\beta(a)\mathcal{F}(a) \to 0$ as $a \to \infty$.*

(iv) *u_0 is essentially bounded on $[0, \infty)$ and $\int_t^\infty \beta(a)\,\mathrm{d}a \to 0$ as $t \to \infty$.*

(v) *$u_0/\mathcal{F}$ is essentially bounded on $[0, \infty)$ and $\int_t^\infty \beta(a)\mathcal{F}(a)\,\mathrm{d}a \to 0$ as $t \to \infty$.*

Define the nonlinearity f by

$$f(x) = g(x) \int_0^c k(a)\,\mathrm{d}a.$$

Then the convolution equation (VIE) has the form (IE) in Toolbox B.6 with the probability measure

$$P(\Omega) = \frac{\int_\Omega \beta(a)\mathcal{F}(a)\,\mathrm{d}a}{\int_0^c k(a)\,\mathrm{d}a},$$

and we have the following result from Theorem B.40.

Theorem 16.2. *Let $g : [0, \infty) \to [0, \infty)$ be continuous and $g(x) > 0$ for all $x > 0$. Assume that the associated function f has a unique fixed point $x^* > 0$ (with 0 possibly being another fixed point) and that f^2 has no fixed point in $(0, \infty)$ other than x^*. Furthermore, assume that there exist x_1 and x_2 such that*

$$0 < x_1 < x^* < x_2 < \infty, \qquad f(x_1) > x_1, \qquad f(x_2) < x_2.$$

If $x_\diamond$ satisfies one of the assumptions in Lemma 16.1, all nonnegative solutions x of (VIE) which are bounded and bounded away from 0 satisfy

$$x(t) \to x^*, \quad t \to \infty.$$

16.1 A Demographic Model with a Juvenile and an Adult Stage

In Chapter 11 we model a population with a juvenile (larval, for example) and an adult stage,

$$N(t) = L(t) + A(t),$$

with N, L, A denoting the sizes of the total, juvenile, and adult populations, respectively. We assume there that both the juvenile and the adult stages have exponentially distributed durations and survivorships. Here we assume that the duration and survivorship of the adult stage are exponentially distributed, but the juvenile stage has arbitrarily distributed duration and survivorship. This leads to a differential equation for the size of the adult population,

$$A' = C - \alpha A \tag{16.4}$$

with α being the per capita mortality rate and C the (living) output from the juvenile stage. We assume that the input into the juvenile stage is given by

$$B(t) = g(A(t)). \tag{16.5}$$

Following Section 13.1 we obtain for the juvenile population that

$$L(t) = \int_0^\infty u(t,a)\,\mathrm{d}a$$

with u given by (16.1) and the sojourn function $\mathcal{F} = \mathcal{F}_\mathrm{s} \cdot \mathcal{F}_\mathrm{d}$ being the product of the survival function $\mathcal{F}_\mathrm{s}$ and the duration function $\mathcal{F}_\mathrm{d}$. By Section 13.6, the (living) output of the juvenile stage is given by

$$\begin{aligned} C(t) &= C_1(t) + C_0(t), \\ C_1(t) &= -\int_0^t B(t-a)\mathcal{F}_\mathrm{s}(a)\mathcal{F}_\mathrm{d}(\mathrm{d}a), \\ C_0(t) &= -\int_t^\infty \frac{u_0(a-t)}{\mathcal{F}(a-t)}\mathcal{F}_\mathrm{s}(a)\mathcal{F}_\mathrm{d}(\mathrm{d}a). \end{aligned}$$

Integrating (16.4),

$$\begin{aligned} A(t) &= A_1(t) + A_0(t) + A(0)\mathrm{e}^{-\alpha t}, \\ A_1(t) &= \int_0^t C_1(t-s)\mathrm{e}^{-\alpha s}\,\mathrm{d}s, \\ A_0(t) &= \int_0^t C_0(t-s)\mathrm{e}^{-\alpha s}\,\mathrm{d}s. \end{aligned}$$

Substituting the expression for C_1 and setting $B(t) = 0$ for $t < 0$, we obtain

$$A_1(t) = -\int_0^t \left(\int_0^t B(t-s-a)\mathcal{F}_\mathrm{s}(a)\mathcal{F}_\mathrm{d}(\mathrm{d}a) \right) \mathrm{e}^{-\alpha s}\,\mathrm{d}s.$$

Changing the order of integration,

$$A_1(t) = -\int_0^t \left(\int_0^t B(t-s-a)\mathrm{e}^{-\alpha s}\,\mathrm{d}s \right) \mathcal{F}_\mathrm{s}(a)\mathcal{F}_\mathrm{d}(\mathrm{d}a).$$

Substituting $t - s - a = r$ and recalling that $B(r)$ is zero for negative r,

$$A_1(t) = -\int_0^t \left(\int_0^{t-a} B(r)\mathrm{e}^{\alpha(r+a-t)}\,\mathrm{d}r \right) \mathcal{F}_\mathrm{s}(a)\mathcal{F}_\mathrm{d}(\mathrm{d}a).$$

Changing the order of integration again,

$$A_1(t) = -\int_0^t B(r)\left(\int_0^{t-r} e^{\alpha(r+a-t)}\mathcal{F}_s(a)\mathcal{F}_d(da)\right)dr.$$

Substituting $t - r = s$,

$$A_1(t) = \int_0^t B(t-s)k(s)\,ds,$$

$$k(s) = -\int_0^s e^{\alpha(a-s)}\mathcal{F}_s(a)\mathcal{F}_d(da).$$

Notice that

$$\int_0^\infty k(s)\,ds = \frac{p}{\alpha}$$

with

$$p = -\int_0^\infty \mathcal{F}_s(a)\mathcal{F}_d(da)$$

providing the probability of surviving the juvenile stage and $1/\alpha$ being the expectation of adult life. Recalling that $B(t) = g(A(t))$, we have obtained that $x(t) = A(t)$ satisfies (VIE) with $x_\diamond(t) = A_0(t) + A(0)e^{-\alpha t}$.

In order to find conditions for $x_\diamond(t) \to 0$ as $t \to \infty$, we substitute the expression for C_0 and set $u_0(a) = 0$ for $a < 0$:

$$A_0(t) = -\int_0^t\left(\int_0^\infty \frac{u_0(a-t+s)}{\mathcal{F}(a-t+s)}\mathcal{F}_s(a)\mathcal{F}_d(da)\right)e^{-\alpha s}\,ds.$$

Changing the order of integration,

$$A_0(t) = -\int_0^\infty\left(\int_0^t \frac{u_0(a-t+s)}{\mathcal{F}(a-t+s)}e^{-\alpha s}\,ds\right)\mathcal{F}_s(a)\mathcal{F}_d(da).$$

Substituting $a - t + s = r$ and recalling that $u_0(a) = 0$ for $a < 0$,

$$A_0(t) = -\int_0^\infty\left(\int_{a-t}^a \frac{u_0(r)}{\mathcal{F}(r)}e^{-\alpha(r+t-a)}\,dr\right)\mathcal{F}_s(a)\mathcal{F}_d(da).$$

Changing the order of integration again (notice that $a - t \leqslant r \leqslant a$ if and only if $r \leqslant a \leqslant r + t$),

$$A_0(t) = -\int_0^\infty \frac{u_0(r)}{\mathcal{F}(r)}\left(\int_r^{r+t} e^{-\alpha(r+t-a)}\mathcal{F}_s(a)\mathcal{F}_d(da)\right)dr.$$

We have the estimate

$$|A_0(t)| \leqslant -\int_0^\infty \frac{|u_0(r)|}{\mathcal{F}_d(r)}\left(\int_r^{r+t} e^{-\alpha(r+t-a)}\mathcal{F}_d(da)\right)dr.$$

Integrating by parts,

$$|A_0(t)| \leqslant \int_0^\infty \frac{|u_0(r)|}{\mathcal{F}_{\mathrm{d}}(r)} \Big(\mathrm{e}^{-\alpha t} \mathcal{F}_{\mathrm{d}}(r) - \mathcal{F}_{\mathrm{d}}(r+t) + \alpha \int_r^{r+t} \mathrm{e}^{-\alpha(r+t-a)} \mathcal{F}_{\mathrm{d}}(a) \,\mathrm{d}a \Big) \,\mathrm{d}r.$$

After a substitution,

$$|A_0(t)| \leqslant \mathrm{e}^{-\alpha t} \int_0^\infty |u_0(r)| \,\mathrm{d}r + \int_0^\infty |u_0(r)| \Big(\alpha \int_0^t \frac{\mathcal{F}_{\mathrm{d}}(t-a+r)}{\mathcal{F}_{\mathrm{d}}(r)} \mathrm{e}^{-\alpha a} \,\mathrm{d}a \Big) \,\mathrm{d}r.$$

This reveals that $A_0(t) \to 0$ as $t \to \infty$ if $\mathcal{F}_{\mathrm{d}}(a) \to 0$ as $a \to \infty$. Similarly as in Theorem 16.2 we have the following result.

Theorem 16.3. *Let $g : [0, \infty) \to [0, \infty)$ be continuous and $g(x) > 0$ for all $x > 0$. Assume that the function f,*

$$f(x) = \frac{p}{\alpha} g(x),$$

has a unique fixed point $x^* > 0$ (with 0 possibly being another fixed point) and that f^2 has no fixed point in $(0, \infty)$ other than x^*. Furthermore, assume that there exist x_1 and x_2 such that

$$0 < x_1 < x^* < x_2 < \infty, \qquad f(x_1) > x_1, \qquad f(x_2) < x_2.$$

If $\mathcal{F}_{\mathrm{d}}(a) \to 0$ as $a \to \infty$, all nonnegative model solutions with A being bounded and bounded away from 0 satisfy

$$A(t) \to x^*, \quad t \to \infty.$$

16.2 A Differential Delay Equation

If the juvenile stage has exponentially distributed duration and survivorship, the system has the form

$$L' = g(A) - (\mu + \nu)L, \qquad A' = \nu L - \alpha A$$

and we have convergence of $A(t), t \to \infty$ under much less restrictive conditions, namely whenever $g(A)/A$ is strictly decreasing and tends to 0 as $A \to \infty$ (cf. Theorem 11.6). For comparison let us consider a juvenile stage of fixed length τ. Then (cf. Section 13.6)

$$C(t) = B(t-\tau)\mathcal{F}_{\mathrm{s}}(\tau), \quad t > \tau,$$

and we obtain the delay differential equation

$$A'(t) = pg(A(t-\tau)) - \alpha A(t), \quad t > \tau, \tag{DDE}$$

where $p = \mathcal{F}_s(\tau)$ is again the probability of surviving the juvenile stage. In the following we consider p to be independent of τ. A possible justification is that the effect of the length of the juvenile period on the probability of surviving it is compensated for by other factors. If g is a Ricker-type nonlinearity, the following theorem holds.

Theorem 16.4. *Let the function g be of Ricker type,*

$$g(A) = \beta A \mathrm{e}^{-\gamma A},$$

with positive constants γ, β. Consider $\mathcal{R}_0 = p\beta/\alpha$. Then the following trichotomy holds.

(i) *If $\mathcal{R}_0 \leqslant 1$, then, for all $\tau > 0$, every nontrivial nonnegative solution of (DDE) converges towards 0.*

(ii) *If $\mathcal{R}_0 \in (1, \mathrm{e}^2]$, there exists a unique equilibrium $A^* > 0$ such that, for all $\tau > 0$, every nonnegative solution of (DDE) which is not identically zero converges to A^*.*

(iii) *If $\mathcal{R}_0 > \mathrm{e}^2$, there exists a unique equilibrium $A^* > 0$ and some $\tau_0 > \tau$ such that, for every $\tau > \tau_0$, there exists a nontrivial positive periodic solution oscillating about A^*.*

The periodic solutions in (iii) can be chosen to be *slowly oscillating*, i.e., their periods are larger than 2τ.

Proof. Since g is bounded on $[0, \infty)$, every nonnegative solution of (DDE) exists for all positive times and is bounded on $[0, \infty)$.

(i) Let $p\beta \leqslant \alpha$ and A a solution of (DDE). Then $A' \leqslant 0$ and A decreases to a limit A^* as $t \to \infty$. Since g is bounded and A is bounded, A' is bounded. It follows that A'' is bounded. By Barbalat's lemma (Corollary A.17), $A'(t) \to 0$ as $t \to \infty$. From (DDE), $0 = pA^*\beta \mathrm{e}^{-\gamma A^*} - \alpha A^*$, which implies that $A^* = 0$.

If $p\beta > \alpha$, there exists a unique $A^* > 0$ such that $p\beta \mathrm{e}^{-\gamma A^*} = \alpha$. Setting $A = cx$ and choosing $c > 0$ appropriately transforms (DDE) into

$$\frac{1}{\alpha}\frac{\mathrm{d}x}{\mathrm{d}t} = f(x(t-\tau)) - x, \qquad f(x) = \mathrm{e}^{r(1-x)}, \quad r = \ln\frac{p\beta}{\alpha}.$$

Scaling the time as $t = \tau s$,

$$\nu\frac{\mathrm{d}x}{\mathrm{d}s} = f(x(t-1)) - x, \qquad f(x) = \mathrm{e}^{r(1-x)}, \quad r = \ln\frac{p\beta}{\alpha}, \quad \nu = \frac{1}{\tau\alpha}.$$

(ii) Let $0 < r \leqslant 2$. Then (DDE) can be rewritten in the form of (IE) in Toolbox B.6 with f satisfying condition (i) in Corollary 9.9 and Assumption 9.4.

Persistence theory (Toolbox A.5) implies that all nonnegative solutions that are not identically 0 are bounded away from 0 and so converge to 1 as $t \to \infty$ by Theorem B.40.

(iii) Let $r > 2$. Then $f'(1) < -1$ and $(x-1)[f(x)-1] < 0$ for $x \neq 1$. Then there exists some $\nu_0 > 0$ such that (DDE) has a periodic solution with period greater than 2 for each $\nu \in (0, \nu_0)$ (Hadeler, Tomiuk, 1977; see also Ivanov, Sharkovsky, 1991, Theorem 6.1). □

Part (iii) of Theorem 16.4 shows that global stability of model solutions may fail for juvenile periods of fixed duration, while it holds for juvenile periods with exponentially distributed duration and survivorship.

Bibliographic Remarks

Some other types of simple nonlinear age-structured models have been considered by Kot (2001, Chapter 24). A fully developed mathematical theory of nonlinear demographics is presented in Webb (1985). A variety of nonlinear demographic models are presented by Cushing (1998). Demographic models with discrete age-structure are discussed in Caswell (1989) and Cushing (1998).

(DDE) is an example of a delay differential equations, more generally, a functional differential equation; see Hale (1977), Kuang (1993), Diekmann, van Gils, et al. (1995), Takeuchi (1996), Brauer, Castillo-Chavez (2001, Chapter 3), and the references in these books. Cooke et al. (1999) have analyzed the model in Theorem 16.4 for the case that $p = \mathrm{e}^{-d_1 \tau}$ depends on τ.

PART 3

Host–Parasite Population Growth: Epidemiology of Infectious Diseases

Chapter Seventeen

Background

From an anthropocentric point of view, infectious disease dynamics are an important special case of multi-species population dynamics, as human populations often play an integral role. Since humans have become top predators in the trophic chain and cultivate their own prey, there are hardly any predator–prey models that explicitly involve human populations. Similarly, while humans still compete with rodents and insects for plant food resources, the associated models typically contain the food resource, the pest species, and possible antagonists, but not human populations as explicit players. This is different in the spread of infectious diseases, i.e., host–parasite population dynamics, where humans typically take the role of the host population and the parasites are microparasites like viruses, bacteria, and fungi, or macroparasites like various types of worms, or protozoa like trypanosoma or plasmodia, which are somewhere in between. Sometimes, other host species are involved as final or intermediate hosts, as vectors or reservoirs. In this book we focus on microparasitic diseases, which can be modeled more easily than macroparasitic diseases, because it is not necessary to model the parasite population explicitly, but only its effects on the host population by dividing the latter according to the various stages of infection. See the bibliographic remarks for references concerning the modeling of macroparasitic diseases. In Chapter 23 we present a framework that is general enough to include not only multiple groups within one host population, but also several host populations, some of which may be vectors or reservoirs.

Infectious diseases have afflicted humankind severely throughout the ages, and despite medical and sanitary progress, they continue to do so today. They are a continuing threat through emerging infectious diseases (HIV, Ebola, Hanta and Lassa viruses, dengue, Chagas' disease) and reemerging diseases with drug-resistant strains (tuberculosis, staph, strep, malaria, diarrheal diseases) and/or insecticide-resistant vectors (malaria).

Many of the emerging infectious diseases (the various hemorrhagic fevers, but not HIV) manifest themselves as epidemics. Their spread occurs on a much faster time scale than the demographic processes in the host population. Epidemic outbreaks typically collapse after some time by exhausting the reservoir of susceptibles. All this happens sufficiently fast that the demographic processes (in particular natural death and birth) can be neglected in the mathematical model. Epidemic models will be considered in Chapters 18–20.

In many of the reemerging diseases and in HIV, as well as in many of the so-called childhood diseases, the demographic processes are fast enough to influence the course of the disease. In particular, the reservoir of susceptibles is refilled by new births or loss of disease immunity before the endemic outbreak collapses completely. So, the disease can establish itself in the population, i.e., it becomes *endemic*. Endemic models will be considered in Chapters 21–23.

17.1 Impact of Infectious Diseases in Past and Present Time

> *The angel of the Lord went out and struck down a hundred and eighty-four thousand men in the Assyrian camp; when morning dawned, they all were dead. So, Sennacherib king of Assyria broke camp, went back to Nineveh and stayed there.*
>
> 2 Kings, 19:35–36 and Isaiah, 37:36–38

Imagine how history could have changed if the Assyrians had conquered Jerusalem in 701 B.C. instead of being stopped by what presumably was an epidemic outbreak of an infectious disease. Jerusalem was spared the fate of the 'ten lost tribes' of the northern kingdom of Israel and was not conquered before 597 B.C., by the Babylonians, and could be resettled around 520 B.C. by the returnees from Babylonian captivity.

In Greece, during the Peloponnesian War, the plague of Thucydides, presumably a mixture of smallpox, spotted fever, and maybe typhoid, put an end to the golden age of Athens in 429 B.C. The epidemic killed one-third of the population, including the political leader Pericles and all his sons.

Epidemics of unknown infectious diseases leading to a population decline weakened the Roman empire during the years A.D. 165 180 and A.D. 251 266.

Around A.D. 542 the East Roman Empire experienced the outbreak of the plague of Justinian, which interfered with the plans of reunifying the old Roman Empire. The outbreaks occurred in several waves until A.D. 750. Some historians believe that the demographic and economic consequences are partially responsible for the relatively little resistance which East Rome could offer against the Muslim forces that spread from Arabia from A.D. 634.

In the 14th century the Plague (black death) killed some 25 million out of a population of 100 million in Europe (rough estimates). In England it stopped population increase for several centuries (McNeill, 1976):

year	1086	1348	1377	1430	1603	1690
millions	1.1	3.7	2.2	2.1	3.8	4.1

The black death swept through England in its first big wave from 1346 to 1350. Other waves occurred in the 1360s and 1370s. In 1337, at the onset of the 100 Years War, England had started its efforts to conquer France, which finally failed in 1453. Without the black death interfering the English might have succeeded.

China experienced a population decline from an estimated 123 million around A.D. 1200 to an estimated 65 million in 1393, which was presumably due to a combination of war (with the Mongols) and plague. According to Chinese chroniclers, up to two-thirds of the population in eight different and distant provinces died from the plague in the years around 1353 and 1354. McNeill (1976) argues that the Mongols were also hit by the plague and that the losses from the disease made an essential contribution to the decline of the Mongol empire after 1350 (starting with their withdrawal from China and the establishment of the Ming dynasty in 1368).

Smallpox has been estimated to have been responsible for as much as 20% of mortalities in Sweden in the early 18th century (Razzell, 1974). Smallpox may also have caused other major epidemics that claimed a quarter to a third of the affected populations: China A.D. 49, Rome A.D. 165, Cyprus A.D. 251–266, Greece A.D. 312, Japan A.D. 552, Mecca A.D. 569–571, Arabia A.D. 683, Europe (various sites) A.D. 700–800 (Garrett, 1994, p. 625).

When the Spaniards, under Cortez, invaded Mexico, they brought not only swords and cannons, but (unintentionally) smallpox as an even more formidable weapon. The Spanish soldiers were rather resistant to this disease, which they presumably had survived during their childhood. The Aztecs had no immunity at all and lost half of their 3.5 million population (Bailey, 1975). More than the cannons, the invulnerability of the Spaniards to smallpox convinced the Aztecs that the invaders were irresistible supermen. The population decrease in central Mexico continued even after the conquest, and in 1568 numbers were down to a tenth of what they were before the invasion. The population also declined in the far south of Mexico. Chance (1989, Table 11, p. 69) gives the following estimated population sizes (in hundreds) for the Alcaldía Mayor of Villa Alta, a region in the northern part of Oaxaca, which comprised about 110 Native American villages, most of which had populations well under one thousand people:

year	1520	1532	1548	1568	1595	1622	1703	1742	1781	1789	1820
estimate	3469	3154	959	315	214	208	364	491	471	428	385

The decline was also due to military conquest and forced or voluntary resettlement; however, until at least 1550, the Spanish had only ephemeral control over the region and few Spaniards or blacks inhabited the area during the 16th century, so much of the depopulation during that period can be attributed to epidemic disease (Chance, 1989, p. 68). The second decline in the late 18th century was again due to smallpox, typhoid, yellow fever or unknown infectious diseases (Chance, 1989, pp. 73–86).

Smallpox and measles, which came with the European colonists, reduced the Native American population in North America by more than 90%. At its very lowest, the population size of Native Americans in the Americas has been estimated to be less than one-twentieth of what it was before Columbus.

While, in the words of Simpson (1980), the Europeans started to conquer America by waging unpremeditated biological warfare in the late 15th century, syphilis spread through Europe to a degree not noticed before. It is still under debate whether this was the Americas striking back (mainly from the Caribbean) or whether syphilis was lurking in Europe (Wills, 1996) at that time.

(For more historical remarks concerning the impact of epidemics on human history, see McNeill (1976), Marks, Beatty (1976), Simpson (1980), Anderson, May (1991, Section 1.1), Garrett (1994, Section 9.1), and Wills (1996).)

The first attempt to build the Panama canal failed for various reasons: among them were yellow fever and malaria, from which more than 50 000 people associated with the construction died between 1881 and 1888 (Koch, 1987).

Before the second (successful) attempt was seriously started in 1907, the mosquito vectors of these diseases were eradicated in the construction zone in a three-year battle. Thousands of tons of crude oil were used to spray the swamps in which the larvae lived (Poirier, 1962; Desowitz, 1997).

The influenza epidemic in 1918–1919 killed 20 million people worldwide (12.5 million in India, 548 000 in the United States). The epidemics in 1957 and 1968 together caused 98 000 deaths in the United States; the costs of treatment, insurance benefits, etc., of the 1968/1969 epidemic in the United States have been estimated to be $4600 million. (For more details see Cliff et al. (1986).)

In the developing world, various forms of gastroenteritis and pneumonia cause some 18 million deaths per year (mainly among young children) (Hassell et al., 1982).

Rubella is usually a mild illness, but congenital rubella syndrome occurs in at least 20% of infants born to women who acquired the infection during the first trimester of pregnancy. The last rubella epidemic in 1964–1965 in the United States resulted in an estimated 11 000 wasted pregnancies and 20 000 children with congenital defects (see Hethcote (1989b) for details and references).

Infectious diseases terminated the lives of the following famous people.

Cholera: Karl von Clausewitz, Friedrich Hegel.

Malaria: Alaric, Oliver Cromwell, Jacob I (Mary Stuart's son), Raffaello Santi (Raphael), Michelangelo da Caravaggio.

Plague: Matthias Grünewald, Holbein the Younger, Tiziano Vecellio (Titian).

Smallpox: Marcus Aurelius, Cuitiáhuac (king of the Aztecs), Louis XV, Ramses V.

Syphilis: Francis I of France, Friedrich Nietzsche, Heinrich Heine, Charles Baudelaire, Guy de Maupassant.

Typhoid: Georg Büchner, Wilhelm Hauff.

Tuberculosis: Francis of Assisi, Frederic Chopin, Aldous Huxley, Nicolò Paganini, Madame Pompadour, Friedrich Schiller, Carl Maria von Weber.

(See Koch, 1987, pp. 201, 202; Marks, Beatty, 1976, p. 133; Wills, 1996, pp. 189, 190.)

The lists of syphilis victims is presumably longer (see the just-mentioned references), but when this disease is involved it seems to be difficult to determine the exact cause of death. Therefore, I have listed only the names which appear in at least two sources.

Mathematicians rarely seem to quite make it into the hall of fame (except among their peers). Since this book is written by a mathematician, I supplement the above list (without claiming completeness).

Tuberculosis: Niels H. Abel, G. F. Bernhard Riemann.

Smallpox: Carl G. J. Jacobi.

Influenza: Sonja Kowalewski, Ernst E. Kummer, Karl W. T. Weierstrass.

Pneumonia: George Boole, René Descartes.

(See Bell, 1937.)

Pierre de Fermat survived the last great outbreak of plague in Toulouse in 1652/1653, which killed approximately 10% of the population, but his health may have been permanently affected (Barner, 2001).

So far we have concentrated on the fatal consequences of infectious diseases. But morbidity also has a tremendous impact by diminishing the quality of life and disrupting the social life and the economy of the affected societies.

In the 1970s the following (mainly worm) diseases were estimated to affect many millions of the world population (Bailey, 1975):

Disease	Millions affected
schistosomiasis	200
filiariasis	250
hookworm	450
trachoma	450

45 million people are at risk from sleeping sickness (human trypanosomiasis), though only 10 000 new cases are reported per year.

Animal trypanosomiasis has a double effect on the rural economies in Africa today. Firstly, it directly causes the death or increased morbidity of countless domestic livestock, mainly cattle. Secondly, and more importantly, the disease precludes the keeping of domestic livestock in large tracts of country which are often very suitable for grazing (reserving them for a number of wild animals which are the original hosts of the disease without suffering too much from it).

Animal trypanosomiasis and its fly vector, tsetse, stopped the horse-borne Fulani invasion in the 19th century before it reached the great population centers of the Benue and Niger river valleys (Desowitz, 1981, "The Fly That Would Be King"). More about trypanosomiasis can be found in Jordan (1986).

Similarly, yellow fever prevented British and French colonists from containing full control over West Africa (Garrett, p. 66). Schistosoma japonicum may have kept Taiwan from being taken by communist troops in 1948 (Desowitz, 1981, Section 8).

Smallpox was eradicated in a campaign that lasted 11 years, involved about 100 highly trained professionals and thousands of local health workers and staff, and cost $300 million. To eliminate malaria, $430 billion dollars was spent in a series of failed attempts between 1958 and 1963, and an additional $793 million was spent by the United States between 1964 and 1981 (Garrett, 1994, p. 47). In 1993, mortality due to malaria was at a historic all-time high in Africa (Garrett, 1994, p. 443).

Humankind had hardly celebrated the eradication of smallpox in the late 1970s when a new disease showed its fearsome face: AIDS, the acquired immunodeficiency syndrome, which is caused by HIV (human immunodeficiency virus). Worldwide, 168 176 cases of AIDS had been reported by 1 August 1988; the estimated numbers at that time were 250 000, and some 10 million were believed to be already infected with the virus. In the United States, 66 464 AIDS cases were reported to the CDC (Centers for Disease Control, Atlanta) from 1981 until 4 July 1988. Of these, 37 535 had died, including more than 80% of the patients diagnosed before 1985. A total of 21 846 cases of AIDS were reported in 1987, about 44 000 in 1991. (For more details see Heyward et al. (1988), Mann et al. (1988), and Hethcote, Van Ark (1992).) In 2000, 36 million adults and children were estimated to live with the virus worldwide (McGeary, 2001).

AIDS has become one of the major health problems confronting the countries of Central and East Africa. In 1988, close to half of all patients in the medical wards of Kinshasa (Zaire), Nairobi (Kenya), and Butare (Rwanda) were infected with HIV, as were 10–25% of the women of childbearing age (Mann et al., 1988). Studies of HIV seropositivity among blood donors in Central African countries suggested levels as high as 9% in Zaire, 11% in Uganda, 15% in Rwanda, and 18% in Zambia (May, Anderson, 1989 (citing Quinn, personal communication)). In the early 1990s HIV/AIDS overtook Malaria in several African states as the number one illness requiring hospitalization (Garrett, 1994, p. 485). Cumulative numbers of AIDS cases and reported HIV infection rates can be found in Garrett (1994, pp. 691–694). In 2000, an estimated 25.3 million people lived with the virus in sub-Saharan Africa; 17 million Africans were estimated to have died since the late 1970s, and 3.8 million to have been newly infected during the preceding year (McGeary, 2001).

Besides the AIDS virus, other formerly unknown microbes triggered epidemic outbreaks which, due to the efforts of so-called virus hunters, could fortunately be contained: Rift Valley virus (1930, Kenya; 1977, Egypt), Junín virus (1953, Argentina), Mapucho virus (1963, Bolivia), Marburg virus (1967; Germany, Yugoslavia), Lassa virus (1969, Nigeria), Ebola virus (1976, Zaire), Fort Dix virus (1976, United States), Legionella bacterium (1976, United States), Sin Nombre (hanta-)virus (1993, United States). Other newly identified infectious diseases are Lyme disease (1975), toxic shock syndrome (1978), hepatitis C (1989), and hepatitis E (1990).

17.2 Epidemiological Terms and Principles

Infectious diseases are *parasitic diseases* with parasites being transmitted from one *host* to another. We understand the term *parasite* to mean an organism which lives in and depends on (and exploits and more or less damages) another organism, its host. Parasites are to be discriminated from *parasitoids*, mainly insects which lay their eggs into other animals. We divide parasites into *microparasites* (viruses, bacteria, fungi, rickettsiae) and *macroparasites* (flukes, tapeworms, roundworms, filiariae, and other worms) with protozoa (unicellular organisms: trypanosoma, plasmodia) forming some kind of intermediate. Microparasites typically reproduce directly in the host (at high rates). Macroparasites often have a sexual and an asexual phase of reproduction which occur in different host species. Whereas the number of microparasites in a host is countless, the number of macroparasites varies a lot and may be quite small.

Infectious diseases can be classified in many ways.

- Microparasitic diseases versus macroparasitic diseases.
- Directly transmitted diseases (smallpox), vector-borne diseases (sleeping sickness), environmental diseases (Arizona valley fever, a fungal disease), diseases with intermediate hosts (schistosomiasis).
- Vertically transmitted diseases (from mother to unborn child), horizontally transmitted diseases (air-borne transmission, for example), diseases with both transmission modes.
- Epidemic diseases (occurring in single outbreaks: plague, for example) versus endemic diseases (common childhood diseases, for example).

From the modeling point of view, the most incisive classification is into microparasitic and macroparasitic diseases; the first class is typically modeled by *prevalence* models, while the second, at least in a certain modeling stage, requires *density* models. Prevalence models structure the host population according to disease stage (for example, dividing them into *susceptible* (S), *exposed* (E), *infective* (I), and *recovered* (R) individuals), while density models further stratify the infective part of the population according to the parasite load of the hosts.

An *infective* individual has the disease and is able to transmit it.

A *susceptible* individual is an infection-free individual that can be infected by direct or indirect contact with an infective individual.

At infection a susceptible individual becomes an *exposed* individual; it has been infected, but is not yet able to transmit the infection. In other words, it is in the *latent period*.

A *recovered individual* has gone through an infection and recovered from the disease, and is completely or partially *immune* to reinfection. After a while it may or may not (depending on the disease) lose its immunity and return to the susceptible part of the population.

The *latent period* should not be confused with the *incubation period*. The first starts at infection and ends when the infected individual becomes infectious itself. The second also starts at infection, but ends with the first appearance of disease symptoms. For most diseases the latent period is shorter than the infectious period: the infected individual becomes infectious itself before it shows symptoms of the disease.

The classification into *epidemic* and *endemic* diseases also leads to a classification into *epidemic* and *endemic* models. Epidemic models deal with singular outbreaks of the disease which are short enough that the natural turnover of the population can be neglected. Endemic models deal with diseases which afflict populations over a long time range and therefore take account of the natural dynamics of the population.

Glossary

carrier	an individual that is infective but develops no symptoms
endemic	from Greek: demos $\simeq$ people, en $\simeq$ in; something (in particular a disease) widespread among many people for a long time
epidemic	from Greek: demos $\simeq$ people, epi $\simeq$ onto; something (in particular a disease) that is transferred onto people; a sudden, rapid spread of a disease
epizootic	epidemic affecting animal populations
exposed	infected, not yet infective

immunity	resistance to the disease which may or may not be acquired by previous exposure
incidence (rate)	number of new cases per unit of time
incubation period	period between the moment of being infected and the appearance of symptoms (typically the incubation period is longer than the latent period)
infective	capable of transmitting the disease, infectious
infectious disease	contagious disease, communicable disease
infective (infectious) period	period of being infective
latent period	period between the moment of being infected and the moment of becoming infectious
prevalence	number of infectives
susceptible	capable of contracting the disease
vector	intermediate host that carries the infectious agents over larger distances, e.g., tsetse fly for trypanosomiasis (sleeping disease) or anopheles mosquito for malaria

A more complete glossary of epidemiological terms and principles can be found in Anderson, May (1982a).

Bibliographic Remarks

The books by McNeill (1976), Marks, Beatty (1976), Simpson (1980), Desowitz (1981, 1997), Garrett (1994), Preston (1995), Wills (1996), and Peters (1997) provide useful and interesting nonmathematical background reading (see also Hethcote, 2000, p. 599). Chance's (1989) book focuses on the ethnohistory of colonial Oaxaca (Mexico), but contains interesting epidemiological information on the side, such as the population numbers of 110 Native American villages from 1548 to 1970 together with references to epidemics that hit some of them.

I have not even tried to sketch a history of mathematical modeling in epidemics here. Instead I refer to the classic by Bailey (1975), to Section 1.2 in Anderson, May (1991), and to the papers by Dietz (1988, 1995), Dietz, Schenzle (1985), Hethcote (2000), and Dietz, Heesterbeek (2002).

Textbooks completely devoted to the mathematical modeling of epidemics have been written by Busenberg, Cooke (1993), Capasso (1993), and Diekmann, Heesterbeek (2000). The mathematical biology texts by Hoppensteadt (1982), Edelstein-Keshet (1987), Murray (1989), and Brauer, Castillo-Chavez (2001) contain sections on epidemics, as does the volume edited by Levin, Hallam, Gross (1989).

Research monographs have been written by Hethcote, Van Ark (1992), Hethcote, Yorke (1984), Mode, Sleeman (2000), Nåsell (1985), and Tan (2000). The book by Scott, Duncan (1998) combines demography and infectious diseases. The book by Nowak and May (2000) concentrates on the dynamics of virus populations within the body.

The monograph by Anderson, May (1991) compiles a wealth of data, graphs, models, references, and other very useful information; for the mathematical details, however, one should read the original papers.

Conference and workshop proceedings abound: recent ones have been edited by Arino et al. (1995), Grenfell, Dobson (1995), Mollison (1995), Isham, Medley (1996), and Castillo-Chavez et al. (2002a,b). The proceedings of the Dahlem workshop edited by Anderson, May (1982a) and of the IIASA workshops edited by Dieckmann et al. (2002) are unique in their coherence. The proceedings edited by Martelli et al. (1996), Arino et al. (1997), and Horn et al. (1998) do not entirely, but to a substantial part, comprise contributions to mathematical epidemiology. Multiauthored volumes devoted to epidemic modeling have been edited by Anderson (1982a) and Scott, Smith (1994). See Hethcote (2000, p. 600) for more. Material on the modeling of macroparasitic diseases can be found in Anderson (1982a), Anderson, May (1982a), Nåsell (1985), Anderson, May (1991, Part II), Scott, Smith (1994), Diekmann, Heesterbeek (2000), and Castillo-Chavez et al. (2002b). Recent surveys of mathematical epidemiology are contained in the articles by Hethcote (2000) and Brauer, van den Driessche (2002).

Chapter Eighteen

The Simplified Kermack–McKendrick Epidemic Model

In the next three chapters we essentially consider *epidemic* models as they have been suggested and analyzed by Kermack, McKendrick (1927–1939). In this chapter all infective individuals are assumed to be equally infective, whereas in Chapter 20 the infectivity will depend on the time that has passed since the moment of infection. In Chapter 19 we will generalize the standard *mass-action law* of infection by introducing a population-size-dependent contact rate.

18.1 A Model with Mass-Action Incidence

The host population is split up into susceptible, infective, and removed individuals whose numbers are denoted by S, I, and R. Susceptible individuals that are infected become infective immediately; there is no latent period. Individuals are removed from the infective stage by death, recovery or quarantine. Individuals that survive the disease become permanently immune. So, once an individual has been infected, it will never become susceptible again. Schematically, this type of epidemic is represented as

$$S \longrightarrow I \longrightarrow R.$$

Model Equations

$$\left.\begin{aligned} \dot{S} &= -\sigma S I, \\ \dot{I} &= \sigma S I - \rho I, \\ \dot{R} &= \rho I. \end{aligned}\right\} \tag{18.1}$$

Explanation

Since all demographic processes are neglected and there is no return to the susceptible population, the size of the susceptible population only changes by infection. Actually, the time derivative $\dot{S}$ equals the negative infection rate. The infection rate (incidence) is assumed to be proportional to the product of the sizes of the susceptible and the infective populations. This is in analogy to the mass-action law of chemical reactions. The infective population size increases by the infection rate and decreases by the removal rate ρI.

The model contains two proportionality constants: σ and ρ. ρ is the per capita removal rate, the rate at which an average individual leaves the infective stage. σ is the rate at which the disease is contracted from an average individual, provided it is infective. Note that ρ has the dimension $1/\text{time}$ and σ has the dimension $(1/\text{time})(1/\text{individual})$.

Scaling (Nondimensionalization) of the System

If possible, it is always a good idea to express the model equations in nondimensional variables. This not only reduces the number of parameters so that the equations look much simpler, but also shows which parameter combinations are really important and whether they are small or large (application of perturbation techniques).

We introduce the new (scaled) dependent variables

$$x = \frac{\sigma}{\rho} S, \qquad y = \frac{\sigma}{\rho} I, \qquad z = \frac{\sigma}{\rho} R. \tag{18.2}$$

Then the system (18.1) takes the form

$$\dot{x} = -\rho xy, \qquad \dot{y} = \rho(xy - y), \qquad \dot{z} = \rho y. \tag{18.3}$$

Finally, we also scale time and introduce the dimensionless time

$$\tau = \rho t. \tag{18.4}$$

As

$$\frac{\mathrm{d}x}{\mathrm{d}t} = \frac{\mathrm{d}\tau}{\mathrm{d}t}\frac{\mathrm{d}x}{\mathrm{d}\tau} = \rho\frac{\mathrm{d}x}{\mathrm{d}\tau}$$

we obtain

$$x' = -xy, \qquad y' = (x - 1)y, \qquad z' = y. \tag{18.5}$$

Here $'$ denotes the derivative with respect to dimensionless time.

Interpretation of the New Time Scale. Replacement Ratios

In order to get back from the dimensionless time τ to the real time $t = (1/\rho)\tau$, we have to identify $1/\rho$ with some epidemiologically meaningful entity. To this end we consider a special case of (18.1), namely $\sigma = 0$. The equation for the infectives then takes the form

$$\dot{I} = -\rho I.$$

The solution

$$I(t) = I_0 \mathrm{e}^{-\rho t}$$

tells us how many infectives out of an initial number I_0 are still infective at time t. This means that

$$e^{-\rho t} = \frac{I(t)}{I_0}$$

gives us the fraction of initial infectives that are still infective at time t. In other words,

$$e^{-\rho t}$$

is the probability of being still infective after a time span t. From this we learn that our model makes the implicit assumption that the length of the infectivity period is negatively exponentially distributed (which may be an unacceptable simplification) and that the *average length of the infectivity period* is given by

$$\int_0^\infty e^{-\rho t}\, dt = \frac{1}{\rho}$$

(cf. Sections 2.2 and 12.7).

Once we know how to get back to real time, the dimensionless time will also be denoted by t.

After having identified $1/\rho$ with the average length of the infectivity period we can interpret $x = (\sigma/\rho)S$. As σ is the average rate at which a susceptible individual contracts the disease from an infective individual, x gives the total number of susceptibles one average infective individual can infect if the number of susceptibles is S all the time.

We call x the *replacement ratio* of the disease at susceptible number S. x_0, the replacement ratio for the initial number of susceptibles S_0, is called the *initial replacement ratio.*

If $x = (\sigma/\rho)S = 1$, one average infective individual can replace itself provided that the number of susceptibles is S all the time. Hence ρ/σ can be interpreted as the number of susceptibles needed in order that one average infective individual can replace itself.

18.2 Phase-Plane Analysis of the Model Equations. The Epidemic Threshold Theorem

Since by their very meaning x, y are nonnegative (use Proposition A.1 or Lemma A.2 for a rigorous argument), we realize from (18.5) that x is a decreasing function of time. There are two cases to be discriminated.

Subthreshold Case: $x_0 \leqslant 1$

Let us assume that x_0, the scaled number of initial susceptibles, does not exceed one. As x is decreasing, x stays below 1, so y is decreasing, as we see from the second equation in (18.5). Actually, there is no epidemic outbreak.

Superthreshold Case: $x_0 > 1$

In this case, y, the scaled number of infectives, initially increases until x has decreased to 1. So, there is an epidemic outbreak which, if measured in terms of y, reaches its maximum when $x = 1$. x continues decreasing and becomes less than 1, so y also starts decreasing. As it does so at an exponential rate, y converges to 0 as time tends to infinity.

The Final Size of the Epidemic

Actually, it is possible to write down an energy functional which remains constant during the course of the epidemic. Adding the first two equations in (18.5) and using the first equation again we find that

$$x' + y' = -y = \frac{x'}{x} = (\ln x)'.$$

Integration of this relation yields

$$\begin{aligned} x + y - \ln x &= \text{const.} = x_0 + y_0 - \ln x_0 \\ &= 1 + y_{\max} \end{aligned} \tag{18.6}$$

with the last relation only holding in the superthreshold case. Recall that y reaches its maximum when $x = 1$. The relation (18.6) remains valid if time tends to infinity, so

$$\begin{aligned} x_\infty - \ln x_\infty &= x_0 + y_0 - \ln x_0 \\ &= 1 + y_{\max} \end{aligned} \tag{18.7}$$

with the last equality being meaningful in the superthreshold case only. x_∞ denotes the final size of the (scaled) susceptible population.

There is another functional which remains constant, namely

$$x + y + z = x_0 + y_0 + z_0, \tag{18.8}$$

because the derivative of the left-hand side is zero, as one realizes from adding the equations in (18.5). This is intuitive because the epidemic does not change the total population size; remember that we still count the fatalities as part of the population.

The scaled number z of removed individuals is increasing with time and has a limit z_∞ as $t \to \infty$, the final size of the removed population. As the relation (18.8) is also preserved in the time limit and the infectives vanish asymptotically, we have

$$z_\infty = x_0 + y_0 + z_0 - x_\infty = y_0 + z_0 + x_0(1 - \mathrm{e}^{z_0 - z_\infty}). \tag{18.9}$$

The second equality follows from integrating the first and the third equation in (18.5).

The Epidemic Threshold Theorem and the Initial Replacement Ratio

Collecting our results we can formulate the following version of the celebrated epidemic threshold theorem by Kermack, McKendrick (1927).

Theorem 18.1. *There is no epidemic outbreak if $x_0 \leqslant 1$. There is an epidemic outbreak if $x_0 > 1$. In either case the final size of the susceptible population is described by (18.7), the final size of the removed population by (18.9).*

We recall that $x_0 = (\sigma/\rho)S_0$ is the *initial replacement ratio* of the disease, i.e., the total number of secondary cases one average infective can produce if the number of susceptibles remained at its initial size S_0. So, Theorem 18.1 contains the intuitive statement that there is an epidemic outbreak if and only if an average infective individual produces more than one secondary case. In the epidemiological literature,

$$x_0 = \frac{\sigma}{\rho} S_0$$

is sometimes called the *net reproductive number* (Metz, Diekmann, 1986), the *initial replacement number* (Hethcote, 1989) or—wrongly, because x_0 is dimensionless—the *initial reproductive rate* of the epidemic.

The practical importance of the initial replacement ratio is the following. In order to prevent epidemic outbreaks it is necessary (and sufficient) to vaccinate so many people that the initial replacement ratio becomes less than one. Note that it is not necessary to vaccinate everybody in order to exclude epidemic outbreaks.

18.3 The Final Size of the Epidemic. Alternative Formulation of the Threshold Theorem

In relation (18.7), $x_\infty - \ln x_\infty = x_0 + y_0 - \ln x_0$, the function

$$f(x) = x - \ln x \tag{18.10}$$

plays a prominent role. The derivatives of f are

$$f'(x) = 1 - \frac{1}{x}, \qquad f''(x) = \frac{1}{x^2}.$$

Hence f is strictly convex for $0 < x < \infty$, strictly decreasing for $0 < x < 1$ and strictly increasing for $x > 1$. f has a global minimum at 1, $f(1) = 1$. Furthermore, $f(x) \to \infty$ if $x \to 0, \infty$.

Lemma 18.2.

(a) $x_\infty \leqslant 1$.

(b) x_∞ *is a continuous function both of* x_0 *and* y_0.

Proof. (a) As $x_\infty \leqslant x_0$ (the final number of susceptibles does not exceed the initial number), the statement is trivial if $x_0 \leqslant 1$. So, we can assume that $x_0 > 1$. Then the equation $f(x) = f(x_0) + y_0$ has two solutions: one bigger than x_0 and one smaller than x_0. As $x_\infty \leqslant x_0$, x_∞ has to be the smaller solution. As $y_0 > 0$ and f is an increasing function of $x \geqslant 1$, x_∞ has to be smaller than 1.

(b) $f(x_0) + y_0$ is a continuous function of both x_0 and y_0. As a continuous strictly decreasing function on $(0, 1)$, f can be continuously inverted on $(0, 1]$. This yields the assertion. □

We now give an alternative formulation of the threshold Theorem 18.1.

Theorem 18.3.

(a) *If* $x_0 \leqslant 1$, *then* $x_\infty \to x_0$ *whenever* $y_0 \to 0$.

(b) *If* $x_0 > 1$, *then* $x_\infty \nearrow x_\infty^0$ *as* $y_0 \to 0$, *where* x_∞^0 *is the unique solution in* $(0, 1)$ *of the equation*

$$x_\infty^0 - \ln x_\infty^0 = x_0 - \ln x_0.$$

Furthermore, $x_\infty < x_\infty^0$, *when* $y_0 > 0$.

Epidemiologically, this means that in the subthreshold case a small number of initial infectives will reduce the susceptible population only a bit, generating only a small number of secondary cases. In the superthreshold case very few initial infectives can trigger a considerable epidemic.

Proof. As x_∞ is a continuous function of y_0, as we have seen in Lemma 18.2 (b),

$$x_\infty^0 = \lim_{y_0 \to 0} x_\infty$$

exists and satisfies

$$x_\infty^0 - \ln x_\infty^0 = x_0 - \ln x_0.$$

Moreover, $x_\infty^0 \leqslant 1$ by Lemma 18.2 (a). If $x_0 \leqslant 1$, $x_\infty^0 = x_0$ because f strictly decreases for $x < 1$. If $x_0 > 1$, then $x_\infty^0 < 1$. In order to show that $x_\infty < x_\infty^0$, we rewrite (18.7) as

$$x_\infty \mathrm{e}^{-x_\infty} = x_0 \mathrm{e}^{-x_0} \mathrm{e}^{-y_0},$$

notice that

$$x_\infty^0 \mathrm{e}^{-x_\infty^0} = x_0 \mathrm{e}^{-x_0},$$

and keep in mind that $x_\infty^0, x_\infty \leqslant 1$. The statement now follows from the strict increase of $x\mathrm{e}^{-x}$ on $[0, 1]$. □

For the superthreshold case $x_0 > 1$, we recognize from the properties of f and from (18.7) that $y_{\max}$ is a strictly increasing function of y_0. The limit of $y_{\max}$ for $y_0 \to 0$ is strictly greater than 0, however. Typically, an epidemic outbreak is triggered by a very small number of initial infectives, so the situation is approximately described by the limiting equation of (18.7), i.e., $y_0 = 0$,

$$x_\infty - \ln x_\infty = x_0 - \ln x_0 = 1 + y_{\max}, \tag{18.11}$$

with $x_0 > 1$ and $x_\infty < 1$ (see Theorem 18.3). Expanding the logarithm one obtains the following relation between the initial and final susceptibles and the maximum number of infectives provided that $x_0 - 1$ is positive, but small.

Theorem 18.4. *If $x_0 - 1$ is positive, but small, then*

$$y_{\max} \approx \tfrac{1}{2}(x_0 - 1)^2 \approx \tfrac{1}{2}(1 - x_\infty)^2.$$

We strive for a more exact relation between x_0 and x_∞.

Theorem 18.5. *If $x_0 > 1$ and $y_0 > 0$ is sufficiently small, we have*

$$\frac{x_0 - 1}{x_0} < 1 - x_\infty < x_0 - 1.$$

Theorem 18.5 relates the final reduction of the susceptibles below the threshold to the surplus of initial susceptibles above the threshold (in terms of the replacement ratio). It is equivalent to the following estimates for the final size of the susceptible population.

Theorem 18.6. *If $x_0 > 1$ and $y_0 > 0$ is sufficiently small, we have*

$$2 - x_0 < x_\infty < \frac{1}{x_0}.$$

The first estimate in Theorem 18.6 is only useful if $x_0 < 2$. The second estimate gives a good first idea of how far an epidemic can reduce the susceptible population. The second estimate holds not only in the limiting case $y_0 = 0$, but for any $y_0 > 0$ because x_∞ depends in a monotone decreasing way on y_0.

We prove Theorem 18.6 in several steps. By Theorem 18.3 we only have to consider the case $y_0 = 0$. The first estimate follows from the next lemma, whose proof is left as an exercise.

Lemma 18.7. *For f in (18.10),*

$$f(1 + \xi) < f(1 - \xi), \quad 0 < \xi \leqslant 1.$$

This lemma implies the first estimate in Theorem 18.6, namely

$$f(x_\infty) = f(x_0) = f(1 + x_0 - 1) < f(1 - (x_0 - 1)) = f(2 - x_0).$$

As f is strictly decreasing for $0 \leqslant x \leqslant 1$, we have $x_\infty > 2 - x_0$.

In order to prove the second estimate in Theorem 18.6 we set $u_0 = \ln x_0$, $u_\infty = \ln x_\infty$ in (18.11) and obtain the relation

$$\mathrm{e}^{u_0} - u_0 = \mathrm{e}^{u_\infty} - u_\infty.$$

The function

$$g(u) = \mathrm{e}^u - u, \quad -\infty < u < \infty,$$

has similar properties to the function f in (18.11): g is strictly convex, strictly decreasing in $u < 0$, and strictly increasing in $u > 0$. The statement of Theorem 18.6 is equivalent to $u_\infty + u_0 < 0$, which follows from the fact that $\mathrm{e}^u - u > \mathrm{e}^{-u} + u$ for $u > 0$.

The details are left as an exercise to the reader.

Better Estimates and an Approximation Procedure for the Final Size of the Susceptible Population

We now try to obtain better estimates and an approximation procedure for x_∞.

To this end we exponentiate formula (18.11). After some rearrangement,

$$x_\infty = x_0 \mathrm{e}^{x_\infty - x_0}. \tag{18.12}$$

In other words, x_∞ is a fixed point of the function F:

$$F(x) = x_0 \mathrm{e}^{x - x_0}. \tag{18.13}$$

F is a strictly monotone increasing function. As

$$[2 - x_0]_+ < x_\infty < \frac{1}{x_0}$$

by Theorem 18.6 (recall that $r_+ := \max\{r, 0\}$) this implies the following estimates by induction

$$\tilde{x}_n < x_\infty = F(x_\infty) < \check{x}_n, \quad n \in \mathbb{N}, \tag{18.14}$$

with

$$\left.\begin{aligned} \tilde{x}_1 &= [2 - x_0]_+, & \tilde{x}_{n+1} &= F(\tilde{x}_n), \\ \check{x}_1 &= 1/x_0, & \check{x}_{n+1} &= F(\check{x}_n). \end{aligned}\right\} \tag{18.15}$$

Do the estimates improve if n increases? One realizes by induction that this is indeed the case, i.e., the sequence $\tilde{x}_n$ is increasing and the sequence $\check{x}_n$ is decreasing, if

$$F(\tilde{x}_1) > \tilde{x}_1, \qquad F(\check{x}_1) < \check{x}_1.$$

The first inequality is obvious if $x_0 \geqslant 2$. If $1 < x_0 \leqslant 2$, then by taking logarithms, the first inequality is equivalent to

$$2(1 - x_0) > \ln(2 - x_0) - \ln x_0,$$

which follows from comparing the derivatives of both sides. Taking logarithms again and setting $u_0 = \ln x_0$ the second estimate is equivalent to

$$0 > 2u_0 + \mathrm{e}^{-u_0} - \mathrm{e}^{u_0}, \quad u_0 > 0,$$

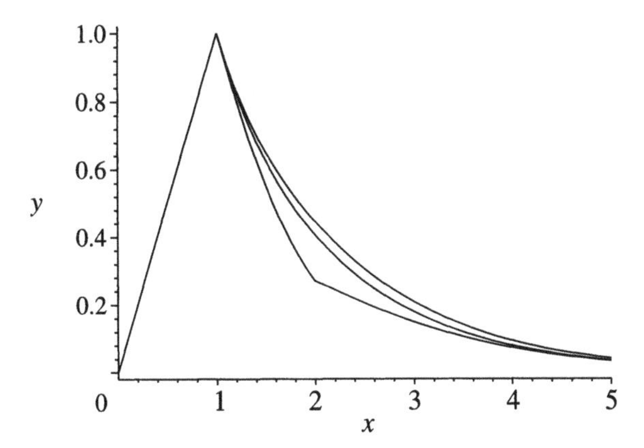

Figure 18.1. The final size of an epidemic for an infinitesimally small number of initial infectives. The variables $x = x_0$ and $y = x_\infty$ denote the initial and the final replacement ratio of the disease, which are proportional to the initial and the final size of the susceptible population. The figure illustrates the threshold phenomenon, as the final replacement ratio equals the initial ratio, as long as the initial ratio is below the threshold ($x = 1$), and then remains well below 1 for $x > 1$. The figure also illustrates the estimates in Theorem 18.8. For $x > 1$, the actual final ratio is given by the middle curve; the other two curves correspond to the lower and upper estimate, respectively.

which follows from differentiating the right-hand side twice. As the sequences $\tilde{x}_n$ and $\check{x}_n$ are monotone and bounded, they converge. By the recursive definition of the sequences their limits are fixed points of the continuous function F. But F has only one fixed point smaller than one, namely x_∞. Hence both sequences $\tilde{x}_n$ and $\check{x}_n$ converge monotone towards x_∞ and their differences provide an upper estimate of the goodness of approximation. Already the first iteration gives a reasonable estimate.

Theorem 18.8. *Under the assumptions of Theorem 18.5, for a sufficiently small number of initial infectives,*

$$x_0 e^{[2-x_0]_+ - x_0} < x_\infty < x_0 e^{(1/x_0)-x_0}.$$

The second estimate actually holds for an arbitrary number of initial infectives.

These estimates are very good for $x_0 \to 1$ and $x_0 \to \infty$. They are worst for $x_0 = 2$, where the quotient of the upper and the lower estimates is the largest.

It may be worthwhile to rewrite the above estimates in terms of the original variables and the initial replacement ratio:

$$x_0 = \frac{\sigma}{\rho} S_0.$$

Corollary 18.9. *If* $x_0 = (\sigma/\rho)S_0 > 1$ *and the number of initial infectives is sufficiently small, the final size of the susceptible population can be estimated* *by*

$$e^{[2-x_0]_+-x_0} < \frac{S_\infty}{S_0} < e^{(1/x_0)-x_0}.$$

The second estimate also holds if the number of initial infectives is not small.

We can rewrite these inequalities in a more suggestive way,

$$e^{2(1-x_0)} < \frac{S_\infty}{S_0} < e^{(1+1/x_0)(1-x_0)}, \quad 1 < x_0 \leqslant 2,$$
$$e^{-x_0} < \frac{S_\infty}{S_0} < e^{1/x_0}e^{-x_0}, \quad x_0 \geqslant 2,$$

which implies

$$\frac{S_\infty}{S_0} \approx e^{2(1-x_0)} \approx 1 - 2(x_0 - 1), \quad x_0 \to 1,$$
$$\frac{S_\infty}{S_0} \approx e^{-x_0}, \quad x_0 \to \infty.$$

The second relation reveals how truly devastating epidemics are if the number of initial susceptibles is well above the threshold. The inverse problem of determining the initial replacement ratio, x_0, from the fraction $p = S_\infty/S_0$ of initial susceptibles that are still susceptible after the epidemic can be solved precisely. From (18.12) one obtains

$$x_0 = \frac{\ln p}{p-1}, \quad p = \frac{S_\infty}{S_0}.$$

Theorem 18.6 has an interesting interpretation in terms of the number of removed individuals. Recall that the number of infectives reaches its maximum exactly at the moment when the replacement ratio reaches the threshold value 1. Following Bailey (1975) we call that moment the central epoch and denote it by $t_{\max}$. Remember that the estimate in Theorem 18.6 also holds if $y_0 > 0$, because x_∞ depends on y_0 in a decreasing manner as we have seen at the beginning of this subsection. Integrating the first equation in (18.5) we find

$$1 = x_0 e^{-z(t_{\max})}, \qquad x_\infty = e^{z_{\max}-z_\infty}.$$

Theorem 18.6 now implies the following corollary.

Corollary 18.10.

$$z_\infty - z(t_{\max}) \geqslant z(t_{\max}).$$

This means that more individuals are removed after the central epoch than before.

From Theorem 18.8 we see that this asymmetry becomes quite large if the number of initial susceptibles is well above the threshold. According to Bailey (1975) this type of asymmetry is often found in actual infectious diseases.

Exercises

18.1. Consider an infectious disease with an average infectious period of length $1/\rho = 7$ days and a per capita rate of contraction of disease of $\sigma = 1/490\,000$ per day per individual. Assume that the population initially consists of 14 million susceptible people and that the number of initial infectives is so small that it can be neglected.

(a) Determine the initial replacement ratio x_0.

(b) Determine the maximum number of infectives.

(c) Determine the minimum fraction of the population that has to be vaccinated in order to prevent an epidemic outbreak.

18.2. Consider an infectious disease with the same parameters ρ, σ as in Exercise 18.1. Assume that there are 21 000 susceptible individuals left after an epidemic. Assume that the number of initial infectives that triggered the epidemic is so small that it can be neglected.

Determine the maximum number of infectives.

18.3. Consider a city of 1 000 000 susceptible inhabitants. Let the average length of the infectious period be $1/\rho = 14$ days. Let the infection rate be $\sigma = 1/1\,000\,000$ per day per individual. Give an upper estimate for the final size S_∞ of the susceptible population.

18.4. Consider a town of 10 000 susceptible inhabitants. Let the average length of the infectious period be 3 days. Let the infection rate be $\sigma = 1/20\,000$ per day per individual. Assume that the number of initial infectives is very small.

(a) Give upper and lower estimates for the final size S_∞ of the susceptible population.

(b) Give an approximation for the maximum number of infectives.

18.5. Consider the function $f(x) = x - \ln x$. Show that

$$f(1+\xi) \leqslant f(1-\xi), \quad 0 \leqslant \xi \leqslant 1$$

(cf. Lemma 18.7).

18.6. (Calculator or computer.) Use the iteration (18.15) to derive the following estimates of x_∞ for various choices of x_0:

(a) $x_0 = 1.5, 0.623 < x_\infty < 0.628$;

(b) $x_0 = 2, 0.406 < x_\infty < 0.407$; and

(c) $x_0 = 5, 0.0348 < x_\infty < 0.0349$.

How many iterations do you need in the various cases in order to obtain these estimates?

Chapter Nineteen

Generalization of the Mass-Action Law of Infection

The infection law σSI has been challenged in various ways. For instance, S and I have been replaced by powers S^α and I^β. See Capasso (1993) for the analysis of some of these models (mainly for endemic models) and Liu (1995) for a survey.

19.1 Population-Size Dependent Contact Rates

Another generalization of the mass-action law of infection replaces σSI by

$$\eta C(N) S \frac{I}{N}$$

(see Dietz, 1982, and Castillo-Chavez et al., 1989, for examples). Here $C(N)$ denotes the number of contacts an average individual makes per unit of time in a population of size N. To be precise, N is the size of the part of the population which is available for contact, i.e., dead and quarantined individuals are not counted. The fraction I/N gives the probability that a given contact actually occurs with an infective individual. η is the average chance that a contact between a susceptible and an infective individual actually leads to an infection. We recover our old model with mass-action incidence if $C(N)$ is proportional to N. If C is constant, we have the so-called *standard incidence*. Data from England and the United States suggest that $C(N)/N$ is a decreasing function of N. Dietz (1982) suggests the saturating contact rate

$$C(N) = \frac{\zeta N}{1 + \alpha N},$$

which is proportional to N for small N and almost independent of N for large N; in other words, this contact rate provides an interpolation between mass-action incidence for small N and standard incidence for large N. Alternatively, one could use $C(N) = (\zeta/\alpha)(1 - e^{-\alpha N})$, in analogy to Roughgarden's (1979) suggestion of a predator's saturating functional response (in lieu of the usual Michaelis–Menten–Monod–Holling type-II response which corresponds to Dietz's suggestion). Anderson (1982b) proposes to set

$$C(N) = \zeta N^\alpha$$

with $0 \leqslant \alpha < 1$. His suggestion is not based on conceptual arguments, but on the fact that this ansatz gives reasonable fits to the data, with α being very small. Actually, the analysis of the model does not change in comparison to the simple Kermack–McKendrick model with mass-action type infection if the size N of available individuals is not changed in the course of the epidemic. But this will necessarily happen if the disease causes fatalities or individuals are quarantined.

If the transmission of the disease requires a close contact between susceptible and infective individuals (sexually transmitted diseases, for example), the contact rate can be modeled by a complex-formation approach (cf. Heesterbeek, Metz, 1993; Diekmann, Heesterbeek, 2000, Exercise 10.12), and the per capita rate of contacts as a function of population size is given by

$$C(N) = \frac{\kappa_1 N}{1 + \kappa_2 N + \sqrt{1 + 2\kappa_2 N}}. \tag{19.1}$$

This contact function has similar qualitative features to the one suggested by Dietz (1982) that we mentioned above.

19.2 Model Modification

We must now be careful in dividing the removed population: we discriminate between the class R of recovered (and immune) individuals and a class D of individuals that died from the disease. We do not assume quarantine. The sizes of the respective classes are denoted by the same symbols. As in Chapter 18, S, I, and N denote the sizes of the susceptible, infective, and total population, respectively. In contrast to Chapter 18, individuals that have died from the disease are no longer considered part of the population; the respective per capita death rate is denoted by δ.

Model Equations

$$\begin{aligned} N &= S + I + R, \\ \dot{S} &= -\eta f(N) S I, \\ \dot{I} &= \eta f(N) S I - (\rho + \delta) I, \\ \dot{R} &= \rho I, \\ \dot{D} &= \delta I, \\ f(N) &= \frac{C(N)}{N}. \end{aligned}$$

It is reasonable to make the following assumptions for the contact rate.

Contact rate assumptions. *$f(N) = C(N)/N$ is a monotone nonincreasing function of N. Furthermore, $C(N) > 0$, whenever $N > 0$, $C(0) = 0$.*

We nondimensionalize time by setting $t = \tau/\delta$. $1/\delta$ is the expectation of remaining life after the moment of infection. We set

$$\sigma = \frac{\eta}{\delta}, \qquad \xi = \frac{\rho}{\delta}.$$

With a prime denoting differentiation with respect to dimensionless time, the model takes the form

$$\begin{aligned} N &= S + I + R, \\ S' &= -\sigma f(N) S I, \\ I' &= \sigma f(N) S I - (\xi + 1) I, \\ R' &= \xi I, \\ D' &= I, \\ f(N) &= \frac{C(N)}{N}. \end{aligned}$$

Adding the S, I, R equations we obtain

$$N' = -I.$$

Next we define

$$u = \sigma S, \qquad v = \sigma I, \qquad w = \sigma N, \qquad f(N) = g(w)$$

and obtain dependent variables which are also dimensionless. The mathematically relevant part of the system then takes the following form:

$$\left.\begin{aligned} u' &= -g(w) u v, \\ v' &= g(w) u v - (1 + \xi) v, \\ w' &= -v. \end{aligned}\right\} \tag{19.2}$$

We keep in mind that

$$u + v \leqslant w.$$

19.3 The Generalized Epidemic Threshold Theorem

We rewrite the scaled system in the following form:

$$\left.\begin{aligned} \frac{u'}{u} &= g(w) w', \\ v' &= -u' + (1 + \xi) w', \\ w' &= -v. \end{aligned}\right\} \tag{19.3}$$

We integrate the u and v equations:

$$\left.\begin{aligned} u &= u_0 \exp\left(-\int_w^{w_0} g(\tilde{w})\,\mathrm{d}\tilde{w}\right), \\ v &= v_0 + u_0 - u + (1+\xi)(w - w_0). \end{aligned}\right\} \tag{19.4}$$

We substitute these relations into the w equation in (19.3):

$$w' = u_0 \exp\left(-\int_w^{w_0} g(x)\,\mathrm{d}x\right) + (1+\xi)(w_0 - w) - u_0 - v_0. \tag{19.5}$$

We derive the following estimate for w from (19.4):

$$w \geqslant \frac{\xi}{1+\xi} w_0. \tag{19.6}$$

Recall that $u_0 + v_0 \leqslant w_0$. We know from (19.3) that w is nonincreasing in time—its derivative is nonpositive—and so has a limit w_∞ as $t \to \infty$ which also satisfies (19.6). It follows from (19.2) that u and w are bounded and so is $v \leqslant w$. By (19.2), v' is bounded and so is w''. Hence $w'(t) \to 0$, $t \to \infty$ by Corollary A.18.

From (19.5) and the third equation in (19.3) we draw the following conclusions for the limiting values of the epidemic:

$$\left.\begin{aligned} (1+\xi)(w_0 - w_\infty) &= v_0 + u_0\left(1 - \exp\left(-\int_{w_\infty}^{w_0} g(\tilde{w})\,\mathrm{d}\tilde{w}\right)\right), \\ v_\infty &= 0, \\ u_\infty &= u_0 \exp\left(-\int_{w_\infty}^{w_0} g(\tilde{w})\,\mathrm{d}\tilde{w}\right), \\ u_0 - u_\infty &= -v_0 + (1+\xi)(w_0 - w_\infty). \end{aligned}\right\} \tag{19.7}$$

The last two relations in (19.7) are the limits of (19.4) for $t \to \infty$.

In order to prove an epidemic threshold theorem we introduce

$$x_0 = \frac{u_0 g(w_0)}{1+\xi}. \tag{19.8}$$

In the original parameters we have

$$x_0 = \frac{\sigma S_0}{1+\xi} f(N_0) = \frac{\eta}{\rho+\delta} C(N_0) \frac{S_0}{N_0}.$$

So, x_0 is the initial replacement ratio in the usual interpretation, namely the total number of secondary cases one average infective individual produces at the beginning of the epidemic.

Theorem 19.1 (epidemic threshold theorem).

(a) *Let* $x_0 \leqslant 1$. *Then* $u_\infty \to u_0$, $w_\infty \to w_0$ *if* $v_0 \to 0$.

(b) *Let* $x_0 > 1$. *Then* u_∞, w_∞ *stay bounded away from* u_0, w_0, *respectively, if* $0 < v_0 \to 0$.

Recall that u, w are the scaled sizes of the susceptible and the total population. So, in the subthreshold case $x_0 \leqslant 1$, the epidemic will be arbitrarily small if the number of initial infectives is sufficiently low. In the superthreshold case $x_0 > 1$, the epidemic will have a certain minimum size however small the number of initial infectives is.

Proof. It follows from (19.7) that we have to prove the statements only for w.

(a) Assume the contrary. From the first equation in (19.7) we find some $w < w_0$ such that

$$(1+\xi)(w_0 - w) \leqslant u_0(1 - \exp(-g(w_0)(w_0 - w))).$$

Here we have used that g is nonincreasing and $1-e^{-r}$ is increasing. As $1-e^{-r} < r$, $r > 0$, we have

$$(1+\xi)(w_0 - w) < u_0 g(w_0)(w_0 - w).$$

But, by (19.8), this implies $x_0 > 1$, in contradiction to our assumption.

(b) Assume the contrary again. From the first equation in (19.7) and from (19.6) we find a sequence $w_j < w_0$, $w_j \to w_0$, $j \to \infty$ such that

$$1+\xi \geqslant \frac{u_0}{w_0 - w_j}\left(1 - \exp\left(-\int_{w_j}^{w_0} g(\tilde{w})\,\mathrm{d}\tilde{w}\right)\right).$$

Taking the limit $j \to \infty$ we obtain

$$1+\xi \geqslant u_0 g(w_0),$$

in contradiction to the assumption $x_0 > 1$. □

In order to describe the limit of w_∞ as $0 < w_0 \searrow u_0$, provided it exists, we need to make additional assumptions. Notice that $w_0 \searrow u_0$ is equivalent to I_0, $R_0 \to 0$. One possibility is addressed in Exercise 19.1 below.

Exercises

19.1. Assume

$$x_0 = \frac{u_0 g(w_0)}{1+\xi} > 1.$$

(a) Prove that there exists a solution $w^* \in (0, u_0)$ of the equation

$$(1+\xi)(u_0 - w^*) = u_0\left(1 - \exp\left(-\int_{w^*}^{u_0} g(\tilde{w})\,\mathrm{d}\tilde{w}\right)\right). \quad (19.9)$$

(b) Assume in addition that the equation

$$\frac{\mathrm{d}}{\mathrm{d}x}\frac{1}{g(x)} = 1 \quad (19.10)$$

has at most one root. Prove that (19.9) has a unique solution $w^* < w_0$.

Hint: if there are two solutions of (19.9) which are strictly smaller than w_0, then the right-hand side of (19.9) considered as a function of w must have an inflection point.

(c) Make the same additional assumption as in (b). Prove that the solutions w_∞ of the first equation in (19.7) converge to the solution w^* of (19.9) as $0 < w_0 \searrow u_0$.

(Remember that $w_0 \searrow u_0$ implies that $v_0 \searrow 0$, because $u_0 + v_0 \leqslant w_0$.)

19.2. Make the additional assumption in Exercise 19.1 (b) and assume $\xi = 0$, i.e., nobody recovers from the disease.

(a) Show that $w^* = 0$ (i.e., the epidemic eradicates the population) if and only if

$$\int_0^1 g(x)\,\mathrm{d}x = \infty.$$

(b) Give an example for a contact function C for which the associated g satisfies

$$\int_0^1 g(x)\,\mathrm{d}x = \infty.$$

19.3. Consider the two contact functions C which have been mentioned, $C(N) = \zeta N^\alpha$ and (19.1). Check whether the associated function g satisfies condition (19.10) in Exercise 14.1(b).

Chapter Twenty

The Kermack–McKendrick Epidemic Model with Variable Infectivity

We continue our consideration of whether the results obtained for the simplified Kermack–McKendrick model survive the addition of more realism. This time we take into account that the infectivity of an infected individual may not be constant during the time after infection. First of all it may be zero as long as the individual stays in the latent period. It has been suggested for HIV, for example, that the infectivity of an infected individual is zero for a short period, then has an early peak which is ended by the immune response and is followed by a rather long period of low infectivity. Finally, the immune system is exhausted and the infectivity rises again.

We mention that variable infectivity has already been considered by Kermack, McKendrick (1927–1939) themselves.

20.1 A Stage-Age Structured Model

In order to take account of the just-mentioned effects, we introduce a second independent variable $b \geqslant 0$ denoting the time that has passed since infection. b is also referred to as the *infection age*. The average rate σ at which a susceptible individual contracts the disease from an infective individual (per unit of time) is now a function of the infection age b, i.e.,

$$\sigma = \sigma(b).$$

Instead of a constant rate ρ at which an infective individual is removed from the infective class, we prescribe the probability of not yet being removed at infection age b,

$$P(b),$$

where $P : [0, \infty) \to [0, \infty)$ is the sojourn function of the infected stage, i.e., P is monotone nonincreasing and $P(0) = 1$. We also assume that the expected infection age at removal, D, is finite,

$$D = \int_0^\infty P(b)\, \mathrm{d}b < \infty$$

(cf. Sections 12.1 and 12.2). We keep the classes S and R of susceptible and removed individuals but replace the class I of *infectious* individuals by the class of *infected* individuals. This class is stratified according to infection age such that

$$i(t, b)$$

denotes the "number" of infected individuals with infection age b at time t. More correctly, $i(t, \cdot)$ is the infection age density of the infected individuals at time t such that

$$\int_{b_1}^{b_2} i(t, b)\, \mathrm{d}b$$

gives the number of infected individuals with infection age between b_1 and b_2.

We remember from Section 13.1 that

$$i(t, b) = \begin{cases} B(t-b)P(b), & t > b \geqslant 0, \\ i_0(b-t)\dfrac{P(b)}{P(b-t)}, & b > t \geqslant 0, \end{cases}$$

where i_0 is the initial density of infected individuals and $B(t)$ is the input into the infected class at time t. The input B into the infected class is the incidence of the disease, i.e., the rate of infections, which, for an epidemic model, is

$$B(t) = -S'(t).$$

As in Chapter 18, we model B via a mass-action law,

$$B(t) = S(t)\phi(t),$$

where $\phi(t)$ is the infective force at time t, i.e., the rate at which an average susceptible individual is infected per unit of time. The infective force is obtained by an integral over the infected individuals which is weighed by σ:

$$\phi(t) = \int_0^\infty \sigma(b) i(t, b)\, \mathrm{d}b.$$

We end up with the following model equations.

Model Equations

$$\left.\begin{aligned} \phi(t) &= \int_0^\infty \sigma(b) i(t, b)\, \mathrm{d}b, \\ S' &= -\phi S, \\ i(t, b) &= \begin{cases} -S'(t-b)P(b), & t > b \geqslant 0, \\ i_0(b-t)\dfrac{P(b)}{P(b-t)}, & b > t \geqslant 0, \end{cases} \\ R'(t) &= -\int_0^\infty i(t, b)\, \mathrm{d}P(b). \end{aligned}\right\} \tag{20.1}$$

The last line of the equation expresses that the rate of the change in the removed class equals the output of the infected class (see Section 13.6). We complete our model by initial conditions

$$S(0) = S_0, \qquad R(0) = 0. \tag{20.2}$$

The relation $i(0, b) = i_0(b)$ is already incorporated in (20.1).

We refer to Problem 20.4 of the comprehensive exercises for Chapters 18–20 for an attempt to combine variable infectivity and the contact function introduced in Chapter 19.

20.2 Reduction to a Scalar Integral Equation

Our aim consists of deriving an integral equation for S. From (20.1),

$$\begin{aligned}\phi(t) &= \int_0^t \sigma(b)(-S'(t-b))P(b)\,\mathrm{d}b + \int_t^\infty \sigma(b) i_0(b-t)\frac{P(b)}{P(b-t)}\,\mathrm{d}b \\ &= \int_0^t \sigma(b)(-S'(t-b))P(b)\,\mathrm{d}b + \int_0^\infty \sigma(t+b) i_0(b)\frac{P(b+t)}{P(b)}\,\mathrm{d}b.\end{aligned}$$

We integrate the S equation in (20.1):

$$\begin{aligned}\ln S_0 - \ln S(t) &= \int_0^t \phi(r)\,\mathrm{d}r \\ &= \int_0^t \int_0^r \sigma(b)(-S'(r-b))P(b)\,\mathrm{d}b\,\mathrm{d}r \\ &\quad + \int_0^t \int_0^\infty i_0(b)\sigma(b+r)\frac{P(b+r)}{P(b)}\,\mathrm{d}b\,\mathrm{d}r.\end{aligned}$$

We change the order of integration:

$$\begin{aligned}\ln S_0 - \ln S(t) &= \int_0^t \sigma(b)\left(\int_b^t (-S'(r-b))\,\mathrm{d}r\right)P(b)\,\mathrm{d}b \\ &\quad + \int_0^\infty \int_0^t i_0(b)\sigma(b+r)\frac{P(b+r)}{P(b)}\,\mathrm{d}r\,\mathrm{d}b.\end{aligned}$$

Using the fundamental theorem of calculus, we arrive at one scalar integral equation for S:

$$\begin{aligned}\ln S_0 - \ln S(t) &= \int_0^t \sigma(b)(S_0 - S(t-b))P(b)\,\mathrm{d}b \\ &\quad + \int_0^\infty \int_0^t i_0(b)\sigma(b+r)\frac{P(b+r)}{P(b)}\,\mathrm{d}r\,\mathrm{d}b.\end{aligned}$$

We recognize from (20.1) that S is a nonincreasing function of time and the limit $S_\infty = \lim_{t\to\infty} S(t)$ exists. S_∞ can be interpreted as the final size of the susceptible population. Taking the limit in the last equation—actually we may use the Lebesgue–Fatou lemma (Toolbox B.2)—we obtain

$$\ln S_0 - \ln S_\infty = (S_0 - S_\infty)\int_0^\infty \sigma(b)P(b)\,\mathrm{d}b + \int_0^\infty i_0(b)\sigma(b+r)\frac{\int_0^\infty P(b+s)\,\mathrm{d}s}{P(b)}\,\mathrm{d}b. \quad (20.3)$$

We realize that this is the same equation as (18.7): we set

$$\kappa = \int_0^\infty \sigma(b)P(b)\,\mathrm{d}b, \quad (20.4)$$

$$x_\infty = S_\infty\kappa, \qquad x_0 = S_0\kappa, \quad (20.5)$$

and

$$y_0 = \int_0^\infty i_0(b)\sigma(b+r)\frac{\int_0^\infty P(b+s)\,\mathrm{d}s}{P(b)}\,\mathrm{d}b. \quad (20.6)$$

With this notation, (20.3) becomes

$$\ln x_0 - x_0 = \ln x_\infty - x_\infty + y_0, \quad (20.7)$$

which is identical to (18.7) and can be analyzed along the lines of Chapter 18, yielding the same results. Notice that y_0 is some kind of infective force exerted by the individuals that were already infected at the beginning. The initial replacement ratio x_0 determining whether or not there is an epidemic outbreak now takes the form

$$x_0 = S_0\kappa,$$

with κ from (20.4), and can again be interpreted as the total number of susceptibles one average infective individual can infect at the beginning of the epidemic.

Theorem 20.1. *The epidemic threshold Theorems 18.3–18.6 and 18.8 and Corollary 18.9 hold verbatim.*

Corollary 18.10 does not generalize to the case of variable infectivity, however.

Exercises

20.1. Let our time unit be 1 day. Let the sojourn function for the infected class be given by a removal rate,

$$P(b) = \exp\left(-\int_0^b \rho(s)\,\mathrm{d}s\right),$$

with

$$\rho(b) = \begin{cases} 0, & 0 \leqslant b \leqslant 1, \\ 1, & b > 1. \end{cases}$$

Let the per capita infection rate σ be given by

$$\sigma(c) = \begin{cases} 0, & 0 \leqslant c \leqslant 1, \\ 1/20\,000 \text{ per day per individual}, & c > 1. \end{cases}$$

Let the initial number of susceptibles be

$$S_0 = 200\,000.$$

Remark: this models a latent period which is exactly 1 day long for everybody.

(a) Determine x_0.

(b) Give lower and upper estimates for S_∞ assuming that the number of initial infectives is very low.

Comprehensive Exercises for Chapters 18–20

20.2. (a) Build an *epidemic* model including a latent period, i.e., SEIR with S, E, I, and R denoting the susceptible, exposed (i.e., in the latent period), infective, and removed individuals, respectively. Assume the mass-action law for infection. Assume that the length of the latent period is negatively exponentially distributed, i.e., if there were no influx of new infections into the latent class we would have

$$\dot{E} = -\gamma E.$$

Borrow the ideas from Chapter 18.

(b) Derive a formula for the final size of the susceptible population and prove a threshold theorem. You may use the techniques of Chapter 18 or of Chapter 20.

20.3. Consider the following *epidemic* model for a disease in which an infective individual passes two stages of different infectivity:

$$\begin{aligned} \dot{S} &= -S(\sigma_1 I_1 + \sigma_2 I_2), \\ \dot{I}_1 &= -\dot{S} - \rho_1 I_1, \\ \dot{I}_2 &= \rho_1 I_1 - \rho_2 I_2, \\ \dot{R} &= \rho_2 I_2. \end{aligned}$$

Derive a formula for the final size of the susceptible population. Prove a threshold theorem. Use the technique of Chapter 20 and define $\phi = \sigma_1 I_1 + \sigma_2 I_2$.

20.4. (a) Formulate a model combining the models in Chapters 19 and 20. Note that

$$N(t) = S(t) + \int_0^\infty i(t, c)\, \mathrm{d}c + R(t).$$

(b) Try to generalize the analysis of Chapter 19 to the model you have formulated. Notice that you need to assume that

$$\delta(c) = \delta_0\, \sigma(c)$$

with a constant $\delta_0 > 0$. Does this assumption make sense?

Bibliographic Remarks

The historical reception of the work by Kermack, McKendrick (1927–1939) is a strange one. For a long time, most people were only aware of the "simplified Kermack–McKendrick" model in Chapter 18, while Kermack and McKendrick actually deal with the "general Kermack–McKendrick" model with variable infectivity in Chapter 20. With the passionate exclamation "This should stop," Diekmann, Heesterbeek, Metz (1995) point out that the first three papers by Kermack, McKendrick (1927, 1932, 1933) have been reprinted and that there is no excuse for this kind of ignorance anymore. The paper by Diekmann, Heesterbeek, Metz also contains a survey and references to the various developments resulting from the work by Kermack, McKendrick, like the introduction of more heterogeneity; these are discussed in detail in Diekmann, Heesterbeek (2000). The impact on studying the spatial spread of epidemics is discussed by Metz, van den Bosch (1995); the spatial spread of epidemics including their speed of spread and the involvement of different subpopulations are comprehensively treated by Rass, Radcliffe (n.d.). While we have only considered the final size of an epidemic here, one can also analyze its complete time development, which is described by the nonlinear renewal theorem of epidemics (see Metz, Diekmann (1986, Section IV.4.1) for an outline and references). For a discussion of standard versus mass-action incidence and for pertinent references we refer to Hethcote (2000, Section 2.1).

Chapter Twenty-One

SEIR ($\rightarrow S$) Type Endemic Models for "Childhood Diseases"

Many infectious diseases do not occur in single outbreaks, but are around all the time, though their prevalence may fluctuate. Their models must therefore take the vital dynamics if not the complete demographics of the respective host populations into consideration.

Some important examples of endemic diseases are the so-called *childhood diseases*, which typically take the following course in an individual. After infection there is a time period—the latent period—in which the infection gets established in the individual, but the individual is not yet infectious itself. After the latent period, the individual enters the *infectious period*, during which it is able to transmit the disease to other individuals. The infection triggers an immune reaction in the body which enables the individual to recover from the disease. It stops being infectious and is immune to reinfection for a certain period of time—the *immunity period*. Immunity may be permanent or temporary. In the second case the individual becomes susceptible again at the end of the immunity period. In developed countries few people die from childhood diseases; consequently, for the dynamical picture of the disease, disease-related deaths can be neglected. (This is different for developing countries.)

In diseases like chicken pox, diphtheria, measles, mumps, poliomyelitis, rubella, scarlet fever, and whooping cough, immunity is practically permanent or lasts for a period of time which is (at least) very large in comparison to the periods of latency and infectivity. The latent period can be short compared with the infective period (scarlet fever: 1–2 days compared with 14–21 days), similarly long (measles: 6–9 days compared with 6–7 days) or considerably longer (mumps: 12–18 days compared with 4–8 days) (see Anderson, May (1982b) and Table 3.1 in Anderson, May (1991)). Regarding these numbers there is actually little justification for neglecting the latent period, except maybe in the cases of scarlet fever and poliomyelitis.

Childhood diseases have fascinated mathematical modelers for quite a while because of the periodic pattern of their occurrence (see Hethcote, Levin (1989) for a survey). Though these diseases are endemic (i.e., present all the time), there are major outbreaks rather regularly. The time between two outbreaks is called the *interepidemic period* (Anderson, May, 1982).

For *measles* in England and Wales (1855–1979), the interepidemic period lasted for 2.2 years on average, with a range of 2–4 years. In New York City (1928–1968) the average was 2.2 years as well, with a range of 2–3 years.

For *mumps* in Baltimore, MD (1928–1973), the interepidemic period averaged 3.0 years, with a range of 2–4 years. In New York City (1928–1967) it averaged 3.0 years, with a range of 2–6 years.

For *scarlet fever* in England and Wales (1897–1978), the interepidemic period averaged 4.4 years, with a range of 3–6 years. (See Anderson, May (1982b) and also Table 6.1 in Anderson, May (1991).)

Many models have been proposed to explain these patterns. See Dietz, Schenzle (1985) and Hethcote, Levin (1989). We restrict ourselves here to a basic model which has already been investigated, for example, by Dietz (1976) and Anderson, May (1982b).

21.1 The Model and Its Well-Posedness

Model Assumptions

The disease is assumed to take the course explained above, and so the host population is split up into susceptible, exposed (i.e., in the latent period), infective, and recovered (immune) individuals. The respective numbers are denoted by S, E, I, and R. The course of the disease can schematically be represented as

$$S \to E \to I \to R \ (\to S).$$

The bracket indicates that immunity may or may not be permanent. We assume that every infectious individual is equally infective. Everybody is assumed to be born susceptible. The per capita birth rate, μ, is supposed to be the same as the per capita death rate such that the population size does not change.

Model Equations

$$\left.\begin{aligned} N &= S + E + I + R,\\ \frac{\mathrm{d}}{\mathrm{d}t} S &= \mu N - \mu S - \sigma S I + \rho R,\\ \frac{\mathrm{d}}{\mathrm{d}t} E &= -\mu E + \sigma S I - \eta E,\\ \frac{\mathrm{d}}{\mathrm{d}t} I &= -\mu I + \eta E - \gamma I,\\ \frac{\mathrm{d}}{\mathrm{d}t} R &= -\mu R + \gamma I - \rho R. \end{aligned}\right\} \tag{21.1}$$

μ is the per capita birth and mortality rate, i.e., $1/\mu$ is the average life expectancy. η is the rate at which individuals leave the latent period, i.e., $1/\eta$ is the mean length of the latent period. γ, ρ are the rates at which individuals leave the infective and recovered classes, respectively, i.e., $1/\gamma$, $1/\rho$ are the mean lengths of the infective and recovered periods. σ is somewhat difficult to interpret, $1/\sigma I$ is the mean length of time it takes an average susceptible individual to become infected if it is exposed to I infective individuals. μ, η, σ, γ are positive constants, ρ is nonnegative allowing for the fact that immunity may be permanent for certain diseases but not for others.

Well-Posedness of the Model

We show that unique solutions exist to system (21.1) which also have the properties suggested by the epidemiological interpretation.

Theorem 21.1. *For any S_0, E_0, I_0, $R_0 \geqslant 0$ there exists a unique solution S, E, I, R, respectively, defined on $[0,\infty)$, to (21.1) satisfying*

$$S(0) = S_0, \qquad E(0) = E_0, \qquad I(0) = I_0, \qquad R(0) = R_0$$

and

$$S(t), E(t), I(t), R(t) \geqslant 0, \quad N(t) = N(0) \quad \forall t \geqslant 0.$$

Proof. The system can be written in the form

$$x' = F(x)$$

with $x = (S, E, I, R)$ and $F(x) = (F_1(x), \ldots, F_4(x))$. We want to apply Theorem A.4. It is easily checked that, for every $j = 1, \ldots, 4$, $F_j(x) \geqslant 0$ if $x \in [0,\infty)^4$ and $x_j = 0$.

Now let x be a solution with values in $\mathbb{R}^n_+$, defined on some interval $[0, b)$. It is sufficient to show that x is bounded on $[0, b)$. Adding the right-hand sides of (21.1) we notice that

$$\frac{\mathrm{d}}{\mathrm{d}t} N = 0.$$

This implies that $N(t) = N(0)$ for all $0 \leqslant t < b$.

As S, E, I, R are nonnegative and add up to N, they are bounded by $N(0)$.

The assertion now follows from Theorem A.4. □

Scaling of the System

We first make the dependent variables S, E, I, R dimensionless. Here we take advantage of the fact that $N = S + E + I + R$ is constant and introduce the fractions of susceptible, exposed, infective, and recovered individuals:

$$u = \frac{S}{N}, \qquad x = \frac{E}{N}, \qquad y = \frac{I}{N}, \qquad z = \frac{R}{N}. \tag{21.2}$$

We obtain the following system:

$$\left.\begin{aligned} 1 &= u + x + y + z, \\ \frac{\mathrm{d}}{\mathrm{d}t}u &= \mu - \mu u - \sigma N u y + \rho z, \\ \frac{\mathrm{d}}{\mathrm{d}t}x &= -\mu x + \sigma N u y - \eta x, \\ \frac{\mathrm{d}}{\mathrm{d}t}y &= -\mu y + \eta x - \gamma y, \\ \frac{\mathrm{d}}{\mathrm{d}t}z &= -\mu z + \gamma y - \rho z. \end{aligned}\right\} \tag{21.3}$$

Scaling in this way we could condense N, σ into the compound parameter σN. By scaling time as well, we can eliminate one further parameter. As we want to look at the cases where η is either large or small compared to γ and where ρ is either zero or strictly positive and where μ is small compared to η, γ we choose to scale time such that σN becomes 1. This is achieved by introducing the new (dimensionless) time τ satisfying

$$\tau = \sigma N t. \tag{21.4}$$

Then

$$\frac{\mathrm{d}}{\mathrm{d}t} = \frac{\mathrm{d}\tau}{\mathrm{d}t}\frac{\mathrm{d}}{\mathrm{d}\tau} = \sigma N \frac{\mathrm{d}}{\mathrm{d}\tau}.$$

The disadvantage of this choice consists of the fact that the new time unit $\tau = 1$ corresponds to $t = 1/\sigma N$. $1/\sigma N$ can be interpreted as the mean time it takes for an average susceptible to be infected in the case in which the whole population is infective. So, one has no immediate guess what $1/\sigma N$ may be quantitatively.

Let differentiation with respect to the new (dimensionless) time be denoted by a prime, i.e., $' = \mathrm{d}/\mathrm{d}\tau$. For simplicity the new time τ is again called t. Equation (21.3) takes the form

$$\left.\begin{aligned} 1 &= u + x + y + z, \\ u' &= \nu(1-u) - uy + \varphi z, \\ x' &= uy - (\xi + \nu)x, \\ y' &= \xi x - (\theta + \nu)y, \\ z' &= \theta y - (\varphi + \nu)z. \end{aligned}\right\} \tag{21.5}$$

The new (dimensionless) parameters $\nu, \xi, \theta, \varphi$ are given by

$$\nu = \frac{\mu}{\sigma N}, \qquad \xi = \frac{\eta}{\sigma N}, \qquad \theta = \frac{\gamma}{\sigma N}, \qquad \varphi = \frac{\rho}{\sigma N}. \tag{21.6}$$

We can use the relation $u + x + y + z = 1$ to eliminate one of the equations; we choose the equation for u. This will facilitate the discussion of the qualitative behavior of solutions:

$$\left.\begin{aligned} x' &= (1 - x - y - z)y - (\xi + \nu)x, \\ y' &= \xi x - (\theta + \nu)y, \\ z' &= \theta y - (\varphi + \nu)z. \end{aligned}\right\} \tag{21.7}$$

21.2 Equilibrium States and the Basic Replacement Ratio

When looking for equilibria (equilibrium solutions, stationary solutions, steady states, time-independent solutions), the system (21.5) takes the form

$$\left.\begin{aligned} 1 &= u + x + y + z, \\ 0 &= \nu(1 - u) - uy + \varphi z, \\ 0 &= uy - (\xi + \nu)x, \\ 0 &= \xi x - (\theta + \nu)y, \\ 0 &= \theta y - (\varphi + \nu)z. \end{aligned}\right\} \tag{21.8}$$

We solve the fourth equation for x and the fifth equation for y:

$$\left.\begin{aligned} x &= \frac{\nu + \theta}{\xi} y, \\ y &= \frac{\nu + \varphi}{\theta} z. \end{aligned}\right\} \tag{21.9}$$

We substitute the expression for x in (21.9) into the third equation of (21.8) and obtain

$$0 = uy - (\nu + \xi)\frac{\nu + \theta}{\xi} y. \tag{21.10}$$

The Disease-Free Equilibrium

One solution of (21.10) is $y^\circ = 0$. If we substitute this into (21.9) and the first or second equation in (21.8), we obtain

$$u^\circ = 1, \qquad x^\circ = y^\circ = z^\circ = 0. \tag{21.11}$$

This is the so-called *disease-free* equilibrium at which the whole population consists of susceptible individuals.

The Endemic Equilibrium

The second solution of (21.10) is

$$u^* = \frac{\nu + \xi}{\xi}(\nu + \theta). \tag{21.12}$$

We substitute y in (21.9) into the second equation of (21.8):

$$0 = \nu(1 - u^*) - u^* \frac{\nu + \varphi}{\theta} z^* + \varphi z^*.$$

We solve this equation for z^* in terms of u^*:

$$z^* = \frac{\theta \nu (1 - u^*)}{(\nu + \varphi) u^* - \varphi \theta}. \tag{21.13}$$

Notice from (21.12) that $u^* > \theta$. Hence $z^* > 0$ if $u^* < 1$, and $z^* = 0$ if $u^* = 1$. Once z^* has been determined we find x^*, y^* from (21.9). x^*, y^* are positive or zero, respectively, if z^* is positive or zero. Hence this equilibrium makes epidemiological sense if and only if $u^* \leqslant 1$. If $u^* = 1$, we again obtain the disease-free equilibrium. If $u^* < 1$, this second equilibrium is called the *endemic equilibrium*. Notice that there is no third equilibrium.

The Basic Replacement Ratio

Apparently, there is an endemic equilibrium only under certain circumstances. In order to understand these better we define the *basic replacement ratio* $\mathcal{R}_0$ by

$$\mathcal{R}_0 = \frac{1}{u^*}.$$

The name will be explained later. From (21.12) we obtain

$$\mathcal{R}_0 = \frac{\xi}{\nu + \xi} \frac{1}{\nu + \theta}. \tag{21.14}$$

We substitute (21.6) in order to obtain an expression in the original parameters:

$$\mathcal{R}_0 = \frac{\eta}{\mu + \eta} \frac{\sigma N}{\mu + \gamma}. \tag{21.15}$$

$\mathcal{R}_0$ plays an important role in the existence of an endemic equilibrium.

Theorem 21.2.

(a) *Let $\mathcal{R}_0 \leqslant 1$. Then there exists only one epidemiologically meaningful equilibrium of the system (21.5), namely the disease-free equilibrium $u^\circ = 1, x^\circ = y^\circ = z^\circ = 0$.*

(b) *Let $\mathcal{R}_0 > 1$. Then there exist two epidemiologically meaningful equilibria of the system (21.5), the disease-free equilibrium, and the endemic equilibrium*

$$u^* = \frac{1}{\mathcal{R}_0}, \qquad z^* = \frac{\theta \nu (\mathcal{R}_0 - 1)}{\nu + \varphi - \varphi \theta \mathcal{R}_0},$$

$$y^* = \frac{\nu + \varphi}{\theta} z^*, \qquad x^* = \frac{\nu + \theta}{\xi} y^*.$$

Interpretation

We now seek an interpretation of $\mathcal{R}_0$ which justifies the term *basic replacement ratio* and makes Theorem 21.2 plausible.

$\mu + \gamma$ is the rate at which individuals are leaving the infective class by recovering from the disease or by dying a natural death. So,

$$\tau_{\mathrm{I}} = \frac{1}{\mu + \gamma} \tag{21.16}$$

is the mean sojourn time spent in the infective class (during one infection cycle; cf. Sections 2.2 and 12.7). We call τ_{I} the mean infectious sojourn time. The mean infectious sojourn time has to be distinguished from the average duration $D_{\mathrm{I}} = 1/\gamma$ of the infectious period. The first takes death during the infectious period into account, the second does not. In most childhood diseases the two notions practically coincide, since the per capita mortality rate is very small compared with the rate of recovering from the disease.

$\mu + \eta$ is the rate at which individuals are leaving the exposed class by becoming infectious themselves or by dying a natural death. Hence

$$\tau_{\mathrm{E}} = \frac{1}{\mu + \gamma} \tag{21.17}$$

is the mean sojourn time spent in the exposed class. η is the per capita rate of leaving the exposed class by becoming infectious. Hence

$$P_{\mathrm{E}} = \frac{\eta}{\mu + \eta}.$$

is the probability of making it through the exposed class alive (cf. Sections 2.3 and 13.6). So we can rephrase $\mathcal{R}_0$ as

$$\mathcal{R}_0 = P_{\mathrm{E}} \tau_{\mathrm{I}} \sigma N, \tag{21.18}$$

with P_{E} being the probability of surviving the latent period and τ_{I} the mean infectious sojourn time. σN can be interpreted as the instantaneous rate at which one average infective individual infects others provided that the whole population (except him/herself) is susceptible. Hence $\tau_{\mathrm{I}} \sigma N$ is the average number of individuals to which one infective individual can pass the disease during the whole infectious period provided that the whole population is susceptible. Those being infected make it to the infectious period themselves with the probability P_{E} of surviving the latent period. So, $\mathcal{R}_0$ gives the average number of infective individuals by which one average infective individual replaces itself—provided that the whole population is susceptible. So, we term $\mathcal{R}_0$ the *basic replacement ratio*. In other words, $\mathcal{R}_0$ is the size of the second generation of infectives going back to one infective individual in an otherwise totally susceptible population. The term *ratio* reflects the fact that we consider the infective *offspring* of only one individual, the term *basic* reflects the fact that the whole population except that one infective individual is assumed to be susceptible.

We are now ready to make Theorem 21.2 plausible. If an average infective individual cannot replace itself by infecting others in a totally susceptible population, then the disease cannot establish itself and the disease-free equilibrium is the only possible equilibrium. If an average infective individual can replace itself by more than just one individual, then the disease can survive and an endemic equilibrium exists. So far we have proved these statements only as far as the equilibrium states are concerned. The dynamical picture will be addressed later.

Terminological Remarks

Dietz (1976) has termed $\mathcal{R}_0$ the *reproduction rate* of the infection. Dietz (1975) used this term before for an analogous entity in a model for arbovirus diseases. Hethcote (1976) used the term *replacement number* for the case where the population is not necessarily totally susceptible; for a totally susceptible population he used the term *contact number*. Anderson, May (1982b) employ *reproductive rate* in the case where the population is not totally susceptible; for a totally susceptible population they speak about *intrinsic reproductive rate*. At the Dahlem Conference (Berlin 1982), Anderson (1982) used *basic reproductive rate* for $\mathcal{R}_0$, whereas Dietz (1982) preferred *basic reproduction rate*. At the end of this workshop the importance of $\mathcal{R}_0$ was celebrated in a limerick (May, 1982):

> The deeper understanding Faust sought,
> Could not from the Devil be bought,
> But now we are told,
> By theorists bold,
> All we need to know is $\mathcal{R}_0$.

Finally, Diekmann, Heesterbeek, Metz (1990), in a universal approach to $\mathcal{R}_0$, coined the expression *basic reproduction ratio*.

I like *ratio* better than *number* because $\mathcal{R}_0$ gives the size of the second generation of infectives generated by one typical infective individual. Both ratio and number are more appropriate than rate (as is now generally accepted) because $\mathcal{R}_0$ is dimensionless. I prefer *replacement* over *reproduction* or even *reproductive*, because, in a model with varying population size, one would also like to speak about the basic reproduction ratio of the population, namely the number of offspring produced by an average individual during its total lifetime. Finally, I find *contact number* slightly confusing because not every contact leads to an infection even if it occurs between an infectious and a susceptible individual. *Contact number*, in the context of SEIR models, further ignores that $\mathcal{R}_0$ also incorporates the probability of surviving the latent period. So, I borrow *basic* from Dietz and Anderson, May, *replacement* from Hethcote, and *ratio* from Diekmann, Heesterbeek, Metz, and go with *basic replacement ratio.*

21.3 The Disease Dynamics in the Vicinities of the Disease-Free and the Endemic Equilibrium: Local Stability and the Interepidemic Period

Scaling and eliminating the equation for the susceptibles, we take the following equations as a starting point (cf. (21.5) and (21.7)):

$$\left.\begin{aligned} x' &= (1-x-y-z)y - (\xi+\nu)x, \\ y' &= \xi x - (\theta+\nu)y, \\ z' &= \theta y - (\varphi+\nu)z. \end{aligned}\right\} \tag{21.19}$$

We write

$$\boldsymbol{x} = \begin{pmatrix} x \\ y \\ z \end{pmatrix}$$

and recover (21.19) in the form (A.5) with

$$F(\boldsymbol{x}) = \begin{pmatrix} (1-x-y-z)y - (\xi+\nu)x \\ \xi x - (\theta+\nu)y \\ \theta y - (\varphi+\nu)z \end{pmatrix}.$$

We follow Toolbox A.4. Let $\boldsymbol{x}^\bullet = (x^\bullet, y^\bullet, x^\bullet)$ be an equilibrium. Then

$$\mathrm{D}F(\boldsymbol{x}^\bullet) = \begin{pmatrix} -(y^\bullet+\xi+\nu) & u^\bullet - y^\bullet & -y^\bullet \\ \xi & -(\theta+\nu) & 0 \\ 0 & \theta & -(\varphi+\nu) \end{pmatrix}. \tag{21.20}$$

Here we have set $u^\bullet = 1 - x^\bullet - y^\bullet - z^\bullet$.

The Disease-Free Equilibrium

The disease-free equilibrium is given by

$$u^\circ = 1, \qquad x^\circ = y^\circ = z^\circ = 0.$$

Hence

$$\mathrm{D}F(\boldsymbol{x}^\circ) = \begin{pmatrix} -(\xi+\nu) & 1 & 0 \\ \xi & -(\theta+\nu) & 0 \\ 0 & \theta & -(\varphi+\nu) \end{pmatrix}.$$

Since this matrix is of block from, we immediately identify one eigenvalue as

$$\lambda_1 = -(\varphi + \nu).$$

The other two eigenvalues coincide with the eigenvalues of the matrix

$$\begin{pmatrix} -(\xi + \nu) & 1 \\ \xi & -(\theta + \nu) \end{pmatrix}.$$

The trace of this matrix is negative, hence $\lambda_2 + \lambda_3 < 0$. From the determinant (which equals the product of the eigenvalues) we obtain that

$$\lambda_2\lambda_3 = (\xi + \nu)(\theta + \nu) - \xi = \xi\left(\frac{1}{\mathcal{R}_0} - 1\right)$$

(see (21.14)). Hence λ_1 and λ_2 are both strictly negative if $\mathcal{R}_0 < 1$. One eigenvalue is strictly positive if $\mathcal{R}_0 > 1$. We draw the following conclusion from Lemma A.28.

Theorem 21.3. *The disease-free equilibrium is locally asymptotically stable if $\mathcal{R}_0 < 1$ and unstable if $\mathcal{R}_0 > 1$.*

Recall that the endemic equilibrium exists if and only if $\mathcal{R}_0 > 1$. Hence the disease-free equilibrium is unstable if the endemic equilibrium exists.

The Endemic Equilibrium

Let $\boldsymbol{x}^* = (x^*, y^*, z^*)$ be the endemic equilibrium of (21.19) with $x^*, y^*, z^* > 0$ and $u^* = 1 - x^* - y^* - z^* > 0$. In order to determine the eigenvalues of $DF(\boldsymbol{x}^*)$ we calculate the characteristic polynomial

$$\begin{aligned}\det(\lambda \boldsymbol{I} - DF(\boldsymbol{x}^*)) = (y^* + \xi + \nu + \lambda)(\theta + \nu + \lambda)(\varphi + \nu + \lambda) \\ + y^*\xi\theta + (y^* - u^*)\xi(\varphi + \nu + \lambda).\end{aligned}$$

$\boldsymbol{I}$ denotes the identity matrix. A bit of algebra yields

$$\begin{aligned}&\det(\lambda \boldsymbol{I} - DF(\boldsymbol{x}^*)) \\ &\quad = \lambda^3 + \lambda^2(y^* + \xi + 3\nu + \theta + \varphi) \\ &\qquad + \lambda([y^* + \xi + \nu][\theta + \nu] + [\varphi + \nu][y^* + \xi + 2\nu + \theta] + [y^* - u^*]\xi) \\ &\qquad + (y^*[\xi + \theta + \nu] - u^*\xi + [\xi + \nu][\theta + \nu])(\varphi + \nu) + y^*\xi\theta.\end{aligned}$$

We recall (21.12): $\xi u^* = (\nu + \xi)(\nu + \theta)$. Hence the characteristic polynomial can be written as

$$\det(\lambda \boldsymbol{I} - DF(\boldsymbol{x}^*)) = \lambda^3 + a\lambda^2 + b\lambda + c \tag{21.21}$$

with

$$\left.\begin{aligned} a &= y^* + \xi + 3\nu + \theta + \varphi, \\ b &= y^*(\theta + \xi + \varphi + 2\nu) + (\varphi + \nu)(\theta + \xi + 2\nu), \\ c &= y^*((\theta + \xi + \nu)(\varphi + \nu) + \xi\theta). \end{aligned}\right\} \tag{21.22}$$

Local Asymptotic Stability of the Endemic Equilibrium

We notice that the coefficients of the characteristic polynomial satisfy $a, b, c > 0$ and $ab > c$ if $y^* > 0$, i.e., $\mathcal{R}_0 > 1$. Hence the Routh–Hurwitz criterion is satisfied (see Toolbox A.4). All roots of the characteristic polynomial—the eigenvalues of $DF(\boldsymbol{x}^*)$—have strictly negative real part. So, we can apply Lemma A.28.

Theorem 21.4. *Let the basic replacement number $\mathcal{R}_0$ strictly exceed* 1. *Then the endemic equilibrium is locally asymptotically stable.*

Damped Oscillations Around the Endemic Equilibrium

Though we already know that the endemic equilibrium is locally asymptotically stable, we are interested in the way in which the solutions converge to it, i.e., whether the endemic equilibrium is a node or rather a spiral point. In the latter case the solutions converge to the endemic equilibrium oscillating around it in a damped fashion. We look at the roots of the characteristic equation more closely in order to see whether some of them have nonzero imaginary part.

As $a, b, c > 0$, the characteristic polynomial has either three negative roots or one negative root λ_0 and two complex conjugate roots λ_+, λ_-.

If the endemic equilibrium exists, then

$$\mathcal{R}_0 = \frac{\xi}{\nu + \xi} \frac{1}{\nu + \theta} > 1.$$

This means that $\theta < 1$, where $1/\theta$ is the scaled length of the infectious period (death disregarded). Since, for childhood diseases, the length of the infectious period is typically very short compared with the life expectancy $L = 1/\nu$, we have that ν/θ is very small and hence $\nu \ll 1$ is very small.

We assume that the length of the immune period is of the same order of magnitude as the life expectancy, i.e., $\nu + \varphi \ll 1$.

In the limiting case $\nu = \varphi = 0$, we have $y^* = 0$ and so $b = 0$ and $c = 0$, and the characteristic polynomial takes the form

$$\lambda^3 + (\xi + \theta)\lambda^2 = 0.$$

Thus it has the double root 0 and the root $\lambda_0 = -(\xi + \theta)$. In order to see what happens to the double root 0 if $\nu + \varphi > 0$ is small, we set

$$\lambda = \epsilon w, \quad \epsilon = (\nu + \varphi)^{1/2}. \tag{21.23}$$

This is motivated by the theory of Puiseux series (Kato, 1976, Chapter 2, Section 1.2). From Theorem 21.2 (b) we remember that

$$y^* = \epsilon^2 \frac{z^*}{\theta}. \tag{21.24}$$

The characteristic equation then takes the form

$$\epsilon w^3 + \tilde{a} w^2 + \epsilon \tilde{b} w + \tilde{c} = 0 \tag{21.25}$$

with

$$\left.\begin{aligned} \tilde{a} &= a = \xi + \theta + \epsilon^2\left(\frac{z^*}{\theta} + 1 + 2\tilde{\nu}\right), \quad \tilde{\nu} = \frac{\nu}{\nu + \varphi} < 1, \\ \tilde{b} &= \left(\frac{z^*}{\theta} + 1\right)(\theta + \xi) + \epsilon^2\left(\frac{z^*}{\theta}(1 + \tilde{\nu}) + 2\tilde{\nu}\right), \\ \tilde{c} &= z^*\xi + \epsilon^2 \frac{z^*}{\theta}(\theta + \xi + \nu). \end{aligned}\right\} \tag{21.26}$$

For $\epsilon = 0$, (21.25) boils down to

$$\tilde{a} w^2 + \tilde{c} = 0,$$

which is solved by

$$w_\pm = \pm \mathrm{i}\tilde{\omega}, \quad \tilde{\omega} = \left(\frac{\tilde{c}}{\tilde{a}}\right)^{1/2} = \left(\frac{z^*\xi}{\xi + \theta}\right)^{1/2} + O(\epsilon^2). \tag{21.27}$$

The expression $O(\epsilon^2)$ stands for a term whose absolute value can be estimated by a constant times ϵ^2. We now look for the solution to (21.25) in the form

$$w = w_\pm + \epsilon w_1. \tag{21.28}$$

This yields

$$(w_\pm)^2 + 2\tilde{a} w_1 + \tilde{b} = O(\epsilon).$$

Hence

$$w_1 = -\frac{\tilde{b} - \tilde{\omega}^2}{2\tilde{a}} + O(\epsilon) = -\frac{1}{2}\left(\frac{z^*}{\theta} + 1 - \frac{z^*\xi}{(\xi + \theta)^2}\right) + O(\epsilon). \tag{21.29}$$

We combine (21.23), (21.27), (21.28), and (21.29), and obtain the following expression for the three roots $\lambda = \lambda_{\pm}, \lambda_0$ of the characteristic polynomial:

$$\lambda_{\pm} = \epsilon\left(\pm \mathrm{i}\left(\frac{z^*\xi}{\xi+\theta}\right)^{1/2} - \epsilon\frac{1}{2}\left(\frac{z^*}{\theta} + 1 - \frac{z^*\xi}{(\xi+\theta)^2}\right)\right) + O(\epsilon^3),$$
$$\lambda_0 = -(\xi+\theta) + O(\nu+\varphi).$$

The expression for λ_0 is found from the fact that $-(\xi+\theta)$ is a root for $\nu+\varphi = 0$ and from the implicit function theorem. We recall (21.24) again:

$$\left.\begin{aligned} \lambda_{\pm} &= \pm \mathrm{i}\left(\frac{y^*\xi\theta}{\xi+\theta}\right)^{1/2} - \frac{1}{2}\left(y^*\left(1 - \frac{\theta\xi}{(\xi+\theta)^2}\right) + \nu + \varphi\right) \\ &\qquad + O((\nu+\varphi)^{3/2}), \\ \lambda_0 &= -(\xi+\theta) + O(\nu+\varphi). \end{aligned}\right\} \tag{21.30}$$

For the case $\varphi = 0$, these formulas have been derived by Dietz (1976) and, more rigorously, by Schwartz, Smith (1983). In comparing the formulas here to those in Schwartz, Smith (1983, Lemma 1.1), one should notice that there z^* has been expanded with respect to ϵ. We prefer the representation (21.30) because, by following Dietz (1976), it will help us to find our way back to the real world.

In the vicinity of the endemic equilibrium, the solutions to (21.19) approximately behave as

$$\boldsymbol{x}(t) = \boldsymbol{x}^* + \mathrm{e}^{-\delta t}(\cos(\omega t)\boldsymbol{a} + \sin(\omega t)\boldsymbol{b}) + \mathrm{e}^{\lambda_0 t}\boldsymbol{c}$$

with appropriate vectors $\boldsymbol{a}$, $\boldsymbol{b}$, $\boldsymbol{c}$ and

$$\left.\begin{aligned} \delta &= \frac{1}{2}\left(y^*\left(1 - \frac{\theta\xi}{(\xi+\theta)^2}\right) + \nu + \varphi\right) + O((\nu+\varphi)^{3/2}), \\ \omega &= \left(\frac{y^*\xi\theta}{\xi+\theta}\right)^{1/2} + O((\nu+\varphi)^{3/2}), \\ \lambda_0 &= -(\xi+\theta) + O(\nu+\varphi). \end{aligned}\right\} \tag{21.31}$$

The Length of the Interepidemic Period

As we have argued above, in childhood diseases, $\nu + \varphi$ is typically very small, hence δ is considerably smaller than $|\lambda_0|$, so $\mathrm{e}^{-\delta}$ is the relevant damping factor of the solution.

The function

$$\mathrm{e}^{-\delta t}\Big(\cos(\omega t)\boldsymbol{a} + \sin(\omega t)\boldsymbol{b}\Big)$$

has the *quasi-period*

$$T = \frac{2\pi}{\omega}.$$

Here *quasi-period* is defined to be the length of the time intervals between two subsequent local maxima (or minima) in cases where these lengths turn out to be the same for all pairs of subsequent local maxima. Since this part of the solution dominates the other, we can approximate the length of the *interepidemic period*—the time span between two peaks of the disease prevalence—by the quasi-period. We fit ω into the formula for T:

$$T \approx 2\pi \left(\frac{\xi + \theta}{y^*\xi\theta}\right)^{1/2}. \tag{21.32}$$

We recall that

$$D_{\rm E} = \frac{1}{\xi}, \qquad D_{\rm I} = \frac{1}{\theta} \tag{21.33}$$

are the average durations of the latent and infectious periods. As has been observed by Dietz (1975),

$$D_{\rm S} = \frac{1}{y^*} \tag{21.34}$$

can be interpreted as the *average duration of the susceptible period*—or the *expected age at first infection*—provided the disease dynamics operate close to the endemic equilibrium. With this notation we obtain the following formula (cf. Dietz 1976; Anderson, May 1982b):

$$T \approx 2\pi (D_{\rm S}(D_{\rm E} + D_{\rm I})))^{1/2}. \tag{21.35}$$

Indeed, if the disease dynamics operate at the endemic equilibrium, the dynamics of a cohort C of susceptibles—a group of susceptibles that entered the susceptible class at the same time—can be described (in scaled time) by

$$C' = -(y^* + \nu)C.$$

Compare the equation for u' in (21.5). If we neglect those who die, we obtain $C' = -y^*C$. This means that the probability of still being susceptible at age a is given by

$$\frac{C(a)}{C(0)} = \mathrm{e}^{-y^*a},$$

i.e., the length of the susceptibility period is exponentially distributed and its average duration (in scaled time) is $1/y^*$. As such a cohort can well consist of newborn individuals, we can identify $D_{\rm S}$ with the expected age at first infection.

For many childhood diseases, formula (21.35) leads to a surprisingly good agreement between calculated and observed values for the interepidemic period (see Anderson, May 1991, Table 6.1).

It may also be interesting to know how long it takes for the amplitude of the damped oscillations to become half its previous size. This time H—let us call it *half-value time* in analogy to other processes—is approximately given by

$$\tfrac{1}{2} = \mathrm{e}^{-\delta H},$$

i.e., by (21.31),

$$H = 2 \ln 2 \frac{D_\mathrm{S}}{1 - (D_\mathrm{E} D_\mathrm{I}/(D_\mathrm{E} + D_\mathrm{I})^2) + (D_\mathrm{S}/\tau_\mathrm{R})}. \tag{21.36}$$

Here

$$\tau_\mathrm{R} = \frac{1}{\nu + \varphi} \tag{21.37}$$

is the mean immune sojourn time, i.e., the average time that recovered individuals are immune (and alive). If immunity is permanent, τ_R coincides with the life expectancy.

The Special Case of Permanent Immunity

The basic replacement number $\mathcal{R}_0$ is a key epidemiological parameter because it indicates the effort one has to make to eradicate the disease. Provided that the disease dynamics operate at the endemic equilibrium, one could use the relation

$$\mathcal{R}_0 = \frac{1}{u^*} \tag{21.38}$$

to estimate $\mathcal{R}_0$. But the size of the fraction of susceptibles seems difficult to measure. In the case of permanent immunity, $\varphi = 0$, Theorem 21.2 (b) displays the simple relationship

$$y^* = \nu(\mathcal{R}_0 - 1).$$

We recall that $D_\mathrm{S} = 1/y^*$ is the average duration of the susceptible period, in this case the expected age at infection, if the disease dynamics operate at the endemic equilibrium, and that

$$L = \frac{1}{\nu}$$

is the life expectancy. We solve for $\mathcal{R}_0$:

$$\mathcal{R}_0 = 1 + \frac{L}{D_\mathrm{S}}. \tag{21.39}$$

This formula will be revisited in the context of age-structured endemic models (Section 22.5).

21.4 Some Global Results: Extinction, Persistence of the Disease; Conditions for Attraction to the Endemic Equilibrium

In Section 21.2 we have identified the basic replacement ratio $\mathcal{R}_0$ and studied its relation with the existence of an *endemic* equilibrium state: if $\mathcal{R}_0 \leqslant 1$, the disease dynamics have only one equilibrium state, the disease-free equilibrium; if $\mathcal{R}_0 > 1$, there is also a unique endemic equilibrium, i.e., an equilibrium state at which the disease is present in the population. This does not really give information about the dynamic behavior of the disease in these two cases. Intuition tells us that the disease dies out if $\mathcal{R}_0 \leqslant 1$, because an average infective individual is only able to replace itself if everybody else is susceptible. If $\mathcal{R}_0 > 1$, the disease should be able to persist in the population.

Various Forms of Disease Persistence

In this context one would say that the disease *persists weakly* in the population if the infective part does not vanish asymptotically, i.e.,

$$y^\infty = \limsup_{t\to\infty} y(t) > 0$$

for every solution with $x(0) + y(0) > 0$. Though the disease does not die out if it persists weakly, it may get arbitrarily close to extinction every now and then, and the odds are that stochastic effects (which are not incorporated in our model) will finally extinguish it. This cannot happen so easily if the infective fraction of the population is bounded away from 0 for large times, i.e.,

$$y_\infty = \liminf_{t\to\infty} y(t) > 0.$$

In this case the disease is called *strongly persistent.*

Uniform strong disease persistence means that $y_\infty > \epsilon > 0$, where ϵ does not depend on the initial data, provided that $x(0) + y(0) > 0$, i.e., the disease is actually present initially.

A useful technical concept is *uniform weak persistence*, which means $y^\infty > \epsilon > 0$, where ϵ does not depend on the initial data, provided that $x(0) + y(0) > 0$.

Remember that, after appropriate scaling, our epidemic model takes the form

$$\left.\begin{aligned}
1 &= u + x + y + z,\\
u' &= \nu(1-u) - uy + \varphi z,\\
x' &= uy - (\xi + \nu)x,\\
y' &= \xi x - (\theta + \nu)y,\\
z' &= \theta y - (\varphi + \nu)z.
\end{aligned}\right\} \tag{21.40}$$

We introduce the new dependent variable

$$v = x + \alpha y, \tag{21.41}$$

where α will be determined appropriately (cf. Simon, Jacquez, 1992). Then

$$v' = uy - (\xi + \nu)x + \alpha[\xi x - (\theta + \nu)y] = y[u - \alpha(\theta + \nu)] + (\alpha\xi - \xi - \nu)x.$$

In order to eliminate x from the equation, we now choose

$$\alpha = \frac{\xi + \nu}{\xi}$$

and obtain

$$v' = y\left(u - \frac{1}{\mathcal{R}_0}\right),$$

where

$$\mathcal{R}_0 = \frac{\xi}{(\xi + \nu)(\theta + \nu)}$$

is the basic replacement ratio expressed in the scaled parameters (see (21.14)). We now rewrite system (21.40) in the dependent variables u, v, y, z:

$$\left.\begin{aligned} 1 &= u + v - \frac{\nu}{\xi}y + z, \\ u' &= \nu(1 - u) - uy + \varphi z, \\ v' &= y\left(u - \frac{1}{\mathcal{R}_0}\right), \\ y' &= \xi v - (2\nu + \xi + \theta)y, \\ z' &= \theta y - (\nu + \varphi)z. \end{aligned}\right\} \tag{21.42}$$

Furthermore,

$$0 \leqslant x + y \leqslant v \leqslant 1 + \frac{\nu}{\xi}, \tag{21.43}$$

because u, x, y, z are nonnegative and add up to 1.

The following lemma is crucial for the following discussion. Recall that v^∞, v_∞, etc., are the limit superior (inferior) of $v(t)$ for $t \to \infty$.

Lemma 21.5.

(a)

$$y^\infty \leqslant \frac{\xi}{\nu + \theta}x^\infty, \qquad z^\infty \leqslant \frac{\theta}{\nu + \varphi}y^\infty,$$

$$y^\infty \leqslant \frac{\xi}{2\nu + \xi + \theta}v^\infty, \qquad x^\infty \leqslant \frac{1}{\nu + \xi}y^\infty.$$

(b)

$$y_\infty \geqslant \frac{\xi}{\nu+\theta} x_\infty, \qquad z_\infty \geqslant \frac{\theta}{\nu+\varphi} y_\infty, \qquad y_\infty \geqslant \frac{\xi}{2\nu+\xi+\theta} v_\infty.$$

In particular, if x or v converge for $t \to \infty$, so do y, z, u.

Proof. We only show the first estimate in (a). The others are proved in exactly the same way.

We apply the method of fluctuations in Toolbox A.3. By Proposition A.22 there exists a sequence $t_n \to \infty$ $(n \to \infty)$ such that

$$y(t_n) \to y^\infty, \quad y'(t_n) \to 0, \quad n \to \infty.$$

From the y-equation in (21.40),

$$0 \leftarrow y'(t_n) = \xi x(t_n) - (\theta + \nu) y(t_n).$$

Hence

$$(\theta + \nu) y^\infty \leqslant \xi x^\infty.$$

□

Extinction of the Disease

We show that the disease dies out if the basic replacement ratio is smaller than 1.

Theorem 21.6. *Let $\mathcal{R}_0 \leqslant 1$. Then the disease dies out in the population, i.e., the solutions u, x, y, z of system (21.40) converge towards the disease-free equilibrium $u^\circ = 1$, $x^\circ = y^\circ = z^\circ = 0$ as time tends to infinity.*

Proof. Let $\mathcal{R}_0 \leqslant 1$. Since $u \leqslant 1$ and $y \geqslant 0$, we have from the v-equation in (21.42) that $v' \leqslant 0$. Thus $v(t)$ converges for $t \to \infty$. It follows from Lemma 21.5 that y, z, u converge as well. By Corollary A.19, the limits form an equilibrium of (21.42) with nonnegative coordinates. As the disease-free equilibrium is the only one with nonnegative coordinates if $\mathcal{R}_0 \leqslant 1$, the assertion follows. □

Uniform Strong Persistence of the Disease

After some preparation, we will obtain that the disease uniformly strongly persists in the population if the basic replacement ratio is larger than 1.

Lemma 21.7. *Let $\mathcal{R}_0 > 1$. If $v(0) > 0$, $v(t)$ does not converge to 0 as $t \to \infty$ and $u_\infty \leqslant 1/\mathcal{R}_0$.*

Proof. Suppose that $v(t) \to 0$, $t \to \infty$. By Lemma A.25, we can find a sequence $t_n \to \infty$ with $v(t_n) \to 0$, $n \to \infty$, $v'(t_n) < 0$. If $v(t)$ tends to zero, so do $y(t)$, $z(t)$ by Lemma 21.5. It follows from the u-equation in (21.42) that $u(t) \to 1$, $t \to \infty$. As $\mathcal{R}_0 > 1$, $u(t_n) - (1/\mathcal{R}_0) > 0$ for sufficiently large n, i.e., $v'(t_n) \geqslant 0$, in contradiction to a previous result.

If $u_\infty > 1/\mathcal{R}_0$, then $v'(t) \geqslant 0$ for sufficiently large t and $v(t)$ converges for $t \to \infty$ and so do y, z, u by Lemma 21.5. As the system does not converge towards the disease-free equilibrium, it converges toward the endemic equilibrium. In particular, $u_\infty \leqslant 1/\mathcal{R}_0$, a contradiction. □

Actually, it is possible to find a lower bound for v^∞ which does not depend on the initial data.

Lemma 21.8. *Let $\mathcal{R}_0 > 1$. If $v(0) > 0$,*

$$v^\infty \geqslant x^\infty \geqslant \left(1 + \frac{\xi}{\nu + \theta} + \frac{\theta}{\nu}\frac{\xi}{\nu + \theta}\right)^{-1}\left(1 - \frac{1}{\mathcal{R}_0}\right) =: \epsilon_0.$$

Notice that $\epsilon_0 > 0$ does not depend on φ because $\mathcal{R}_0$ does not.

Proof. We combine the estimates in Lemma 21.5:

$$x^\infty + y^\infty + z^\infty \leqslant \left(1 + \frac{\xi}{\nu + \theta} + \frac{\theta}{\nu}\frac{\xi}{\nu + \theta}\right)x^\infty.$$

By Lemma 21.7,

$$x^\infty + y^\infty + z^\infty \geqslant (1 - u)^\infty = 1 - u_\infty \geqslant 1 - \frac{1}{\mathcal{R}_0}.$$

The last inequality follows from Lemma 21.7. The statement now follows by combining the two inequalities. □

Proposition 21.9. *Let $\mathcal{R}_0 > 1$. Then there exists some $\epsilon > 0$ such that $x^\infty, y^\infty, z^\infty \geqslant \epsilon$ for all solutions of (21.40) with $u(0), x(0), y(0), z(0) \geqslant 0$, $u(0) + x(0) + y(0) + z(0) = 1$, $x(0) + y(0) > 0$.*

In particular, the disease persists uniformly weakly in the population.

Proof. By definition of v, $v(0) > 0$. The assertion now follows from Lemmas 21.8 and 21.5. □

Proposition 21.10. *Let $\mathcal{R}_0 > 1$. Then there exists some $\epsilon > 0$ such that $y_\infty, z_\infty \geqslant \epsilon$ for all solutions of (21.40) with $u(0), x(0), y(0), z(0) \geqslant 0$, $u(0) + x(0) + y(0) + z(0) = 1$, $x(0) + y(0) > 0$.*

Proof. We apply the persistence theory in Toolbox A.5. The set $u, x, y, z \geqslant 0$, $u+x+y+z = 1$ is forward invariant for the solutions of (21.40). Equivalently, the set $x, y, z \geqslant 0, x+y+z \leqslant 1$ is forward invariant for the solutions of (21.7). Furthermore, if $x(0) > 0$, then $x(t) > 0$ for all $t \geqslant 0$. This means that the solutions of (21.7) induce a semiflow on

$$X = \{(x, y, z) \in \mathbb{R}^3_+;\ x > 0,\ y \geqslant 0,\ z \geqslant 0,\ x + y + z \leqslant 1\}.$$

Set

$$\rho(x, y, z) = x.$$

Then the semiflow induced by our system on X is uniformly weakly ρ-persistent by Proposition 21.9. Furthermore, the intersections of ρ-rings $\{\delta \leqslant x \leqslant \epsilon\}$ with X are compact in X. Hence the compactness condition (C) in Theorem A.32 is satisfied with $B = X$. Thus the semiflow is uniformly strongly ρ-persistent by Theorem A.32. This implies that there exists some $\epsilon > 0$ such that $x_\infty \geqslant \epsilon$ for all solutions with $x(0) > 0$. Hence $v_\infty \geqslant \epsilon$. Now apply Lemma 21.5 (b) to obtain the statements for z_∞ and y_∞.

The case where $y(0) > 0$ is handled analogously with $\rho(x, y, z) = y$. This yields $y_\infty \geqslant \epsilon > 0$ with ϵ not depending on the initial data. The statement concerning z_∞ follows from Lemma 21.5 (b). □

A Global Attraction Result for the Endemic Equilibrium

Theorem 21.11. *Let $\mathcal{R}_0 > 1$, $x(0) + y(0) > 0$. Then the disease dynamics converge towards the endemic equilibrium, as time tends to infinity, provided that*

$$(\nu + \xi + \theta)(\nu + \varphi) > \theta\xi.$$

Proof. It follows from Lemma 21.5 that the solution of (21.42) converges if v converges. By Corollary A.19, the limit is an equilibrium and, by Proposition 21.9, the endemic equilibrium.

Let us suppose that $v(t)$ does not converge as time tends to infinity. By Lemma A.20 we find a sequence $t_n \to \infty$, $n \to \infty$, such that $v(t_n) \to v^\infty$, $u(t_n) = 1/\mathcal{R}_0 = u^*$. It follows from the first equation in (21.42) that

$$v^\infty \leqslant 1 - u^* + \frac{\nu}{\xi} y^\infty - z_\infty.$$

By Lemma 21.5,

$$v^\infty \leqslant 1 - u^* + \frac{\nu}{\xi} \frac{\xi}{2\nu + \xi + \theta} v^\infty - \frac{\theta}{\nu + \varphi} \frac{\xi}{2\nu + \xi + \theta} v_\infty.$$

We rearrange terms:

$$\frac{\nu + \xi + \theta}{2\nu + \xi + \theta} v^\infty \leqslant 1 - u^* - \frac{\theta}{\nu + \varphi} \frac{\xi}{2\nu + \xi + \theta} v_\infty.$$

Similarly, we can derive the following estimate:

$$\frac{\nu+\xi+\theta}{2\nu+\xi+\theta}v_\infty \geqslant 1-u^* - \frac{\theta}{\nu+\varphi}\frac{\xi}{2\nu+\xi+\theta}v^\infty.$$

We subtract the two inequalities and simplify:

$$(\nu+\xi+\theta)(v^\infty - v_\infty) \leqslant \frac{\theta\xi}{\nu+\varphi}(v^\infty - v_\infty).$$

By assumption,

$$\nu+\xi+\theta > \frac{\theta\xi}{\nu+\varphi},$$

and it follows that $v^\infty = v_\infty$, a contradiction, because we have supposed that v does not converge. □

In particular, we have convergence towards the endemic equilibrium if the condition in Theorem 21.11 is satisfied for $\nu = 0$. For this case the condition can be rewritten as

$$\frac{1}{\theta}+\frac{1}{\xi} \geqslant \frac{1}{\varphi}.$$

We recall that

$$D_{\mathrm{E}} = \frac{1}{\xi}, \qquad D_{\mathrm{I}} = \frac{1}{\theta}, \qquad D_{\mathrm{R}} = \frac{1}{\varphi}$$

are the average durations of the latent, infectious, and immune period, respectively.

Corollary 21.12. *Let $\mathcal{R}_0 > 1$, $x(0)+y(0) > 0$. Then the disease dynamics converge towards the endemic equilibrium, as time tends to infinity, provided that the average duration of the immune period is smaller than the sum of the average durations of the latent and the infectious periods.*

For almost all childhood diseases this condition is not satisfied. If immunity is permanent, which seems to hold for most childhood diseases, convergence towards the endemic equilibrium has been shown by Li, Muldowney (1995).

Theorem 21.13. *Let $\mathcal{R}_0 > 1$, $x(0)+y(0) > 0$, $\varphi = 0$. Then the disease dynamics converge towards the endemic equilibrium, as time tends to infinity.*

We sketch the proof (for details see Li, Muldowney (1995)). Let

$$X = \{(u,x,y,z);\ u,x,y,z \geqslant 0,\ x+y>0,\ u+x+y+z=1\}.$$

X is forward invariant under the solutions of (21.40). By Proposition 21.10 all ω-limit sets of solutions starting in X are contained in X. System (21.40), for $\varphi = 0$, becomes a competitive three-dimensional ordinary differential equation system after dropping the equation for z' and replacing x by

$-x$. (See Toolbox A.2 for the definition of a competitive system.) Since three-dimensional competitive ordinary differential equation systems have the Poincaré–Bendixson property (see Hirsch, 1990; Smith, 1995, Theorem 4.1), all minimal sets (see Toolbox A.6) are equilibria or periodic orbits of (21.40). Using compound matrices (Proposition A.39, taken from Muldowney (1990)), one can show that all periodic orbits are locally asymptotically stable (Li, Muldowney, 1995, Theorem 3.2). Since the endemic equilibrium is locally asymptotically stable by Theorem 3.2 and is the only equilibrium in X, we have that all minimal sets are locally asymptotically stable. By Theorem A.37, there exists a globally asymptotically stable minimal subset of X which, of course, is the only minimal set in X. Since the endemic equilibrium is a minimal set, it coincides with this globally asymptotically stable minimal set.

The compound matrix approach also provides that the condition in Theorem 21.11 can be replaced either by $\phi + \theta + \nu > \xi$ (sufficiently short immune or infectious periods, compared to the latent period) or by another, more complicated, condition which covers the case of very long immune periods (Li, Muldowney, van den Driessche, 1999). This suggests that the disease dynamics converge to the endemic equilibrium for immune periods of arbitrary length. A proof is still elusive though.

Exercises

21.1. Model the spread of an infectious disease in an endemic situation by dividing a population of size N into susceptible individuals, S, infective individuals, I, and recovered individuals, R. The transition between the various parts can be schematically described by

$$S \to I \to R\ (\to S).$$

Assume that the per capita birth rate equals the per capita mortality rate, μ. Assume that the disease is nonlethal. Assume that the various per capita rates do not depend on age or time.

(a) Formulate the model. Identify a basic replacement ratio $\mathcal{R}_0$ and establish its relation to the existence of an endemic equilibrium.

(b) Show that if $\mathcal{R}_0 \leqslant 1$, then all solutions with nonnegative initial data converge to the disease-free equilibrium.

(c) Show that if $\mathcal{R}_0 > 1$, then there exists a uniquely determined endemic equilibrium and all solutions with nonnegative initial data and $I(0) > 0$ converge to the endemic equilibrium.

Hint: see Toolbox A.2 (cf. Chapters 10 and 11).

21.2. Same as 1, but divide the population of size N into susceptible individuals, S, exposed individuals, E, and infective individuals, I. The transition between the various parts can be schematically described by

$$S \to E \to I \ (\to S).$$

Bibliographic Remarks

The agreement between the quasi-periods of the damped oscillations found in Section 21.3 and the interepidemic periods observed in childhood diseases can be improved if the exponential length distributions of the latent and the infectious periods are replaced by more realistic distributions (Feng, Thieme, 2000b). See our discussion in Section 12.7. Then, in formula (21.35), the expected duration of the infectious period, D_{I}, needs to be replaced by the average expectation of remaining duration (cf. Section 12.4).

Many possible mechanisms have been suggested that may cause or contribute to the damped oscillations observed in this model being tipped into the undamped oscillations (recurrent outbreaks), which are observed in measles and other childhood diseases:

- seasonal variation in the per capita infection rate (London, Yorke, 1973a; Grossman et al., 1977; Grossman, 1980; Schwartz, Smith, 1983; Smith, 1983a,b; Schaffer, 1985; Olsen et al., 1988);
- interaction of seasonal variation and age-structure (Schenzle, 1984);
- stochastic fluctuations (Bartlett, 1957, 1960; London, Yorke, 1973b; Nåsell, n.d.);
- Allee type effects in the infection rate (Liu et al., 1987);
- dose-dependent lengths of the latent period (Liu, 1993); and
- isolation of infectives with symptoms (Feng, Thieme, 1995, 2000b; Feng, 1999; Wu, Feng, 2000; Hethcote et al., 2002).

See Hethcote et al. (1981), Dietz, Schenzle (1985), Hethcote, Levin (1989), Anderson, May (1991, Section 6.5), and Liu (1995) for a further discussion of mechanisms that can generate undamped oscillations in endemic models.

The analysis of the global large-time behavior of the SEIR(S) model becomes much easier if one class (E, for example) is dropped. See Capasso (1993), which is also a good source for models where the incidence σSI is replaced by $\sigma S^{\alpha} I^{\beta}$, $\alpha, \beta > 0$. Actually, the Li, Muldowney (1995) global stability result also holds for this type of incidence provided that $\beta \leqslant 1$. It also holds if vertical transmission is added to the model (Li et al., 2001).

The model analysis becomes somewhat different if disease fatalities and more complex vital dynamics are included. Since the population size is no longer constant, the choice of the right contact function again becomes an issue (cf. Chapter 19). There are many various modeling possibilities; see Hethcote (1994), for example, for a survey which makes the point that there are more than 1001 possible epidemic (or rather endemic) models. Among other things, new sources of undamped disease oscillations appear (e.g., Pugliese, 1990; Diekmann, Kretzschmar, 1991; Gao et al., 1995).

Chapter Twenty-Two

Age-Structured Models for Endemic Diseases and Optimal Vaccination Strategies

Vaccination is an important way of controlling the spread of infectious diseases, both in epidemic outbreaks and in endemic situations. While in an epidemic outbreak it is important to vaccinate as many individuals as possible as fast as possible, in an endemic situation it is a relevant question to determine at what age or ages individuals are most effectively vaccinated. The question of optimal vaccination ages cannot be answered, of course, without incorporating age-structure into the model. In the following, we formulate an age-structured SEIR model with so-called separable mixing in the age classes and with vaccination, and we relate the existence of an endemic equilibrium and the stability of the disease-free equilibrium to the size of the *net replacement ratio* (we say *net* rather than *basic* because we include vaccination). For disease dynamics that are much faster than the demographics, like those of most childhood diseases, we derive approximate formulas for the net replacement ratio which relate it to the *average duration of susceptibility* and the *average age at infection*. Finally, we show that optimal vaccination strategies (provided that they are subject to only one constraint, like a maximum amount of money that can be spent) can be found among one-age or two-age strategies where individuals are immunized at one or two specific ages.

22.1 A Model with Chronological Age-Structure

We make some simplifying assumptions. We first assume that the disease causes so few fatalities that they can be neglected. Furthermore, we assume that the population has reached an equilibrium state and that the population neither increases nor decreases. We assume that the disease confers permanent immunity and that there is no vertical transmission of the disease. As for the vaccination, we assume that susceptible and recovered individuals are vaccinated indiscriminately and that the vaccination provides permanent immunity to the disease.

Let $S(t, a)$, $E(t, a)$, $I(t, a)$, $R(t, a)$ denote the respective age-densities of the susceptible, infective, and recovered parts of the population, where a denotes age and t denotes time. Then the dynamics of the various parts of the population are described by the following partial differential equations and boundary conditions (see Section 13.2).

$$\left.\begin{aligned}
\frac{\partial}{\partial t}S(t,a)+\frac{\partial}{\partial a}S(t,a)&=-(\mu(a)+v(a))S(t,a)-S(t,a)\phi(t,a),\\
\frac{\partial}{\partial t}E(t,a)+\frac{\partial}{\partial a}E(t,a)&=-(\mu(a)+\eta(a))E(t,a)+S(t,a)\phi(t,a),\\
\frac{\partial}{\partial t}I(t,a)+\frac{\partial}{\partial a}I(t,a)&=-(\mu(a)+\gamma(a))I(t,a)+\eta(a)E(t,a),\\
\frac{\partial}{\partial t}R(t,a)+\frac{\partial}{\partial a}R(t,a)&=-(\mu(a)+v(a))R(t,a)+\gamma(a)I(t,a),\\
S(t,0)=B,\qquad E(t,0)=0,\qquad &I(t,0)=0,\qquad R(t,0)=0,\\
\phi(t,a)=\int_0^\infty &k(a,b)I(t,b)\,\mathrm{d}b.
\end{aligned}\right\}\tag{22.1}$$

In addition, we have initial conditions

$$\begin{aligned}
S(0,a)&=S_0(a), & E(0,a)&=E_0(a),\\
I(0,a)&=I_0(a), & R(0,a)&=R_0(a).
\end{aligned}$$

As usual, $\mu(a)$ is the age-dependent per capita mortality rate, while $v(a)$ is the age-dependent per capita vaccination rate. The function ϕ is the per capita rate of susceptibles to be infected at age a at time t. The integral kernel $k(a,b)$ involves the contact rate between individuals of age a and age b as well as the respective transmission probabilities. $\eta(a)$ is the age-dependent rate at which individuals leave the exposed class and $\gamma(a)$ is the age-dependent per capita rate at which individuals recover from the disease. B is the constant influx of newborn individuals into the susceptible part of the population.

In order to have a terminology that is both more general and simpler, we introduce the probabilities $\mathcal{F}(a)$ of being still alive at age a, the probabilities $V(a)$ of still not being vaccinated at age a, the probabilities $H(a)$ of still being in the exposed class at age a in case one was infected directly at birth, and the probabilities $G(a)$ of still being infective at age a if one was already infective at birth:

$$\left.\begin{aligned}
\mathcal{F}(a)&=\exp\biggl(-\int_0^a\mu(b)\,\mathrm{d}b\biggr),\\
V(a)&=\exp\biggl(-\int_0^a v(b)\,\mathrm{d}b\biggr),\\
G(a)&=\exp\biggl(-\int_0^a\gamma(b)\,\mathrm{d}b\biggr),\\
H(a)&=\exp\biggl(-\int_0^a\eta(b)\,\mathrm{d}b\biggr).
\end{aligned}\right\}\tag{22.2}$$

Furthermore, we introduce

$$u(t,a) = \frac{S(t,a)}{\mathcal{F}(a)V(a)}, \qquad x(t,a) = \frac{E(t,a)}{\mathcal{F}(a)H(a)}, \qquad y(t,a) = \frac{I(t,a)}{\mathcal{F}(a)G(a)}.$$

Then

$$\begin{aligned}
&\frac{\partial}{\partial t}u(t,a) + \frac{\partial}{\partial a}u(t,a) \\
&\quad = \left(\frac{\partial}{\partial t}S(t,a) + \frac{\partial}{\partial a}S(t,a)\right)\frac{1}{\mathcal{F}(a)V(a)} + (\mu(a)+\upsilon(a))\frac{S(t,a)}{\mathcal{F}(a)V(a)} \\
&\quad = -u(t,a)\phi(t,a).
\end{aligned}$$

Furthermore,

$$\begin{aligned}
&\frac{\partial}{\partial t}x(t,a) + \frac{\partial}{\partial a}x(t,a) \\
&\quad = \left(\frac{\partial}{\partial t}E(t,a) + \frac{\partial}{\partial a}E(t,a)\right)\frac{1}{\mathcal{F}(a)H(a)} + (\mu(a)+\eta(a))\frac{E(t,a)}{\mathcal{F}(a)H(a)} \\
&\quad = u(t,a)\frac{V(a)}{H(a)}\phi(t,a).
\end{aligned}$$

Finally,

$$\begin{aligned}
&\frac{\partial}{\partial t}y(t,a) + \frac{\partial}{\partial a}y(t,a) \\
&\quad = \left(\frac{\partial}{\partial t}I(t,a) + \frac{\partial}{\partial a}I(t,a)\right)\frac{1}{\mathcal{F}(a)G(a)} + (\mu(a)+\gamma(a))\frac{I(t,a)}{\mathcal{F}(a)G(a)} \\
&\quad = \frac{\eta(a)E(t,a)}{\mathcal{F}(a)G(a)} = -\frac{H'(a)}{G(a)}x(t,a).
\end{aligned}$$

Summary

We can rewrite the problem in terms of u, x and y:

$$\left.\begin{aligned}
\frac{\partial}{\partial t}u(t,a) + \frac{\partial}{\partial a}u(t,a) &= -u(t,a)\phi(t,a), \\
\frac{\partial}{\partial t}x(t,a) + \frac{\partial}{\partial a}x(t,a) &= \frac{V(a)}{H(a)}u(t,a)\phi(t,a), \\
\frac{\partial}{\partial t}y(t,a) + \frac{\partial}{\partial a}y(t,a) &= -\frac{H'(a)}{G(a)}x(t,a), \\
u(t,0) = B, \qquad x(t,0) &= 0, \qquad y(t,0) = 0, \\
\phi(t,a) &= \int_0^\infty k(a,b)\mathcal{F}(b)G(b)y(t,b)\,\mathrm{d}b,
\end{aligned}\right\} \tag{22.3}$$

with initial conditions

$$u(0,a) = \frac{S_0(a)}{\mathcal{F}(a)V(a)}, \qquad x(0,a) = \frac{E_0(a)}{\mathcal{F}(a)H(a)}, \qquad y(0,a) = \frac{I_0(a)}{\mathcal{F}(a)G(a)}.$$

Problem (22.3) also makes sense if V, G, and $\mathcal{F}$ cannot be expressed in terms of per capita rates, but are arbitrary nonnegative nonincreasing functions with $V(0) = 1, G(0) = 1, \mathcal{F}(0) = 1$. $\mathcal{F}$ is the survival function. We call the function V a *vaccination profile*. H and G can be interpreted as generalized duration functions of the latent and infectious periods, respectively, i.e., $H(a+\cdot)/H(a)$ and $G(a+\cdot)/G(a)$ are the duration functions of the latent and infectious periods for an individual that has entered the respective period at age a. For instance, if $b > a$, $H(b)/H(a)$ is the probability that an individual that entered the exposed period at age a (i.e., was infected at age a) is still in the exposed period at age b.

The natural space for $S(t,\cdot)$, $I(t,\cdot)$, $R(t,\cdot)$ is $L^1[0, a^\sharp)$, where $a^\sharp \in (0, \infty]$ is the maximum age of individuals, i.e., $\mathcal{F}(a) > 0$ for all $a \in [0, a^\sharp)$ and $\mathcal{F}(a) = 0$ for all $a > a^\sharp$. This means that we should look for solutions u, x, y of (22.2) such that

$$\int_0^\infty \mathcal{F}(a)V(a)u(t,a)\,\mathrm{d}a < \infty,$$

$$\int_0^\infty \mathcal{F}(a)H(a)x(t,a)\,\mathrm{d}a < \infty,$$

$$\int_0^\infty \mathcal{F}(a)G(a)y(t,a)\,\mathrm{d}a < \infty,$$

with the initial conditions obeying analogous conditions. Realistically, $a^\sharp < \infty$, but mathematically it is reasonable to include $a^\sharp = \infty$ in order to be able to interpret endemic models without age-structure as special cases. Moreover, an upper integration limit ∞ is often easier to manipulate than a finite integration limit. We need to introduce the convention, however, that the initial data associated with (22.3) satisfy $u(0,a) = x(0,a) = y(0,a) = 0$ whenever $a > a^\sharp$, and that $\mathcal{F}(a)/\mathcal{F}(b) = 0$ whenever $a \geqslant b$ and $\mathcal{F}(b) = 0$.

The most direct way of showing that the system (22.3) is well-posed (i.e., has unique solutions that continuously depend on the initial data) consists of integrating the partial differential equations along characteristic lines and reducing the system to an integral equation in ϕ which is solved by the contraction mapping principle.

A special case that is often considered because it is much simpler than the general case is *separable mixing* between the age classes, i.e., the infection kernel k factors as

$$k(a,b) = k_1(a)k_2(b).$$

Then

$$\phi(t,a) = k_1(a)\psi(t),$$

with

$$\psi(t) = \int_0^\infty k_2(b) I(t,b)\,\mathrm{d}b = \int_0^\infty k_2(a)\mathcal{F}(a)G(a)y(t,a)\,\mathrm{d}a. \tag{22.4}$$

We call $\psi(t)$ the *infective force* at time t. We will eventually restrict our consideration to the case of separable mixing and refer to the bibliographic remarks for some literature on the general case.

An Equivalent Model of Integral Equations

In the following we reformulate model (22.3) as a system of integral equations. This presents the most elementary way of establishing the well-posedness of the model, though we will not do this here. Furthermore, it will be our basis for discussing extinction and persistence of the disease (Section 22.3). The reader who is not interested in technical details is advised to skip this part.

The partial differential equation for u in (22.3) can be rewritten by formula (13.8) in Section 13.2,

$$u(t,a) = \begin{cases} B\exp\left(-\int_0^a \phi(t-r,a-r)\,\mathrm{d}r\right), & t > a, \\ u(0,a-t)\exp\left(-\int_0^t \phi(t-r,a-r)\,\mathrm{d}r\right), & a > t \geqslant 0. \end{cases}$$

By Exercise 13.1 in Section 13.2,

$$x(t,a) = \int_0^{t\wedge a} \frac{V(a-s)}{H(a-s)} u(t-s,a-s)\phi(t-s,a-s)\,\mathrm{d}s + x(0,a-t),$$

where $x(0,s) := 0$ for $s < 0$ and $t \wedge a = \min\{t,a\}$. By the same exercise,

$$y(t,a) = -\int_0^{t\wedge a} \frac{H'(a-r)}{G(a-r)} x(t-r,a-r)\,\mathrm{d}r + y(0,a-t).$$

Substituting $a-r=s$, we see that this expression can be rewritten as a Stieltjes integral and that differentiability of H is not needed:

$$y(t,a) = -\int_{[a-t]_+}^{a} \frac{x(t+s-a,s)}{G(s)}\,\mathrm{d}H(s) + y(0,a-t)$$

(see Toolbox B.1). If x is not continuous, we can still write this formula as a Lebesgue–Stieltjes integral (see Toolbox B.4). r_+ is the positive part of r, which is r for $r \geqslant 0$ and 0 for $r \leqslant 0$. Fitting this expression into ϕ we obtain

$$\phi(t,a) = \int_0^t k(a,b)\mathcal{F}(b)G(b)y_1(t,b)\,\mathrm{d}b + \int_t^\infty k(a,b)\mathcal{F}(b)G(b)y_0(t,b)\,\mathrm{d}b,$$

where $y_1(t, a) = y(t, a)$ for $t > a$ and $y_0(t, a) = y(t, a)$ for $t < a$. Hence

$$y_1(t, b) = -\int_0^b \frac{x_1(t+s-b, s)}{G(s)} \, \mathrm{d}H(s), \quad t > b,$$

$$x_1(t+s-b, s) = \int_0^s \frac{V(s-r)}{H(s-r)} u(t-b+s-r, s-r) \times \phi(t-b+s-r, s-r) \, \mathrm{d}r.$$

Substituting $s - r = a$ in the expression for x_1 and fitting it into the expression for y_1 yields

$$y_1(t, a) = -\int_0^b \left(\int_0^s \frac{V(a)}{H(a)} u(t-b+a, a) \phi(t-b+a, a) \, \mathrm{d}a \right) \frac{1}{G(s)} \, \mathrm{d}H(s).$$

Furthermore,

$$y_0(t, b) = -\int_{b-t}^b \frac{x_0(t+s-b, s)}{G(s)} \, \mathrm{d}H(s) + y(0, b-t), \quad t < b,$$

$$x_0(t+s-b, s) = \int_0^{t+s-b} \frac{V(s-r)}{H(s-r)} u(t-b+s-r, s-r) \times \phi(t-b+s-r, s-r) \, \mathrm{d}r + x(0, b-t).$$

Substituting $s - r = a$ in the expression for x_0 and fitting it into the expression for y_0 yields

$$y_0(t, b) = -\int_{b-t}^b \left(\int_{b-t}^s \frac{V(a)}{H(a)} u(t-b+a, a) \phi(t-b+a, a) \, \mathrm{d}a + x(0, b-t) \right) \times \frac{1}{G(s)} \, \mathrm{d}H(s) + y(0, b-t), \quad t < b.$$

This system can be solved by using Banach's fixed-point theorem (contraction-mapping theorem) in a similar way to the integral equation in Toolbox B.6. In the following we concentrate on the special case of separable mixing, $\phi(t, a) = k_1(a)\psi(t)$ with ψ as in (22.4):

$$\psi(t) = \int_0^t k_2(b)\mathcal{F}(b)G(b)y_1(t, b) \, \mathrm{d}b + \int_t^\infty k_2(b)\mathcal{F}(b)G(b)y_0(t, b) \, \mathrm{d}b,$$

$$y_1(t, a) = -\int_0^b \left(\int_0^s \frac{V(a)}{H(a)} u(t-b+a, a) \times \psi(t-b+a)k_1(a) \, \mathrm{d}a \right) \frac{1}{G(s)} \, \mathrm{d}H(s),$$

$$y_0(t,b) = -\int_{b-t}^{b}\left(\int_{b-t}^{s}\frac{V(a)}{H(a)}u(t-b+a,a)\times\psi(t-b+a)k_1(a)\,\mathrm{d}a\right)\frac{1}{G(s)}\,\mathrm{d}H(s) - x(0,b-t)\int_{b-t}^{b}\frac{1}{G(s)}\,\mathrm{d}H(s) + y(0,b-t), \quad t<b.$$

Changing the order of integration,

$$y_1(t,a) = -\int_0^b \psi(t-b+a)\frac{V(a)}{H(a)}k_1(a)\left(\int_a^b u(t-b+a,a)\frac{1}{G(s)}\,\mathrm{d}H(s)\right)\mathrm{d}a,$$

and substituting $b-a=r$,

$$y_1(t,a) = -\int_0^b \psi(t-r)\frac{V(b-r)}{H(b-r)}\times k_1(b-r)u(t-r,b-r)\left(\int_{b-r}^{b}\frac{1}{G(s)}\,\mathrm{d}H(s)\right)\mathrm{d}r.$$

Fitting this expression into the one for ψ and changing the order of integration, we obtain the following integral equation for ψ:

$$\psi(t) = \int_0^t \psi(t-r)\left[\int_r^t u(t-r,b-r)K(b,b-r)\,\mathrm{d}b\right]\mathrm{d}r + \psi_2(t)+\psi_3(t)+\psi_4(t), \quad (22.5)$$

with

$$K(b,a) = -k_2(b)\mathcal{F}(b)\frac{V(a)}{H(a)}k_1(a)\left(\int_a^b \frac{G(b)}{G(s)}\,\mathrm{d}H(s)\right). \quad (22.6)$$

The functions ψ_j are given by

$$\psi_2(t) = -\int_t^\infty k_2(b)\mathcal{F}(b)G(b)\times\left[\int_{b-t}^{b}\left(\int_{b-t}^{s}\frac{V(a)}{H(a)}u(t-b+a,a)\times\psi(t-b+a)k_1(a)\,\mathrm{d}a\right)\frac{1}{G(s)}\,\mathrm{d}H(s)\right]\mathrm{d}b,$$

$$\psi_3(t) = -\int_t^\infty k_2(b)\mathcal{F}(b)G(b)x(0,b-t)\left(\int_{b-t}^{b}\frac{1}{G(s)}\,\mathrm{d}H(s)\right)\mathrm{d}b,$$

$$\psi_4(t) = \int_t^\infty k_2(b)\mathcal{F}(b)G(b)y(0,b-t)\,\mathrm{d}b.$$

Substituting $b = r + t$ and possibly changing the order of integration,

$$\left.\begin{aligned}
\psi_2(t) &= -\int_0^\infty k_2(r+t)\mathcal{F}(r+t)\Bigg[\int_r^{r+t} \frac{V(a)}{H(a)} u(a-r,a)\psi(a-r)k_1(a) \\
&\qquad\qquad \times \left(\int_a^{r+t} \frac{G(r+t)}{G(s)}\,\mathrm{d}H(s)\right)\mathrm{d}a\Bigg]\,\mathrm{d}r, \\
\psi_3(t) &= -\int_0^\infty k_2(r+t)\mathcal{F}(r+t)G(r+t)x(0,r) \\
&\qquad\qquad \times \left(\int_r^{r+t} \frac{1}{G(s)}\,\mathrm{d}H(s)\right)\mathrm{d}b, \\
\psi_4(t) &= \int_0^\infty k_2(r+t)\mathcal{F}(r+t)G(r+t)y(0,r)\,\mathrm{d}r.
\end{aligned}\right\} \tag{22.7}$$

Recall that

$$u(t,a) = \begin{cases} B\exp\left(-\int_0^a \psi(t-r)k_1(a-r)\,\mathrm{d}r\right), & t > a > 0, \\ u(0,a-t)\exp\left(-\int_0^t \psi(t-r)k_1(a-r)\,\mathrm{d}r\right), & a > t \geqslant 0. \end{cases} \tag{22.8}$$

The system (22.5), (22.7), and (22.8) presents an alternative formulation of the partial differential equations model (22.3). As mentioned before, it can be used to prove well-posedness of the model along the lines of Toolbox B.6, though we will not go into this here. It also presents a starting point to discuss extinction and persistence of the disease (Section 22.3).

22.2 Disease-Free and Endemic Equilibrium: the Replacement Ratio

The equations (22.3) for equilibrium states (which do not depend on time) take the form

$$\left.\begin{aligned}
u'(a) = -u(a)\phi(a), \quad x'(a) = u(a)\frac{V(a)}{H(a)}\phi(a), \quad y'(a) = -x(a)\frac{H'(a)}{G(a)}, \\
u(0) = B, \quad x(0) = 0, \quad y(0) = 0, \\
\phi(a) = \int_0^\infty k(a,b)\mathcal{F}(b)G(b)y(b)\,\mathrm{d}b.
\end{aligned}\right\} \tag{22.9}$$

There is one obvious equilibrium state, the disease free-state:

$$u^\diamond(a) = B, \qquad x^\diamond(a) = 0, \qquad y^\diamond(a) = 0.$$

In order to find an endemic equilibrium state, we integrate the first equation,

$$u^*(a) = B \exp\left(-\int_0^a \phi^*(b)\,\mathrm{d}b\right),$$

and the second equation,

$$x^*(a) = \int_0^a u^*(b)\frac{V(b)}{H(b)}\phi^*(b)\,\mathrm{d}b.$$

Finally, we integrate the third equation,

$$\begin{aligned} y^*(a) &= -\int_0^a \frac{x(s)}{G(s)}\,\mathrm{d}H(s) \\ &= -\int_0^a \left(\int_0^s u^*(b)\frac{V(b)}{H(b)}\phi^*(b)\,\mathrm{d}b\right)\frac{1}{G(s)}\,\mathrm{d}H(s). \end{aligned}$$

We restrict our consideration to the case of separable mixing, $k(a, b) = k_1(a)k_2(b)$. Then

$$\left.\begin{aligned} \phi^*(a) &= k_1(a)\psi^*, \\ \psi^* &= \int_0^\infty k_2(a)\mathcal{F}(a)G(a)y^*(a)\,\mathrm{d}a, \\ u^*(a) &= B\exp(-K_1(a)\psi^*), \qquad K_1(a) = \int_0^a k_1(b)\,\mathrm{d}b, \\ x^*(a) &= \int_0^a u^*(b)\frac{V(b)}{H(b)}k_1(b)\psi^*\,\mathrm{d}b, \\ y^*(a) &= -\int_0^a \left(\int_0^s u^*(b)\frac{V(b)}{H(b)}\psi^* k_1(b)\,\mathrm{d}b\right)\frac{1}{G(s)}\,\mathrm{d}H(s). \end{aligned}\right\} \tag{22.10}$$

Substituting the expression for y^* into ψ^*,

$$\psi^* = -\int_0^\infty k_2(a)\mathcal{F}(a)\left(\int_0^a \left(\int_0^s u^*(b)\frac{V(b)}{H(b)}\psi^* k_1(b)\,\mathrm{d}b\right)\frac{G(a)}{G(s)}\,\mathrm{d}H(s)\right)\mathrm{d}a.$$

Apparently, for an endemic equilibrium to exist, it is necessary and sufficient that $\psi^* > 0$. We divide the last relation by ψ^* and obtain

$$1 = -\int_0^\infty k_2(a)\mathcal{F}(a)\left(\int_0^a \left(\int_0^s u^*(b)\frac{V(b)}{H(b)}k_1(b)\,\mathrm{d}b\right)\frac{G(a)}{G(s)}\,\mathrm{d}H(s)\right)\mathrm{d}a.$$

Substituting the formula for u^* in (22.10), we obtain

$$\begin{aligned}1 = &- \int_0^\infty k_2(a)\mathcal{F}(a) \\ &\times \left(\int_0^a \left(\int_0^s B\exp(-K_1(b)\psi^*)\frac{V(b)}{H(b)}k_1(b)\,\mathrm{d}b\right)\frac{G(a)}{G(s)}\,\mathrm{d}H(s)\right)\mathrm{d}a \\ =:&\ \mathcal{R}(\psi^*).\end{aligned} \tag{22.11}$$

We see that $\mathcal{R}$ is a nonincreasing function which is strictly decreasing if $\mathcal{R}(0) > 0$. Moreover, we notice that

$$\exp(-K_1(b)\psi^*)k_1(b) \to 0, \quad \psi^* \to \infty, \quad b > 0.$$

Lebesgue's theorem of dominated convergence implies that $\mathcal{R}(\psi^*) \to 0$ as $\psi^* \to \infty$. Moreover, $\mathcal{R}(\psi^*)$ is a continuous function of ψ^*. The intermediate value theorem implies the following result.

Theorem 22.1. *There exists an endemic equilibrium if and only if*

$$\begin{aligned}1 &< \mathcal{R}(0) \\ &= -B\int_0^\infty k_2(a)\mathcal{F}(a)\left(\int_0^a \left(\int_0^s \frac{V(b)}{H(b)}k_1(b)\,\mathrm{d}b\right)\frac{G(a)}{G(s)}\,\mathrm{d}H(s)\right)\mathrm{d}a.\end{aligned}$$

The endemic equilibrium (when it exists) is uniquely determined.

In order to interpret $\mathcal{R}(\psi^*)$ we remember that

$$B\exp(-K_1(b)\psi^*)V(b) = u^*(b)V(b) = \frac{S^*(b)}{\mathcal{F}(b)},$$

where we use the symbol S^* in order to indicate that this is the age-density of the susceptible part at endemic equilibrium. Then

$$\begin{aligned}\mathcal{R}(\psi^*) &= -\int_0^\infty k_2(a)\left(\int_0^a \left(\int_0^s S^*(b)\frac{\mathcal{F}(a)G(a)}{\mathcal{F}(b)H(b)}k_1(b)\,\mathrm{d}b\right)\frac{\mathrm{d}H(s)}{G(s)}\right)\mathrm{d}a \\ &= \int_0^\infty S^*(b)k_1(b)\left(\int_b^\infty k_2(a)\frac{\mathcal{F}(a)}{\mathcal{F}(b)}\left[\int_b^a \frac{G(a)}{G(s)}\frac{\mathrm{d}H(s)}{H(b)}\right]\mathrm{d}a\right)\mathrm{d}b.\end{aligned}$$

We introduce

$$M(b) = -\int_b^\infty k_2(a)\frac{\mathcal{F}(a)}{\mathcal{F}(b)}\left[\int_b^a \frac{G(a)}{G(s)}\frac{\mathrm{d}H(s)}{H(b)}\right]\mathrm{d}a.$$

Remember that $\mathcal{F}(a)/\mathcal{F}(b)$ can be interpreted as the probability of still being alive at age $a > b$ provided one was alive at age b, while $G(a)/G(s)$ is the probability of still being infective at age $a > b$ when one entered the infective class at age b. $-H'(s)/H(b)$ is the rate of leaving the exposed class at age $s > b$ under the provision that one has entered it at age b. Hence

$$\frac{\mathcal{F}(a)}{\mathcal{F}(b)} \int_b^a \frac{G(a)}{G(s)} \frac{\mathrm{d}H(s)}{H(b)}$$

is the probability of being infective and alive at age $a > b$ under the provision that one was infected at age b. So, $M(b)$ can be interpreted as the expected infectivity of an individual that enters the infective class at age b.

Now introduce

$$f(b) = \frac{S^*(b)k_1(b)}{\int_0^\infty S^*(a)k_1(a)\,\mathrm{d}a}.$$

Then $f(b)$ can be interpreted as the conditional probability density of being infected at age b under the condition of being infected at all. Recall that S^* is the age-density of the susceptible part of the population and that k_1 is the age-dependent susceptibility of a susceptible individual.

We can now rewrite $\mathcal{R}(\psi^*)$ as

$$\mathcal{R}(\psi^*) = \left(\int_0^\infty S^*(a)k_1(a)\,\mathrm{d}a\right) \int_0^\infty f(b)M(b)\,\mathrm{d}b.$$

The second integral can be interpreted as the average infectivity of one typical infective individual, while the first integral describes the size of the reservoir of susceptible individuals weighed by the appropriate age-dependent susceptibilities.

Hence $\mathcal{R}(\psi^*)$ gives the average number of secondary cases one infectious individual produces in a population under a constant infective force ψ^* and is called the *replacement ratio* at infective force ψ^*.

If $\psi^* = 0$, the population is disease free, and $\mathcal{R}(0)$ gives the average number of secondary cases one infectious individual produces if it is introduced into an otherwise disease-free population. $\mathcal{R}(0)$ is called the *basic replacement ratio* if there is no vaccination, and the *net replacement ratio* if there is vaccination. We will often write $\mathcal{R}_0$ instead of $\mathcal{R}(0)$ and, if we want to stress the dependence of the net replacement ratio on the vaccination profile, $\mathcal{R}_0(V)$. The basic replacement ratio will sometimes be denoted by $\mathcal{R}_0(\emptyset)$.

22.3 The Net Replacement Ratio, and Disease Extinction and Persistence

In the previous section, we have linked the net replacement ratio to the existence of an endemic equilibrium. This does not yet mean that the size of the net replacement ratio decides the dynamic fate of the disease, extinction or persistence, though it gives an important clue. Usually, in the literature, the net replacement ratio is linked to the local stability of the endemic equilibrium:

local asymptotic stability if $\mathcal{R}_0 < 1$, and instability if $\mathcal{R}_0 > 1$. The formal calculations, which derive a characteristic equation and determine the location of their roots, can also be done in our case. However, rather sophisticated dynamical systems considerations are required that guarantee that the location of the roots of the characteristic equation in the left complex half-plane implies the stability of the equilibrium. After all, we have an infinite-dimensional system. Here we present a global alternative which provides stronger statements and only requires elementary, though lengthy, manipulations of integrals which, however, are not much more involved than those needed for the local analysis. We will show that the disease becomes extinct, i.e., in terms of the infective force ψ, that $\psi(t) \to 0$ as $t \to \infty$, if $\mathcal{R}_0 < 1$ and some nonrestrictive technical conditions are satisfied. We will also show that, if $\mathcal{R}_0 > 1$, the disease either vanishes after finite time, or the infective force ψ satisfies $\limsup_{t\to\infty} \psi(t) \geqslant \epsilon$ with $\epsilon > 0$ being independent of the initial data. The latter means that the infective force is either bounded away from 0 or keeps bouncing back to a certain positive level.

The proofs of these statements are based on the integral equations we derived at the end of Section 22.1 and are quite technical. The reader who is satisfied with understanding the role of $\mathcal{R}_0$ as explained above and is not interested in the mathematical details may like to skip the rest of this section (and perhaps return later).

We start by considering the age-distribution of the total population, $N(t, a) = S(t, a) + E(t, a) + I(t, a) + R(t, a)$. Adding the equations in (21.1),

$$\frac{\partial}{\partial t}N(t,a) + \frac{\partial}{\partial a}N(t,a) = -\mu(a)N(t,a), \quad N(t,0) = B.$$

Using the survivorship $\mathcal{F}$ instead, this is equivalent to

$$\frac{\partial}{\partial t}\frac{N(t,a)}{\mathcal{F}(a)} + \frac{\partial}{\partial a}\frac{N(t,a)}{\mathcal{F}(a)} = 0, \quad N(t,0) = B,$$

which extends the equation to nondifferentiable survival functions. Integrating,

$$N(t,a) = \begin{cases} B\mathcal{F}(a), & t > a, \\ \dfrac{N(0,a-t)\mathcal{F}(a)}{\mathcal{F}(a-t)}, & t < a. \end{cases}$$

Since $I(t, a) \leqslant N(t, a)$ we have from (22.4) that

$$\begin{aligned}\psi(t) &= \int_0^\infty k_2(b)I(t,b)\,\mathrm{d}b \leqslant \int_0^\infty k_2(b)N(t,b)\,\mathrm{d}b \\ &= \int_0^t k_2(b)B\mathcal{F}(b)\,\mathrm{d}b + \int_0^\infty \frac{N(0,b)\mathcal{F}(b+t)}{\mathcal{F}(b)}\,\mathrm{d}b.\end{aligned}$$

This shows that ψ is bounded. By (22.8),

$$u(t,a) \leqslant \begin{cases} B, & t > a, \\ u(0, a-t), & a > t. \end{cases}$$

By (22.5) and (22.7),

$$\psi(t) = \psi_1(t) + \psi_2(t) + \psi_3(t) + \psi_4(t)$$

with

$$\psi_1(t) \leqslant \int_0^t \psi(t-r)\left[\int_r^t BK(b, b-r)\,\mathrm{d}b\right]\mathrm{d}r,$$

and

$$\begin{aligned} \psi_2(t) \leqslant &- \int_0^\infty k_2(r+t)\mathcal{F}(r+t) \\ &\times \left[\int_r^{r+t} \frac{V(a)}{H(a)} u(0,r)\bar{\psi}k_1(a)\left(\int_a^{r+t} \frac{G(r+t)}{G(s)}\,\mathrm{d}H(s)\right)\mathrm{d}a\right]\mathrm{d}r, \\ \psi_3(t) = &- \int_0^\infty k_2(r+t)\mathcal{F}(r+t)x(0,r)\left(\int_r^{r+t} \frac{G(r+t)}{G(s)}\,\mathrm{d}H(s)\right)\mathrm{d}b, \\ \psi_4(t) = &\int_0^\infty k_2(r+t)\mathcal{F}(r+t)G(r+t)y(0,r)\,\mathrm{d}r, \end{aligned}$$

where $\bar{\psi}$ is the supremum of ψ. Notice that, by changing the order of integration, change of variables, and (22.6),

$$\begin{aligned} &\int_0^\infty \left[\int_r^\infty BK(b, b-r)\,\mathrm{d}b\right]\mathrm{d}r \\ &\quad = B\int_0^\infty \left(\int_0^b K(b, b-r)\,\mathrm{d}r\right)\mathrm{d}b \\ &\quad = B\int_0^\infty \left(\int_0^b K(b,a)\,\mathrm{d}a\right) \\ &\quad = -B\int_0^\infty k_2(b)\mathcal{F}(b)\left[\int_0^b \frac{V(a)}{H(a)}k_1(a)\left(\int_a^b \frac{G(b)}{G(s)}\,\mathrm{d}H(s)\right)\mathrm{d}a\right]\mathrm{d}b \\ &\quad = \mathcal{R}_0. \end{aligned}$$

Since ψ is bounded, we can apply the Lebesgue–Fatou lemma to ψ_1 (see Toolbox B.2). Using the notation $\psi^\infty = \limsup_{t\to\infty} \psi(t)$, we obtain

$$\psi_1^\infty \leqslant \mathcal{R}_0\psi^\infty.$$

From the decrease of H,

$$-\left(\int_a^{r+t} \frac{G(r+t)}{G(s)}\,\mathrm{d}H(s)\right) \leqslant H(a),$$

and

$$\psi_2(t) \leqslant \int_0^\infty k_2(r+t)\mathcal{F}(r+t)\left[\int_r^{r+t} V(a)u(0,r)\bar{\psi}k_1(a)\,\mathrm{d}a\right]\mathrm{d}r,$$

$$\psi_3(t) \leqslant \int_0^\infty k_2(r+t)\mathcal{F}(r+t)x(0,r)H(r)\,\mathrm{d}r.$$

Since $u(0,r) = S_0(r)/\mathcal{F}(r)V(r)$,

$$\psi_2(t) \leqslant \int_0^\infty k_2(r+t)\mathcal{F}(r+t)\left[\int_r^{r+t} V(a)\frac{S_0(r)}{\mathcal{F}(r)V(r)}\bar{\psi}k_1(a)\,\mathrm{d}a\right]\mathrm{d}r.$$

Since V is nonincreasing,

$$\psi_2(t) \leqslant \int_0^\infty k_2(r+t)\frac{\mathcal{F}(r+t)}{\mathcal{F}(r)}\bar{\phi}S_0(r)\left[\int_r^{r+t} k_1(a)\,\mathrm{d}a\right]\mathrm{d}r.$$

Similarly, since $x(0,a) = E_0(a)/(\mathcal{F}(a)H(a))$ and H is nonincreasing,

$$\psi_3(t) \leqslant \int_0^\infty k_2(r+t)\frac{\mathcal{F}(r+t)}{\mathcal{F}(r)}E_0(r)\,\mathrm{d}r.$$

Finally,

$$\psi_4(t) = \int_0^\infty k_2(r+t)\frac{\mathcal{F}(r+t)G(r+t)}{\mathcal{F}(r)G(r)}I_0(r)\,\mathrm{d}r.$$

After these preparations we can show that the disease dies out if an average infective individual does not replace itself when introduced into an otherwise disease-free population.

Theorem 22.2. *Let the net replacement ratio satisfy $\mathcal{R}_0 < 1$, the functions k_1 and k_2 be bounded and $t\mathcal{F}(t) \to 0$ as $t \to \infty$. Then $\psi(t) \to 0$ as $t \to 0$, and the disease dies out.*

Proof. It follows from the Lebesgue–Fatou lemma that $\psi_j(t) \to 0$ as $t \to \infty$ for $j = 2, 3, 4$. Hence $\psi^\infty = \psi_1^\infty \leqslant \mathcal{R}_0\psi^\infty$. Since $\mathcal{R}_0 < 1$, this is only possible if $\psi^\infty = 0$. □

Disease Persistence, or Endemicity

To stress the decisive role of $\mathcal{R}_0$ in disease extinction versus disease persistence, we show $\mathcal{R}_0 > 1$ implies uniform weak disease persistence (or endemicity) under reasonable assumptions. The disease is called uniformly weakly persistent (or endemic) if there exists some $\epsilon > 0$, which does not depend on the initial data S_0, E_0, I_0, R_0, such that the infective force ψ satisfies

$$\psi^\infty := \limsup_{t\to\infty} \psi(t) > \epsilon,$$

unless ψ is 0 almost everywhere on $[0, \infty)$. By (22.5),

$$\psi(t) \geqslant \int_0^t \psi(t-r)\left[\int_r^t u(t-r, b-r)K(b, b-r)\,\mathrm{d}b\right]\mathrm{d}r.$$

Let ψ_s denote the translation $\psi_s(t) = \psi(t+s)$. Then

$$\begin{aligned}\psi_s(t) &\geqslant \int_0^{t+s} \psi(t+s-r)\left[\int_r^{t+s} u(t+s-r, b-r)K(b, b-r)\,\mathrm{d}b\right]\mathrm{d}r \\ &\geqslant \int_0^t \psi_s(t-r)\left[\int_r^t u(t+s-r, b-r)K(b, b-r)\,\mathrm{d}b\right]\mathrm{d}r.\end{aligned}$$

Assume that $\psi(t) \leqslant \epsilon$ for $t \geqslant s$. By (22.8), $u(t+s-r, b-r) \geqslant B\mathrm{e}^{-\epsilon(b-r)}$ for $b \leqslant t$, and

$$\psi_s(t) \geqslant \int_0^t \psi_s(t-r)\left[\int_r^t B\mathrm{e}^{-\epsilon(b-r)}K(b, b-r)\,\mathrm{d}b\right]\mathrm{d}r.$$

In particular,

$$\psi_s(t+T) \geqslant \int_0^{t+T} \psi_s(t+T-r)\left[\int_r^{t+T} B\mathrm{e}^{-\epsilon(b-r)}K(b, b-r)\,\mathrm{d}b\right]\mathrm{d}r.$$

Noticing that $\psi_s(t+T) = \psi_{s+T}(t)$,

$$\psi_{s+T}(t) \geqslant \int_0^t \psi_{s+T}(t-r)\left[\int_r^{r+T} B\mathrm{e}^{-\epsilon(b-r)}K(b, b-r)\,\mathrm{d}b\right]\mathrm{d}r.$$

Taking Laplace transforms,

$$\widehat{\psi_{s+T}}(\lambda) \geqslant \widehat{\psi_{s+T}}(\lambda)\alpha(\epsilon, \lambda, T),$$

$$\alpha(\epsilon, \lambda, T) = \int_0^\infty \mathrm{e}^{-\lambda r}\left[\int_r^{r+T} B\mathrm{e}^{-\epsilon(b-r)}K(b, b-r)\,\mathrm{d}b\right]\mathrm{d}r.$$

Since ψ is bounded, the Laplace transform of ψ_{s+T} is defined on $(0, \infty)$. By (22.6) and Theorem 18.1 we notice that $\alpha(0, 0, \infty) = \mathcal{R}_0 > 1$; so $\alpha(\epsilon, \lambda, T) > 1$ if $\epsilon, \lambda > 0$ are chosen small enough and T large enough. Then $\widehat{\psi_{s+T}}(\lambda) = 0$, which means that ψ_{s+T} is 0 a.e. (almost everywhere) on $[0, \infty)$. In other words, ψ *is eventually* 0, i.e., there exists some $r > 0$ such that $\psi(t) = 0$ for a.a. (almost all) $t \geqslant r$. We have shown the following theorem.

Theorem 22.3. *Let $\mathcal{R}_0 > 1$. Then there exists some $\epsilon > 0$, which is independent of the initial data, such that the following alternative holds for the infective force ψ. Either ψ is eventually 0 or $\psi^\infty = \limsup_{t\to\infty} \psi(t) > \epsilon$.*

Our next aim consists of deriving conditions such that every infective force ψ that is eventually 0 is 0 a.e. on $[0, \infty)$. Recall that ψ is bounded and let $\bar{\psi}$

be the supremum. Then $u(t-r, b-r) \geqslant B e^{-\bar{\psi}(b-r)}$ for $t \geqslant b$ and

$$\psi(t) \geqslant \int_0^t \psi(t-r) \left[\int_r^t B e^{-\bar{\psi}(b-r)} K(b, b-r) \, db \right] dr.$$

Let ψ be nonnegative and not 0 a.e. on $[0, \infty)$. Assume that ψ is eventually 0. Then there exists some $T > 0$ such that $\psi(t) = 0$ for a.a. $t > T$. We choose $T > 0$ as small as possible. Then

$$0 = \int_T^\infty \psi(t) \, dt$$

and

$$\psi(t) \geqslant \int_0^t \psi(t-r) \left[B e^{-\bar{\psi} T} \int_r^t K(b, b-r) \, db \right] dr, \quad 0 \leqslant t \leqslant T.$$

This implies

$$0 = \int_T^\infty \left(\int_0^t \psi(t-r) \left[\int_r^t K(b, b-r) \, db \right] dr \right) dt.$$

Changing the order of integration and substituting $b - r = a$,

$$0 = \int_0^\infty \left(\int_{r \vee T}^\infty \psi(t-r) \left[\int_0^{t-r} K(a+r, a) \, da \right] dt \right) dr,$$

where $r \vee T = \max\{r, T\}$. Substituting $t - r = s$,

$$0 = \int_0^\infty \left(\int_{0 \vee (T-r)}^\infty \psi(s) \left[\int_0^s K(a+r, a) \, da \right] ds \right) dr.$$

Recalling that $\psi = 0$ on (T, ∞) and changing the order of integration again,

$$0 = \int_0^T \psi(s) \left(\int_{T-s}^\infty \left[\int_0^s K(a+r, a) \, da \right] dr \right) ds.$$

By the choice of T, for every $\delta \in (0, T)$, ψ does not vanish a.e. on $(0, T - \delta)$. This means that there exists some $s \in (T - \delta, T)$ such that

$$\int_{T-s}^\infty \left[\int_0^s K(a+r, a) \, da \right] dr = 0.$$

Hence, for all $\delta \in (0, T)$,

$$\int_\delta^\infty \left[\int_0^{T-\delta} K(a+r, a) \, da \right] dr = 0.$$

Taking the limit of $\delta \to 0$,

$$\int_0^\infty \left[\int_0^T K(a+r, a) \, da \right] dr = 0.$$

Changing the order of integration another time and making the appropriate substitution

$$\int_0^T \left[\int_a^\infty K(b,a)\,\mathrm{d}b\right] \mathrm{d}a = 0.$$

Recalling the definition of K in (22.6),

$$0 = \int_0^T \left(\int_a^\infty k_2(b)\mathcal{F}(b)\frac{V(a)}{H(a)}k_1(a)\left(\int_a^b \frac{G(b)}{G(s)}\,\mathrm{d}H(s)\right)\mathrm{d}b\right)\mathrm{d}a.$$

Changing the order of integration and substituting $s = r + a$:

$$0 = \int_0^T \left(\int_a^\infty k_2(b)\mathcal{F}(b)\frac{V(a)}{H(a)}k_1(a)\left(\int_0^{b-a} \frac{G(b)}{G(s+a)}H(a+\mathrm{d}s)\right)\mathrm{d}b\right)\mathrm{d}a.$$

Let us assume that the lengths of the exposed and infectious periods do not depend on the age of the infected individual, then

$$\frac{H(a+s)}{H(a)} = \tilde{H}(s), \qquad \frac{G(a+s)}{G(a)} = \tilde{G}(s)$$

do not depend on a with $\tilde{H}$ and $\tilde{G}$ being the duration functions of the exposed and infectious periods, respectively. The last integral takes the form

$$0 = \int_0^T V(a)k_1(a)\left(\int_a^\infty k_2(b)\mathcal{F}(b)\Gamma(b-a)\,\mathrm{d}b\right)\mathrm{d}a$$

with

$$\Gamma(b) = \int_0^b \tilde{G}(b-s)\,\mathrm{d}\tilde{H}(s).$$

After another substitution,

$$0 = \int_0^T V(a)k_1(a)\left(\int_0^\infty k_2(b+a)\mathcal{F}(b+a)\Gamma(b)\,\mathrm{d}b\right)\mathrm{d}a.$$

Since

$$\int_0^\infty k_2(b+a)\mathcal{F}(b+a)\Gamma(b)\,\mathrm{d}b \to \int_0^\infty k_2(b)\mathcal{F}(b)\Gamma(b)\,\mathrm{d}b, \quad a \to 0,$$

we have the following result.

Theorem 22.4. *Let $\mathcal{R}_0 > 1$. Assume that $k_1(\cdot)V(\cdot)$ does not vanish a.e. in a neighborhood of 0 and that the durations of the exposed and infectious periods are not age dependent and*

$$\int_0^\infty k_2(b)\mathcal{F}(b)\Gamma(b)\,\mathrm{d}b > 0.$$

Then the disease persists uniformly weakly in the sense that every infective force ψ that is not 0 *a.e. satisfies*

$$\limsup_{t \to \infty} \psi(t) > \epsilon$$

with $\epsilon > 0$ not depending on the initial data.

Proof. Under these assumptions we have

$$\int_0^T V(a)k_1(a)\left(\int_0^\infty k_2(b+a)\mathcal{F}(b+a)\Gamma(b)\,\mathrm{d}b\right)\mathrm{d}a > 0$$

for all $T > 0$, so, by contraposition of the preceding arguments, ψ is not eventually 0. The assertion now follows from Theorem 22.3. □

22.4 Cost of Vaccinations and Optimal Age Schedules

Vaccination (or immunization) campaigns try to give a population as much protection from the disease as possible, within budgetary limits. We will restrict our deliberations to the protection of an uninfected population. We will take the point of view that the degree of protection is reciprocally related to the net replacement ratio, the average number of secondary cases one typical infected person produces when introduced into an uninfected, but to some degree immunized, population.

Cost of Vaccination

As mentioned before, we restrict our consideration to the cost of vaccinating a population in the disease-free equilibrium. The cost per age class depends on the vaccination rate and on other factors like access to the age class (it may be less expensive to vaccinate school children because one can take advantage of the school system). Let $p(a)$ denote the per capita cost of vaccinating at age a. We first consider the special case where the vaccination strategy can be described by a vaccination rate v, $v(a)V(a) = -V'(a)$ (recall (22.2)):

$$\begin{aligned} C &= \int_0^\infty v(a)p(a)S^\diamond(a)\,\mathrm{d}a \\ &= \int_0^\infty v(a)p(a)B\mathcal{F}(a)V(a)\,\mathrm{d}a \\ &= -\int_0^\infty p(a)B\mathcal{F}(a)V'(a)\,\mathrm{d}a. \end{aligned}$$

Closer inspection of this formula shows that C is actually a rate, the cost of vaccinations per time unit. If, more generally, the vaccination strategy is described by a vaccination profile V, then the vaccination cost is given by the Stieltjes integral

$$C = C(V) = -\int_0^\infty p(a)B\mathcal{F}(a)\,\mathrm{d}V(a).$$

Optimal Vaccination Schedules

In designing an optimal vaccination schedule, the cost of the vaccination campaign is typically balanced against the degree of protection of the (uninfected) population, which (as we have agreed) is reciprocally related to the net replacement ratio.

Recall from (22.11), with $\psi^* = 0$, that the net replacement ratio is

$$\begin{aligned}\mathcal{R}_0 &= \mathcal{R}(0)\\ &= -B\int_0^\infty k_2(a)\mathcal{F}(a)\bigg(\int_0^a\bigg(\int_0^s \frac{V(b)}{H(b)}k_1(b)\,\mathrm{d}b\bigg)\frac{G(a)}{G(s)}\,\mathrm{d}H(s)\bigg)\,\mathrm{d}a.\end{aligned}$$

In order to stress the dependence of the net replacement ratio on the vaccination profile V, we write $\mathcal{R}_0(V)$. Changing the order of integration yields

$$\begin{aligned}\mathcal{R}_0 &= \mathcal{R}_0(V)\\ &= -B\int_0^\infty\bigg(\int_b^\infty k_2(a)\mathcal{F}(a)\bigg(\int_b^a \frac{G(a)}{G(s)}\frac{\mathrm{d}H(s)}{H(b)}\bigg)\,\mathrm{d}a\bigg)k_1(b)V(b)\,\mathrm{d}b.\end{aligned}$$

Set

$$\xi(b) = -Bk_1(b)\int_b^\infty k_2(a)\mathcal{F}(a)\bigg(\int_b^a \frac{G(a)}{G(s)}\frac{\mathrm{d}H(s)}{H(b)}\bigg)\,\mathrm{d}a. \tag{22.12}$$

Then

$$\mathcal{R}_0 = \int_0^\infty \xi(b)V(b)\,\mathrm{d}b = \int_0^\infty \xi(b)\,\mathrm{d}b + \int_0^\infty\bigg(\int_b^\infty \xi(a)\,\mathrm{d}a\bigg)\,\mathrm{d}V(b).$$

We introduce

$$W(a) = 1 - V(a).$$

$W(a)$ is the probability of being vaccinated up to age a. We call W a *vaccination distribution*. We have $W(0) = 0$, W is nondecreasing and takes values between 0 and 1. Particular vaccination distributions are those where everybody is vaccinated at one age, let us say $b \geqslant 0$,

$$W_b(a) = \begin{cases} 0, & 0 \leqslant a < b,\\ 1, & a > b.\end{cases}$$

If $0 \leqslant q \leqslant 1$, qW_b represents a vaccination strategy which is concentrated at age b and has a fraction q of the population vaccinated.

The costs and the net replacement ratio can be expressed in terms of a vaccination distribution W as

$$C = C(W) = \int_0^\infty p(a) B \mathcal{F}(a)\, \mathrm{d}W(a),$$

$$\mathcal{R}_0 = \mathcal{R}_0(W) = \int_0^\infty \xi(b)\, \mathrm{d}b - \int_0^\infty \left(\int_b^\infty \xi(a)\, \mathrm{d}a \right) \mathrm{d}W(b).$$

We need to make sure that ξ is integrable and the integral over ξ is finite. To this end we assume that the expectation of remaining life at age a, $L(a)$, is a bounded function of a, i.e., there exists some $\epsilon > 0$ such that

$$\frac{\mathcal{F}(a+b)}{\mathcal{F}(b)} \leqslant \frac{1}{\epsilon} \mathrm{e}^{-\epsilon a} \quad \forall a, b \geqslant 0.$$

Recall Proposition 12.1. Furthermore, assume that k_1, k_2, p are Borel measurable and bounded. Then

$$0 \leqslant \xi(b) \leqslant -\text{const.} \times B k_1(b) \mathcal{F}(b) \int_b^\infty \mathrm{e}^{-\epsilon(a-b)} \left(\int_b^a \frac{\mathrm{d}H(s)}{H(b)} \frac{G(b)}{G(s)} \right) \mathrm{d}a$$
$$\leqslant \text{const.} \times \mathcal{F}(b).$$

Summarizing, we can write

$$\left.\begin{aligned} C = C(W) &= \int_0^\infty g(a)\, \mathrm{d}W(a), \\ g(a) &= p(a) B \mathcal{F}(a), \\ \mathcal{R}_0 = \mathcal{R}_0(W) &= f(0) - \int_0^\infty f(a)\, \mathrm{d}W(a), \\ f(a) &= \int_a^\infty \xi(b)\, \mathrm{d}b, \end{aligned}\right\} \tag{22.13}$$

where ξ is the integrable function given by (22.12). Notice that f, g are continuous nonnegative functions and that f is nonincreasing and $f(a), g(a) \to 0$ as $a \to \infty$. Furthermore,

$$\frac{f(a)}{g(a)} \leqslant \text{const.} \times \frac{L(a)}{p(a)}, \tag{22.14}$$

with

$$L(a) = \frac{\int_a^\infty \mathcal{F}(b)\, \mathrm{d}b}{\mathcal{F}(a)}$$

being the expected remaining life at age a. In the important special case where the vaccination cost does not depend on age, g is nonincreasing.

Notice that C is a linear functional of W, while $\mathcal{R}_0(W)$ is the sum of a linear functional and a constant. This means that $\mathcal{R}_0$ is affine, i.e.,

$$\mathcal{R}_0(sW_1 + (1-s)W_2) = s\mathcal{R}_0(W_1) + (1-s)\mathcal{R}_0(W_2) \quad \forall s \in \mathbb{R},$$

where W_1, W_2 are arbitrary vaccination distributions.

Remember that W is a nondecreasing function with values between 0 and 1. The latter constraint can be expressed as

$$\int_0^\infty \mathrm{d}W(a) = W(\infty) - 1 \leqslant 0.$$

Following Hadeler, Müller (1996) and Müller (1998) we now consider two optimization problems. We can minimize the net replacement ratio $\mathcal{R}_0(W)$ for a given maximum cost c,

$$C(W) - c \leqslant 0,$$

or we can minimize the cost $C(W)$ under the constraint of a given maximum replacement ratio ρ,

$$\mathcal{R}_0(W) - \rho \leqslant 0.$$

Mathematically, either optimization problem has two constraints.

One possible, extreme, vaccination strategy in this model consists of vaccinating everybody at birth (we ignore the complication that the immune system has not yet been established). This leads to the vaccination distribution W_0 with

$$W_0(0) = 0, \qquad W_0(a) = 1, \quad a > 0. \tag{22.15}$$

Then $\mathcal{R}_0(W_0) = 0$. Hence, for every $\rho > 0$, there exists some W of the form sW_0, $0 < s < 1$, such that

$$\mathcal{R}_0(W) - \rho < 0, \quad W(\infty) - 1 < 0.$$

This makes the problem of minimizing the cost under the constraint of a given maximum replacement ratio feasible in so far as there exists a vaccination strategy that meets the constraint.

Another extreme strategy consists of not vaccinating at all, i.e.,

$$W = W_\infty \equiv 0. \tag{22.16}$$

Then $C(W_\infty) = 0$ and, for every $c > 0$,

$$C(W_\infty) - c < 0, \quad W_\infty(\infty) - 1 < 0.$$

This means that the problem of minimizing the net replacement ratio under given maximum costs is feasible as there always exists a vaccination strategy that meets the cost constraint.

The problem of minimizing $\mathcal{R}_0$ under given maximum cost c is only nontrivial if

$$C(W_0) > c,$$

because otherwise we can adopt the strategy of vaccinating everybody at birth. If $C(W_0) > c$, a vaccination distribution W with $\mathcal{R}_0(W) > 0$ and $C(W) < c$ is not an optimal strategy. Indeed, let

$$W_s = sW + (1 - s)W_0.$$

Then $C(W_s) = sC(W) + (1 - s)C(W_0)$ and thus there exists some $s \in (0, 1)$ such that $C(W_s) = c$. Since $\mathcal{R}_0$ is affine,

$$\mathcal{R}_0(W_s) = s\mathcal{R}_0(W) + (1 - s)\mathcal{R}_0(W_0) = s\mathcal{R}_0(W) < \mathcal{R}_0(W).$$

W_0 is the only vaccination strategy W with $\mathcal{R}_0(W) = 0$, unless there is an interval $[0, a]$ on which f is constant. Assume that a is maximal. Then any strategy W with the property $W(b) = 1$ for all $b > a$ has $\mathcal{R}_0(W) = 0$, and there are no other ones.

Similarly, the problem of minimizing C under a given maximum net replacement ratio ρ is only nontrivial if

$$\mathcal{R}_0(W_\infty) = f(0) > \rho,$$

because otherwise we can adopt the strategy of not vaccinating at all. If $\mathcal{R}_0(W_\infty) > \rho$, a strategy W with $C(W) > 0$ and $\mathcal{R}_0(W) < \rho$ is not optimal. Indeed, let

$$W_s = sW + (1 - s)W_\infty.$$

Since $\mathcal{R}_0$ is affine and C is linear,

$$\mathcal{R}_0(W_s) = s\mathcal{R}_0(W) + (1 - s)\mathcal{R}_0(W_\infty) = \rho$$

for some $s \in (0, 1)$ and

$$C(W_s) = sC(W) + (1 - s)C(W_\infty) = sC(W) < C(W).$$

In the following we concentrate on strategies that maximize the protection of the population, i.e., minimize the net replacement ratio. We assume that this problem is nontrivial, $C(W_0) > c$, which means that it is not affordable to vaccinate everybody at age 0.

Optimal One-Age Strategies for Maximum Protection

If we assume that the per capita cost function p is strictly positive on $[0, a^\sharp)$ and

$$\lim_{a \to a^\sharp} \frac{L(a)}{p(a)} = 0,$$

where $a^\sharp$ is the maximum lifetime and $L(a)$ the expected remaining life at age a, we can develop a criterion for a one-age strategy (where everybody is vaccinated at the same age) to be an optimal strategy. We consider minimizing the net replacement ratio under given costs. Equivalently, we maximize

$$\Xi(W) = \int_0^\infty f(a)\, \mathrm{d}W(a) = \int_0^{a^\sharp} \frac{f(a)}{g(a)} g(a)\, \mathrm{d}W(a).$$

Since f/g is continuous and converges to 0 as $a \to a^\sharp$ by (22.14), the intermediate value theorem for integrals implies that

$$\int_0^\infty f(a)\, \mathrm{d}W(a) = \frac{f(b)}{g(b)} C(W)$$

for some $b \in [0, a^\sharp)$. If $W = qW_{a^\diamond}$ is a one-age strategy concentrated at age $a^\diamond$, we have

$$\int_0^\infty f(a)\, \mathrm{d}W(a) = \frac{f(a^\diamond)}{g(a^\diamond)} C(qW_{a^\diamond}).$$

So, the one-age strategy concentrated at $a^\diamond$ where f/g takes its maximum is optimal, provided that the costs are c or more if everybody is vaccinated, $C(W_{a^\diamond}) \geqslant c$. The percentage of people vaccinated is then chosen such that the cost ceiling c is exactly matched, $q = c/C(W_{a^\diamond})$. If the costs of vaccinating everybody at $a^\diamond$ is smaller than the ceiling c, this one-age strategy is not optimal, and, as we have seen before, a better strategy can be obtained by combining it with vaccinations at age 0. Let us ponder whether in this case, namely $C(W_{a^\diamond}) < c$, there exists another one-age strategy that is optimal, i.e., an optimal strategy concentrated at a with $(f(a)/g(a)) < (f(a^\diamond)/g(a^\diamond))$. For such a strategy we would have

$$\Xi(qW_a) = \frac{f(a)}{g(a)} c = f(a)q$$

with $q \in (0, 1]$ being the fraction of individuals vaccinated, and

$$c = C(qW_a) = qg(a).$$

Since $\Xi(W_{a^\diamond}) = f(a^\diamond)$ and f is nonincreasing, $a < a^\diamond$. If $q < 1$, $W = q(1-\epsilon)W_a + \epsilon(c/g(a^\diamond))W_{a^\diamond}$ is a two-age vaccination strategy for sufficiently small $\epsilon > 0$ which has also cost c, but provides a lower net replacement ratio because

$$\begin{aligned}\Xi(W) &= f(a)q(1-\epsilon) + f(a^\diamond)\frac{\epsilon c}{g(a^\diamond)} \\ &= \frac{f(a)}{g(a)} c(1-\epsilon) + \frac{f(a^\diamond)}{g(a^\diamond)} \epsilon c > \frac{f(a)}{g(a)} c = \Xi(qW_a).\end{aligned}$$

So, if the cost of vaccinating everybody at age $a^\diamond$ is smaller than the cost ceiling c, the only optimal one-age strategies are those where everybody is vaccinated when reaching the same specific age $a \leqslant a^\diamond$.

Optimal Two-Age Strategies for Maximum Protection

It can be shown that there exists an optimal strategy which is concentrated at either one age or two ages, i.e., W is a step function with at most two discontinuities, or equivalently the associated Stieltjes measure is a one- or two-point measure. See Hadeler, Müller (1996), Müller (1998), or Theorem B.28 in Toolbox B.4. Let us try to figure out what such an optimal two-age strategy looks like. We assume that $g(a^\diamond) < c$, where $a^\diamond$ is an age where f/g takes its maximum; otherwise we have an optimal one-age vaccination strategy. For the two-age strategy we have

$$\Xi(W) = \frac{f(a_1)}{g(a_2)} c_1 + \frac{f(a_2)}{g(a_2)} c_2,$$
$$c = c_1 + c_2, \qquad c_1 = q_1 g(a_1) > 0, \qquad c_2 = q_2 g(a_2) > 0,$$
$$1 \geqslant q_1 + q_2, \qquad q_1, q_2 > 0.$$

We have $g(a_j) \geqslant c$ for $j = 1$ or $j = 2$; we assume $g(a_2) \geqslant c$ without restriction of generality. Substituting the second equation into the first, we see that

$$\Xi(W) = \frac{f(a_1)}{g(a_1)} c_1 + \frac{f(a_2)}{g(a_2)} (c - c_1).$$

This shows that $\Xi(W)$ can be increased and $C(W)$ can be decreased by replacing a_2 by $a^\diamond$. The second follows from $g(a_2) \geqslant c > g(a^\diamond)$. So, we can assume that an optimal strategy has the form

$$W = q_1 W(a) + q_2 W(a^\diamond).$$

We necessarily have $g(a) > c$; otherwise $C(W) < c$ because $g(a^\diamond) < c$. Since f is nonincreasing, this implies that $a < a^\diamond$; otherwise $f(a) \leqslant f(a^\diamond)$ and, with $q = q_1 + q_2$,

$$\Xi(W) \leqslant \Xi(qW_{a^\diamond}), \qquad C(W) = c > C(qW_{a^\diamond}),$$

and we can find a better two-age strategy by combining vaccinations at age 0 and at age $a^\diamond$. So,

$$\Xi(W) = \frac{f(a)}{g(a)} c_1 + \frac{f(a^\diamond)}{g(a^\diamond)} c_2,$$
$$c = C(W) = c_1 + c_2, \quad c_1 = g(a) q_1, \quad c_2 = g(a^\diamond) q_2,$$

where

$$a < a^\diamond, \qquad g(a) > c > g(a^\diamond), \qquad q_1 + q_2 \leqslant 1.$$

Obviously, $\Xi(W)$ becomes as large as possible if we choose c_2 as large as the constraints allow. Now, with $q = q_1 + q_2 \leqslant 1$,

$$c_2 = c - c_1 = c - g(a)q_1 = c - g(a)(q - q_2)$$
$$= c - g(a)\left(q - \frac{c_2}{g(a^\diamond)}\right) = c - g(a)q + c_2\frac{g(a)}{g(a^\diamond)}.$$

Solving for c_2,

$$c_2\left(\frac{g(a)}{g(a^\diamond)} - 1\right) = g(a)q - c,$$

from which we learn that c_2 becomes as large as possible if $q = q_1 + q_2 = 1$. This means that everybody is finally vaccinated, and the fraction vaccinated at the second age is

$$q_2 = \frac{c_2}{g(a^\diamond)} = \frac{g(a) - c}{g(a) - g(a^\diamond)}.$$

In principle, we could substitute the formula for q_2 and $q_1 = 1 - q_2$ into

$$\Xi(W) = q_1 f(a) + q_2 f(a^\diamond)$$

and look for an optimal $a < a^\diamond$ satisfying $g(a) > c$. But this does not seem to provide any additional interpretive insight.

Let us summarize. *There exists an optimal vaccination strategy which is concentrated at one or two ages, depending on the functions f and g in the expressions for the net replacement ratio and the cost functional. If f/g takes its maximum at an age $a^\diamond$ and the cost of vaccinating everybody at age $a^\diamond$, denoted by $c^\diamond$, is as much as or more than the cost ceiling c ($c^\diamond \geqslant c$), then the vaccination strategy concentrated at $a^\diamond$ is optimal, and the fraction of individuals vaccinated is $c/c^\diamond$.*

If for all ages $a^\diamond$ where f/g takes its maximum, the strategy where everybody is vaccinated at age $a^\diamond$ costs less than c, then there exists either an optimal one-age strategy where everybody is vaccinated when reaching the same specific age $a \leqslant a^\diamond$, or an optimal two-age strategy which has the following features: at the second vaccination age, f/g takes its maximum and everybody is vaccinated who has not been vaccinated at the first age.

At the level of generality we have employed here, the function f is too complex for interpretation. In the next section, we will exploit the different time scales between demographics and endemics to derive an approximate formula for $\mathcal{R}_0(V)$, and we will revisit the description of the optimal ages at vaccination at its end.

The optimization problem considered here has two constraints: the natural constraint $W \leqslant 1$, which is inherent to the problem because the maximal fraction that can be vaccinated is 1; and the economic constraint of a cost ceiling.

If other constraints are added that cannot be incorporated into the previous ones, the number of ages at which an optimal vaccination strategy is concentrated may increase and can equal the number of constraints, but without exceeding it (see Toolbox B.4, Theorem B.28).

22.5 Estimating the Net Replacement Ratio: Average Duration of Susceptibility and Average Age at Infection. Optimal Vaccination Schedules Revisited

In order to facilitate the estimation of the net replacement ratio from epidemiological data and to further interpret the optimal one- or two-age vaccination schedules discussed in the last section, we will derive approximate formulas for the net replacement ratio, $\mathcal{R}_0(V)$, in infectious diseases which operate on a much faster time scale than the demographic dynamics of the affected population (like most childhood diseases). If $\mathcal{R}_0(V) > 1$ and ψ^* is the infective force at endemic equilibrium, we will find that

$$\mathcal{R}_0(V) \approx \frac{\int_0^\infty V(b)k_1(b)k_2(b)\mathcal{F}(b)\,\mathrm{d}b}{\int_0^\infty \mathrm{e}^{-K_1(b)\psi^*}V(b)k_1(b)k_2(b)\mathcal{F}(b)\,\mathrm{d}b}, \qquad K_1(b) = \int_0^b k_1(a)\,\mathrm{d}a.$$

We will introduce the following epidemiological concepts: *average age at infection*, A_{I}, and *average duration of susceptibility*, D_{S}. These two concepts have sometimes not been clearly distinguished in the literature; the average duration of susceptibility ignores the fact that an individual can die before it can be infected.

Since, at endemic equilibrium, the probability of still being susceptible (i.e., neither infected nor vaccinated) at age b is given by $\mathrm{e}^{-\psi^* K_1(b)}V(b) = u^*(b)V(b)$, the average duration of susceptibility satisfies

$$D_{\mathrm{S}} = \int_0^\infty \mathrm{e}^{-\psi^* K_1(b)}V(b)\,\mathrm{d}b$$

(see Section 12.2). In an unvaccinated population, if k_1 does not depend on age, we have $D_{\mathrm{S}} = 1/\psi^* k_1$, and D_{S} can be estimated by fitting the so-called *simple catalytic curve* $1 - u^*(a) = 1 - \mathrm{e}^{-a/D_{\mathrm{S}}}$ to serological profiles giving the fraction of people in different age groups with antibodies to the antigens of the infectious disease (Dietz, 1975; Anderson, May, 1991, Section 3.2.3).

At endemic equilibrium, the infection rate at age a is given by $S^*(a)\phi^*(a)$; so, the average age at infection is

$$A_{\mathrm{I}} = \frac{\int_0^{a^\sharp} aS^*(a)\phi^*(a)\,\mathrm{d}a}{\int_0^{a^\sharp} S^*(a)\phi^*(a)\,\mathrm{d}a}.$$

Since ages at infection can be recorded, the average age of infection can (at least in principle) be determined for many childhood diseases.

As a preview, here are approximations of the basic replacement ratio (i.e., without vaccinations). They hold under suitable assumptions which will be spelled out later (e.g., k_1 and k_2 are not allowed to depend on age). If $L = \int_0^\infty \mathcal{F}(a)\, da$ denotes the life expectancy, we will find that

$$\mathcal{R}_0(\emptyset) \approx \frac{L}{D_{\mathrm{S}}} \frac{1}{\mathcal{F}(D_{\mathrm{S}})} \quad \text{and} \quad \mathcal{R}_0(\emptyset) \approx \frac{L}{A_{\mathrm{I}}} \frac{\mathcal{F}(2A_{\mathrm{I}})}{\mathcal{F}(A_{\mathrm{I}})^2}.$$

Once an estimate for the basic replacement ratio is known, one can forecast an estimate for the net replacement ratio under vaccination:

$$\frac{\mathcal{R}_0(V)}{\mathcal{R}_0(\emptyset)} \approx \frac{\int_0^\infty V(b)\mathcal{F}(b)\, \mathrm{d}b}{L}.$$

Notice that the numerator on the right-hand side gives the expectation of unvaccinated life.

Approximation of the Net Replacement Ratio

By (22.11) and a change of the order of integration, the replacement ratio at constant infective force ψ^* is given by

$$\mathcal{R}(\psi^*) = -B \int_0^\infty \mathcal{F}(b) k_2(b) \times \left(\int_0^b \mathrm{e}^{-K_1(a)\psi^*} k_1(a) \frac{V(a)}{H(a)} \left[\int_a^b \frac{G(b)}{G(s)}\, \mathrm{d}H(s) \right] da \right) db.$$

After changing the order of integration again,

$$\begin{aligned}\mathcal{R}(\psi^*) &= -B \int_0^\infty \mathrm{e}^{-K_1(a)\psi^*} V(a) k_1(a) \times \left(\int_a^\infty k_2(b)\mathcal{F}(b) \left[\int_a^b \frac{G(b)}{G(s)} \frac{\mathrm{d}H(s)}{H(a)} \right] db \right) da \\ &= -B \int_0^\infty \mathrm{e}^{-K_1(a)\psi^*} V(a) k_1(a) \times \left(\int_0^\infty k_2(b+a)\mathcal{F}(b+a) \left[\int_a^{b+a} \frac{G(b+a)}{G(s)} \frac{\mathrm{d}H(s)}{H(a)} \right] db \right) da.\end{aligned}$$

We consider the special case that the lengths of the exposed and infectious stages do not depend on age, i.e.,

$$\frac{G(b+a)}{G(a)} = \tilde{G}(b), \qquad \frac{H(b+a)}{H(a)} = \tilde{H}(b)$$

with nonincreasing functions $\tilde{G}$, $\tilde{H}$, $\tilde{G}(0) = 1 = \tilde{H}(0)$, $\tilde{G}(\infty) = 0 = \tilde{H}(\infty)$. Then

$$-\int_a^{b+a} \frac{G(b+a)}{G(s)} \frac{\mathrm{d}H(s)}{H(a)} = -\int_0^b \frac{G(b+a)}{G(a+s)} \mathrm{d}_s \frac{H(a+s)}{H(a)}$$
$$= -\int_0^b \tilde{G}(b-s)\, \mathrm{d}\tilde{H}(s)$$
$$=: \Gamma(b).$$

All integrals in this formula are Stieltjes integrals (see Toolbox B.1); the symbol d_s in the second integral means that the Stieltjes integral is taken with respect to the variable s. This allows us to write more compactly

$$\mathcal{R}(\psi^*) = B \int_0^\infty \mathrm{e}^{-K_1(a)\psi^*} V(a) k_1(a) \left(\int_0^\infty k_2(b+a) \mathcal{F}(b+a) \Gamma(b)\, \mathrm{d}b \right) \mathrm{d}a.$$

A similar proof as the one of Lemma B.29 shows that

$$\int_0^\infty \Gamma(b)\, \mathrm{d}b = \int_0^\infty \tilde{G}(b)\, \mathrm{d}b = D_\mathrm{I},$$

with D_I denoting the average length of the infectious period (average duration of the infectious stage).

We will now assume that the vital dynamics of the population, the vaccination rates, and the change of susceptibility and infectivity will act on a much slower time scale than the disease; in particular we will assume that the average length of the infectious period will be be much smaller than the life expectancy, L:

$$D_\mathrm{I} \ll L := \int_0^\infty \mathcal{F}(a)\, \mathrm{d}a.$$

In order to express this assumption mathematically, we normalize $\mathcal{F}$ and Γ

$$\mathcal{F}(a) = \mathcal{F}_0(a/L), \qquad \Gamma(a) = \Gamma_0(a/D_\mathrm{I}),$$

such that

$$\int_0^\infty \mathcal{F}_0(s)\, \mathrm{d}s = 1 = \int_0^\infty \Gamma_0(s)\, \mathrm{d}s,$$

and assume that

$$V(a) = V_0(a/L), \qquad k_1(a) = k_1^\circ(a/L), \qquad k_2(a) = k_2^\circ(a/L).$$

By substitution,

$$K_1(b) = \int_0^b k_1(s)\, \mathrm{d}s = \int_0^b k_1^\circ(s/L)\, \mathrm{d}s = L K_1^\circ(b/L)$$

with

$$K_1^\circ(r) = \int_0^r k_1^\circ(s)\,\mathrm{d}s.$$

If B is the constant input of newborn individuals and N is the total population size, then, at equilibrium, $N = BL$ (recall Theorem 13.2). Using these time scales,

$$\mathcal{R}(\psi^*) = \frac{N}{L}\int_0^\infty \mathrm{e}^{-K_1^\circ(a/L)L\psi^*} V_0\Big(\frac{a}{L}\Big)k_1^\circ\Big(\frac{a}{L}\Big) \times \bigg(\int_0^\infty k_2^\circ\Big(\frac{b+a}{L}\Big)\mathcal{F}_0\Big(\frac{b+a}{L}\Big)\Gamma_0\Big(\frac{b}{D_\mathrm{I}}\Big)\,\mathrm{d}b\bigg)\,\mathrm{d}a.$$

Substituting $a = rL$ and $b = sD_\mathrm{I}$ and introducing

$$\epsilon = \frac{D_\mathrm{I}}{L} \ll 1,$$

we obtain

$$\mathcal{R}(\psi^*) = ND_\mathrm{I}\int_0^\infty \mathrm{e}^{-K_1^\circ(r)L\psi^*} V_0(r)k_1^\circ(r) \times \bigg(\int_0^\infty k_2^\circ(\epsilon s + r)\mathcal{F}_0(\epsilon s + r)\Gamma_0(s)\,\mathrm{d}s\bigg)\,\mathrm{d}r. \tag{22.17}$$

Since the life expectancy is much larger than the mean length of the infectious period, i.e., $\epsilon \ll 1$, and since the integral over Γ_0 equals 1, we expect that

$$\mathcal{R}(\psi^*) \approx ND_\mathrm{I}\int_0^\infty \mathrm{e}^{-K_1^\circ(r)L\psi^*} V_0(r)k_1^\circ(r)k_2^\circ(r)\mathcal{F}_0(r)\,\mathrm{d}r.$$

Indeed the following lemma holds.

Lemma 22.5.

$$\int_0^\infty \mathrm{e}^{-K_1^\circ(r)L\psi^*} V_0(r)k_1^\circ(r)\bigg(\int_0^\infty k_2^\circ(\epsilon s + r)\mathcal{F}_0(\epsilon s + r)\Gamma_0(s)\,\mathrm{d}s\bigg)\,\mathrm{d}r - \int_0^\infty \mathrm{e}^{-K_1^\circ(r)L\psi^*} V_0(r)k_1^\circ(r)k_2^\circ(r)\mathcal{F}_0(r)\,\mathrm{d}r \longrightarrow 0, \quad \epsilon \to 0, \quad \textit{uniformly for } L, \quad \psi^* \geqslant 0.$$

Proof. Let $\Delta(\epsilon, L\psi^*)$ denote the difference in the statement of the theorem. Then

$$|\Delta(\epsilon, L\psi^*)| \leqslant \int_0^c \mathrm{e}^{-K_1^\circ(r)L\psi^*} V_0(r)k_1^\circ(r)\mathcal{F}_0(r)h_\epsilon(r)\,\mathrm{d}r$$

with

$$h_\epsilon(r) = \int_0^\infty k_2^\circ(\epsilon s + r)\bigg[1 - \frac{\mathcal{F}_0(\epsilon s + r)}{\mathcal{F}_0(r)}\bigg]\Gamma_0(s)\,\mathrm{d}s.$$

Here $c \in (0, \infty]$ has been chosen such that $\mathcal{F}_0(r) > 0$ for $r < c$ and $\mathcal{F}_0(r) = 0$ for $r > c$. Recall that $\mathcal{F}_0$ is monotone nonincreasing. Hence

$$0 \leqslant 1 - \frac{\mathcal{F}_0(\epsilon s + r)}{\mathcal{F}_0(r)} \leqslant 1,$$

and

$$|\Delta(\epsilon, L\psi^*)| \leqslant \int_0^c V_0(r) k_1^\circ(r) \mathcal{F}_0(r) h_\epsilon(r)\, \mathrm{d}r.$$

Notice that this expression no longer contains L and ψ^*. Since $\mathcal{F}_0$ is monotone nonincreasing, it is continuous everywhere except at countably many points. By the theorem of dominated a.e. convergence (e.g., Berberian, 1965, Section 30; McDonald, Weiss, 1999, Proposition 4.11; Aliprantis, Burkinshaw, 1998, Theorem 22.11), we have

$$h_\epsilon(r) \longrightarrow 0, \quad \epsilon \to 0 \quad \forall r \geqslant 0.$$

Furthermore,

$$0 \leqslant h_\epsilon(r) \leqslant \sup k_2^\circ \quad \forall r \geqslant 0.$$

Since k_1° and V are bounded and $\mathcal{F}$ is integrable, we apply the theorem of dominated convergence one more time and obtain

$$\Delta(\epsilon, L\psi^*) \to 0, \quad \epsilon \to 0, \quad \text{uniformly in } L, \quad \Psi^* \geqslant 0.$$

□

Proposition 22.6. *Let $\mathcal{R}_0 > 1$ and let ψ^* be the infective force at endemic equilibrium. Then*

$$\mathcal{R}_0 \approx \frac{\int_0^\infty V(b) k_1(b) k_2(b) \mathcal{F}(b)\, \mathrm{d}b}{\int_0^\infty \mathrm{e}^{-K_1(b)\psi^*} V(b) k_1(b) k_2(b) \mathcal{F}(b)\, \mathrm{d}b}, \qquad \epsilon := \frac{D_\mathrm{I}}{L} \to 0.$$

Proof. Since $1 = \mathcal{R}(\psi^*)$, from (22.17),

$$\begin{aligned}\mathcal{R}_0 &= \frac{\mathcal{R}_0}{\mathcal{R}(\psi^*)} \\ &= \frac{\int_0^\infty V_0(r) k_1^\circ(r) (\int_0^\infty k_2^\circ(\epsilon s + r) \mathcal{F}_0(\epsilon s + r) \Gamma_0(s)\, \mathrm{d}s)\, \mathrm{d}r}{\int_0^\infty \mathrm{e}^{-K_1^\circ(r) L\psi^*} V_0(r) k_1^\circ(r) (\int_0^\infty k_2^\circ(\epsilon s + r) \mathcal{F}_0(\epsilon s + r) \Gamma_0(s)\, \mathrm{d}s)\, \mathrm{d}r}.\end{aligned}$$

By Lemma 22.5 (notice that the convergence is uniform in ψ^*, which may depend on ϵ),

$$\mathcal{R}_0 \longrightarrow \frac{\int_0^\infty V_0(r) k_1^\circ(r) k_2^\circ(r) \mathcal{F}_0(r)\, \mathrm{d}r}{\int_0^\infty \mathrm{e}^{-K_1^\circ(r) L\psi^*} V_0(r) k_1^\circ(r) k_2^\circ(r) \mathcal{F}_0(r)\, \mathrm{d}r}, \qquad \epsilon \searrow 0.$$

Substituting $r = b/L$ provides the assertion. □

We finally assume that k_1 and k_2 are constant. Then

$$\mathcal{R}_0 \approx \frac{\int_0^\infty V(b)\mathcal{F}(b)\,\mathrm{d}b}{\int_0^c \mathrm{e}^{-k_1\psi^* b} V(b)\mathcal{F}(b)\,\mathrm{d}b}.$$

By (22.10),

$$\mathcal{R}_0 \approx \frac{\int_0^\infty V(b)\mathcal{F}(b)\,\mathrm{d}b}{\int_0^\infty u^*(b)V(b)\mathcal{F}(b)\,\mathrm{d}b} =: \tilde{\mathcal{R}}_0 \qquad (22.18)$$

with

$$u^*(b) = \frac{S^*(b)}{S^*(0)} = \mathrm{e}^{-k_1\psi^* b}$$

being the probability of not having yet been infected at age b.

To interpret formula (22.18) we notice that $V(b)\mathcal{F}(b)$ is the joint probability of still being alive and unvaccinated at age b, hence

$$\int_0^\infty V(b)\mathcal{F}(b)\,\mathrm{d}b$$

is the expected age at which people are either vaccinated or die; in other words, it is the *expectation of unvaccinated life* (see Section 12.2). Similarly, $u^*(a)V(b)\mathcal{F}(b)$ is the joint probability of still being alive, susceptible, and unvaccinated. Hence

$$\int_0^\infty u^*(b)V(b)\mathcal{F}(b)\,\mathrm{d}b$$

is the average age at which, at the endemic equilibrium, individuals leave the susceptible class by being infected, vaccinated or by dying; in other words, it is the expected sojourn in the susceptible class at endemic equilibrium.

Average Duration of Susceptibility and Average Age at Infection in an Unvaccinated Population

In order to link formula (22.18) to epidemiological concepts for which data can be collected, we introduce *average age at infection*, A_I, and *average duration of susceptibility*, D_S. We assume in this subsection that there are no vaccinations and that the per capita susceptibility, k_1, and the per capita infectivity, k_2, do not depend on age. As we pointed out at the beginning of this section, these assumptions imply that

$$D_\mathrm{S} = \frac{1}{k_1\psi^*}.$$

By (22.18), with $V \equiv 1$,

$$\frac{1}{\mathcal{R}_0} \approx \frac{\int_0^\infty e^{-b/D_S} \mathcal{F}(b)\, db}{\int_0^\infty \mathcal{F}(b)\, db} =: \frac{1}{\tilde{\mathcal{R}}_0}. \tag{22.19}$$

We find an approximation for the numerator in (22.19) using Proposition B.36 with $\varphi(t) = \mathcal{F}(t)$ and $g(t) = e^{-t/D_S}$. Assume that $\mathcal{F}$ is twice continuously differentiable on $(0, a^\sharp)$, where $a^\sharp \in (0, \infty]$ denotes the maximum age. Then

$$\int_0^\infty e^{-b/D_S} \mathcal{F}(b)\, db = \mathcal{F}(E) \int_0^{a^\sharp} e^{-b/D_S}\, db + \text{rest}$$

with

$$E = \frac{\int_0^{a^\sharp} b e^{-b/D_S}\, db}{\int_0^{a^\sharp} e^{-b/D_S}\, db}$$

and

$$|\text{rest}| \leqslant \Lambda \left[L^{-1} \int_0^{a^\sharp} e^{-b/D_S}\, db \right]^2 \frac{1}{2} \frac{\int_0^{a^\sharp} b^2 e^{-b/D_S}\, db}{(\int_0^{a^\sharp} e^{-b/D_S}\, ds)^3}.$$

Here Λ is the bound of the second derivative of the normalization of $\mathcal{F}$, $\mathcal{F}_0(s) = \mathcal{F}(sL)$. Substituting these results into (22.19),

$$\frac{1}{\mathcal{R}_0} \approx \mathcal{F}(E) \frac{D_S}{L} \int_0^{a^\sharp/D_S} e^{-s}\, ds,$$

$$E = D_S \frac{\int_0^{a^\sharp/D_S} s e^{-s}\, ds}{\int_0^{a^\sharp/D_S} e^{-s}\, ds},$$

and

$$|\text{rest}| \leqslant \Lambda \left[\frac{D_S}{L} \right]^2 \frac{1}{1 - e^{-a^\sharp/D_S}}.$$

If we understand the discussion in Section 3.2.3 of Anderson, May (1991) correctly, they actually mean D_S when they speak about average age of infection, so we assume that $a^\sharp/D_S \geqslant 10$ and $L/D_S \geqslant 5$ for most childhood diseases (cf. Anderson, May, 1991, Table 6.1). Then $E \approx D_S$ and we can further approximate

$$\mathcal{R}_0 \approx \frac{1}{\mathcal{F}(D_S)} \frac{L}{D_S}. \tag{22.20}$$

As mentioned after Proposition B.36, the approximation is particularly good if $\mathcal{F}$ is linear on $(0, a^\sharp)$, i.e., if

$$\mathcal{F}(a) = \begin{cases} p - \alpha a, & 0 < a < a^\sharp, \\ 0, & a > a^\sharp, \end{cases}$$

where $\alpha a^\sharp \leqslant p$, $0 < p \leqslant 1$. If the survivorship is exponentially distributed, formula (22.20) and the Taylor expansion provide

$$\mathcal{R}_0 \approx \mathrm{e}^{D_\mathrm{S}/L} \frac{L}{D_\mathrm{S}} = \frac{L}{D_\mathrm{S}} + 1 + \tfrac{1}{2}\mathrm{e}^{\zeta} \frac{D_\mathrm{S}}{L}, \quad \text{with some } 0 < \zeta < \frac{D_\mathrm{S}}{L}.$$

Actually, one can show that the equality $\tilde{R}_0 = (L/D_\mathrm{S}) + 1$ holds (Exercise 22.2).

The average duration of susceptibility has often been identified with, but should be discriminated from, the average age at infection, A_I. Since, at endemic equilibrium, the infection rate at age a is given by $S^*(a)\phi^*(a)$, the average age at infection is

$$A_\mathrm{I} = \frac{\int_0^{a^\sharp} aS^*(a)\phi^*(a)\,\mathrm{d}a}{\int_0^{a^\sharp} S^*(a)\phi^*(a)\,\mathrm{d}a}.$$

Recalling that $S^*(a) = u^*(a)V(a)\mathcal{F}(a)$ and $\phi^*(a) = k_1(a)\psi^*$,

$$A_\mathrm{I} = \frac{\int_0^{a^\sharp} au^*(a)V(a)\mathcal{F}(a)k_1(a)\,\mathrm{d}a}{\int_0^{a^\sharp} u^*(a)V(a)\mathcal{F}(a)k_1(a)\,\mathrm{d}a}.$$

Let us assume that the per capita susceptibility does not depend on age and that there are no vaccinations. Then

$$A_\mathrm{I} = \frac{\int_0^{a^\sharp} a\mathrm{e}^{-a/D_\mathrm{S}}\mathcal{F}(a)\,\mathrm{d}a}{\int_0^{a^\sharp} \mathrm{e}^{-a/D_\mathrm{S}}\mathcal{F}(a)\,\mathrm{d}a}.$$

Using (22.18),

$$A_\mathrm{I} = \frac{\int_0^{a^\sharp} a\mathrm{e}^{-a/D_\mathrm{S}}\mathcal{F}(a)\,\mathrm{d}a}{\int_0^{a^\sharp} \mathcal{F}(a)\,\mathrm{d}a}\tilde{\mathcal{R}}_0.$$

So,

$$\frac{A_\mathrm{I}L}{\tilde{\mathcal{R}}_0} = \int_0^{a^\sharp} a\mathrm{e}^{-a/D_\mathrm{S}}\mathcal{F}(a)\,\mathrm{d}a = D_\mathrm{S}^2 \int_0^{a^\sharp/D_\mathrm{S}} b\mathrm{e}^{-b}\mathcal{F}(bD_\mathrm{S})\,\mathrm{d}b. \tag{22.21}$$

Applying Proposition B.36, with $\varphi(t) = \mathcal{F}(tD_\mathrm{S})$ and $g(t) = t\mathrm{e}^{-t}$,

$$\int_0^{a^\sharp/D_\mathrm{S}} b\mathrm{e}^{-b}\mathcal{F}(bD_\mathrm{S})\,\mathrm{d}b = \mathcal{F}\left(D_\mathrm{S}\frac{\int_0^{a^\sharp/D_\mathrm{S}} b^2\mathrm{e}^{-b}\,\mathrm{d}b}{\int_0^{a^\sharp/D_\mathrm{S}} b\mathrm{e}^{-b}\,\mathrm{d}b}\right) + \text{rest}$$

with

$$|\text{rest}| \leqslant \Lambda\left[\frac{\int_0^{a^\sharp/D_S} b\mathrm{e}^{-b}\,\mathrm{d}s}{\int_0^{a^\sharp/D_S} \mathcal{F}(bD_S)\,\mathrm{d}s}\right]^2 \frac{1}{2}\frac{\int_0^\infty b^3\mathrm{e}^{-b}\,\mathrm{d}b}{(\int_0^{a^\sharp/D_S} b\mathrm{e}^{-b}\,\mathrm{d}b)^3}$$
$$\leqslant \Lambda\left[\frac{D_S}{L}\right]^2 \frac{3}{\int_0^{a^\sharp/D_S} b\mathrm{e}^{-b}\,\mathrm{d}b}.$$

Using the fact that $a^\sharp/D_S$ is large and D_S/L is small, we obtain

$$\int_0^{a^\sharp/D_S} b\mathrm{e}^{-b}\mathcal{F}(bD_S)\,\mathrm{d}b \approx \mathcal{F}(2D_S).$$

Substituting this into (22.21) and using (22.18),

$$\frac{A_I L}{\mathcal{R}_0} \approx D_S^2\mathcal{F}(2D_S).$$

Since $\mathcal{R}_0 D_S \approx L/\mathcal{F}(D_S)$ by (22.20),

$$D_S \approx A_I \frac{\mathcal{F}(D_S)}{\mathcal{F}(2D_S)} > A_I.$$

Substituting this back into the equation for $\mathcal{R}_0$ yields

$$\mathcal{R}_0 \approx \frac{L}{A_I}\frac{\mathcal{F}(2D_S)}{\mathcal{F}(D_S)^2}.$$

Mortality between D_S and $2D_S$ is presumably low, so D_S and A_I should not be very different. Since mortality between A_I and D_S and $2A_I$ and $2D_S$ should also be low, $\mathcal{F}(D_S) \approx \mathcal{F}(A_I)$ and $\mathcal{F}(2D_S) \approx \mathcal{F}(2A_I)$. So,

$$\mathcal{R}_0 \approx \frac{L}{A_I}\frac{\mathcal{F}(2A_I)}{\mathcal{F}(A_I)^2}. \tag{22.22}$$

Both for an exponentially distributed survivorship and for fixed life length (assuming $2A_I \leqslant a^\sharp = L$), this specializes to

$$\mathcal{R}_0 \approx \frac{L}{A_I}.$$

Since the approximation in Proposition B.36 is actually an equality if the function ϕ is linear on the interval (a, b) in Proposition B.36, this approximation is very good for a survivorship that is linear on $(0, a^\sharp)$, in particular for the step survival function. We even have the equality $\tilde{\mathcal{R}}_0 = L/A_I$ for an exponentially distributed survivorship (Exercise 22.2). However, if $\mathcal{F}$ is a step survival function with infant mortality (see Section 12.5), then

$$\mathcal{R}_0 \approx \frac{L}{A_I}\frac{1}{p},$$

where p is the probability of surviving infancy, provided that, realistically, the average age of infection is beyond infancy and smaller than half the maximum age.

Comparison of Net Replacement Ratios for Different Vaccination Strategies

Lemma 22.5 can also be used to derive approximate formulas that compare the net replacement ratios for different vaccination strategies. The same proof as for Proposition 22.6 provides the following result.

Proposition 22.7. *Let V_1 and V_2 be two vaccination profiles and let $\mathcal{R}_0(V_1)$ and $\mathcal{R}_0(V_2)$ denote the related net replacement ratios. Then*

$$\frac{\mathcal{R}_0(V_1)}{\mathcal{R}_0(V_2)} \approx \frac{\int_0^\infty V_1(b)k_1(b)k_2(b)\mathcal{F}(b)\,\mathrm{d}b}{\int_0^\infty V_2(b)k_1(b)k_2(b)\mathcal{F}(b)\,\mathrm{d}b}, \quad \frac{D_\mathrm{I}}{L} \to 0.$$

Let $\mathcal{R}_0(\emptyset)$ denote the basic replacement ratio, without vaccination, and assume that k_1 and k_2 are constant.

Corollary 22.8.

$$\frac{\mathcal{R}_0(V)}{\mathcal{R}_0(\emptyset)} \approx \frac{\int_0^\infty V(b)\mathcal{F}(b)\,\mathrm{d}b}{L}, \quad \frac{D_\mathrm{I}}{L} \to 0.$$

Recall that the numerator on the right-hand side gives the expectation of unvaccinated life. This formula becomes more suggestive if we express it as follows:

$$\begin{aligned}\int_0^\infty V(b)\mathcal{F}(b)\,\mathrm{d}b &= -\int_0^\infty V(b)\frac{\mathrm{d}}{\mathrm{d}b}\left(\int_b^\infty \mathcal{F}(a)\,\mathrm{d}a\right)\mathrm{d}b \\ &= L + \int_0^\infty \mathcal{F}(b)L(b)\,\mathrm{d}V(b),\end{aligned}$$

where

$$L(b) = \frac{\int_b^\infty \mathcal{F}(a)\,\mathrm{d}a}{\mathcal{F}(b)}$$

is the expected remaining life at age b (Section 12.4). Thus

$$\frac{\mathcal{R}_0(V)}{\mathcal{R}_0(\emptyset)} = 1 + \int_0^\infty \mathcal{F}(b)\frac{L(b)}{L}\,\mathrm{d}V(b).$$

Optimal Ages at Vaccination Revisited

Let us return to the problem of minimizing the net replacement ratio constrained by a cost ceiling, which we considered in the last section. Equivalently, we can minimize $\mathcal{R}(V)/\mathcal{R}(\emptyset)$. By Proposition 22.7 and integration by parts, this amounts to minimizing

$$\int_0^\infty V(b)k_1(b)k_2(b)\mathcal{F}(b)\,\mathrm{d}b = \int_0^\infty k_1(b)k_2(b)\mathcal{F}(b)\,\mathrm{d}b - \int_0^\infty f(a)\,\mathrm{d}W(a)$$

with

$$f(a) = \int_a^\infty k_1(b)k_2(b)\mathcal{F}(b)\,\mathrm{d}b, \quad W = 1 - V.$$

This means that our optimization problem is equivalent to maximizing

$$\Xi(W) = \int_0^\infty f(a)\,\mathrm{d}W(a)$$

under the constraint

$$c \geqslant C(W) = \int_0^\infty g(a)\,\mathrm{d}W(a), \quad g(a) = p(a)\mathcal{F}(a).$$

At the end of the last section we found that the ages in an optimal one- or two-age strategy are chosen at or before those ages where f/g takes its maximum,

$$\begin{aligned} \frac{f(a)}{g(a)} &= \frac{\int_a^\infty k_1(b)k_2(b)\mathcal{F}(b)\,\mathrm{d}b}{p(a)\mathcal{F}(a)} \\ &= \frac{\int_0^\infty k_1(a+r)k_2(a+r)\mathcal{F}(r \mid a)\,\mathrm{d}r}{p(a)}. \end{aligned}$$

Recall that $\mathcal{F}(r \mid a)$ is the probability of living for at least another time span of length r provided one has survived to age a (see Section 12.4). If the susceptibility k_1 and the infectivity k_2 are constant,

$$\frac{f(a)}{g(a)} = \text{const.} \times \frac{L(a)}{p(a)},$$

so the ages at vaccination should be chosen at or before those ages which maximize the quotient of expected remaining life and of per capita cost. If the per capita cost of vaccination does not depend on age, the ages at vaccination in an optimal one- or two-age strategy should be chosen at or before those ages which maximize the expected remaining life. In this connection it is interesting that the figure in Smith, Keyfitz (1977, p. 5) shows curves of expected remaining life from various centuries which peak somewhere between three and eight years. It can be argued, however, that a reduction in infant mortality has made expected remaining life take its maximum at an earlier age since then.

Exercises and Selected Solutions

Vaccination Strategies

22.1. Let $c > 0$ and let W^* be a vaccination strategy that minimizes $R_0(W)$ under the constraint $C(W) \leqslant c$. Let

$$\rho^* = \mathcal{R}_0(W^*) > 0.$$

Show that W^* minimizes $C(W)$ under the side condition $\mathcal{R}_0(W) \leqslant \rho^*$.

Solution. We first notice that $c < C(W_0)$. Otherwise, since $\mathcal{R}_0(W_0) = 0$, W^* with $\mathcal{R}_0(W^*) > 0$ does not minimize $\mathcal{R}_0(W)$ under the side condition $C(W) \leqslant c$.

Now suppose that W^* does not minimize $C(W)$ under the side condition $\mathcal{R}(W) \leqslant \rho^*$. Then there exists a vaccination distribution W such that

$$C(W) < C(W^*) \leqslant c, \quad \mathcal{R}_0(W) \leqslant \rho^*.$$

Let W_s be the vaccination distribution

$$W_s = (1-s)W + sW_0, \quad s \in [0, 1].$$

Since C is linear and $\mathcal{R}_0$ is affine, we have

$$C(W_s) = (1-s)C(W) + sC(W_0) < (1-s)c + sC(W_0),$$
$$\mathcal{R}_0(W_s) = (1-s)\mathcal{R}_0(W) + s\mathcal{R}_0(W_0) = (1-s)\mathcal{R}_0(W) \leqslant (1-s)\rho^*.$$

Remember $C(W) < c < C(W_0)$ and set

$$s = \frac{c - C(W)}{C(W_0) - C(W)}.$$

Then $0 < s < 1$ and

$$C(W_s) = c, \qquad \mathcal{R}_0(W_s) \leqslant (1-s)\rho^* < \rho^* = \mathcal{R}_0(W^*),$$

because $\rho^* > 0$. Thus the strategy W^* does not minimize $\mathcal{R}_0(W)$ under the constraint $C(W) \leqslant c$, a contradiction.

Approximating the Basic Replacement Ratio

22.2. Consider an exponentially distributed survivorship, $\mathcal{F}(a) = e^{-a/L}$, with life expectancy L. Show for $\tilde{\mathcal{R}}_0$ in (22.19) and A_I in (22.21):

(a) $\tilde{\mathcal{R}}_0 = (L/D_S) + 1$;

(b) $\tilde{\mathcal{R}}_0 = (L/A_I)$.

Another Way of Approximating the Basic Replacement Ratio

22.3. Recall that the net replacement ratio at constant infective force ψ^* is given by

$$\mathcal{R}(\psi^*) = -B \int_0^\infty \mathcal{F}(b) k_2(b) \times \left(\int_0^b \mathrm{e}^{-K_1(a)\psi^*} k_1(a) \frac{V(a)}{H(a)} \left[\int_a^b \frac{G(b)}{G(s)} \,\mathrm{d}H(s) \right] \mathrm{d}a \right) \mathrm{d}b.$$

We consider the special case in which we have per capita rates of leaving the exposed and the infectious stage that do not depend on the age, i.e.,

$$G(b) = \mathrm{e}^{-\gamma b}, \qquad H(a) = \mathrm{e}^{-\eta a}$$

with constants $\gamma, \eta > 0$. Furthermore, we assume that there is no vaccination, i.e., $V \equiv 1$, and that k_1 and k_2 are also age independent.

Show that

$$\mathcal{R}(\psi^*) = \frac{\eta B}{\gamma - \eta} k_1 k_2 \left(\frac{1}{\eta - k_1 \psi^*} \left[\int_0^\infty \mathcal{F}(b) \mathrm{e}^{-k_1 \psi^* b} \,\mathrm{d}b - \int_0^\infty \mathcal{F}(b) \mathrm{e}^{-\eta b} \,\mathrm{d}b \right] - \frac{1}{\gamma - k_1 \psi^*} \left[\int_0^\infty \mathcal{F}(b) \mathrm{e}^{-k_1 \psi^* b} \,\mathrm{d}b - \int_0^\infty \mathcal{F}(b) \mathrm{e}^{-\gamma b} \,\mathrm{d}b \right] \right).$$

By interpreting these terms and arguing that the average durations of the infectious and of the exposed period are much smaller than the mean age of either dying or being infected, give a new derivation of formula (22.19) in Section 22.5:

$$\mathcal{R}_0 \approx \frac{L}{\int_0^\infty \mathcal{F}(b) \mathrm{e}^{-b/D_\mathrm{S}} \,\mathrm{d}b},$$

where L is the life expectancy and D_S the average duration of susceptibility.

Solution. We have

$$-\int_a^b \frac{G(b)}{G(s)} \,\mathrm{d}H(s) = \int_a^b \mathrm{e}^{\gamma(s-b)} \eta \mathrm{e}^{-\eta s} \,\mathrm{d}s$$
$$= \eta \mathrm{e}^{-\gamma b} \int_a^b \mathrm{e}^{(\gamma-\eta)s} \,\mathrm{d}s$$
$$= \mathrm{e}^{-\gamma b} \frac{\eta}{\gamma - \eta} (\mathrm{e}^{(\gamma-\eta)b} - \mathrm{e}^{(\gamma-\eta)a}).$$

Hence

$$-\frac{1}{H(a)}\int_a^b \frac{G(b)}{G(s)}\,\mathrm{d}H(s) = \frac{\eta}{\gamma-\eta}(\mathrm{e}^{\eta(a-b)} - \mathrm{e}^{\gamma(a-b)}).$$

Thus

$$\mathcal{R}(\psi^*) = B\int_0^\infty \mathcal{F}(b)k_2 \times \left(\int_0^b \mathrm{e}^{-k_1\psi^* a}k_1\frac{\eta}{\gamma-\eta}[\mathrm{e}^{\eta(a-b)} - \mathrm{e}^{\gamma(a-b)}]\,\mathrm{d}a\right)\mathrm{d}b.$$

Hence

$$\mathcal{R}(\psi^*) = \frac{\eta B}{\gamma-\eta}k_1k_2\int_0^\infty \mathcal{F}(b)\left(\left[\int_0^b \mathrm{e}^{(\eta-k_1\psi^*)a}\,\mathrm{d}a\right]\mathrm{e}^{-\eta b} - \left[\int_0^b \mathrm{e}^{(\gamma-k_1\psi^*)a}\,\mathrm{d}a\right]\mathrm{e}^{-\gamma b}\right)\mathrm{d}b.$$

Thus

$$\mathcal{R}(\psi^*) = \frac{\eta B}{\gamma-\eta}k_1k_2\int_0^\infty \mathcal{F}(b)\left(\frac{1}{\eta-k_1\psi^*}[\mathrm{e}^{-k_1\psi^* b} - \mathrm{e}^{-\eta b}] - \frac{1}{\gamma-k_1\psi^*}[\mathrm{e}^{-k_1\psi^* b} - \mathrm{e}^{-\gamma b}]\right)\mathrm{d}b.$$

Hence

$$\mathcal{R}(\psi^*) = \frac{\eta B}{\gamma-\eta}k_1k_2 \times\left(\frac{1}{\eta-k_1\psi^*}\left[\int_0^\infty \mathcal{F}(b)\mathrm{e}^{-k_1\psi^* b}\,\mathrm{d}b - \int_0^\infty \mathcal{F}(b)\mathrm{e}^{-\eta b}\,\mathrm{d}b\right] - \frac{1}{\gamma-k_1\psi^*}\left[\int_0^\infty \mathcal{F}(b)\mathrm{e}^{-k_1\psi^* b}\,\mathrm{d}b - \int_0^\infty \mathcal{F}(b)\mathrm{e}^{-\gamma b}\,\mathrm{d}b\right]\right). \tag{22.23}$$

Recall that $\mathrm{e}^{-k_1\psi^* b}$ is the probability of not being infected in the time interval $[0, b]$ and

$$D_\mathrm{S} = \frac{1}{k_1\psi^*} = \int_0^\infty \mathrm{e}^{-k_1\psi^* b}\,\mathrm{d}b$$

is the average duration of the susceptibility period, D_S.

Hence $\mathcal{F}(b)\mathrm{e}^{-k_1\psi^* b}$ is the joint probability of neither dying nor being infected in the interval from birth to age b, and

$$\int_0^\infty \mathcal{F}(b)\mathrm{e}^{-k_1\psi^* b}\,\mathrm{d}b$$

can be interpreted as the expected age at which an individual dies or is infected. Now

$$\int_0^\infty \mathcal{F}(b)\mathrm{e}^{-\eta b}\,\mathrm{d}b < \int_0^\infty \mathrm{e}^{-\eta b}\,\mathrm{d}b = \frac{1}{\eta} = D_{\mathrm{E}}$$

is smaller than the average duration of the exposed period, D_{E}, and

$$\int_0^\infty \mathcal{F}(b)\mathrm{e}^{-\gamma b}\,\mathrm{d}b < \int_0^\infty \mathrm{e}^{-\gamma b}\,\mathrm{d}b = \frac{1}{\gamma} = D_{\mathrm{I}}$$

is smaller than the average duration of the infectious period, D_{I}. Since, for most diseases, the average durations of the infectious period and the exposed period are much smaller than the expected age at death or infection, they can be neglected in (22.23) if ψ^* is the infective force at the endemic equilibrium:

$$\begin{aligned} 1 &= \mathcal{R}(\psi^*) \\ &\approx \frac{\eta B}{\gamma - \eta} k_1 k_2 \left(\frac{1}{\eta - k_1\psi^*} - \frac{1}{\gamma - k_1\psi^*} \right) \int_0^\infty \mathcal{F}(b)\mathrm{e}^{-k_1\psi^* b}\,\mathrm{d}b \\ &= \frac{\eta B}{\gamma - \eta} k_1 k_2 \left(\frac{D_{\mathrm{E}}}{1 - (D_{\mathrm{E}}/D_{\mathrm{S}})} - \frac{D_{\mathrm{I}}}{1 - (D_{\mathrm{I}}/D_{\mathrm{S}})} \right) \int_0^\infty \mathcal{F}(b)\mathrm{e}^{-b/D_{\mathrm{S}}}\,\mathrm{d}b. \end{aligned}$$

Again the average duration of susceptibility is very large in comparison to the average durations of the exposed and the infectious periods, i.e.,

$$D_{\mathrm{E}} \ll D_{\mathrm{S}}, \qquad D_{\mathrm{I}} \ll D_{\mathrm{S}}.$$

Hence

$$1 = \mathcal{R}(\psi^*) \approx \frac{\eta B}{\gamma - \eta} k_1 k_2 (D_{\mathrm{E}} - D_{\mathrm{I}}) \int_0^\infty \mathcal{F}(b)\mathrm{e}^{-b/D_{\mathrm{S}}}\,\mathrm{d}b. \quad (22.24)$$

For $\psi^* = 0$ we have from (22.23) that

$$\begin{aligned} \mathcal{R}_0 = \frac{\eta B}{\gamma - \eta} k_1 k_2 \bigg(&\frac{1}{\eta}\left[\int_0^\infty \mathcal{F}(b)\,\mathrm{d}b - \int_0^\infty \mathcal{F}(b)\mathrm{e}^{-\eta b}\,\mathrm{d}b\right] \\ &- \frac{1}{\gamma}\left[\int_0^\infty \mathcal{F}(b)\,\mathrm{d}b - \int_0^\infty \mathcal{F}(b)\mathrm{e}^{-\gamma b}\,\mathrm{d}b\right]\bigg). \end{aligned}$$

Now

$$L = \int_0^\infty \mathcal{F}(b)\,\mathrm{d}b$$

is the life expectancy, which again is much larger than the average durations of the infectious and exposed periods. Hence

$$\mathcal{R}_0 \approx \frac{\eta B}{\gamma - \eta} k_1 k_2 \left(\frac{1}{\eta} - \frac{1}{\gamma} \right) L = \frac{\eta B}{\gamma - \eta} k_1 k_2 (D_{\mathrm{E}} - D_{\mathrm{I}}) L.$$

From the last equation and from (22.24) we obtain

$$\mathcal{R}_0 = \frac{\mathcal{R}_0}{\mathcal{R}(\psi^*)} \approx \frac{L}{\int_0^\infty \mathcal{F}(b) \mathrm{e}^{-b/D_{\mathrm{S}}}\,\mathrm{d}b}.$$

This is formula (22.19).

Bibliographic Remarks

A survey on endemic models with chronological age-structure (till 1991) can be found in Thieme (1991); an up-to-date overview is presented in Hethcote (2000). Capasso (1993) and Iannelli (1994) contain some material and references concerning the SIS case. An early age-structured epidemic model is due to Bernoulli (1766), who estimated the gain in life expectancy if smallpox was eliminated as a cause of death. See Dietz, Heesterbeek (2002) for the historical account and a mathematical generalization.

The local stability of the endemic equilibrium is discussed by Inaba (1990), Thieme (1991), Andreasen (1995), and Cha et al. (2000). The endemic equilibrium can be unstable if the per capita susceptibility decreases in age (Andreasen, 1995), or the per capita infectivity is concentrated in a very narrow age-window (Thieme, 1991). Hethcote (2000) gives a nice Lyapunov(–LaSalle) argument for the global stability of the disease-free equilibrium if $\mathcal{R}_0 < 1$, and its instability if $\mathcal{R}_0 > 1$. He also discusses the concepts of average duration of susceptibility and average age at infection, and he seems to have been the first to observe that the two should be discriminated. The relation of these two concepts to $\mathcal{R}_0$ is derived for exponentially distributed survivorships and survival step functions. His model is more general in so far as it allows exponential population growth and maternal antibodies, and more restrictive as the lengths of the latent and infectious periods are exponentially distributed. Hadeler, Müller (1996) use the Kuhn–Tucker theory (see Zeidler, 1985, for example) to show that generically all optimal vaccination strategies are one- or two-age strategies. In Toolbox B.3 we follow the approach by Müller (1998), though we have recast it in a measure-theoretic framework.

Assuming separable mixing is a huge (over)simplification. Without separable mixing, even uniqueness of the endemic equilibrium is a difficult problem. For sufficient conditions see Inaba (1990) and Cha, Iannelli, Milner (1998). Also the optimality of one- or two-age vaccinations cannot be shown by the techniques used in these notes, because the net replacement ratio $\mathcal{R}_0(V)$ becomes the spectral radius of some linear operator (Müller, 1998), and it is no longer clear whether it is an affine function of the vaccination strategy V. For estimation of the net replacement ratio in the case of nonseparable mixing and for additional references, see Anderson, May (1991), Greenhalgh, Dietz (1994), and Farrington, Kanaan, Gay (2001). Vaccination programs may have aims other than minimizing the net replacement ratio; some target age-specific risk groups, like women of child-bearing age in the case of rubella (see, for example, Hethcote, 1989b; Greenhalgh, 1990; Anderson, May (1991); and the references therein).

Age-structure helps to explain the negative binomial distribution of macroparasites in their hosts (Dietz, 1982); for the analysis of the resulting models see Hadeler, Dietz (1984) and the literature cited there.

Age-structure is only one example of structure in endemic models; others are spatial structure (Keeling, Grenfell, 1998), temporal variation (Grossman et al., 1977; Grossman, 1980; Schwartz, Smith, 1983; Smith, 1983a,b; Schaffer, 1985; Olsen et al., 1988; Thieme, 2000), multiple groups (see Chapter 23 and its bibliographic remarks), and difference in activity levels (Thieme, 1985). Several structures can be combined; Schenzle (1984), for example, considers an age-structured model with temporal variation. See the proceedings edited by Mollison (1995) as a recent source of more information. For modeling contact structures in heterogeneous populations, see also Castillo-Chavez, Velasco-Hernandez, Fridman (1994).

Chapter Twenty-Three

Endemic Models with Multiple Groups or Populations

In this chapter we present a general framework for handling the spread of nonfatal diseases in heterogeneous populations that have constant size. The heterogeneity of the population can result from the existence of several groups that behave differently with respect to the disease. There may even be different populations such as a host population (humans, for example) and a vector population (flies or mosquitoes, for example). One aim of this chapter is to see whether and how concepts like the basic replacement ratio, endemic equilibria, disease persistence (endemicity), which have turned out to be important in homogeneous populations, generalize to the multigroup setting. That this is largely possible is due to the powerful theory of nonnegative matrices associated with the names of Perron and Frobenius (Toolbox A.8). A new concept will arise: the basic replacement matrix or (epidemic) second-generation matrix, and its spectral radius. The spectral radius takes the role of the basic replacement ratio both as far as epidemiological interpretation and mathematical properties are concerned. While this is theoretically satisfying, we will investigate several special cases which show that there is no general simple way of determining the basic replacement ratio from the basic replacement matrix.

The framework of our model will be general enough to allow the disease to take different courses in every group, leading to different subdivisions of infected individuals according to disease stage. We give a list of the types of disease courses included:

$$
\begin{aligned}
&S \to E \to I \to R \ (\to S),\\
&S \to E \to I[1] \to I[2] \ (\to S),\\
&S \to E[1] \to E[2] \to I \ (\to S),\\
&S \to I[1] \to I[2] \to I[3] \ (\to S),\\
&S \to I[1] \to I[2] \to R[1] \ (\to S),\\
&S \to I[1] \to R[1] \to R[2] \ (\to S).
\end{aligned}
$$

The first line represents a disease course like that considered in Chapter 21, with transitions from susceptible to exposed to infectious to removed. The second line represents a disease course involving two infectious stages, the third a disease course with two exposed stages, etc. Every course of the disease which leaves out one of the classes is also admitted as long as there is at least one infective class. The model allows individuals to return from every infected (exposed, infective, and recovered) class into the susceptible class. While the formulation of our model and some results will hold for arbitrarily many infected classes, one important result (local asymptotic stability of strongly endemic equilibria) will hold in general for at most three infected classes only.

We include both vaccination of newly recruited individuals and random vaccination, which brings susceptible individuals into an additional class, the class of vaccinated individuals, V, from which they may or may not return to the susceptible class, e.g.,

$$\begin{array}{l} \to V\ (\to S) \\ \quad\ \uparrow \\ \to S \to E \to I \to R\ (\to S). \end{array}$$

We do not merge the class of vaccinated individuals into one of the recovered classes because it is entirely possible that the average length of protection through vaccination is different from the average length of protection after recovery.

Since our model does not keep track of the age of individuals, it does not account for age-specific vaccination (except at age 0).

23.1 The Model

We assume that the population consists of m groups, the sizes of which are denoted by N_j. We should make clear that by population we mean the epidemiologically relevant part of the population. (In the case of a sexually transmitted disease, for example, the epidemiologically relevant part of the population is the sexually active part of the population.) This means that an individual does not necessarily enter a population group by birth, but by some general form of recruitment. Each group is split into susceptible individuals, S_j, infected individuals, J_j, and vaccinated individuals, V_j:

$$N_j = S_j + J_j + V_j. \tag{23.1}$$

Note that J_j here denotes the size of the infected (not infective!) part of group j with the understanding that the infected part includes the exposed, infective, and removed (recovered) individuals. The infected stage is divided into n substages, called *infection stages*,

$$J_j = \sum_{l=1}^{n} J_j[l], \tag{23.2}$$

with $J_j[l]$ denoting the number of infected individuals in group j that are in the lth infected class (or infection stage). There may be situations (host-vector diseases) where each subgroup has a different number of infection stages, but it is always possible to introduce dummy stages and make the number of stages equal. We consider the following system:

$$\left.\begin{aligned} S_j' &= (1-p_j)\Lambda_j - (\mu_j+\alpha_j)S_j - S_j\phi_j + \rho_{j0}V_j + \sum_{l=1}^{n}\rho_{jl}J_j[l], \\ V_j' &= p_j\Lambda_j + \alpha_j S_j - (\mu_j+\rho_{j0})V_j, \\ J_j'[1] &= S_j\phi_j - (\mu_j+\rho_{j1}+\eta_{j1})J_j[1], \\ J_j'[l] &= \eta_{j,l-1}J_j[l-1] - (\mu_j+\rho_{jl}+\eta_{jl})J_j[l], \quad l=2,\ldots,n, \\ \phi_j &= \sum_{k=1}^{m} C_{jk}(N_1,\ldots,N_m)\left(\sum_{l=1}^{n}\xi_{jk}[l]\frac{J_k[l]}{N_k}\right). \end{aligned}\right\} \tag{23.3}$$

$\Lambda_j > 0$ is the rate at which new individuals are recruited into the jth group; $p_j \in [0, 1)$ is the proportion which is vaccinated immediately upon recruitment.

$\mu_j > 0$ denotes the per capita mortality rate in the jth group.

$\alpha_j \geqslant 0$ is the per capita rate at which susceptibles in the jth group are vaccinated at random.

ϕ_j is the infective force in group j, i.e., the per capita rate at which susceptible individuals are infected in the jth group.

$\rho_{jl} \geqslant 0$ denotes the per capita rate at which individuals in the jth group that are in the lth infection stage return to the susceptible class.

$\eta_{jl} \geqslant 0$ denotes the per capita rate at which individuals in the jth group move from the lth to the $(l+1)$th stage.

$C_{jk}(N_1, \ldots, N_m)$ is the average probability of an individual in the jth group coming into contact with an individual in the kth group. This probability depends on the sizes of the various groups.

The fractions $J_k[l]/N_k$ give the probability that a given contact with an individual in group k actually occurs with an infected individual in the lth stage; $\xi_{jk}[l]$ is the rate at which this contact leads to an infection (provided that the other individual is susceptible, of course).

If the different groups belong to the same population, each group presumably has the same infection stages. In this case the stage transition rates satisfy $\eta_{jl} > 0$ for all $j = 1, \ldots, m$ and $l = 1, \ldots, n-1$. If the various groups are from different populations, the number of infection stages may be different for each group. In that case, if n_j is the number of infection stages for population j, we let n be the maximum of the numbers n_j, $j = 1, \ldots, m$. If $n_j < n$, we introduce dummy stages for $l = 1 + n_j, \ldots, n$ and set $\eta_{jl} = 0$ for $l = n_j, \ldots, n$, while $\eta_{jl} > 0$ is assumed for $l = 1, \ldots, n_j - 1$. Since the respective stage transition rates are zero this way, nobody actually gets into the dummy stages.

As long as we consider nonfatal diseases and diseases that do not permanently or temporarily remove individuals from the active population, we do not need to know how the contact probabilities C_{jk} depend on the group sizes in order to discuss the dynamics of the disease. This question only becomes relevant for the calculation of the basic replacement ratio (see Chapter 21). The group sizes converge towards a fixed value, as can be seen by adding the differential equations, resulting in

$$N_j' = \Lambda_j - \mu_j N_j,$$

which implies that

$$N_j(t) \to N_j^* := \frac{\Lambda_j}{\mu_j}, \quad t \to \infty. \tag{23.4}$$

This leads us to consider the limiting system

$$\left.\begin{aligned} S_j' &= (1 - p_j)\mu_j N_j^* - (\mu_j + \alpha_j)S_j - S_j\phi_j + \rho_{j0}V_j + \sum_{l=1}^{n} \rho_{jl} J_j[l], \\ V_j' &= p_j \mu_j N_j^* + \alpha_j S_j - (\mu_j + \rho_{j,0})V_j, \\ J_j'[1] &= S_j \phi_j - (\mu_j + \rho_{j,1} + \eta_{j,1})J_j[1], \\ J_j'[l] &= \eta_{j,l-1} J_j[l-1] - (\mu_j + \rho_{jl} + \eta_{jl})J_j[l], \quad l = 2, \ldots, n, \\ \phi_j &= \sum_{k=1}^{m} C_{jk}(N_1^*, \ldots, N_m^*)\left(\sum_{l=1}^{n} \xi_{jk}[l]\frac{J_k[l]}{N_k^*}\right). \end{aligned}\right\} \tag{23.5}$$

Moreover, we have

$$S_j(t) + V_j(t) + \sum_{l=1}^{n} J_j[l](t) = N_j^* \quad \forall t \geqslant 0.$$

It is easily seen that equilibria of the systems (23.3) and (23.5) are in a one-to-one correspondence and have the same local stability properties if the functions C_{jk} are differentiable.

Under quite general assumptions the theory of asymptotically autonomous systems allows us to reduce the global large-time behavior of system (23.3) to that of system (23.5) (see Thieme, 1992; Castillo-Chavez, Thieme, 1995; Mischaikow, Smith, Thieme, 1995).

It is convenient to introduce the fractions

$$u_j = \frac{S_j}{N_j^*}, \qquad v_j = \frac{V_j}{N_j^*}, \qquad x_{jl} = \frac{J_j[l]}{N_j^*}.$$

In order to keep the indices for x apart, we should keep in mind that the first index denotes the group and the second the infection stage. We will use j and k in $\{1, \ldots, m\}$ as group indices and $l \in \{1, \ldots, n\}$ as infection stage indices.

We obtain the following system

$$\left.\begin{aligned}
u_j' &= (1-p_j)\mu_j - (\mu_j+\alpha_j)u_j - u_j\phi_j + \rho_{j,0}v_j + \sum_{l=1}^{n}\rho_{jl}x_{jl},\\
v_j' &= p_j\mu_j + \alpha_j u_j - (\mu_j+\rho_{j,0})v_j,\\
x_{j1}' &= u_j\phi_j - (\mu_j+\rho_{j,1}+\eta_{j,1})x_{j1},\\
x_{jl}' &= \eta_{j,l-1}x_{j,l-1} - (\mu_j+\rho_{jl}+\eta_{jl})x_{jl}, \quad l=2,\dots,n,\\
\phi_j &= \sum_{k=1}^{m}\sum_{l=1}^{n}\sigma_{jk}[l]x_{kl}.
\end{aligned}\right\} \tag{23.6}$$

with

$$\sigma_{jk}[l] = C_{jk}(N_1^*,\dots,N_m^*)\xi_{jk}[l]. \tag{23.7}$$

Moreover,

$$u_j(t) + v_j(t) + \sum_{l=1}^{n} x_{jl}(t) = 1 \quad \forall t \geqslant 0, \tag{23.8}$$

The following existence and uniqueness result follows from standard ordinary differential equation theory and Proposition A.1.

Theorem 23.1. *Let u_j°, v_j°, x_{jl}°, $j=1,\dots,m$, $l=1,\dots,n$, be nonnegative and*

$$u_j^\circ + v_j^\circ + \sum_{l=1}^{n} x_{jl}^\circ = 1 \quad \forall j=1,\dots,m.$$

Then there exists a unique nonnegative solution of system (23.5) with initial data u_j°, v_j°, x_{jl}° and

$$u_j + v_j + \sum_{l=1}^{n} x_{jl} = 1 \quad \text{on } [0,\infty) \quad \forall j=1,\dots,m.$$

Relation (23.8) makes the u_j equation in (23.6) redundant. We also use it to rewrite this system as follows:

$$\begin{aligned}
1 &= u_j + v_j + \sum_{l=1}^{n} x_{jl},\\
v_j' &= p_j\mu_j + \alpha_j - \alpha_j\left(\sum_{l=1}^{n} x_{jl}\right) - (\mu_j+\rho_{j,0}+\alpha_j)v_j,\\
x_{j1}' &= u_j\phi_j - (\mu_j+\rho_{j,l}+\eta_{j,l})x_{j1},\\
x_{jl}' &= \eta_{j,l-1}x_{j,l-1} - (\mu_j+\rho_{jl}+\eta_{jl})x_{jl}, \quad l=2,\dots,n,\\
\phi_j &= \sum_{k=1}^{m}\sum_{l=1}^{n}\sigma_{jk}[l]x_{kl}.
\end{aligned}$$

In order to deal with this system further, it is convenient to condense the notation by setting

$$x_{j0} = v_j$$

and

$$\left.\begin{aligned} \nu_{j0} &= \mu_j + \rho_{j,0} + \alpha_j, \\ \nu_{jl} &= \mu_j + \rho_{jl} + \eta_{jl}, \quad l = 1, \dots, n. \end{aligned}\right\} \tag{23.9}$$

Then system (23.6) takes the form

$$\left.\begin{aligned} 1 &= u_j + \sum_{l=0}^{n} x_{jl}, \\ x'_{j0} &= p_j \mu_j + \alpha_j - \alpha_j \left(\sum_{l=1}^{n} x_{jl} \right) - \nu_{j0} x_{j0}, \\ x'_{j1} &= u_j \phi_j - \nu_{j1} x_{j1}, \\ x'_{jl} &= \eta_{j,l-1} x_{j,l-1} - \nu_{jl} x_{jl}, \quad l = 2, \dots, n, \\ \phi_j &= \sum_{k=1}^{m} \sum_{l=1}^{n} \sigma_{jk}[l] x_{kl}. \end{aligned}\right\} \tag{23.10}$$

We should keep in mind though that

$$\nu_{j0} > \alpha_j, \qquad \nu_{jl} > \eta_{jl}, \quad l = 1, \dots, n. \tag{23.11}$$

Exercise. Prove Theorem 23.1.

23.2 Equilibrium Solutions

The equilibrium solutions of system (23.10) satisfy the following system:

$$\left.\begin{aligned} u_j^* &= 1 - \sum_{l=0}^{n} x_{jl}^*, \\ 0 &= p_j \mu_j + \alpha_j - \alpha_j \left(\sum_{l=1}^{n} x_{jl}^* \right) - \nu_{j0} x_{j0}^*, \\ 0 &= u_j^* \phi_j^* - \nu_{j1} x_{j1}^*, \\ 0 &= \eta_{j,l-1} x_{j,l-1}^* - \nu_{jl} x_{jl}^*, \quad l = 2, \dots, n, \\ \phi_j^* &= \sum_{k=1}^{m} \sum_{l=1}^{n} \sigma_{jk}[l] x_{kl}^*. \end{aligned}\right\} \tag{23.12}$$

We first look for a disease-free equilibrium, with no infected individuals, i.e., all

$$x_{jl}^{\diamond} = 0, \quad l = 1, \dots, n.$$

This implies

$$\left.\begin{aligned} x_{j0}^{\diamond} &= \frac{p_j \mu_j + \alpha_j}{\nu_{j0}} = v_j^{\diamond}, \\ u_j^{\diamond} &= \frac{(1 - p_j)\mu_j + \rho_{j,0}}{\nu_{j0}}. \end{aligned}\right\} \tag{23.13}$$

In order to find another possible equilibrium, we first obtain from the second equation in (23.12) that

$$x_{j0}^{*} = \frac{p_j \mu_j + \alpha_j}{\nu_{j0}} - \frac{\alpha_j}{\nu_{j0}} \sum_{l=1}^{n} x_{jl}^{*}.$$

Now, from the first equation in (23.12),

$$u_j^{*} = 1 - x_{j0}^{*} - \sum_{l=1}^{n} x_{jl}^{*} = 1 - \frac{p_j \mu_j + \alpha_j}{\nu_{j0}} - \left(1 - \frac{\alpha_j}{\nu_{j0}}\right) \sum_{l=1}^{n} x_{jl}^{*}.$$

Since $\nu_{j0} = \mu_j + \alpha_j + \rho_{j0}$,

$$\left.\begin{aligned} u_j^{*} &= u_j^{\diamond} - \zeta_j \sum_{l=1}^{n} x_{jl}^{*}, \\ \zeta_j &= \frac{\mu_j + \rho_{j,0}}{\nu_{j0}}. \end{aligned}\right\} \tag{23.14}$$

From the other equations in (23.12) we obtain that

$$\begin{aligned} x_{j1}^{*} &= u_j^{*} \sum_{k=1}^{m} \sum_{l=1}^{n} \tilde{\sigma}_{jk}[l] x_{kl}^{*}, \\ x_{jl}^{*} &= \gamma_{j,l} x_{j,l-1}^{*}, \quad l = 2, \dots, n, \end{aligned}$$

with

$$\tilde{\sigma}_{jk}[l] = \frac{\sigma_{jk}[l]}{\nu_{j1}}, \qquad \gamma_{j,l} = \frac{\eta_{j,l-1}}{\nu_{jl}}. \tag{23.15}$$

Hence

$$x_{jl}^{*} = \left(\prod_{q=2}^{l} \gamma_{jq}\right) x_{j1}^{*}. \tag{23.16}$$

Set

$$z_j = x_{j1}^{*}.$$

Then

$$z_j = u_j^* \sum_{k=1}^{m} \kappa_{jk} z_k \tag{23.17}$$

with

$$\kappa_{jk} = \tilde{\sigma}_{jk}[1] + \sum_{l=2}^{n} \tilde{\sigma}_{jk}[l] \left(\prod_{q=2}^{l} \gamma_{kq} \right). \tag{23.18}$$

By (23.14),

$$z_j = (u_j^\diamond - \beta_j z_j) \sum_{k=1}^{m} \kappa_{jk} z_k \tag{23.19}$$

with

$$\beta_j = \zeta_j \left[1 + \sum_{l=1}^{n} \left(\prod_{q=2}^{l} \gamma_{j,q} \right) \right]. \tag{23.20}$$

We solve for z_j:

$$z_j = \frac{u_j^\diamond \sum_{k=1}^{m} \kappa_{jk} z_k}{1 + \beta_j \sum_{k=1}^{m} \kappa_{jk} z_k}. \tag{23.21}$$

An equilibrium is called endemic if the disease is present in some groups and strongly endemic if the disease is present in all groups. Noticing from (23.16) that $x_j^*[1] = 0$ implies $x_{jl}^* = 0$ for all $l = 1, \ldots, n$, we make the following definition.

Definition 23.2. A nonnegative equilibrium solution u_j^*, x_{jl}^* to (23.8) is called *endemic* if $z_j = x_{j1}^* > 0$ for some j. It is called a *strongly endemic* equilibrium if $z_j = x_{j1}^* > 0$ for all $j = 1, \ldots, m$. We also call $z_j = x_{j1}^*$ an endemic or strongly endemic equilibrium if u_j^*, x_{jl}^* has this property.

Our previous discussions can be summarized as follows.

Lemma 23.3. *If u_j^*, x_{jl}^* is an endemic equilibrium, then z with $z_j = x_{j1}^*$ is a positive vector satisfying (23.19), or equivalently (23.21). On the other hand, if z is a positive vector satisfying (23.21), then there exists an endemic equilibrium u_j^*, x_{jl}^* that is uniquely determined by $x_{j1}^* = z_j$.*

Uniqueness of Strongly Endemic Equilibria

In order to guarantee that the disease-free equilibrium and at most one endemic equilibrium are the only equilibria possible, we introduce the following definition.

Definition 23.4. A matrix $(a_{jk})_{m\times m}$, $m \geqslant 2$, is called *irreducible* if the following holds. For any proper nonempty subset P of $\{1,\ldots,m\}$ there are $k \in P$, $j \notin P$ such that $a_{jk} \neq 0$.

A 1×1 matrix is called irreducible if it is not the 0 matrix.

Epidemiologically speaking, the irreducibility of the contact matrix means that the disease finally affects all groups.

Theorem 23.5. *Let the matrix $(\kappa_{jk})_{m\times m}$ be irreducible. Then any endemic equilibrium is strongly endemic.*

Proof. Let P contain those j with $z_j > 0$. As the equilibrium is different from the disease-free equilibrium, P is nonempty. Suppose that P does not coincide with $\{1,\ldots,m\}$. As $(\kappa_{jk})_{m\times m}$ is irreducible, there are $k \in P$, $j \notin P$ such that

$$\kappa_{jk} > 0.$$

Hence $\kappa_{jk} z_k > 0$, by the definition of P. Equation (23.21) implies that $z_j > 0$, i.e., $j \in P$, a contradiction, because we already have that $j \notin P$. □

Theorem 23.6. *Let z be an endemic equilibrium and $\tilde{z}$ a strongly endemic equilibrium. Then $z \leqslant \tilde{z}$. In particular, there exists at most one strongly endemic equilibrium. If the matrix κ is irreducible, there exists at most one endemic equilibrium.*

Proof. Define

$$\xi = \max_{j=1,\ldots,m} \frac{z_j}{\tilde{z}_j}.$$

Since z is nonnegative and not the zero vector, we have $\xi > 0$ and

$$z_k \leqslant \xi \tilde{z}_k \quad \forall k = 1,\ldots,m. \tag{23.22}$$

Furthermore,

$$z_j = \xi \tilde{z}_j \quad \text{for some } j. \tag{23.23}$$

We consider this coordinate j. By (23.22),

$$\sum_{k=1}^{m} \kappa_{jk} z_k \leqslant \sum_{k=1}^{m} \kappa_{jk} \xi \tilde{z}_k = \xi \sum_{k=1}^{m} \kappa_{jk} \tilde{z}_k.$$

As the function

$$x \mapsto \frac{u_j^\diamond x}{1 + \beta_j x}$$

is increasing, we have from (23.23) that

$$z_j \leqslant \frac{u_j^\diamond \xi \sum_{k=1}^{m} \kappa_{jk} \tilde{z}_k}{1 + \beta_j \xi \sum_{k=1}^{m} \kappa_{jk} \tilde{z}_k}.$$

Now we suppose that $\xi > 1$. By (23.21) and $\tilde{z}_j > 0$,

$$\sum_{k=1}^{m} \kappa_{jk}\tilde{z}_k > 0,$$

and

$$z_j < \frac{\xi u_j^\diamond \sum_{k=1}^{m} \kappa_{jk}\tilde{z}_k}{1 + \beta_j \sum_{k=1}^{m} \kappa_{jk}\tilde{z}_k} = \xi\tilde{z}_j = z_j.$$

This contradiction implies $\xi \leqslant 1$, so by (23.22),

$$z_k \leqslant \tilde{z}_k, \quad k = 1, \ldots, m.$$

If z is also a strongly endemic equilibrium, we can switch the roles of z and $\tilde{z}$ in this proof and obtain equality. The last statement follows from Theorem 23.5. □

Let us explore the irreducibility of κ in terms of the original parameters. By (23.15) and (23.18),

$$\kappa_{jk} = \sum_{l=1}^{n} \tilde{\sigma}_{jk}[l]\left(\prod_{q=2}^{l} \gamma_{kq}\right) = \sum_{l=1}^{n} \frac{\sigma_{jk}[l]}{\nu_{j1}}\left(\prod_{q=2}^{l} \frac{\eta_{k,q-1}}{\nu_{kq}}\right).$$

By (23.7),

$$\kappa_{jk} = \sum_{l=1}^{n} C_{jk}(N_1^*, \ldots, N_m^*)\frac{\xi_{jk}[l]}{\nu_{j1}}\left(\prod_{q=2}^{l} \frac{\eta_{k,q-1}}{\nu_{kq}}\right).$$

Reorganizing terms,

$$\kappa_{jk} = C_{jk}(N_1^*, \ldots, N_m^*)\left[\sum_{l=1}^{n} \frac{\xi_{jk}[l]}{\nu_{j1}}\left(\prod_{q=2}^{l} \frac{\eta_{k,q-1}}{\nu_{kq}}\right)\right]. \tag{23.24}$$

Proposition 23.7. *The matrix κ_{jk} is irreducible if and only if the contact rate matrix $C_{jk}(N_1^*, \ldots, N_m^*)$ is irreducible for all $N_1, \ldots, N_m > 0$, and*

$$\sum_{l=1}^{n_j} \xi_{jk}[l] > 0 \quad \forall j, k = 1, \ldots, m.$$

Recall that $\xi_{jk}[l]$ is the probability that the contact between a susceptible individual in group j with an infected individual in group k and stage l actually leads to an infection. So, the second condition in Proposition 23.7 means that, for each group, at least one of the infection stages is an infectious stage. Recall that $\eta_{j,l}$ is the per capita rate with which an infected individual in group j moves from stage l to stage $l+1$ and that $\eta_{jl} > 0$ for $l = 1, \dots, n_j - 1$.

Existence of Endemic Equilibria

Apparently, an endemic equilibrium exists if the mapping

$$G : \mathbb{R}_+^m \to \mathbb{R}_+^m,$$

$$G_j(z) = \frac{u_j^\diamond \sum_{k=1}^m \kappa_{jk} z_k}{1 + \beta_j \sum_{k=1}^m \kappa_{jk} z_k}$$

has a nonzero fixed point. Let

$$\bar{z}_j = \frac{u_j^\diamond}{\beta_j}.$$

Notice that

$$G(\bar{z}) \leqslant \bar{z}.$$

Define the matrix $\tilde{\kappa}$ by

$$\tilde{\kappa}_{jk} = u_j^\diamond \kappa_{jk}.$$

Notice that the matrix $\tilde{\kappa}$ is irreducible if and only if the matrix κ is irreducible.

We recall that the spectral radius of the matrix $\tilde{\kappa} = (\tilde{\kappa}_{jk})$, $\mathcal{R}_0 = r(\tilde{\kappa})$, is defined as

$$\mathcal{R}_0 = \sup\{|\lambda|; \lambda \text{ eigenvalue of } \tilde{\kappa}\}.$$

Theorem 23.8. *Let $\mathcal{R}_0$ be the spectral radius of the matrix $\tilde{\kappa}$. There exists no endemic equilibrium if $\mathcal{R}_0 < 1$ or if $\mathcal{R}_0 \leqslant 1$ and κ is irreducible. If $\mathcal{R}_0 > 1$, there exists an endemic equilibrium (which is strongly endemic if κ is irreducible).*

For the proof of Theorem 23.8 we need some facts about nonnegative matrices (see Toolbox A.8).

Proof of Theorem 23.8. Let us first assume that there exists an endemic equilibrium and show that this implies $\mathcal{R}_0 \geqslant 1$ or even $\mathcal{R}_0 > 1$ if the matrix κ is irreducible. If there exists an endemic equilibrium, there exists a positive vector z satisfying (23.19) (see Lemma 23.3). If the matrix κ is irreducible, z is strictly positive. It follows from (23.19) that

$$z_j \leqslant \sum_{k=1}^m \tilde{\kappa}_{jk} z_k.$$

Theorem A.42 implies that the spectral radius of $\tilde{\kappa}$, $\mathcal{R}_0$, satisfies $\mathcal{R}_0 \geqslant 1$. If κ is irreducible and thus z strictly positive, all these inequalities are strict and $\mathcal{R}_0 > 1$.

Now let $\mathcal{R}_0 > 1$. By Theorem A.41 there exists a positive vector $\tilde{z}$ such that

$$\tilde{z}_j = \mathcal{R}_0 \sum_{k=1}^{m} u_j^\diamond \kappa_{jk} \tilde{z}_k.$$

If $\epsilon > 0$ is chosen small enough, we have

$$\epsilon \tilde{z}_j \leqslant \frac{u_j^\diamond \sum_{k=1}^{m} \kappa_{jk} \epsilon \tilde{z}_k}{1 + \beta_j \sum_{k=1}^{m} \kappa_{jk} \epsilon \tilde{z}_k}.$$

We also choose $\epsilon > 0$ small enough that $\epsilon \tilde{z} \leqslant \bar{z}$. Now we have

$$G(\epsilon \tilde{z}) \geqslant \epsilon \tilde{z}, \qquad G(\bar{z}) \leqslant \bar{z}.$$

Furthermore, G is a monotone increasing operator and so maps the order interval $[\epsilon z, \bar{z}]$ into itself. By Brower's fixed-point theorem, G has a fixed point in the order interval. This fixed point is an endemic equilibrium because $\epsilon \tilde{z}$ is not the zero vector. □

Exercises

23.1. Show that the mapping G in the proof of Theorem 23.8 is monotone increasing on $\mathbb{R}_+^m$ with the canonical order

$$z \leqslant \tilde{z} \longleftrightarrow \tilde{z} - z \in \mathbb{R}_+^m.$$

23.2. Show that an order interval

$$[\tilde{z}, \bar{z}] = \{z \in \mathbb{R}^n;\ \tilde{z} \leqslant z \leqslant \bar{z}\}, \qquad \tilde{z}, \bar{z} \in \mathbb{R}^n, \quad \tilde{z} \leqslant \bar{z},$$

is a convex compact set.

23.3 Local Asymptotic Stability of Strongly Endemic Equilibria

Among other results, we prove in this section that a strongly endemic equilibrium is always locally asymptotically stable if every group has at most three infected stages. This is the best we can expect in this generality because even a one-group model of $S \to I \to R_1 \to R_2 \to R_3 \to S$ type with one infective and three removed or recovered stages (without a vaccinated class) can have an unstable endemic equilibrium (Hethcote, Stech, van den Driessche, 1981; Stech, William, 1981)).

We linearize (23.10) around the strongly endemic equilibrium x_{jl}^*:

$$y'_{j0} = -\alpha_j\left(\sum_{j=1}^{n} y_{jl}\right) - \nu_{j0}y_{j0},$$
$$y'_{j1} = u_j^*\sum_{k=1}^{m}\sum_{l=1}^{n}\sigma_{jk}[l]y_{kl} - \nu_{j1}y_{j1} - \phi_j^*\left(\sum_{l=0}^{n} y_{jl}\right),$$
$$y'_{jl} = \eta_{j,l-1}y_{j,l-1} - \nu_{jl}y_{jl}, \quad l = 2, \dots, n.$$

with

$$\phi_j^* = \sum_{k=1}^{m}\sum_{l=1}^{n}\sigma_{jk}[l]x_{kl}^* = \eta_{j1}\sum_{k=1}^{m}\kappa_{jk}x_{k1}^* > 0.$$

Setting $y_{jl}(t) = y_{jl}^* e^{\lambda t}$, this system becomes

$$\left.\begin{aligned}
\lambda y_{j0}^* &= -\alpha_j\left(\sum_{l=1}^{n} y_{jl}^*\right) - \nu_{j0}y_{j0}^*,\\
\lambda y_{j1}^* &= u_j^*\sum_{k=1}^{m}\sum_{l=1}^{n}\sigma_{jk}[l]y_{kl}^* - \nu_{j1}y_{j1}^* - \phi_j^*\left(\sum_{l=0}^{n} y_{jl}^*\right),\\
\lambda y_{jl}^* &= \eta_{j,l-1}y_{j,l-1}^* - \nu_{jl}y_{jl}^*, \quad l = 2, \dots, n.
\end{aligned}\right\} \tag{23.25}$$

As our system can be arbitrarily big, there is no hope of proving locally asymptotic stability via the Routh–Hurwitz criterion. Hence we work directly with (23.25). The subsequent result follows from Lemma A.28.

Lemma 23.9. *The strongly endemic equilibrium is locally asymptotically stable if the following holds: if* λ, $y_{jl}^* \in \mathbb{C}$ *is a solution to (23.25) with* $y_{jl}^* \neq 0$ *for some* j, l, *then* $\Re\lambda < 0$.

In order to apply Lemma 23.9 we assume that

$$\Re\lambda \geqslant 0.$$

We solve for y_{jl}^* in (23.25) and use (23.15):

$$y_{jl}^* = \frac{\eta_{j,l-1}}{\lambda + \nu_{jl}}y_{j,l-1}^* = \frac{\gamma_{jl}}{1 + (\lambda/\nu_{jl})}y_{j,l-1}^*. \tag{23.26}$$

This yields

$$|y_{jl}^*| = \frac{\gamma_{jl}}{|1 + (\lambda/\nu_{jl})|}|y_{j,l-1}^*|.$$

If $\zeta = \zeta_1 + i\zeta_2$ is a complex number with $\Re\zeta = \zeta_1 \geqslant 0$, then

$$|1 + \zeta|^2 = (1 + \zeta_1)^2 + \zeta_2^2 \geqslant 1.$$

Hence

$$|y_{jl}^*| \leqslant \gamma_{jl}|y_{j,l-1}^*|, \quad l = 2, \dots, n,$$

and

$$|y_{jl}^*| \leqslant \left(\prod_{q=2}^{l} \gamma_{jq}\right)|y_{j1}^*|, \quad l = 2, \dots, n.$$

From the second equation in (23.25),

$$\left(1 + \frac{\lambda}{\nu_{j1}}\right)y_{j1}^* + \frac{\phi_j^*}{\nu_{j1}}\left(\sum_{j=0}^{n} y_{jl}^*\right) = u_j^* \sum_{k=1}^{m}\sum_{l=1}^{n} \tilde{\sigma}_{jk}[l]y_{kl}^*$$

with $\tilde{\sigma}$ defined by (23.15). Taking absolute values,

$$\left|\left(1 + \frac{\lambda}{\nu_{j1}}\right)y_{j1}^* + \frac{\phi_j^*}{\nu_{j1}}\left(\sum_{l=0}^{n} y_{jl}^*\right)\right| \leqslant u_j^* \sum_{k=1}^{m}\sum_{l=1}^{n} \tilde{\sigma}_{jk}[l]|y_{kl}^*| \leqslant u_j^* \sum_{k=1}^{m} \kappa_{jk}|y_{k1}^*|$$

with κ_{jk} being defined as in (23.18). Substituting the following result from the first equation in (23.25),

$$y_{j0}^* = -\frac{\alpha_j}{\lambda + \nu_{j0}}\left(\sum_{l=1}^{n} y_{j1}^*\right),$$

we obtain

$$\left|\left(1 + \frac{\lambda}{\nu_{j1}}\right)y_{j1}^* + \frac{\phi_j^*}{\nu_{j1}}\left[1 - \frac{\alpha_j}{\lambda + \nu_{j0}}\right]\left(\sum_{j=1}^{n} y_{jl}^*\right)\right| \leqslant u_j^* \sum_{k=1}^{m} \kappa_{jk}|y_{k1}^*|.$$

Repeating (23.26),

$$y_{jl}^* = \prod_{q=2}^{l} \frac{\eta_{j,q-1}}{\nu_{jq} + \lambda} y_{j1}^*,$$

which yields

$$\left|1 + \frac{\lambda}{\nu_{j1}} + \frac{\phi_j^*}{\nu_{j1}}\psi_j(\lambda)\right| |y_{j1}^*| \leqslant u_j^* \sum_{k=1}^{m} \kappa_{jk}|y_{k1}^*|$$

with

$$\psi_j(\lambda) = \left[1 - \frac{\alpha_j}{\lambda + \nu_{j0}}\right]\left(1 + \sum_{l=2}^{n}\prod_{q=2}^{l} \frac{\eta_{j,q-1}}{\nu_{jq} + \lambda}\right).$$

Lemma 23.10. *Assume that $\psi_j(\lambda) > 0$ whenever $\Re\lambda = 0$ and $\psi_j(\lambda) \geqslant 0$ whenever $\Re\lambda > 0$. Then the strongly endemic equilibrium is locally asymptotically stable.*

Proof. Suppose that the strongly endemic equilibrium is not locally asymptotically stable. By Lemma 23.9 and the previous considerations, we obtain the two relations above with $\Re\lambda \geqslant 0$. By assumption,

$$\epsilon = \min_{j=1,\dots,m} \Re\left(\frac{\lambda}{\nu_{j1}} + \frac{\phi_j^*}{\nu_{j1}}\psi_j(\lambda)\right) > 0,$$

and

$$\left|1 + \frac{\lambda}{\nu_{j1}} + \frac{\phi_j^*}{\nu_{j1}}\psi_j(\lambda)\right| \geqslant \left|\Re\left(1 + \frac{\lambda}{\nu_{j1}} + \frac{\phi_j^*}{\nu_{j1}}\psi_j(\lambda)\right)\right| \geqslant 1 + \epsilon.$$

Thus

$$(1+\epsilon)|y_{j1}^*| \leqslant u_j^* \sum_{k=1}^{m} \kappa_{jk}|y_{k1}^*|. \tag{23.27}$$

Let

$$\xi = \max_{k=1,\dots,m} \frac{|y_{k1}^*|}{z_k},$$

where z is a solution of (23.17) and $z_k > 0$ for $k = 1, \dots, m$. At least one of the elements y_{k1}^* is strictly positive, hence $\xi > 0$. Furthermore,

$$|y_{k1}^*| \leqslant \xi z_k, \quad k = 1, \dots, m,$$

and

$$|y_{j1}^*| = \xi z_j \quad \text{for some } j.$$

We consider this coordinate j. Then, by (23.27) and (23.17),

$$(1+\epsilon)|y_{j1}^*| \leqslant u_j^* \sum_{k=1}^{m} \kappa_{jk}|y_{k1}^*| \leqslant u_j^* \sum_{k=1}^{m} \kappa_{jk}\xi z_k = \xi z_j = |y_{j1}^*|,$$

a contradiction, as $\epsilon > 0$. □

Lemma 23.11. *Let* $\nu_0 > \alpha$ *and*

$$\psi(\lambda) = \left[1 - \frac{\alpha}{\lambda + \nu_0}\right](1 + \chi(\lambda)).$$

Then, if $\Re\lambda \geqslant 0$, *we have* $\Re\psi(\lambda) > 0$ *whenever*

$$1 + \Re\chi(\lambda) > 0, \qquad \Im\chi(\lambda) \leqslant 0.$$

Proof. Let $\lambda = \varphi + \mathrm{i}\omega$, $\varphi \geqslant 0$. Then

$$\begin{aligned}\psi(\lambda) &= \left[1 - \alpha\frac{\nu_0 + \varphi - \mathrm{i}\omega}{\omega^2 + (\varphi + \nu_0)^2}\right](1 + \Re\chi(\lambda) + \mathrm{i}\Im\chi(\lambda)) \\ &= \left[\frac{(\nu_0 + \varphi)(\nu_0 + \varphi - \alpha) + \omega^2 + \mathrm{i}\alpha\omega}{\omega^2 + (\varphi + \nu_0)^2}\right](1 + \Re\chi(\lambda) + \mathrm{i}\Im\chi(\lambda)).\end{aligned}$$

Hence

$$\Re\psi(\lambda) = \frac{(\nu_0+\varphi)(\nu_0+\varphi-\alpha)+\omega^2}{\omega^2+(\varphi+\nu_0)^2}(1+\Re\chi(\lambda)) - \frac{\alpha\omega}{\omega^2+\varphi^2+\nu_0^2}\Im\chi(\lambda).$$

Since $\nu_0 > \alpha$, the assertion follows. □

Lemma 23.12. *Let*

$$\chi(\lambda) = \frac{\eta_1}{\nu_1+\lambda}\left(1+\frac{\eta_2}{\nu_2+\lambda}\right).$$

Then $\Im\chi(\lambda) \leqslant 0$ *and* $\Re(1+\chi(\lambda)) > 0$ *whenever* $\Re\lambda \geqslant 0$.

Proof. Let $\lambda = \varphi + i\omega$ with $\varphi \geqslant 0$. Then

$$\chi(\lambda) = \eta_1\frac{\nu_1+\varphi-i\omega}{(\nu_1+\varphi)^2+\omega^2}\left(1+\eta_2\frac{\nu_2+\varphi-i\omega}{(\nu_2+\varphi)^2+\omega^2}\right).$$

Hence $\Im\chi(\lambda) \leqslant 0$ and

$$1+\Re\chi(\lambda) \geqslant 1+\eta_1\Re\left(\frac{\nu_1+\varphi-i\omega}{(\nu_1+\varphi)^2+\omega^2}\eta_2\frac{\nu_2+\varphi-i\omega}{(\nu_2+\varphi)^2+\omega^2}\right).$$

Thus $1+\Re\chi(\lambda)$ is positive if the following expression is positive,

$$\begin{aligned}
&((\nu_1+\varphi)^2+\omega^2)((\nu_2+\varphi)^2+\omega^2)+\eta_1\eta_2\Re((\nu_1+\varphi-i\omega)(\nu_2+\varphi-i\omega))\\
&\quad= ((\nu_1+\varphi)^2+\omega^2)((\nu_2+\varphi)^2+\omega^2)+\eta_1\eta_2(\nu_1+\varphi)(\nu_2+\varphi)-\omega^2\eta_1\eta_2\\
&\quad> \omega^2(\nu_1^2+\nu_2^2-\eta_1\eta_2) \geqslant \omega^2(2\nu_1\nu_2-\eta_1\eta_2).
\end{aligned}$$

Since $\nu_l > \eta_l$, $1+\Re\chi(\lambda) > 0$. □

Theorem 23.13. *Let* $n \leqslant 3$, *i.e., every group has at most three infected stages. Then every strongly endemic equilibrium (if one exists) is locally asymptotically stable.*

Proof. We want to apply Lemma 23.10. The function ψ_j is of the form in Lemma 23.11,

$$\psi_j(\lambda) = \left[1-\frac{\alpha_j}{\lambda+\nu_{j0}}\right](1+\chi_j(\lambda))$$

with χ_j as in Lemma 23.12 and $\nu_{j0} > \alpha_j$ by (23.9). Let $\Re\lambda \geqslant 0$. By Lemma 23.12, $\Im\chi_j(\lambda) \leqslant 0$ and $\Re(1+\chi_j(\lambda)) > 0$. By Lemma 23.11, $\Re\psi_j(\lambda) > 0$. Since this holds for every j, Lemma 23.10 implies our assertion. □

23.4 Extinction or Persistence of the Disease?

In this section we tie the magnitude of $\mathcal{R}_0$ in relation to unity to the question of disease extinction or disease persistence (endemicity). In a first result we show that the disease-free equilibrium is locally asymptotically stable if $\mathcal{R}_0 < 1$. This means that the disease dies out once the disease dynamics get close enough to the disease-free equilibrium.

Theorem 23.14. *If $\mathcal{R}_0 < 1$, the disease-free equilibrium is locally asymptotically stable.*

Proof. We leave the proof as an exercise. Combine the procedure used in discussing the stability of a strongly endemic equilibrium in Section 23.3 with Theorem A.41. □

Remember that $\mathcal{R}_0 < 1$ also precludes the existence of an endemic equilibrium (Theorem 23.8). In order to obtain disease extinction, i.e., global asymptotic stability of the disease-free equilibrium, we need some extra conditions because we have included random vaccinations in our model (cf. Theorem 23.16). We assume that, in every group, the return rate of vaccinated individuals to the susceptible class is not smaller than any of the return rates of infected individuals. It is not clear to me whether this is just a technical condition or whether there is more behind it. Certain vaccination regimes are known to cause subcritical bifurcation of endemic equilibria, which can preclude global stability of the endemic equilibrium even if $\mathcal{R}_0 < 1$ (see the bibliographic remarks). But in our case there are no subcritical bifurcations, as stated by Theorem 23.8.

Theorem 23.15. *Let $\mathcal{R}_0 < 1$ and*

$$\rho_{j0} \geqslant \rho_{jl} \quad \forall j = 1, \ldots, m, \quad l = 1, \ldots, n.$$

Then the disease dies out, i.e.,

$$x_{jl}(t) \to 0, \quad t \to \infty \quad \forall j = 1, \ldots, m, \quad l = 1, \ldots, n.$$

Proof. Recall the following subsystem of (23.10):

$$\left.\begin{aligned} x'_{j1} &= u_j \phi_j - \nu_{j1} x_{j1}, \\ x'_{jl} &= \eta_{j,l-1} x_{j,l-1} - \nu_{jl} x_{jl}, \quad l = 2, \ldots, n, \\ \phi_j &= \sum_{k=1}^{m} \sum_{l=1}^{n} \sigma_{jk}[l] x_{kl}. \end{aligned}\right\} \tag{23.28}$$

Furthermore, substituting the relation (23.8) into the first equation in (23.6) we obtain the inequality

$$u_j' \leqslant (1-p_j)\mu_j - (\mu_j+\alpha_j)u_j + \rho_{j0}(1-u_j) + \sum_{l=1}^{n}(\rho_{jl}-\rho_{j0})x_{jl}$$
$$\leqslant (1-p_j)\mu_j + \rho_{j0} - \nu_{j0}u_j.$$

Recall $\nu_{j0} = \mu_j + \alpha_j + \rho_{j0}$. Fix j. By Lemma A.20, we find a sequence $t_i \to \infty$ $(i \to \infty)$ such that

$$u_j'(t_i) \to 0, \quad u_j(t_i) \to u_j^\infty, \quad i \to \infty.$$

Here we again use the convention that the superscript '∞' denotes the limit superior as $t \to \infty$. From the last inequality we obtain

$$0 \leqslant (1-p_j)\mu_j + \rho_{j0} - \nu_{j0}u_j^\infty.$$

By (23.13),

$$u_j^\infty \leqslant u_j^\diamond.$$

Again by Lemma A.20, we find a sequence $t_i \to \infty$ $(i \to \infty)$ such that

$$x_{j1}'(t_i) \to 0, \quad x_{j1}(t_i) \to x_j^\infty[1], \quad i \to \infty.$$

From the first equation in (23.28),

$$0 \leqslant u_j^\infty \phi_j^\infty - \nu_{j1}x_j^\infty[1].$$

Furthermore,

$$\phi_j^\infty \leqslant \sum_{k=1}^{m}\sum_{l=1}^{n}\sigma_{jk}[l]x_{kl}^\infty.$$

Applying Lemma A.20 to the second equation in (23.28) yields

$$0 \leqslant \eta_{j,l-1}x_{j,l-1}^\infty - \nu_{jl}x_{jl}^\infty, \quad l = 2, \dots, n. \tag{23.29}$$

Substituting these inequalities into each other we obtain

$$x_{j1}^\infty \leqslant \sum_{k=1}^{n}\tilde{\kappa}_{jk}x_{k1}^\infty,$$

where $\tilde{\kappa}_{jk} = u_j^\diamond \kappa_{jk}$ with κ_{jk} from (23.18). Suppose that $x_{j1}^\infty > 0$ for some j. By Theorem A.42, these inequalities imply that the spectral radius of $\tilde{\kappa}$, $\mathcal{R}_0$, is greater than or equal to 1. Since this contradicts our assumptions, we conclude that

$$x_{j1}^\infty = 0 \quad \forall j = 1, \dots, m.$$

By (23.29), $x_{jl}^\infty = 0$ for all $l = 1, \dots, n$, $j = 1, \dots, m$. □

We do not need the extra assumptions in Theorem 23.15 when there are no random vaccinations.

Theorem 23.16. *Let $\mathcal{R}_0 < 1$, and assume that there are no random vaccinations, i.e., all $\alpha_j = 0$. Then the disease dies out, i.e.,*

$$x_{jl}(t) \to 0, \quad t \to \infty \quad \forall j = 1, \dots, m, \quad l = 1, \dots, n.$$

Proof. In this case,

$$x_{j0}(t) \to \frac{p_j \mu_j}{\mu_j + \rho_{j0}}, \quad t \to \infty,$$

and

$$u_j^\infty \leqslant 1 - \frac{p_j \mu_j}{\mu_j + \rho_{j0}} = u_j^\diamond.$$

The rest of the proof is the same as in Theorem 23.15. □

We turn to disease persistence and first establish uniform weak persistence, which will imply uniform strong persistence by Theorem A.32.

Proposition 23.17. *Let $\mathcal{R}_0 > 1$ and κ be irreducible. Then the disease is uniformly weakly persistent in the following sense. There exists some $\epsilon > 0$ such that*

$$\limsup_{t\to\infty} \max_{j=1,\dots,m} x_{j1}(t) \geqslant \epsilon$$

for all solutions of (23.10) with nonnegative initial data and $x_{j1}(t) > 0$ for at least one j and one $t \geqslant 0$. In particular, the disease-free equilibrium is unstable.

Proof. Suppose that $x_{j1}(t) > 0$ for at least one j and one $t \geqslant 0$ and

$$\limsup_{t\to\infty} \max_{j=1,\dots,m} x_{j1}(t) < \epsilon.$$

It follows from the x_{j1} equation in (23.6) that $x_{j1}(t) > 0$ for all sufficiently large $t > 0$. As before, Lemma A.20 provides

$$x_{jl}^\infty \leqslant \frac{\eta_{j,l-1}}{\nu_{jl}} x_{j,l-1}^\infty = \gamma_{jl} x_{j,l-1}^\infty.$$

Hence, for a given $\delta > 0$, by choosing $\epsilon > 0$ small enough, we can achieve that

$$\sum_{l=1}^{n} x_{jl}^\infty < \delta \quad \forall j$$

and

$$\phi_j^\infty < \tfrac{1}{2}\delta \quad \forall j.$$

From the first equation in (23.6), for sufficiently large t,

$$u_j' \geqslant (1-p_j)\mu_j - (\mu_j+\alpha_j)u_j - \tfrac{1}{2}\delta u_j + \rho_{j0}(1-u_j-\delta).$$

By Lemma A.20, there exists a sequence $t_i \to \infty$, $i \to \infty$ such that

$$u_j'(t_i) \to 0, \quad u_j(t_i) \to u_{j\infty}, \quad i \to \infty.$$

Hence

$$u_{j\infty} \geqslant \frac{(1-p_j)\mu_j + \rho_{j0}(1-\delta)}{\mu_j+\alpha_j+\rho_{j0}+\delta/2} > \frac{(1-p_j)\mu_j + \rho_{j0}(1-\delta)}{\nu_{j0}+\delta} =: \theta_j(\delta).$$

Notice from (23.13) that $\theta_j(0) = u_j^\diamond$. For large t, we have from (23.10) that

$$\begin{aligned} x_{j1}' &\geqslant \theta_j(\delta)\phi_j - \nu_{j1}x_{j1}, \\ x_{jl}' &= \eta_{j,l-1}x_{j,l-1} - \nu_{jl}x_{jl}, \quad l = 2,\ldots,n, \\ \phi_j &= \sum_{k=1}^{m}\sum_{l=1}^{n}\sigma_{jk}[l]x_{kl}. \end{aligned}$$

By a shift in time, we can assume that these inequalities hold for all $t \geqslant 0$ and $x_{j1}(0) > 0$ for some j. Since all x_{jl} are bounded, the Laplace transforms

$$\hat{x}_{jl}(\lambda) = \int_0^\infty \mathrm{e}^{-\lambda t}x_{jl}(t)\,\mathrm{d}t$$

are defined for all $\lambda > 0$ and satisfy

$$\begin{aligned} \lambda\hat{x}_{j1}(\lambda) - x_{j1}(0) &\geqslant \theta_j(\delta)\hat{\phi}_j(\lambda) - \nu_{j1}\hat{x}_{j1}(\lambda), \\ \hat{x}_{jl}(\lambda) - x_{jl}(0) &= \eta_{j,l-1}\hat{x}_{j,l-1}(\lambda) - \nu_{jl}\hat{x}_{jl}, \quad l = 2,\ldots,n, \\ \hat{\phi}_j(\lambda) &= \sum_{k=1}^{m}\sum_{l=1}^{n}\sigma_{jk}[l]\hat{x}_{kl}(\lambda). \end{aligned}$$

Solving for $\hat{x}_{jl}(\lambda)$, $\lambda > 0$,

$$\begin{aligned} \hat{x}_{j1}(\lambda) &\geqslant \frac{1}{\lambda+\nu_{j1}}(\theta_j(\delta)\hat{\phi}_j(\lambda) + x_{j1}(0)), \\ \hat{x}_{jl}(\lambda) &\geqslant \frac{\eta_{j,l-1}}{\lambda+\nu_{jl}}\hat{x}_{j,l-1}(\lambda), \quad l = 2,\ldots,n, \\ \hat{\phi}_j(\lambda) &= \sum_{k=1}^{m}\sum_{l=1}^{n}\sigma_{jk}[l]\hat{x}_{kl}(\lambda). \end{aligned}$$

Set $w_j(\lambda) = \hat{x}_{j1}(\lambda)(\lambda+\nu_{j1})$. Then the vector $w(\delta)$ is positive and

$$w(\delta) \geqslant \tilde{\kappa}(\delta)w(\delta)$$

with the matrix

$$\tilde{\kappa}_{jk}(\delta) = \theta_j(\delta)\left[\frac{\sigma_{jk}[1]}{\delta + \nu_{j1}} + \sum_{l=2}^{n} \frac{\sigma_{jk}[l]}{\delta + \nu_{j1}}\left(\prod_{q=2}^{l} \frac{\eta_{k,q-1}}{\delta + \nu_{jq}}\right)\right].$$

Notice that, for $\delta = 0$, this is the irreducible matrix $\tilde{\kappa}$, whose spectral radius is strictly greater than 1. $\tilde{\kappa}(\delta)$ inherits irreducibility from the matrix κ. By Lemma A.47, $r(\tilde{\kappa}(\delta)) \leqslant 1$, where δ is any sufficiently small positive number. Since the spectral radius depends continuously on the matrix, $r(\kappa) \leqslant 1$, a contradiction. □

We turn to uniform strong persistence of the disease in the population. Let n_j be the number of infection stages for group j, i.e., the largest natural number $\tilde{n} \leqslant n$ such that $\eta_{j,l} > 0$ for $l = 1, \ldots, \tilde{n} - 1$.

Theorem 23.18. *Let $\mathcal{R}_0 > 1$ and κ be irreducible. Then the disease is uniformly strongly persistent in the following sense. There exists some $\epsilon > 0$ such that*

$$x_{jl}(t) \geqslant \epsilon \quad \forall j = 1, \ldots, m, \quad l = 1, \ldots, n_j$$

for all solutions of (23.10) with nonnegative initial data and $x_{j1}(t) > 0$ for at least one j and one $t \geqslant 0$.

Proof. We apply Theorem A.34. Let X be the space of all nonnegative vectors $w = ((u_j), (v_j), (x_{jl})) \in \mathbb{R}^{m(n+2)}$ such that

$$\sum_{j=1}^{m}\left(u_j + v_j + \sum_{l=1}^{m} x_{jl}\right) = 1.$$

X is a compact set, so the condition (C)$'$ is trivially satisfied with $B = X$. Set

$$\rho(w) = \max_{j=1,\ldots,m} x_{j1}, \qquad \tilde{\rho}(w) = \min_{j=1,\ldots,m} x_{j1}.$$

Let Φ be the semiflow induced by the solutions of (23.6). Then $\rho(\Phi(t,\cdot)) = \max_j\{x_{j1}(t)\}$ and $\tilde{\rho}(\Phi(t,\cdot)) = \min_j\{x_{j1}(t)\}$. Now assume that $w \in X$, $\rho(w) > 0$. Then $x_{j1}(0) > 0$ for some $j \in \{1, \ldots, m\}$. The differential equation for x_{j1} in (23.6) implies that $x_{j1}(t)$ for all $t \geqslant 0$. So, $\rho(\Phi(t, w)) > 0$ for all $t \geqslant 0$. Now consider a total orbit $w(t)$ of Φ with $\rho(w(t)) > 0$. This corresponds to a solution of (23.6) that is defined for all $t \in \mathbb{R}$. An argument similar to that used before provides some $k \in \{1, \ldots, m\}$ such that $x_{k1}(t) > 0$ for all $t \in \mathbb{R}$. The form of (23.6) implies that $x_{kl}(t) > 0$ for all $t \in \mathbb{R}$, $l = 1, \ldots, n_k$. The irreducibility of the matrix κ implies that $\phi_j(t) > 0$ for all $t \in \mathbb{R}$. Since $u_j(t) > 0$ for all $t \in \mathbb{R}$, we have $x_{j1}(t) > 0$ for all $t \in \mathbb{R}$. Repeating the previous argument this implies $x_{jl}(t) > 0$ for all $t \in \mathbb{R}$ and all j, l. So, $\tilde{\rho}(w(t)) > 0$ for all $t \in \mathbb{R}$. Since Φ is uniformly weakly ρ-persistent by Proposition 23.17, the statement follows from Theorem A.34. □

Exercises

23.3. Prove Theorem 23.14.

23.4. Give the details of the proof of Theorem 23.18.

23.5 The Basic Replacement Matrix, Alias Next-Generation Matrix

So far in this chapter, we have used the symbol $\mathcal{R}_0$ for the spectral radius of some matrix $\tilde{\kappa}$, while we have used it in previous chapters as the symbol for the basic replacement ratio with the interpretation that it is the number of secondary cases that an average infected individual will generate if introduced into a disease-free population. Actually, for multi-group models, we will need to modify this interpretation slightly by saying that the basic replacement ratio is the number of secondary cases an average infected individual produces as long as the disease dynamics are close to the disease-free equilibrium state.

In the previous sections, we have shown that the $\mathcal{R}_0$ notation is justified mathematically, because $\mathcal{R}_0$ completely determines the stability of the disease-free equilibrium (Theorems 23.14 and 23.18), as it did in Chapters 21 and 22. Here we want to provide some epidemiological justification. In a first step we show that $\mathcal{R}_0$ is the spectral radius of a matrix the entries of which can be interpreted as basic replacement ratios from one group to another.

Recall that the eigenvalue problem for the matrix $\tilde{\kappa}$, $\tilde{\kappa}_{jk} = u_j^\diamond \kappa_{jk}$, with κ from (23.18), has the form

$$\lambda z_j = u_j^\diamond \sum_{k=1}^{m} \left[\sum_{l=1}^{n} \frac{\sigma_{jk}[l]}{\nu_{j1}} \left(\prod_{q=2}^{l} \frac{\eta_{k,q-1}}{\nu_{kq}} \right) \right] z_k.$$

Setting $\tilde{z}_j = \nu_{j1} N_j^* z_j$, with N_j^* from (23.4), we see that this eigenvalue problem is equivalent to

$$\lambda \tilde{z}_j = u_j^\diamond \sum_{k=1}^{m} \left[\sum_{l=1}^{n} \frac{N_j^*}{N_k^*} \frac{\sigma_{jk}[l]}{\nu_{k1}} \left(\prod_{q=2}^{l} \frac{\eta_{k,q-1}}{\nu_{kq}} \right) \right] \tilde{z}_k.$$

This means that $\tilde{\kappa}$ has the same eigenvalues (and the same spectral radius) as the matrix $\mathcal{R}^\circ$ with entries

$$\mathcal{R}_{jk}^\circ = u_j^\diamond \frac{N_j^*}{N_k^*} \sum_{l=1}^{n} \frac{\sigma_{jk}[l]}{\nu_{k1}} \left(\prod_{q=2}^{l} \frac{\eta_{k,q-1}}{\nu_{kq}} \right).$$

Recalling that $u_j^\diamond N_j^* = S_j^\diamond$ is the number of susceptibles in group j at the disease-free equilibrium, we rewrite these expressions as

$$\mathcal{R}_{jk}^\circ = \frac{S_j^\diamond}{N_k^*} \sum_{l=1}^{n} \frac{\sigma_{jk}[l]}{\nu_{kl}} P_{kl}$$

with

$$P_{kl} = \left(\prod_{q=1}^{l-1} \frac{\eta_{kq}}{\nu_{kq}} \right). \tag{23.30}$$

Remember that, in group k, η_{kq} is the per capita rate of transition from the infected stage q to stage $q+1$, while $\nu_{kq} = \eta_{kq} + \mu_k + \rho_{kq}$ is the overall rate of leaving stage q, which can occur by either going to the next infected stage, dying, or returning to the susceptible stage. So,

$$\frac{\eta_{kq}}{\nu_{kq}}$$

is the probability of making it through stage q (alive) and getting into the next infected stage. Hence P_{kl} is the probability of getting to the lth infected stage (alive) in group k. Recalling that

$$\sigma_{jk}[l] = C_{jk}(N^*)\xi_{jk}[l], \quad N^* = (N_1^*, \dots, N_m^*),$$

we find

$$\mathcal{R}_{jk}^\circ = \frac{S_j^\diamond}{N_k} C_{jk}(N^*) \sum_{l=1}^{n} \frac{\xi_{jk}[l]}{\nu_{kl}} P_{kl}. \tag{23.31}$$

Remember that $\xi_{jk}[l]$ is the per capita rate at which a susceptible individual in group j contracts the disease from an infected individual in group k and stage l, while $1/\nu_{kl}$ is the mean sojourn time in stage l of an infected individual in group k. So,

$$\frac{\xi_{jk}[l]}{\nu_{kl}}$$

is the number of susceptibles in group j that an infected individual in group k can infect during stage l, provided that immediately after every finished contact a new contact with a susceptible individual from group j is made. Since P_{kl} is the probability of reaching stage l,

$$\sum_{l=1}^{n} \frac{\xi_{jk}[l]}{\nu_{kl}} P_{kl}$$

is the number of susceptible individuals from group j that an infected individual in group k can infect from the moment it is infected to the moment it either dies or returns to the susceptible class, provided that immediately after every finished contact a new contact with a susceptible individual from group j is made.

We finally recall that C_{jk} is the average probability with which a given individual in group j has a contact with an individual in group k. So, $S_j^\diamond C_{jk}$ is the average number of contacts of susceptible individuals in group j with individuals in group k. Hence $(S_j^\diamond / N_k^*) C_{jk}$ is the average probability with which a given individual in group k has a contact with a susceptible individual in group j.

Summarizing, if a given infected individual is introduced into group k of a disease-free population, $\mathcal{R}_{jk}^\circ$ is the product of the average probability of having a contact with a susceptible individual in group j and the average number of susceptible individuals from group j that it can infect from the moment it was infected itself to the moment it either dies or returns to the susceptible class, provided that immediately after every finished contact a new contact with a susceptible individual from group j is made.

In other words, $\mathcal{R}_{jk}^\circ$ is the basic replacement ratio from group k to group j in the sense that it is the mean number of secondary cases an average infected individual in group k produces in group j (cf. Jacquez, Simon, Koopman (1995), where these entries are called the basic replacement numbers for the group j contacts of an infected individual in group k).

We call the matrix $\mathcal{R}^\circ$ formed by the entries $\mathcal{R}_{jk}^\circ$ the *basic replacement matrix* or *next-generation matrix* (where "generation" is to be understood epidemiologically not demographically) in order to tie our considerations to the discussion of the basic replacement ratio in Diekmann, Heesterbeek, Metz (1990, 1995), Diekmann, Dietz, Heesterbeek (1991), and Diekmann, Heesterbeek (2000).

23.6 The Basic Replacement Ratio as Spectral Radius of the Basic Replacement Matrix

In the previous section we have shown that, in this multi-group model (and typically for multi-group epidemic and endemic models in general), we can identify a *basic replacement matrix* (or *second-generation matrix*), $\mathcal{R}^\circ$, the spectral radius of which, $\mathcal{R}_0$, has the following property.

If $\mathcal{R}_0 < 1$, the disease-free equilibrium is locally asymptotically stable. If $\mathcal{R}_0 > 1$, then the disease-free equilibrium is unstable, and the disease persists uniformly strongly in the population (in the sense of Theorem 23.18) provided that $\mathcal{R}^\circ$ is irreducible.

While this shows that $\mathcal{R}_0$ has the mathematical properties which the basic replacement ratio has in single-group models, this does not yet answer our question of whether $\mathcal{R}_0$ is the basic replacement ratio in the sense that it gives the number of secondary cases produced by an average infected individual as long as the disease dynamics still linger close to the disease-free steady state. Following Jacquez, Simon, Koopman (1995), we introduce the *basic replacement ratio of group k* as

$$\mathcal{R}_k^\circ = \sum_{j=1}^{m} \mathcal{R}_{jk}^\circ, \tag{23.32}$$

because this sum gives us the number of secondary cases produced by an average infected individual in group k. Then the (overall) basic replacement ratio can be defined formally as

$$\mathcal{R}_0 = \sum_{k=1}^{m} \mathcal{R}_k^\circ Q_k, \tag{23.33}$$

where Q_k is the probability that an average infected individual belongs to group k. The question is whether a vector $Q = (Q_1, \ldots, Q_m)$ of such probabilities can be found.

One candidate is provided by the Perron–Frobenius theory (Theorem A.45 and Remark A.46), which we rephrase and supplement as follows. The following term will be useful. A *probability vector* is a positive vector whose coordinates sum up to 1.

Theorem 23.19. *Let the basic replacement matrix $\mathcal{R}^\circ$ be irreducible.*

(a) *Then its spectral radius $\mathcal{R}_0$ is an eigenvalue with a unique strictly positive probability eigenvector $Q = (Q_1, \ldots, Q_m)$.*

(b) *Furthermore, $\mathcal{R}_0$ is an eigenvalue of the transposed matrix $\mathcal{R}^{\circ *}$ with a uniquely determined eigenvector Q^* satisfying $\langle Q^*, Q\rangle = 1$.*

We call Q the *Perron–Frobenius vector* associated with the basic replacement matrix $\mathcal{R}^\circ$. Summing the eigenvalue–eigenvector relation for Q,

$$\mathcal{R}_0 Q_j = \sum_{k=1}^{m} \mathcal{R}_{jk}^\circ Q_k$$

provides the crucial relation (23.33) for the basic replacement ratio:

$$\mathcal{R}_0 = \sum_{k=1}^{m} \mathcal{R}_k^\circ Q_k.$$

Supporting evidence that the Perron–Frobenius vector is the average probability distribution of group membership for infected individuals is provided by the theorems of ergodicity and mean ergodicity.

If a nonnegative vector $z(0)$ represents the infected individuals that are introduced into a disease-free population, then the vectors $z(l)$ defined below give an approximation of the lth epidemiological generation of infected individuals:

$$z(l) = (\mathcal{R}^\circ)^l z(0) \tag{23.34}$$

or

$$z_j(l+1) = \sum_{k=1}^{m} \mathcal{R}^\circ_{jk} z_k(l), \quad l \in \mathbb{N}.$$

Then

$$Q(l) = \frac{z(l)}{\sum_{j=1}^{m} z_j(l)} \tag{23.35}$$

is the probability distribution of group membership for the lth generation (in an epidemiological sense) of infected individuals. These are only approximations because the population starts moving away from the disease-free state as soon as the disease is introduced, but this takes the longer the smaller the number of infected individuals is at the beginning.

We recall that a nonnegative square matrix is called *primitive* if one of its powers has strictly positive entries. It is easily seen that a nonnegative matrix is primitive if it is irreducible and all entries in its main diagonal are strictly positive (see Exercise A.1). This means that there is intragroup infection in every group. This is why primitive basic replacement matrices typically occur in models for sexually transmitted diseases in homosexual populations. We have the following asymptotic behavior from Corollary A.50.

Theorem 23.20. *Let the basic replacement matrix $\mathcal{R}^\circ$ be primitive and define a sequence of probability vectors $Q(l)$ by (23.35) and (23.34).*

Then $Q(l) \to Q$ as $l \to \infty$ with Q being the uniquely determined probability vector in Theorem 23.19 (a).

This shows that the unique probability eigenvector Q associated with $\mathcal{R}_0$ is the probability distribution of group membership for an average infected individual after sufficiently many epidemiological generations (provided that the number of infected individuals introduced into the disease-free population is so small that the state of the system is very close to the disease-free equilibrium for a sufficiently long time.

In models for sexually transmitted diseases in heterosexual populations, the second-generation matrix is not primitive. In the simplest possible model that has two groups, one female and one male, the basic replacement matrix has the form

$$\mathcal{R}^\circ = \begin{pmatrix} 0 & \mathcal{R}^\circ_{12} \\ \mathcal{R}^\circ_{21} & 0 \end{pmatrix}$$

because there is no intragroup transmission. This matrix, provided $\mathcal{R}^\circ_{12}$ and $\mathcal{R}^\circ_{21}$ are strictly positive, is irreducible, but not primitive. Theorem 23.20 is no longer valid for nonprimitive matrices, but mean ergodicity still holds for irreducible matrices (Theorem A.52), which means, epidemiologically, that the convergence in Theorem 23.20 holds for the respective averages over the epidemiological generations.

Theorem 23.21. *Let the basic replacement matrix $\mathcal{R}^\circ$ be irreducible. Then*

$$\frac{1}{\ell+1}\sum_{l=0}^{\ell}\frac{z(l)}{\langle z(l), Q^*\rangle} \to Q, \quad \ell \to \infty.$$

Relationships Between the Basic Replacement Ratio and Basic Intergroup Replacement Ratios

Once it is clear that the basic replacement ratio is the spectral radius of the second-generation matrix, one can take advantage of the many relations that have been established between the spectral radius of a nonnegative matrix and its entries. We only list one that is epidemiologically relevant.

Let M be a subset of $\{1, \dots, m\}$. Then we set

$$\mathcal{R}_k(M) = \sum_{j\in M} \mathcal{R}^\circ_{jk}.$$

M represents the subpopulation that is formed by the groups with index in M. $\mathcal{R}_k(M)$ is the basic replacement ratio from group k to subpopulation M, i.e., the number of secondary cases produced in subpopulation M by an average infective in group k. If $M = \{1, \dots, m\}$ is the whole population, we recover $\mathcal{R}_k(M) = \mathcal{R}^\circ_k$, the basic replacement ratio of group k given by (23.32). Furthermore, we set

$$\mathcal{R}_\diamond(M) = \min_{k\in M} \mathcal{R}_k(M), \qquad \mathcal{R}^\diamond(M) = \max_{k\in M} \mathcal{R}_k(M).$$

Proposition 23.22. *Let $\mathcal{R}^\circ$ be a nonnegative matrix. Then*

$$\mathcal{R}_0 \geqslant \mathcal{R}_\diamond(M) \quad \forall M \subseteq \{1, \dots, m\}.$$

Proof. This is a special case of Exercise 23.5(b) below. □

Specializing Proposition 23.22 to the singletons M and to $M = \{1, \dots, m\}$, we obtain the following theorem.

Theorem 23.23. *Let $\mathcal{R}^\circ$ be a nonnegative matrix. Then*

$$\mathcal{R}_0 \geqslant \max_{j=1,\dots,m} \mathcal{R}^\circ_{jj}$$

and

$$\min_{k=1,\dots,m} \mathcal{R}^\circ_k \leqslant \mathcal{R}_0 \leqslant \max_{k=1,\dots,m} \mathcal{R}^\circ_k.$$

The last inequality follows from Remark A.48 and Theorem A.41.

We can conclude that $\mathcal{R}_0 > 1$ if $\mathcal{R}^\circ_{jj} > 1$ for at least one of the intragroup replacement ratios or if the basic replacement ratios of all groups are strictly larger than 1. Conversely, if the basic replacement ratios of all groups are strictly smaller than 1, so is the (overall) basic replacement ratio. We will see in the next section that these results cannot be improved in general; in particular, the average for the basic group replacement ratios is no estimate, from above or below, for the basic replacement ratio.

Exercises

23.5. Let $\mathcal{R}^\circ$ be a nonnegative matrix with spectral radius $\mathcal{R}_0$ and α_j, $j = 1, \dots, m$, be nonnegative numbers.

(a) Show that

$$\mathcal{R}_0 \geqslant \min\left\{\frac{1}{\alpha_j}\sum_{k=1}^{m} \mathcal{R}^\circ_{jk}\alpha_k;\ j = 1, \dots, m,\ \alpha_j > 0\right\}.$$

(b) Show that

$$\mathcal{R}_0 \geqslant \min\left\{\frac{1}{\alpha_k}\sum_{j=1}^{m} \alpha_j \mathcal{R}^\circ_{jk};\ k = 1, \dots, m,\ \alpha_k > 0\right\}.$$

Hint: Theorems A.41 and A.42.

23.6. Let $f_2, \dots, f_m$ be nonnegative numbers, $f_m > 0$, and $g_2, \dots, g_m$ be strictly positive numbers. Let the basic replacement matrix be given by

$$\begin{aligned} \mathcal{R}^\circ_{1k} &= f_k, \quad k = 2, \dots, m, \\ \mathcal{R}^\circ_{k,k-1} &= g_k, \quad k = 2, \dots, m, \\ \mathcal{R}^\circ_{jk} &= 0, \quad \text{otherwise.} \end{aligned}$$

(a) Show that the matrix $\mathcal{R}^\circ$ is irreducible.

(b) Set

$$\tau_0 = \sum_{k=2}^{m} f_k\left(\prod_{j=2}^{k} g_j\right).$$

Show the following equivalences:

$$\begin{aligned} \mathcal{R}_0 > 1 &\iff \tau_0 > 1, \\ \mathcal{R}_0 = 1 &\iff \tau_0 = 1, \\ \mathcal{R}_0 < 1 &\iff \tau_0 < 1. \end{aligned}$$

23.7 Some Special Cases of Mixing

In general it is difficult to find explicit expressions for the basic replacement ratio. This is relatively easy for the following cases, however, which we use to illustrate that the basic replacement ratio is larger than the sum of intragroup replacement ratios in some cases and smaller in others (see the remarks after Theorem 23.25) and that there is no fixed-order relation between $\mathcal{R}_0$ and the average of group replacement ratios either (see the example after Theorem 23.24).

Separable Mixing

We speak about *separable mixing* in multi-group models if the basic replacement matrix factorizes as

$$\mathcal{R}^\circ_{jk} = f_j g_k. \tag{23.36}$$

This has the consequence that

$$\frac{\mathcal{R}^\circ_{jk}}{\mathcal{R}^\circ_k} \text{ is independent of } k \text{ for every } j,$$

i.e., without restriction of generality, $g_k = \mathcal{R}^\circ_k$, where $\mathcal{R}^\circ_k$ is the basic replacement ratio of group k given by (23.32).

Theorem 23.24. *In the case of separable mixing,*

$$\mathcal{R}_0 = \sum_{k=1}^{m} \mathcal{R}^\circ_{kk}.$$

This means that, in the case of separable mixing, the basic replacement ratio is the sum of the intragroup basic replacement ratios.

Proof. The eigenvalue problem $\lambda v = \mathcal{R}^\circ v$ takes the form

$$\lambda v_j = f_j \sum_{k=1}^{m} g_k v_k.$$

Setting $v_j = f_j$, we find that $v = (f_1, \ldots, f_m)$ is an eigenvector associated with the eigenvalue

$$\lambda = \sum_{k=1}^{m} g_k f_k.$$

Hence

$$\mathcal{R}_0 \geqslant \sum_{k=1}^{m} g_k f_k \geqslant 0.$$

Without restriction of generality we can assume that $\mathcal{R}_0 > 0$; otherwise this inequality is already an equality. Let $\lambda \neq 0$ be an eigenvalue with $|\lambda| = \mathcal{R}_0$. Then there exists an eigenvector v such that

$$\lambda v_j = \sum_{k=1}^{m} \mathcal{R}^\circ_{jk} v_k = f_j \sum_{k=1}^{m} g_k v_k.$$

Since $\lambda \neq 0$, v is a nonzero scalar multiple of the vector $(f_1, \dots, f_m)$ and we can assume that v equals this vector. Hence

$$\lambda = \sum_{k=1}^{m} g_k f_k = \sum_{k=1}^{m} \mathcal{R}^\circ_{kk}.$$

Since $\lambda \geqslant 0$, $\mathcal{R}_0 = \lambda$. □

We use separable mixing in order to illustrate that, with m groups, it may happen that all but one basic group-replacement ratio $\mathcal{R}^\circ_k$ exceed one, but the basic replacement ratio is still smaller than 1.

Example. Let

$$g_2 = \cdots g_m = 2, \qquad g_1 = \tfrac{1}{2}, \qquad f_2 = \cdots f_m = \frac{1}{4(m-1)}, \qquad f_1 = \tfrac{3}{4},$$

such that

$$\sum_{j=1}^{m} f_j = 1.$$

Then

$$\mathcal{R}^\circ_1 = \tfrac{1}{2}, \quad \mathcal{R}^\circ_k = g_k = 2, \quad k = 2, \dots, m; \qquad \mathcal{R}_0 = \tfrac{7}{8} < 1.$$

This example also shows that the basic replacement ratio can be substantially smaller than the average of the group replacement ratios, for

$$\frac{1}{m} \sum_{k=1}^{m} R^\circ_k = 2 - \frac{3}{2m}.$$

It can be modified to illustrate that the basic replacement ratio can be substantially larger than the average of the group replacement ratios (Exercise 23.7).

Separable Extra-Group Mixing

Here we assume that

$$\mathcal{R}^\circ_{jk} = f_j g_k, \quad j \neq k, \qquad \mathcal{R}^\circ_{jj} = h_j, \tag{23.37}$$

with $f_j g_j > 0$ and $h_j \geqslant 0$ for all j. Obviously, the second-generation matrix $\mathcal{R}^\circ$ is irreducible. For $h_j = f_j g_j$ we recover the case of separable mixing; if $h_j \geqslant f_j g_j$ we obtain separable mixing with intragroup enhancement (cf. Diekmann, Heesterbeek, Metz, 1995) or preferred (intragroup) mixing (cf. Jacquez, Simon, Koopman, 1995). The case $0 \leqslant h_j \leqslant f_j g_j$ could be called separable mixing with intragroup inhibition.

The eigenvalue problem takes the form

$$\lambda v_j = \sum_{j \neq k=1}^{m} f_j g_k v_k + h_j v_j,$$

or equivalently

$$(\lambda + f_j g_j - h_j) v_j = f_j \sum_{k=1}^{m} g_k v_k.$$

By Theorem A.41, $\mathcal{R}_0$ is associated with a positive eigenvector v. By Theorem 23.23, $\mathcal{R}_0 \geqslant h_j$ for all j. This implies that

$$\mathcal{R}_0 + f_j g_j - h_j > 0$$

and

$$v_j = \frac{f_j}{\mathcal{R}_0 + f_j g_j - h_j} \sum_{k=1}^{m} g_k v_k.$$

We multiply by g_j, sum over j and divide appropriately to obtain

$$1 = \sum_{j=1}^{m} \frac{f_j g_j}{\mathcal{R}_0 + f_j g_j - h_j}.$$

Theorem 23.25. *Consider separable extragroup mixing as explained above (see (23.37)).*

(a) *Then $\mathcal{R}_0 > 1$ if and only if $h_j = \mathcal{R}^\circ_{jj} > 1$ for at least one j, or if*

$$h_j \leqslant 1 \quad \forall j \quad \text{and} \quad 1 < \sum_{j=1}^{m} \frac{f_j g_j}{1 + f_j g_j - h_j}.$$

(b) *Conversely, $\mathcal{R}_0 < 1$ if and only if*

$$h_j < 1 \quad \forall j \quad \text{and} \quad \sum_{j=1}^{m} \frac{f_j g_j}{1 + f_j g_j - h_j} < 1.$$

Proof. By Theorem 23.23, $\mathcal{R}_0 \geqslant h_j$ for all j. Hence $\mathcal{R}_0 > 1$ if $h_j > 1$ for at least one j. Conversely, $\mathcal{R}_0 \leqslant 1$ implies that $h_j \leqslant 1$ for all j, and $\mathcal{R}_0 < 1$ implies $h_j < 1$ for all j. So, in order to complete our discussion, we can assume that $h_j \leqslant 1$ for all j. Then the function

$$\tau(x) = \sum_{j=1}^{m} \frac{f_j g_j}{x + f_j g_j - h_j}$$

is defined for all $x \geqslant \min\{1, \mathcal{R}_0\}$ and strictly monotone decreasing. Furthermore, $\tau(\mathcal{R}_0) = 1$. Hence $\tau(1) < 1$ if $\mathcal{R}_0 < 1$, $\tau(1) = 1$ if $\mathcal{R}_0 = 1$, and $\tau(1) > 1$ if $\mathcal{R}_0 > 1$. □

We obtain an explicit formula for $\mathcal{R}_0$ if the intragroup deviation is constant, i.e., $h_j = f_j g_j + \delta$ with a constant $\delta \in \mathbb{R}$, $\delta > -\min\{f_j g_j\}$. Then $\mathcal{R}_0 = \sum_{k=1}^{m} f_k g_k + \delta$. Notice that

$$\sum_{k=1}^{m} \mathcal{R}^\circ_{kk} = \sum_{k=1}^{m} f_k g_k + m\delta = \mathcal{R}_0 + (m-1)\delta.$$

So, $\mathcal{R}_0$ can be smaller or larger than $\sum_{k=1}^{m} \mathcal{R}^\circ_{kk}$ depending on the sign of δ.

Sexually Transmitted Diseases in Heterosexual Populations

Let us assume that we have $m_\uparrow$ groups of males and $m_\ddagger$ groups of females and that the $m_\uparrow \times m_\ddagger$ matrix $R^\ddagger$ gives the basic replacement ratios from the female groups to the male groups, and that the $m_\ddagger \times m_\uparrow$ matrix $R^\uparrow$ gives the basic replacement ratios from the male groups to the female groups. We assume that there is no disease transmission between male groups and no transmission between female groups. If we count the male groups first, the basic replacement matrix is the $m \times m$ matrix, $m = m_\uparrow + m_\ddagger$,

$$\mathcal{R}^\circ = \begin{pmatrix} 0 & \mathcal{R}^\ddagger \\ \mathcal{R}^\uparrow & 0 \end{pmatrix}. \tag{23.38}$$

Counting the female groups first, we could also take

$$\begin{pmatrix} 0 & \mathcal{R}^\uparrow \\ \mathcal{R}^\ddagger & 0 \end{pmatrix}.$$

It will turn out that both matrices have the same spectral radius. Since the spectral radius of a product of two commuting matrices is the product of their spectral radii, we consider

$$(\mathcal{R}^\circ)^2 = \begin{pmatrix} \mathcal{R}^\ddagger \mathcal{R}^\uparrow & 0 \\ 0 & \mathcal{R}^\uparrow \mathcal{R}^\ddagger \end{pmatrix}.$$

Since $\mathcal{R}^{\uparrow}\mathcal{R}^{\ddagger}$ and $\mathcal{R}^{\ddagger}\mathcal{R}^{\uparrow}$ have the same spectral radius (Lemma A.53 in the toolbox), the spectral radius of $(\mathcal{R}^{\circ})^2$ coincides with the spectral radius of $\mathcal{R}^{\ddagger}\mathcal{R}^{\uparrow}$ (Exercise A.3). This leads to the following result.

Theorem 23.26. *$\mathcal{R}_0$ is the square root of the spectral radius of $\mathcal{R}^{\ddagger}\mathcal{R}^{\uparrow}$.*

If there are just one female group and one male group, then

$$\mathcal{R}^{\circ} = \begin{pmatrix} 0 & \mathcal{R}_1^{\circ} \\ \mathcal{R}_2^{\circ} & 0 \end{pmatrix},$$

with the $\mathcal{R}_j^{\circ}$ being the group replacement ratios. Then $\mathcal{R}_0$ is the geometric mean of the group replacement ratios, which is strictly smaller than the arithmetic mean unless the group replacement ratios are equal.

Let us now assume that there is separable mixing between the male groups on one side and the female groups on the other, i.e.,

$$\mathcal{R}_{ij}^{\ddagger} = f_i^{\ddagger} g_j^{\ddagger}, \qquad \mathcal{R}_{jk}^{\uparrow} = f_j^{\uparrow} g_k^{\uparrow}.$$

Then

$$(\mathcal{R}^{\ddagger}\mathcal{R}^{\uparrow})_{ik} = f_i^{\ddagger}\left(\sum_{j=1}^{m_{\ddagger}} g_j^{\ddagger} f_j^{\uparrow}\right) g_k^{\uparrow}.$$

This is a separable matrix again and its spectral radius is given by Theorem 23.24.

Corollary 23.27. *Consider separable mixing for a sexually transmitted disease in a heterosexual population as described above. Then the basic replacement ratio is given by*

$$(\mathcal{R}_0)^2 = \left(\sum_{j=1}^{m_{\ddagger}} g_j^{\ddagger} f_j^{\uparrow}\right)\left(\sum_{k=1}^{m_{\uparrow}} f_k^{\ddagger} g_k^{\uparrow}\right).$$

These considerations also apply to vector-borne diseases with the vertebrate host corresponding to one sex and the invertebrate vector to the other sex.

Exercises

23.7. Consider a population with m groups, where m is an arbitrary natural number. Find a basic replacement matrix $\mathcal{R}^{\circ}$ with separable mixing such that the basic replacement ratios of group k, $\mathcal{R}_k^{\circ}$, are strictly less than 1 for all but one k, but the basic replacement ratio, $\mathcal{R}_0$, strictly exceeds 1.

23.8. Consider a population with m groups, where m is an arbitrary natural number. Find a basic replacement matrix $\mathcal{R}^{\circ}$ with separable extragroup mixing and no intragroup mixing, i.e., $\mathcal{R}_{jj}^{\circ} = 0$ for all j, such that the basic replacement ratios of group k, $\mathcal{R}_k^\circ$, are strictly less than 1 for all but one k, but the basic replacement ratio, $\mathcal{R}_0$, strictly exceeds 1.

23.9. Consider a population with m groups where m is an arbitrary natural number. Find a basic replacement matrix $\mathcal{R}^\circ$ with separable extragroup mixing and no intragroup mixing, i.e., $\mathcal{R}_{jj} = 0$ for all j, such that the basic replacement ratios of group k, $\mathcal{R}_k^\circ$, are strictly larger than 1 for all but one k, but the basic replacement ratio, $\mathcal{R}_0$, is strictly smaller than 1.

23.10. One of the classical models for the spread of gonorrhea in a heterosexual population divides the population into four groups: men and women, each with symptomatic and asymptomatic infection.

Let the 2×2 matrices $\mathcal{R}^\dagger$ and $\mathcal{R}^\ddagger$ give the respective basic replacement ratios from the male to the female groups and vice versa.

Express the basic replacement ratio in terms of the entries of $\mathcal{R}^\dagger$ and $\mathcal{R}^\ddagger$.

23.11. Let $f_2, \ldots, f_m$ be nonnegative numbers, $f_m > 0$, and $g_2, \ldots, g_m$, $h_1, \ldots, h_m$ strictly positive numbers. Let the basic replacement matrix be given by

$$\begin{aligned} \mathcal{R}_{1k}^\circ &= f_k, & k &= 2, \ldots, m, \\ \mathcal{R}_{k,k-1}^\circ &= g_k, & k &= 2, \ldots, m, \\ \mathcal{R}_{kk} &= h_k, & k &= 1, \ldots, m, \\ \mathcal{R}_{jk}^\circ &= 0, & & \text{otherwise.} \end{aligned}$$

(a) Show that the matrix $\mathcal{R}^\circ$ is primitive.

(b) Show that $\mathcal{R}_0 > 1$ if and only if $h_j > 1$ for at least one j or if

$$h_j < 1 \quad \forall j \quad \text{and} \quad 1 < \frac{1}{1-h_1} \sum_{k=2}^{m} f_k \left(\prod_{j=2}^{k} \frac{g_j}{1-h_j} \right).$$

Remark: cf. Exercise 23.6 from Section 23.6.

Bibliographic Remarks

The interest in multi-group endemic models originally stems from gonorrhea modeling (see Lajmanovich, Yorke (1976) for a pioneering paper, and Hethcote, Yorke (1984) for a serious attempt at converting mathematical insight into practical recommendations for gonorrhea control), but has been boosted, of course, by the HIV/AIDS epidemic/endemic.

By assuming that the groups of the population remain constant in size, we have excluded immediate application of our theory to HIV/AIDS, which induces substantial fatalities. Our approach to local stability of the disease-free equilibrium and to persistence of the disease should still work for AIDS models also. Uniqueness of the endemic equilibrium no longer holds in general (Lin, 1991; Huang, Cooke, Castillo-Chavez, 1992; Jacquez, Simon, Koopman, 1995; van den Driessche, Watmough, 2002), and the assumptions guaranteeing uniqueness become rather complicated (Lin, 1991). Related to this, the disease-free equilibrium may be locally, but not globally, asymptotically stable in multi-group AIDS models. See Simon, Jacquez (1992) for conditions that guarantee extinction of the disease (global stability of the disease-free equilibrium) for various mixing scenarios between the groups of the population. Nonuniqueness of endemic equilibria is often connected with subcritical bifurcation from the disease-free equilibrium. This phenomenon may not only occur in certain multi-group models, but can also by caused by certain vaccination regimes (Hadeler, Castillo-Chavez, 1995; Kribs-Zaleta, Velasco-Hernández 2000; Kribs-Zaleta, Martcheva, 2002). For a survey on other causative mechanisms and more references, see Martcheva, Thieme (n.d.).

Our techniques for studying the local stability of a strongly endemic equilibrium, which were first used in Thieme (1983), can be generalized to AIDS models as well, but lead to much less clean results. They can also be generalized to models with variable per capita infectivity (Hethcote, Thieme, 1985).

We have not touched on global stability of the strongly endemic equilibrium because it seems to be difficult to obtain clean general results. We mention the work by Lin, So (1993), who show global asymptotic stability of the endemic equilibrium for an SIRS endemic model with multiple groups of constant size where intragroup contact rates are substantially larger than intergroup contact rates. Simon, Jacquez (1992, Theorem 3) present a sufficient condition for an endemic model with several infected stages; their assumption of restricted mixing divides the population into isolated groups, however, such that it can no longer be considered a proper multi-group model.

As for the threshold condition for the stability of the disease-free equilibrium, there seem to be two schools of thought: Diekmann, Heesterbeek, Metz (1990, 1995) and Diekmann, Dietz, Heesterbeek (1991) on the one hand, who promote $\mathcal{R}_0$ as the spectral radius of the next-generation matrix; and Simon, Jacquez (1992) and Jacquez, Simon, Koopman (1995) on the other hand, who emphasize the connection to basic group replacement ratios. Our presentation tries to bring them closer together. Infinite-dimensional generalizations of separable mixing and separable extragroup mixing and additional examples in which $\mathcal{R}_0$ can explicitly be determined can be found in Diekmann, Heesterbeek (2000, Section 5.2). For a brief history of $\mathcal{R}_0$ see Heesterbeek (2002).

If the disease changes the sizes of the population and its groups, the dependence of the intergroup contact probabilities C_{jk} on the group sizes becomes very relevant for the dynamics of the disease, and a lot of effort has been expended on modeling these functional relationships. (For surveys and references see Castillo-Chavez, Velasco-Hernandez, Fridman (1994), Diekmann, Heesterbeek, Metz (1995, Section 5), Hethcote (1995), Levin (1995), Jacquez, Simon, Koopman (1995) and Diekmann, Heesterbeek (2000, Chapter 10).) If the C_{jk} are "constant," i.e.,

$$C_{jk}(rN_1, \dots, rN_m) = C_{jk}(N_1, \dots, N_m) \quad \forall r > 0,$$

all epidemiologically reasonable choices of C_{jk} have been classified by Busenberg, Castillo-Chavez (1991). We agree with Diekmann, Heesterbeek, Metz (1995), however, that this approach is purely descriptive and that a mechanistic derivation of the contact functions would be preferable. Castillo-Chavez, Velasco-Hernandez, Fridman (1994) have adapted a mechanistic derivation from Holling's (1966) celebrated modeling of a predator's functional response to prey abundance. In the context of complex formation, this approach neglects, however, that an individual that is bound in a complex is at that point not available for other contacts. This feature is accounted for in the complex-formation approach of Heesterbeek, Metz (1993). Unfortunately, this approach does not seem to lead to explicit solutions for C_{jk} in general, but works quite well for proportional mixing. See Thieme, Yang (2000) for more details of, among other things, estimates and approximations of C_{jk} and for modeling contacts in a heterosexual multi-group population.

The theory of (quasi-)positive matrices and of the associated linear dynamical systems (Toolbox A.8), as rarely as it is taught in standard linear algebra or ordinary differential equations courses, is an immensely powerful tool in population models with some kind of structure like multi-species communities (Hofbauer, Sigmund, 1998; Smith, Waltman, 1995) and models with discrete age- or stage structure (Caswell, 1989, 2001; Cushing, 1998).

PART 4

Toolbox

Appendix A

Ordinary Differential Equations

In Appendix A, I have tried to organize in a somewhat systematic way the tools from ordinary differential equation theory that have been used to analyze the models in the main text. The following books on mathematical biology have useful technical appendixes which can be consulted as well: Capasso (1993), Edelstein-Keshet (1988), Murray (1989), Smith, Waltman (1995), and Farkas (2001).

A.1 Conservation of Positivity and Boundedness

In population dynamics we deal with functions representing population sizes, densities, biomasses, etc., all of which only make sense if they are nonnegative. It therefore becomes a major issue to show that trajectories that start nonnegative stay nonnegative. A very general condition for solutions of ordinary differential equations is given in Proposition B.7 in Smith, Waltman (1995).

Proposition A.1. *Consider a system of differential equations in* $\mathbb{R}^n$,

$$x' = F(t, x),$$

$x(t) = (x_1(t), \dots, x_n(t))$, $F(t, x) = (F_1(t, x), \dots, F_n(t, x))$, *where* $F(t, x)$ *is defined for all* $t \geqslant 0$, $x \in \mathbb{R}^n$. *Assume that* F *has the property that solutions of the initial value problems* $x(t_0) = x^\circ$ *are unique for* $x^\circ \in [0, \infty)^n$, $t_0 \geqslant 0$. *Furthermore, assume that, for all* $j = 1, \dots, n$, $t \geqslant 0$, *we have*

$$F_j(t, x) \geqslant 0 \quad \text{whenever}\, x \in [0, \infty)^n, \quad x_j = 0, \quad t \geqslant 0.$$

Then $x(t) \in [0, \infty)^n$ *for all* $t \geqslant t_0 \geqslant 0$ *for which it is defined, whenever* $x(t_0) \in [0, \infty)^n$.

A much more elementary condition which works without uniqueness, however, is the following.

Lemma A.2. *Consider the system*

$$x' = F(t, x)$$

with a continuous vector field F *from an open subset* D *of* $\mathbb{R}^{n+1}$ *to* $\mathbb{R}^n$. *Assume that the* j*th component of* F *has the form*

$$F_j(t, x) = x_j G_j(t, x)$$

with a continuous scalar-valued function G_j*. If* x *is a solution with* $x_j(t_0) > 0$ *or* $x_j(t_0) \geqslant 0$*, then this respective property holds for all* $t > t_0$ *for which* x *exists.*

Proof. Set

$$g(t) = G_j(t, x(t)).$$

Then

$$x_j(t) = x_j(t_0) \exp\left(\int_{t_0}^{t} g(s)\, \mathrm{d}s\right),$$

and the assertion follows. □

A somewhat more careful analysis provides the following result.

Proposition A.3. *Consider the system*

$$x' = F(t, x)$$

with a continuous vector field F *from a subset* D *of* $\mathbb{R}^{n+1}$ *to* $\mathbb{R}^n$*. Assume that the* j*th component of* F *has the following property:*

$F_j(t, x) = 0$ *whenever* $x_j = 0$*, and for every* $r > 0$ *there exists some* $c > 0$ *such that*

$$\frac{1}{x_j}|F_j(t, x)| \leqslant c \quad \text{whenever } |t| \leqslant r, \quad |x| \leqslant r, \quad (t, x) \in D, \quad x_j > 0.$$

If x *is a solution with* $x_j(t_0) > 0$*, then* $x(t) > 0$ *for all* $t > t_0$ *for which* x *exists.*

Proof. Suppose the statement is not true. Then there exists some $t > t_0$ such that $x_j(s) > 0$ for $s \in [t_0, t)$ and $x_j(t) = 0$. Since x is continuous, there exists some $r > 0$ such that $[t_0, t] \subseteq [-r, r]$ and $|x(t)| \leqslant r$ for $t \in [t_0, t]$. By assumption there exists some $c > 0$ such that

$$\frac{1}{x_j}|F_j(s, x)| \leqslant c, \quad s \in [t_0, t], \quad |x| \leqslant r, \quad x_j > 0.$$

We define

$$G_j(s, x) = \frac{1}{x_j} F_j(s, x), \quad s \in [t_0, t], \quad |x| \leqslant r, \quad (s, x) \in D, \quad x_j > 0.$$

Then

$$x_j'(s) = x_j(s) G_j(s, x(s)), \quad t_0 \leqslant s < t,$$

and

$$|G_j(s, x(s))| \leqslant c, \quad t_0 \leqslant s < t.$$

Hence

$$x_j(s) = x_j(t_0) \exp\left(\int_{t_0}^{s} G_j(r, x(r))\, \mathrm{d}r \right), \quad t_0 \leqslant s < t.$$

Thus

$$x_j(s) \geqslant x_j(t_0) \exp(-(s - t_0)c) \geqslant x_j(t_0) \exp(-(t - t_0)c) > 0, \quad t_0 \leqslant s < t.$$

But this contradicts $x_j(t) = 0$ and the continuity of x. □

Theorem A.4. *Let $\mathbb{R}^n_+ = [0, \infty)^n$ be the cone of nonnegative vectors in $\mathbb{R}^n$. Let $F : \mathbb{R}^{n+1}_+ \to \mathbb{R}^n$ be locally Lipschitz,*

$$F(t, x) = (F_1(t, x), \ldots, F_n(t, x)), \quad x = (x_1, \ldots, x_n),$$

and satisfy

$$F_j(t, x) \geqslant 0 \quad \text{whenever } t \geqslant 0, \quad x \in \mathbb{R}^n_+, \quad x_j = 0.$$

Then, for every $x^\circ \in \mathbb{R}^n_+$, there exists a unique solution of $x' = F(t, x)$, $x(0) = x^\circ$, with values in $\mathbb{R}^n_+$, which is defined on some interval $[0, b)$, $b > 0$. If $b < \infty$, then

$$\limsup_{t \nearrow b} \sum_{j=1}^{n} x_j(t) = \infty.$$

Proof. Extend F from $\mathbb{R}^{n+1}_+$ to $\mathbb{R}^{n+1}$ by

$$F(t, x) = F(t_+, x_+),$$

where

$$t_+ = \max\{t, 0\}, \quad x_+ = (x_{1+}, \ldots, x_{n+1}),$$

are the positive parts of the scalar t and the vector x. One easily checks that

$$\|x_+ - y_+\| \leqslant \|x - y\|$$

for any of the usual norms on $\mathbb{R}^n$. Hence F is a locally Lipschitz continuous vector field on $\mathbb{R}^n$ which still satisfies $F_j(t, x) \geqslant 0$ whenever $t \in \mathbb{R}$, $x \in \mathbb{R}^n_+$, $x_j = 0$.

Now, for any $x_0 \in \mathbb{R}^n$, there exists some $b > 0$ and a uniquely determined solution x of $x' = F(t, x)$ on $[0, b)$ with initial datum x_0 such that

$$\limsup_{t \nearrow b} \|x(t)\| \to \infty, \quad \text{whenever } b < \infty$$

(see Hale, 1980, Theorem I.2.1). Furthermore, the mapping

$$\Phi(t, x_0) = x(t)$$

is continuous in (t, x_0) wherever it is defined (see Hale, 1980, Theorem I.3.1).

Let $x_0 \in \mathbb{R}^n_+$. Then, by Proposition A.1, $x(t) \in \mathbb{R}^n_+$ for all $t \in [0, b)$. This implies the assertion. □

Corollary A.5. *Let $\mathbb{R}^n_+ = [0, \infty)^n$ be the cone of nonnegative vectors in $\mathbb{R}^n$. Let $F : \mathbb{R}^{n+1}_+ \to \mathbb{R}^n$ be locally Lipschitz,*

$$F(t, x) = (F_1(t, x), \ldots, F_n(t, x)), \quad x = (x_1, \ldots, x_n),$$

and satisfy

$$F_j(t, x) \geqslant 0 \quad \text{whenever } t \geqslant 0, \quad x \in \mathbb{R}^n_+, \quad x_j = 0.$$

Let B be a subset of $\mathbb{R}^n$ such that all solutions x of $x' = F(t, x)$ (with values in $\mathbb{R}^n_+$) that start in B stay in B as long as they are defined. Furthermore, let $B \cap \mathbb{R}^n_+$ be bounded.

Then, for every $x^\circ \in B \cap \mathbb{R}^n_+$, there exists a unique solution of $x' = F(t, x)$, $x(0) = x^\circ$, with values in $B \cap \mathbb{R}^n_+$, which is defined on all of $[0, \infty)$.

Proof. Let $x^\circ \in B \cap \mathbb{R}^n_+$. By Theorem A.4, there exists a unique solution of $x' = F(t, x)$, $x(0) = x^\circ$, with values in $\mathbb{R}^n_+$, which is defined on some interval $[0, b)$, $b > 0$. If $b < \infty$, then

$$\limsup_{t \nearrow b} \sum_{j=1}^n x_j(t) = \infty.$$

Since the solution starts in $B \cap \mathbb{R}^n_+$, we have that $x(t) \in B \cap \mathbb{R}^n_+$ for all $t \in [0, \infty)$. Since $B \cap \mathbb{R}^n_+$ is bounded, the solution cannot blow up in finite time, hence $b = \infty$. □

The following elementary consideration is sometimes useful to establish boundedness of solutions on finite intervals.

Lemma A.6. *Let $x : [a, b) \to \mathbb{R}$ be continuous and differentiable on (a, b). Let $t \in (a, b)$ and $\bar{x} = \max_{[a,t]} x$. Then $\bar{x} = x(a)$ or there exists $s \in (a, t]$ such that $\bar{x} = x(s)$ and $x'(s) \geqslant 0$.*

Proof. First notice that x is bounded on $[a, t]$ and the supremum $\bar{x}$ is taken at some element $s \in [0, t]$. If $s = a$, then $\bar{x} = x(a)$. If $s \in (a, t)$, then there is a local maximum at s and $x'(s) = 0$. If $s = t$, then $x(t) \geqslant x(r)$ for all $r \in [a, t]$ and $x'(t) \geqslant 0$. □

A.2 Planar Ordinary Differential Equation Systems

The large-time behavior of solutions of planar autonomous systems of ordinary differential equations is limited by the Poincaré–Bendixson theory. Convergence towards equilibria can be established by the Dulac criterion or by *cooperativity* or *competitiveness* of the system. In special configurations it can also be established by showing that all possible periodic orbits are orbitally stable (this approach is discussed in Toolbox A.6).

Let X be an open subset of $\mathbb{R}^2$ and $F : X \to \mathbb{R}^2$ be continuous,

$$F(x) = (F_1(x), F_2(x)), \quad x = (x_1, x_2) \in X.$$

Furthermore, assume that any solution

$$x' = F(x), \quad x(0) = x^\circ \tag{A.1}$$

on an interval $(-\epsilon, \epsilon)$ is uniquely determined by x°.

A point $z \in X$ with $F(z) = 0$ is called a *critical point* or an *equilibrium*; a point $z \in X$ with $F(z) \neq 0$ is called a *regular* point.

Let x be a solution of (A.1) in X that is defined and bounded on $[0, \infty)$. Then its ω-limit set $\omega(x^\circ)$ can be defined as

$$\omega(x^\circ) = \bigcap_{r \geqslant 0} \overline{\{x(t);\ t \geqslant r\}}.$$

Equivalently, we can say that an element $y \in \mathbb{R}^2$ is contained in $\omega(x^\circ)$ if and only if there exists a sequence (t_j) of positive numbers such that $t_j \to \infty$ and $x(t_j) \to y$ as $j \to \infty$. Similarly, if x is a solution of (A.1) in X that is defined and bounded on $(-\infty, 0]$, then its α-limit set, $\alpha(x^\circ)$, can be defined as

$$\alpha(x^\circ) = \bigcap_{r \geqslant 0} \overline{\{x(t);\ t \leqslant -r\}}.$$

Poincaré–Bendixson Theory

Obviously, $\omega(x^\circ)$ is a subset of $\bar{X}$. We will restrict our attention to the case where $\omega(x^\circ)$ is a subset of X. We follow the presentation in Coddington, Levinson (1955, Section 16.2).

Theorem A.7 (Poincaré–Bendixson dichotomy). *Let x be a solution of (A.1) in X that is defined and bounded on $[0, \infty)$ with $\omega(x^\circ) \subseteq X$. Then $\omega(x^\circ)$ contains an equilibrium or is a periodic orbit.*

An analogous statement holds for $\alpha(x^\circ)$.

Proposition A.8. *Let x be a solution of (A.1) in X that is defined and bounded on $[0, \infty)$ with $\omega(x^\circ) \subseteq X$. If $\omega(x^\circ)$ contains an equilibrium and a regular point y°, then there exists a solution y of $y' = F(y)$ in $\omega(x^\circ)$ that is defined on $\mathbb{R}$ and satisfies $y(0) = y^\circ$ such that both $\omega(y^\circ)$ and $\alpha(y^\circ)$ consist of equilibria.*

This proposition follows from the proof of Theorem 3.1 in Coddington, Levinson (1955). It has the following immediate consequence.

Corollary A.9. *Let x be a solution of (A.1) in X that is defined and bounded on $[0, \infty)$ with $\omega(x^\circ) \subseteq X$. If $\omega(x^\circ)$ contains exactly one equilibrium x^* and a regular point y°, then there exists a solution y of $y' = F(y)$ in $\omega(x^\circ)$ that is defined on $\mathbb{R}$ and satisfies $y(0) = y^\circ$ such that $y(t) \to x^*$ as $t \to \infty$ and $t \to -\infty$.*

In other words, we have a homoclinic orbit through y° that connects the equilibrium x^* to itself. If X only contains a finite number of critical points, a more complete description of the ω-limit set is possible.

Theorem A.10 (Poincaré–Bendixson trichotomy). *Assume that X only contains finitely many equilibria. Let x be a solution of (A.1) in X that is defined and bounded on $[0, \infty)$ with $\omega(x^\circ) \subseteq X$. Then either*

(i) *$\omega(x^\circ)$ consists of an equilibrium, or*

(ii) *$\omega(x^\circ)$ is a periodic orbit, or*

(iii) *$\omega(x^\circ)$ consists of finitely many equilibria and of orbits that connect them.*

Mutatis mutandis, this statement also holds for $\alpha(x^\circ)$.

Remark A.11. If part (iii) of the trichotomy in Theorem A.10 holds, $\omega(x^\circ)$ contains a cyclic chain of equilibria, i.e., there are equilibria $x^1, \dots, x^m, m \geqslant 1$, in $\omega(X^\circ)$ and solutions y^k, $k = 1, \dots, m$, of $y' = F(y)$ in $\omega(x^\circ)$ which are defined on $\mathbb{R}$ such that

$$\begin{aligned} y^k(t) &\to x^k, & t &\to -\infty, & k &= 1, \dots, m, \\ y^k(t) &\to x^{k+1}, & t &\to \infty, & k &= 1, \dots, m-1, \\ y^m(t) &\to x^1, & t &\to \infty. \end{aligned}$$

If $m = 1$, we have a homoclinic orbit connecting the equilibrium to itself.

A proof of this remark can be found in Thieme (1994a), for example. A convenient way of ruling out periodic orbits and cyclic chains of equilibria is the following (see, for example, Hahn, 1967).

Dulac–Bendixson Criterion

Theorem A.12. *Let Z be an open and simply connected subset of $\mathbb{R}^2$, $Z \subseteq X$. Assume that F is continuously differentiable on Z and that there exists a continuously differentiable real-valued function ρ on Z such that the divergence of ρF,*

$$\nabla \cdot (\rho F)(x) = \frac{\partial}{\partial x_1}(\rho F_1)(x) + \frac{\partial}{\partial x_2}(\rho F_2)(x),$$

is either strictly positive almost everywhere on Z or strictly negative almost everywhere on Z. Then Z contains no periodic orbits or cyclic chains of $x' = F(x)$.

A function ρ that makes the divergence of ρF have one sign is called a *Dulac function*. If $\rho \equiv 1$, Theorem A.12 is referred to as the *Bendixson criterion*.

Corollary A.13. *Let Y and Z be subsets of X, Z open and simply connected in $\mathbb{R}^2$, with the following properties:*

- *Every solution of (A.1) in X that is defined and bounded on $[0, \infty)$ has its ω-limit set in Y, and Y contains only finitely many equilibria.*
- *All possible periodic orbits of (A.1) and all cyclic chains (i.e., the equilibria and the connecting orbits) of (A.1) that lie in Y are actually contained in Z.*
- *F is continuously differentiable on Z, and there exists a continuously differentiable real-valued function ρ on Z such that the divergence of ρF,*

$$\nabla \cdot (\rho F)(x) = \frac{\partial}{\partial x_1}(\rho F_1)(x) + \frac{\partial}{\partial x_2}(\rho F_2)(x),$$

  is either strictly positive almost everywhere on Z or strictly negative almost everywhere on Z.

Then every solution of (A.1) in X that is defined and bounded on $[0, \infty)$ converges towards an equilibrium in Y.

Competitive or Cooperative Planar Systems

A system

$$x' = F(x)$$

with continuously differentiable $F : \mathbb{R}^n \to \mathbb{R}^n$, $F(x) = (F_1(x), \dots, F_n(x))$, is called *cooperative* if

$$\frac{\partial F_j}{\partial x_k}(x) \geqslant 0, \quad j \neq k, \quad x \in \mathbb{R}^n.$$

It is called *competitive* if the inequalities are reversed:

$$\frac{\partial F_j}{\partial x_k}(x) \leqslant 0, \quad j \neq k, \quad x \in \mathbb{R}^n.$$

Theorem A.14 (Smith, 1995, Theorem 3.2.2). *Any solution of a competitive or cooperative system in $\mathbb{R}^2$ that is bounded in forward time converges towards an equilibrium as $t \to \infty$.*

A result in the opposite direction has been proved by Zhu (1991) (see Zhu, Smith (1994) for the result).

Existence of a Stable Periodic Orbit

A periodic orbit is called locally (asymptotically) stable if it is locally (asymptotically) stable as a set (see Toolbox A.6). The associated solution is called locally (asymptotically) orbitally stable.

Theorem A.15. *Let Ω be a simply connected open subset of $\mathbb{R}^2$. Let F be a continuously differentiable vector field from Ω to $\mathbb{R}^2$.*

Assume that every solution of $x' = F(x)$ starting in Ω at $t = 0$ is defined for all times $t \geqslant 0$ and that there exists some time $r > 0$ (which may depend on the solution) such that $x(s) \in K$ for $s > r$, where K is a compact subset of Ω that does not depend on the solution x.

Let Ω contain a single equilibrium x^* of F. Assume that x^* is nondegenerate and unstable. Then there exists a locally stable periodic orbit. If F is real analytic, then there exists a locally asymptotically stable periodic orbit.

Recall that an equilibrium point is called degenerate if the associated Jacobian matrix is singular (i.e., has a zero eigenvalue).

A.3 The Method of Fluctuations

In this section we consider bounded differentiable functions $f : (a, \infty) \to \mathbb{R}$ and discuss the relation between the asymptotic behavior of f and f'. We introduce the following notation:

$$f^\infty = \limsup_{t\to\infty} f(t), \qquad f_\infty = \liminf_{t\to\infty} f(t).$$

Theorem A.16. *Let t_n be a sequence in (a, ∞), $t_n \to \infty$ as $n \to \infty$. Assume that*

$$f'(t_n + h) - f'(t_n) \to 0, \quad n \to \infty, \quad h \to 0. \tag{A.2}$$

If $f(t_n) \to f^\infty$ or $f(t_n) \to f_\infty$ as $n \to \infty$, then $f'(t_n) \to 0$ as $n \to \infty$.

Remark. Condition (A.2) means that, for any $\epsilon > 0$, there exists $m \in \mathbb{N}$ and $\delta > 0$ such that

$$|f'(t_n + h) - f'(t_n)| < \epsilon, \quad n > m, \quad |h| < \delta.$$

Condition (A.2) is satisfied if f' is uniformly continuous on some interval $[b, \infty) \subseteq (a, \infty)$.

Proof. It is sufficient to consider the case where $f(t_n)$ converges towards f^∞. The other case can be reduced to this one by a sign change. Let $h > 0$. By the mean value theorem,

$$f(t_n + h) - f(t_n) = f'(t_n + r_n)h = f'(t_n)h + (f'(t_n + r_n) - f'(t_n))h$$

for some $r_n \in (0, h)$. Hence

$$f'(t_n) = (1/h)(f(t_n + h) - f(t_n)) + f'(t_n) - f'(t_n + r_n).$$

Thus

$$f'(t_n) \leqslant (1/h)(f(t_n + h) - f(t_n)) + \sup_{0<r<h} (f'(t_n) - f'(t_n + r)).$$

Hence

$$\begin{aligned}\limsup_{n\to\infty} f'(t_n) \\ &\leqslant (1/h)\Big(\limsup_{t\to\infty} f(t) - f^{\infty}\Big) + \limsup_{n\to\infty} \sup_{0<r<h} (f'(t_n) - f'(t_n + r)) \\ &\leqslant \limsup_{n\to\infty} \sup_{0<r<h} (f'(t_n) - f'(t_n + r)).\end{aligned}$$

Since this holds for every $h > 0$, our assumption implies that

$$\limsup_{n\to\infty} f'(t_n) \leqslant 0.$$

Still let $h > 0$. Again, by the mean value theorem,

$$f(t_n) - f(t_n - h) = f'(t_n - r_n)h = f'(t_n)h + (f'(t_n - r_n) - f'(t_n))h$$

for some $r_n \in (0, h)$. Hence

$$f'(t_n) = (1/h)(f(t_n) - f(t_n - h)) + f'(t_n) - f'(t_n - r_n).$$

Thus

$$f'(t_n) \geqslant (1/h)(f(t_n) - f(t_n - h)) + \inf_{0<r<h} (f'(t_n) - f'(t_n - r)).$$

Hence

$$\begin{aligned}\liminf_{n\to\infty} f'(t_n) \\ &\geqslant (1/h)\Big(f^{\infty} - \limsup_{t\to\infty} f(t)\Big) + \liminf_{n\to\infty} \inf_{0<r<h} (f'(t_n) - f'(t_n - r)) \\ &\geqslant \liminf_{n\to\infty} \inf_{0<r<h} (f'(t_n) - f'(t_n - r)).\end{aligned}$$

Since this holds for every $h > 0$, our assumption implies that

$$\liminf_{n\to\infty} f'(t_n) \geqslant 0.$$

□

We recover a well-known result from Theorem A.16 (Barbalat, 1959).

Corollary A.17. *Assume that $f(t)$ converges as $t \to \infty$. Then $f'(t) \to 0$ if and only if*

$$f'(t+h) - f'(t) \to 0, \quad t \to \infty, \quad h \to 0.$$

Corollary A.18. *Let $x : [0, \infty) \to \mathbb{R}^n$ be differentiable and assume that $x(t)$ converges as $t \to \infty$. Then $x'(t) \to 0$ as $t \to \infty$ if x' is uniformly continuous on $[0, \infty)$.*

Corollary A.19. *Let x be a solution of $x' = F(x)$ that is defined for all forward times and converges to x^* as $t \to \infty$, where F is a continuous vector field on $\mathbb{R}^n$. Then x^* is an equilibrium.*

Proof. Since F is continuous,

$$x'(t+h) - x'(t) = F(x(t+h)) - F(x(t)) \to F(x^*) - F(x^*) = 0, \quad t \to \infty,$$

uniformly in $h \in (0, 1)$. By Corollary A.17,

$$0 = \lim_{t \to \infty} x'(t) = \lim_{t \to \infty} F(x(t)) = F(x^*).$$

□

The following lemma can be found in Hirsch et al. (1985, Lemma 4.2).

Lemma A.20. *Assume that f has no limit for $t \to \infty$. Then there are sequences $s_n, t_n \to \infty$ with the following properties:*

$$f(s_n) \to f_\infty, \quad n \to \infty, \qquad f'(s_n) = 0 \quad \forall n \in \mathbb{N}, \tag{A.3}$$

$$f(t_n) \to f^\infty, \quad n \to \infty, \qquad f'(t_n) = 0 \quad \forall n \in \mathbb{N}. \tag{A.4}$$

If f is twice continuously differentiable, one has in addition that

$$f''(s_n) \geqslant 0, \quad f''(t_n) \leqslant 0, \quad n \in \mathbb{N}.$$

Remark A.21. One can also achieve that, in (A.3), $f'(s_n) > 0$ for all $n \in \mathbb{N}$ or that $f'(s_n) < 0$ for all $n \in \mathbb{N}$. The same applies to (A.4).

Proof. We restrict ourselves to the statement concerning the limit superior. Otherwise we consider $-f$.

Step 1. For every $r > a$ and every $\alpha \in (f_\infty, f^\infty)$ there exists some $t > r$ such that $f(t) \geqslant \alpha$, $f'(t) = 0$. If f is twice differentiable, then we can also achieve that $f''(t) \leqslant 0$.

There also exists some $t > r$ such that $f'(t) < 0$ and some $t > r$ such that $f'(t) > 0$.

Indeed, since $f_\infty < \alpha < f^\infty$ we find some s, u such that $u > s > r$ and $f(s) = \alpha = f(u)$ and $f(t) > \alpha$ for all $t \in [s, u]$. Hence f takes a maximum at some point $t \in (s, u)$ and we have

$$f(t) > \alpha, \quad f'(t) = 0, \quad f''(t) \leqslant 0,$$

with the last property holding if f is twice differentiable.

As for the remark, the mean value theorem also lets us find some $t \in [s, u]$ such that $f'(t) < 0$ and some $t \in [s, u]$ such that $f'(t) > 0$.

Step 2. Finale.

We choose a sequence $\alpha_n \in (f_\infty, f^\infty)$ such that $\alpha_n \to f^\infty$. By Step 1 we find numbers $t_n \to \infty$ as $n \to \infty$ such that

$$f(t_n) > \alpha_n, \quad f'(t_n) = 0, \quad f''(t_n) \leqslant 0,$$

with the last property holding if f is twice differentiable.

Since $t_n \to \infty$, it follows from the definition of the limit superior that

$$\limsup_{n\to\infty} f(t_n) \leqslant f^\infty.$$

Since

$$f(t_n) > \alpha_n \to f^\infty,$$

we have $f(t_n) \to f^\infty$ as $n \to \infty$.

It is clear from Step 1 that we can achieve that $f'(t_n) < 0$ for all $n \in \mathbb{N}$ or $f'(t_n) > 0$ for all $n \in \mathbb{N}$. □

The next result looks like a special case of Theorem A.16, but we do not make any continuity requirements for f'.

Proposition A.22. *There are sequences $s_n, t_n \to \infty$ with the following properties:*

$$\begin{aligned} f(s_n) &\to f_\infty, \quad & f'(s_n) &\to 0, \\ f(t_n) &\to f^\infty, \quad & f'(t_n) &\to 0, \end{aligned}$$

for $n \to \infty$.

Proof. By Lemma A.20 we can assume that $f(t)$ has a finite limit for $t \to \infty$. If the statement of this proposition does not hold, f' does not change sign for sufficiently large t and must be bounded away from 0. But this contradicts the convergence of $f(t)$ to a finite limit. □

Theorem A.23. *Let D be a bounded interval in $\mathbb{R}$ and $g : (t_0, \infty) \times D \to \mathbb{R}$ be bounded and uniformly continuous. Furthermore, let $x : (t_0, \infty) \to D$ be a solution of*

$$x' = g(t, x),$$

which is defined on the whole interval (t_0, ∞). *Then there exist sequences* $s_n, t_n \to \infty$ *such that*

$$\lim_{n\to\infty} g(s_n, x_\infty) = 0 = \lim_{n\to\infty} g(t_n, x^\infty).$$

Proof. As g is bounded, x' is bounded and so x is uniformly continuous on (t_0, ∞). As g is continuous, so is $x'(t)$ on (t_0, ∞). By Proposition A.22, we find sequences $s_n, t_n \to \infty$ with the following properties:

$$\begin{aligned} x(s_n) &\to x_\infty, \quad x'(s_n) = g(s_n, x(s_n)) \to 0, \\ x(t_n) &\to x^\infty, \quad x'(t_n) = g(t_n, x(t_n)) \to 0, \end{aligned}$$

for $n \to \infty$. As g is uniformly continuous, the assertion follows. □

Corollary A.24. *Let the assumptions of Theorem A.23 be satisfied. Then*

(a) $\liminf_{t\to\infty} g(t, x_\infty) \leqslant 0 \leqslant \limsup_{t\to\infty} g(t, x_\infty)$,

(b) $\liminf_{t\to\infty} g(t, x^\infty) \leqslant 0 \leqslant \limsup_{t\to\infty} g(t, x^\infty)$.

We continue with an elementary, but useful, remark.

Lemma A.25. *Let* f *be strictly positive.*

If $f_\infty = 0$, *there exists a sequence* $t_n \to \infty$ *as* $n \to \infty$ *such that* $f(t_n) \to 0$ *and* $f'(t_n) < 0$ *for all* $n \in \mathbb{N}$.

Proof. Let us first assume that $f^\infty = 0$.

Let $r > 0$. We show that there exists some $s \geqslant r$ such that $f'(s) < 0$. (This provides a sequence $t_n \to \infty$, $n \to \infty$, such that $f'(t_n) < 0$. Since $f^\infty = 0$, $f(t_n) \to 0$.)

Since $f(r) > 0$ and $f_\infty = 0$ there exists some $t > r$ such that $f(t) < f(r)$. By the mean value theorem, there exists some $s \in (r, t)$ such that

$$0 > f(t) - f(r) = f'(s)(t - r).$$

Obviously, $f'(s) < 0$.

Let us now assume that $0 = f_\infty < f^\infty$. Then the statement follows from Lemma A.20 and Remark A.21. □

We conclude this section with a result on asymptotically autonomous scalar differential equations.

Theorem A.26. *Let* J *be a closed interval,* $a > 0$ *and* $h : J \to \mathbb{R}$, h *continuous on* J. *Let* x *be a bounded function, defined on* (a, ∞) *with values in* J, *and*

$$x'(t) - h(x(t)) \to 0, \quad t \to \infty.$$

Then the following hold:

(a) *h is zero on (x_∞, x^∞).*

(b) *If h is never identically zero on an open interval, $x(t)$ converges to some $x^* \in J$ as $t \to \infty$ and $h(x^*) = 0$.*

Proof. We first assume that $x_\infty < x^\infty$. Consider $x^\diamond \in (x_\infty, x^\infty)$. Then the solution x keeps crossing $x^\diamond$ as $t \to \infty$, both from above and from below. This means that there are sequences $s_n, t_n \to \infty$, $n \to \infty$, such that

$$x(s_n) = x^\diamond = x(t_n), \qquad x'(s_n) \leqslant 0, \qquad x'(t_n) \geqslant 0.$$

This implies

$$h(x^\diamond) = \lim_{n\to\infty} h(s_n) = \lim_{n\to\infty} x'(s_n) \leqslant 0.$$

Similarly, $h(x^\diamond) \geqslant 0$ and h is identically zero on (x_∞, x^∞).

By contraposition, if h is never identically zero on an open interval, $x_\infty = x^\infty$, and $x(t)$ converges towards some x^*.

By Proposition A.22, there exists a sequence $t_n \to \infty$ such that $x'(t_n) \to 0$ and $x(t_n) \to x^*$. So,

$$h(x^*) = \lim_{n\to\infty} h(x(t_n)) = \lim_{n\to\infty} x'(t_n) = 0.$$

□

A.4 Behavior in the Vicinity of an Equilibrium

Consider a system of ordinary differential equations

$$x' = F(x), \tag{A.5}$$

with $x(t) \in \mathbb{R}^n$ and $F : \mathbb{R}^n \to \mathbb{R}^n$. F is continuously differentiable.

$x^* \in \mathbb{R}^n$ is called an *equilibrium* (*critical point, equilibrium point, steady-state, stationary solution*) of (A.5) if

$$F(x^*) = 0.$$

In this section we collect some results which tell us how a solution to (A.5) behaves in the vicinity of an equilibrium. An important concept is *local stability*.

Definition A.27.

(a) An equilibrium x^* of (A.5) is called *locally stable* if and only if the following holds:

For any $\epsilon > 0$ there exists some $\delta > 0$ such that $\|x(t) - x^*\| < \epsilon$ for $t > 0$ whenever x is a solution to (A.5) and $\|x(0) - x^*\| < \delta$.

(b) A locally stable equilibrium x^* is called *locally asymptotically stable* if and only if there exists some $\delta > 0$ such that $x(t) \to x^*$ for $t \to \infty$ whenever x is a solution to (A.5) and $\|x(0) - x^*\| < \delta$.

(c) An equilibrium is called *unstable* if and only if it is not locally stable, i.e., if and only if the following holds.

There exists some $\epsilon > 0$ and a sequence x_n of solutions to (A.5) and a sequence $t_n > 0$ such that $\|x_n(0) - x^*\| \to 0, n \to \infty$, but $\|x(t_n) - x^*\| > \epsilon$ for all $n \in \mathbb{N}$.

Obviously, it is important to find criteria which guarantee the stability or instability of equilibria. The most important one is the principle of linearized stability. As stability is a local phenomenon, we are interested in how solutions behave in the vicinity of the equilibrium. Hence we expand solutions to (A.5) around the equilibrium:

$$x(t) = x^* + y(t).$$

Using the fact that F is continuously differentiable we obtain

$$\frac{\mathrm{d}}{\mathrm{d}t}y = \mathrm{D}F(x^*)y + \Psi(y)y$$

with $\Psi(y)$ being a matrix such that $\|\Psi(y)\| \to 0$ for $\|y\| \to 0$. $\mathrm{D}F$ denotes the derivative of F. The basic idea consists of assuming (which can actually be proved) that the solution y essentially behaves as the solution z to the linear problem

$$\frac{\mathrm{d}}{\mathrm{d}t}z = \mathrm{D}F(x^*)z.$$

Solutions to the linear problem are found by trying

$$z(t) = \mathrm{e}^{\lambda t} z^*$$

with $z^* \neq 0$. This leads to the problem

$$\mathrm{D}F(x^*)z^* = \lambda z^*, \quad z^* \neq 0.$$

If λ and z^* satisfy this relation, then λ is called an *eigenvalue* of the matrix $\mathrm{D}F(x^*)$ and z^* the associated *eigenvector*. Apparently, any eigenvalue is a solution of the *characteristic equation*

$$\det(\lambda \boldsymbol{I} - \mathrm{D}F(x^*)) = 0$$

with $\boldsymbol{I}$ standing for the identity matrix. Since $z(t) = \mathrm{e}^{\lambda t} z^*$ converges to 0 ($t \to \infty$) if $\Re\lambda < 0$, but diverges to ∞ if $\Re\lambda > 0$, it is suggestive that the eigenvalues of $\mathrm{D}F(x^*)$ determine the stability of the equilibrium x^*. Indeed, the following lemma holds.

Lemma A.28.

(a) *The equilibrium x^* is locally asymptotically stable if all eigenvalues of* $\mathrm{D}F(x^*)$ *have strictly neg ,tive real parts.*

(b) x^* *is unstable if at least one eigenvalue of* $\mathrm{D}F(x^*)$ *has a strictly positive real part.*

These results can be found in standard advanced textbooks on ordinary differential equations (e.g., Hirsch, Smale, 1974, Sections 9.1 and 9.2).

Checking the conditions in Lemma A.28 can be gruesome. Sometimes it helps to remember some facts from linear algebra.

- The sum of all eigenvalues (counting multiplicities) equals the trace of the matrix (sum of the elements in the main diagonal).
- The product of all eigenvalues (counting multiplicities) equals the determinant of the matrix.
- If a matrix

$$A = \begin{pmatrix} B & C \\ 0 & D \end{pmatrix}$$

has block form with square matrices B and D, then the eigenvalues of A are found by determining the eigenvalues of B and D.

The Routh–Hurwitz Criterion

Let the characteristic equation have the form

$$\lambda^n + a_1\lambda^{n-1} + a_2\lambda^{n-2} + \cdots + a_{n-1}\lambda + a_n = 0.$$

Consider the following expressions for characteristic equations with $n \leqslant 4$:

$$\begin{aligned} n = 2: &\quad a_1,\ a_2; \\ n = 3: &\quad a_1,\ a_3,\ a_1a_2 - a_3; \\ n = 4: &\quad a_1,\ a_3,\ a_4,\ a_1a_2a_3 - a_3^2 - a_1^2a_4. \end{aligned}$$

Then all zeros λ have strictly negative real parts if all expressions are strictly positive.

However, if at least one expression is strictly negative, then at least one root λ has a strictly positive real part.

More material that can be useful in discussing the local stability of equilibria can be found in Murray (1989, Appendix 2) and in Smith, Waltman (1995, Appendix A).

A.5 Elements of Persistence Theory

Persistence addresses the question of which components of an ecological system will not become extinct and/or which will remain bounded. Mathematically, persistence is most elegantly treated within the theory of abstract dynamical systems. Dynamical systems or semiflows are induced by ordinary differential equations, functional differential equations, and certain partial differential equations, and they offer a unified analysis of the qualitative behavior of their solutions.

Let X be a metric space with metric d.

A mapping $\Phi : [0, \infty) \times X \to X$ is called a *semiflow* (or *semi-dynamical system*) on X if

$$\Phi_t \circ \Phi_s = \Phi_{t+s}, \quad t, s \geqslant 0, \tag{A.6}$$

where Φ_t represents the mapping $\Phi(t, \cdot) : X \to X$.

The semiflow is called continuous if Φ is a continuous mapping. Generally, one also requires $\Phi(0, x) = x$, but this is not necessary for persistence theory.

Let us describe the two types of semiflows induced by ordinary differential equations that typically show up in our biological models.

Proposition A.29. *Let X be an open subset of $\mathbb{R}^n$ and $F : X \to \mathbb{R}^n$ be continuous. Furthermore, assume that solutions of the differential equation*

$$x' = F(x), \quad x(0) = x_0,$$

are unique for $x_0 \in X$ and exist and stay in X for all $t \geqslant 0$. Then the solutions x of this ordinary differential equation induce a continuous semiflow on X via

$$\Phi(t, x_0) = x(t), \quad x_0 \in X, \quad t \geqslant 0.$$

Moreover, $\Phi(0, x_0) = x_0$.

Proof. In order to prove the semiflow property (A.6) fix $r \geqslant 0$ and set

$$y(t) = \Phi(t + r, x_0), \qquad z(t) = \Phi(t, z_0), \quad z_0 = \Phi(r, x_0).$$

Then

$$y'(t) = \frac{\mathrm{d}}{\mathrm{d}s}\Phi(s, x_0)_{[s=t+r]} = F(\Phi(s, x_0))_{[s=t+r]} = F(y(t)),$$
$$y(0) = \Phi(r, x_0) = z_0,$$

and

$$z'(t) = \frac{\mathrm{d}}{\mathrm{d}t}\Phi(t, z_0) = F(\Phi(t, z_0)) = F(z(t)), \quad z(0) = z_0.$$

Hence y and z are both solutions of the differential equation $x' = F(x)$ with the same initial datum at $t = 0$. Since uniqueness has been assumed, $y(t) = z(t)$

for all $t \geqslant 0$. The continuity of the semiflow follows from Theorem I.3.1 of Hale (1980). □

Corollary A.30. *Let* $\mathbb{R}_+^n = [0, \infty)^n$ *be the cone of nonnegative vectors in* $\mathbb{R}^n$. *Let* $F : \mathbb{R}_+^n \to \mathbb{R}^n$ *be locally Lipschitz,*

$$F(x) = (F_1(x), \dots, F_n(x)), \quad x = (x_1, \dots, x_n),$$

and satisfy

$$F_j(x) \geqslant 0 \quad \text{whenever } x \in \mathbb{R}_+^n, \quad x_j = 0. \tag{A.7}$$

Finally, assume that every solution of

$$x' = F(x)$$

that is defined on an interval $[0, b), 0 < b < \infty$ *(with values in* $\mathbb{R}_+^n$*), is bounded on* $[0, b)$.

Then the solutions x *of this ordinary differential equation induce a continuous semiflow on* $\mathbb{R}_+^n$ *via*

$$\Phi(t, x_0) = x(t), \quad x_0 \in X, \quad t \geqslant 0.$$

Moreover, $\Phi(0, x_0) = x_0$.

Proof. By Theorem A.4, all solutions of $x' = F(x)$ starting from $x_0 \in \mathbb{R}_+^n$ are defined for all forward times and stay in $\mathbb{R}_+^n$. As in Proposition A.29, we show that the mapping Φ defined by the solutions of $x' = F(x)$ is a semiflow on $\mathbb{R}_+^n$. □

Resuming the discussion of persistence, let $\rho : X \to [0, \infty)$ be a nonnegative uniformly continuous functional on X. We assume that the composition

$$\sigma = \rho \circ \Phi \quad \text{or} \quad \sigma(t, x) = \rho(\Phi(t, x))$$

is a continuous mapping from $[0, \infty) \times X$ to $\mathbb{R}$. σ is continuous if Φ is continuous. Notice that we have the relation

$$\sigma(t, \Phi_r(x)) = \sigma(t + r, x), \quad t, r \geqslant 0, \quad x \in X.$$

We introduce the following notation:

$$\sigma^\infty(x) = \limsup_{t\to\infty} \sigma(t, x), \qquad \sigma_\infty(x) = \liminf_{t\to\infty} \sigma(t, x).$$

Definition A.31.

- Φ is called *weakly ρ-persistent* if

$$\sigma^\infty(x) > 0 \quad \forall x \in X, \ \rho(x) > 0.$$

- Φ is called *strongly ρ-persistent* if

$$\sigma_\infty(x) > 0 \quad \forall x \in X,\ \rho(x) > 0.$$

- Φ is called *uniformly weakly ρ-persistent* if there exists some $\epsilon > 0$ such that

$$\sigma^\infty(x) > \epsilon \quad \forall x \in X,\ \rho(x) > 0.$$

- Φ is called *uniformly strongly ρ-persistent* if there exists some $\epsilon > 0$ such that

$$\sigma_\infty(x) > \epsilon \quad \forall x \in X,\ \rho(x) > 0.$$

If no misunderstanding about the functional ρ is possible, we use *persistent* rather than ρ-persistent.

We consider the following compactness condition.

Compactness condition (C). There exists some $\epsilon_0 > 0$ and a closed subset B of X with the following properties:

- For all $x \in \{\rho \leqslant \epsilon_0\}$ we have $\Phi_t(x) \to B, t \to \infty$.
- If $\epsilon_1 \in (0, \epsilon_0)$, the intersection of B with the ring $\{\epsilon_1 \leqslant \rho \leqslant \epsilon_0\}$ is compact.

By $\Phi(t, x) \to B, t \to \infty$, we mean that, for every $\epsilon > 0$, there is some $t > 0$ such that for every $s > t$ some $b \in B$ can be found with $d(\Phi(s, x), b) < \epsilon$.

Theorem A.32. *Let the compactness condition* (C) *hold and assume that* $\sigma(t, x) := \rho(\Phi(t, x)) > 0$ *for all* $t \geqslant 0$ *whenever* $\rho(x) > 0$. *Then uniform weak persistence implies uniform strong persistence.*

Proof. Assume that Φ is uniformly weakly persistent. Then there exists $\epsilon > 0$ such that

$$\sigma^\infty(x) > \epsilon \quad \forall x \in X,\ \rho(x) > 0. \tag{A.8}$$

We can choose ϵ smaller than ϵ_0 in the compactness condition (C).

Now suppose that Φ is not uniformly strongly persistent.

Then there exist sequences $x_j \in X$, $\rho(x_j) > 0$, and $0 < \epsilon_j \to 0$ such that

$$\sigma_\infty(x_j) < \epsilon_j \quad \forall j.$$

Fix $j \in \mathbb{N}$ for a moment such that $\epsilon_j < \epsilon/2$. Then there exists sequences $r_k < s_k$ with $r_k \to \infty$ such that

$$\begin{aligned}\sigma(r_k, x_j) &= \epsilon \quad \forall k \in \mathbb{N},\\ \sigma(s_k, x_j) &\to \sigma_\infty(x_j), \quad k \to \infty,\\ \sigma(s, x_j) &\leqslant \epsilon, \quad r_k \leqslant s \leqslant s_k.\end{aligned}$$

Moreover, by the first part of our compactness assumption (C),

$$\Phi(r_k, x_j) \to B, \quad k \to \infty.$$

Hence we find elements $b_k \in B$ such that

$$d(\Phi(r_k, x_k), b_k) \to 0, \quad k \to \infty.$$

Since ρ is uniformly continuous, the elements b_k are in the intersection of B with the ring $\{\epsilon/2 < \rho < \epsilon_0\}$ for sufficiently large k. After dropping the first elements from the sequence we can assume that

$$b_k \in K := B \cap \{\epsilon/2 \leqslant \rho \leqslant \epsilon_0\} \quad \forall k \in \mathbb{N}.$$

The set K is compact by the second part of assumption (C).

Summarizing, for every $j \in \mathbb{N}$, we can find numbers $s_j > r_j > j$ and an element $b_j \in K$ such that

$$\sigma(r_j, x_j) = \epsilon \quad \forall j \in \mathbb{N}, \tag{A.9}$$

$$\sigma(s_j, x_j) \to 0, \quad j \to \infty, \tag{A.10}$$

$$\sigma(s, x_j) \leqslant \epsilon, \quad r_j \leqslant s \leqslant s_j, \tag{A.11}$$

$$d(\Phi(r_j, x_j), b_j) < 1/j, \quad b_j \in K \quad \forall j \in \mathbb{N}.$$

Since K is compact, we can assume that b_j converges towards an element $x \in \bar{K}$, after choosing a subsequence. Hence we can assume

$$\Phi(r_j, x_j) \to x \in X, \quad j \to \infty.$$

Since ρ is continuous, $\rho(x) = \epsilon$ by (A.9).

Step 1. We claim that the sequence $s_j - r_j$ is unbounded.

We suppose that this sequence is bounded. After choosing a subsequence $s_j - r_j \to s$ for some $s \geqslant 0$. Since σ is continuous and x is the limit of $(\Phi_{r_j}(x_j))$, we have from (A.10) that

$$\sigma(s, x) \leqslant \limsup_{j\to\infty} \sigma(s_j - r_j, \Phi_{r_j}(x_j)) = \lim_{j\to\infty} \sigma(s_j, x_j) = 0.$$

But $\sigma(s, x) > 0$ because $\rho(x) > 0$, a contradiction.

Step 2. The contradiction.

By (A.11), we have

$$\sigma(r, \Phi_{r_j}(x_j)) \leqslant \epsilon, \quad 0 \leqslant r \leqslant s_j - r_j.$$

By Step 1, $s_j - r_j$ is unbounded. After passing to subsequences we can assume that $s_j - r_j$ is increasing and converges to infinity. Then

$$\sigma(r, \Phi_{r_k}(x_k)) \leqslant \epsilon, \quad 0 \leqslant r \leqslant s_j - r_j, \quad k \geqslant j.$$

Since σ is continuous and x the limit of $(\Phi_{r_k}(x_k))$, we have

$$\sigma(r, x) \leqslant \limsup_{k\to\infty} \sigma(r, \Phi_{r_k}(x_k)) \leqslant \epsilon, \quad 0 \leqslant r \leqslant s_j - r_j.$$

Since $s_j - r_j \to \infty$ as $j \to \infty$, this estimate holds for all $r \geqslant 0$, contradicting (A.8). □

As an easy illustration in a perhaps unexpected direction, we show a boundedness result.

Corollary A.33. *Let X be a closed subset of $\mathbb{R}^n$ and Φ a continuous semiflow on X. Assume that there exists some $c > 0$ such that*

$$\liminf_{t\to\infty} \|\Phi_t(x)\| < c \quad \forall x \in X.$$

Then there exists some $c > 0$ such that

$$\limsup_{t\to\infty} \|\Phi_t(x)\| < c \quad \forall x \in X.$$

Proof. Set $\rho(x) = (1 + \|x\|)^{-1}$. The compactness condition (C) is trivially satisfied by choosing $B = X$ because the ρ-rings are compact. Obviously, Φ is uniformly weakly ρ-persistent and so uniformly strongly ρ-persistent. □

Strengthening Persistence

After having established uniform strong ρ-persistence, one might like to prove persistence with respect to some other functional $\tilde{\rho}$, which gives a stronger type of persistence. While ρ is strictly positive, $\tilde{\rho}$ may be nonnegative. The proofs, but not the formulations of the results, require the concept of ω-limit sets.

Let X be a metric space and Φ an autonomous continuous semiflow on X. Let $\rho : X \to [0, \infty)$ be uniformly continuous. We make the following assumption.

Compactness condition (C)′. There exists a closed subset B of X with the following properties:

- For every $x \in X$, $\Phi(t, x) \to B$ as $t \to \infty$.
- For every $\epsilon > 0$ the intersection $B \cap \{\rho \geqslant \epsilon\}$ is compact in X.

Theorem A.34. *Let* (C)′ *hold with attracting set B. Let Φ be uniformly weakly ρ-persistent on X and $\tilde{\rho}$ be a nonnegative continuous functional on X. Assume that $\rho(\Phi(t, x)) > 0$ for all $t \geqslant 0$ whenever $\rho(x) > 0$ and that every bounded total orbit $w : \mathbb{R} \to X$ of Φ satisfies $\tilde{\rho}(w(0)) > 0$ whenever $\rho(w(t)) > 0$ for all $t \in \mathbb{R}$. Then there exists some $\epsilon > 0$ such that*

$$\liminf_{t\to\infty} \tilde{\rho}(\Phi(t, x)) > \epsilon$$

for all $x \in X$ with $\rho(x) > 0$.

We mention that it is sufficient to assume that $\tilde{\rho}$ is lower semi-continuous. We prepare the proof of Theorem A.34 by establishing the following result.

Lemma A.35. *Let* (C) *hold and let* Φ *be uniformly weakly* ρ*-persistent. Then there exists a compact invariant subset* K *of* B *such that* ρ *is strictly positive on* K *and* $\Phi(t, x) \to K$ *as* $t \to \infty$ *for every* $x \in X$ *with* $\rho(x) > 0$.

Proof. Condition (C)$'$ implies condition (C) and so Φ is uniformly strongly persistent by Theorem A.32. Now (C)$'$ implies that every orbit has compact closure and we can consider the union of all ω-limit sets,

$$\tilde{K} := \bigcup_{x \in X} \omega(x) \subseteq B \cap \{\rho \geqslant \epsilon\}$$

for some $\epsilon > 0$. Here the ω-limit sets are defined similarly as in Section A.2: $\omega(x) = \bigcap_{r \geqslant 0} \overline{\{\Phi(t, x); t \geqslant r\}}$. So, $\tilde{K}$ is contained in a compact set and we choose K as the closure of $\tilde{K}$. The invariance of K follows from the invariance of $\tilde{K}$ by standard arguments. □

Proof of Theorem A.34. Consider the compact attracting set K in Lemma A.35. Let $x \in K$. Since K is also invariant, there exists a total orbit $w : \mathbb{R} \to K$ through x. Since ρ is strictly positive on K, $\rho(w(t)) > 0$ for all $t \in \mathbb{R}$. By assumption, $\tilde{\rho}(x) = \tilde{\rho}(w(0)) > 0$. Since $x \in K$ was arbitrary, $\tilde{\rho}$ is strictly positive on K. As $\tilde{\rho}$ is continuous on X and K is compact, there exists $\epsilon > 0$ such that $\tilde{\rho}(x) > \epsilon$ for all $x \in K$. Now let $x \in X$ and $\rho(x) > 0$. Since $\Phi(t, x) \to K$ as $t \to \infty$ by Lemma A.35 and $\tilde{\rho}$ is continuous on X, we have

$$\tilde{\rho}(x) > \epsilon \quad \forall x \in K.$$

Again using the lower semi-continuity of $\tilde{\rho}$,

$$\liminf_{t \to \infty} \tilde{\rho}(\Phi(t, x)) > \epsilon \quad \forall x \in X.$$

□

Bibliographic Remarks

Persistence

The connection between uniform weak and uniform strong persistence seems to have first been noticed by Freedman, Moson (1990). It was elaborated in Thieme (1993), motivated by the observation that the method of fluctuation in Toolbox A.3 provides an efficient elementary tool to prove uniform weak persistence for ordinary (and functional) differential equations. Theorem A.34 is based on ideas by Ruan, Zhao (1999). A more sophisticated approach to uniform weak persistence (which we do not explore here) consists of studying the boundary flow for acyclicity (Thieme, 1993). Much of the literature uses this approach to directly establish strong or uniform strong persistence: see, for example, Butler, Waltman (1986), Butler et al. (1986), and Hale, Waltman (1989). Another approach to persistence uses average Lyapunov functions (e.g., Fonda, 1988; Hofbauer, Sigmund, 1988, 1998; Hutson, 1984; Tang, Kuang, 1996). For surveys on persistence see Hutson, Schmitt (1992) and Waltman (1991). A short introduction and applications and references can also be found in Smith, Waltman (1995). We finally mention that, under appropriate assumptions, uniform strong persistence implies the existence of a *coexistence equilibrium* (or *persistence equilibrium*, *endemic equilibrium* for infectious diseases); see Zhao (1995) for a recent result and for references. We refer to Thieme (2000) and Zhao (2001) for persistence results for non-autonomous semiflows.

Semiflows

Hale (1977) is a classical reference for semiflows generated by functional differential equations. A recent reference is Diekmann, van Gils, et al. (1995), which also carefully explains how semilinear evolution equations in general generate semiflows (thus also covering semilinear partial differential equations). For the latter see also Hale (1988) and Sell, You (2002).

A.6 Global Stability of a Compact Minimal Set

Let X be a metric space with metric d and Φ a continuous semiflow on X.

A subset B of X is called forward invariant (under Φ) if $\Phi_t(B) \subseteq B$ for all $t \geqslant 0$. B is called invariant if $\Phi_t(B) = B$ for all $t \geqslant 0$.

A subset of X is called a *minimal set* if it is closed, invariant and non-empty and contains no proper subset which also has these three properties.

A modification of Lemma I.8.1 in Hale (1980) provides the following consequence of Zorn's lemma.

Lemma A.36. *Every compact forward invariant set contains a compact minimal set.*

An invariant set M is called *locally stable* if for every neighborhood V of M there exists a neighborhood W of M such that $\Phi_t(W) \subseteq V$ for all $t \geqslant 0$. The set M is called *locally asymptotically stable* if there exists a neighborhood U of M such that $\Phi_t(x) \to M$ as $t \to \infty$ for every $x \in U$. The latter, $\Phi_t(x) \to M$ as $t \to \infty$, means that, for every neighborhood W of M, there exists some $r > 0$ such that $\Phi_t(x) \in W$ for all $t > r$.

An invariant set M is called *globally asymptotically stable* if it is locally stable and if $\Phi_t(x) \to M$ as $t \to \infty$ for every $x \in X$.

Theorem A.37. *Assume that X is connected and that every orbit of Φ has compact closure in X and that every compact minimal set is locally asymptotically stable. Then there exists a (uniquely determined) compact minimal set that is globally asymptotically stable.*

Proof. Choose some $x_0 \in X$ and consider the ω-limit set of x_0, $\omega(x_0)$. The set $\omega(x_0)$ is compact and invariant. By Lemma A.36, it contains a compact minimal set, M_0. By assumption, M_0 is locally asymptotically stable. We consider its domain of attraction, X_0:

$$X_0 = \{x \in X : \Phi_t(x) \to M_0,\ t \to \infty\}.$$

X_0 is an open forward invariant set. If $X_0 = X$, the assertion holds. Assume that X_0 is a proper subset of X. Then $\bar{X}_0$ and $X \setminus X_0$ are forward invariant sets and so is $\partial X_0 = \bar{X}_0 \setminus X_0$. Since X is connected, ∂X_0 is not empty. Pick some $x \in \partial X_0$. Then the ω-limit set of x contains a compact minimal set M which is a subset of ∂X_0. By assumption, M is locally asymptotically stable. Hence there exists a neighborhood V of M such that $\Phi_t(y) \to M$ as $t \to \infty$ for all $y \in V$. Since M is a subset of ∂X_0, $V \cap X_0$ is not empty. Hence there exists some $y \in V$ such that $\Phi_t(y) \to M_0$ as $t \to \infty$, a contradiction, because the compact sets M_0 and M can be separated in the metric space X. □

In general, it will be difficult to show that all compact minimal sets are locally asymptotically stable. This task becomes easier if the only compact minimal sets are fixed points and periodic orbits. Semiflows Φ with this property are induced by planar two-dimensional ordinary differential equation systems (see Toolbox A.2), by competitive or cooperative three-dimensional ordinary differential equation systems (Hirsch, 1990; Smith, 1995, Theorem 4.1), by monotone cyclic feedback systems (Mallet-Paret, Smith, 1990), by certain hypercycle systems (Hofbauer, Mallet-Paret, Smith, 1991), by scalar reaction–diffusion systems on a circle (Fiedler, Mallet-Paret, 1989), and by monotone cyclic feedback systems with delay (Mallet-Paret, Sell, 1996).

The stability of periodic orbits of planar ordinary differential equation systems can be determined from the following divergence criterion (Hale, 1980, VI.3, Lemma 3.1). We recall that a periodic solution is called *orbitally stable* if the associated orbit is a locally stable set. A periodic solution is called *locally asymptotically orbitally stable* if the associated orbit is locally asymptotically stable. A periodic solution x is called locally asymptotically orbitally stable *with asymptotic phase* if it is orbitally stable and there exists a neighborhood U of the orbit such that for any solution $y' = F(y)$ with $y(0) \in U$ there exists some $s > 0$ such that

$$|y(t) - x(t+s)| \to 0, \quad t \to \infty.$$

Proposition A.38. *Let $x : \mathbb{R} \to \mathbb{R}^2$ be a periodic solution to a planar ordinary differential equation*

$$x' = F(x)$$

with period $\tau > 0$. Then this solution is locally asymptotically orbitally stable (with asymptotic phase) if

$$\int_0^\tau \operatorname{div} F(x(t))\, \mathrm{d}t < 0.$$

If this integral is strictly positive, the periodic solution is unstable.

For applications of Proposition A.38 see Kuang (1990) and Tang et al. (1997), for example.

The asymptotic stability of a periodic orbit of higher-dimensional ordinary differential equation systems can be studied by using second additive compound matrices.

Proposition A.39 (Muldowney (1990)). *Let $x(t)$ be a periodic solution of the ordinary differential equation $x' = F(x)$ with F being a continuously differentiable vector field on $\mathbb{R}^n$. Then x is orbitally asymptotically stable if the linear system*

$$y = A^{[2]}(t)y$$

is asymptotically stable, where $A(t) = F'(x(t))$ is the Jacobian matrix of F evaluated at $x(t)$ and $A^{[2]}(t)$ is the second compound matrix of $A(t)$.

For more details see Muldowney (1990). If $A = (\alpha_{jk})$ is a 3×3 matrix, then its second compound matrix $A^{[2]}$ is given as (Li, Muldowney, 1995):

$$A^{[2]} = \begin{pmatrix} \alpha_{11} + \alpha_{22} & \alpha_{23} & -\alpha_{13} \\ \alpha_{32} & \alpha_{11} + \alpha_{33} & \alpha_{12} \\ -\alpha_{31} & \alpha_{21} & \alpha_{22} + \alpha_{33} \end{pmatrix}.$$

A.7 Hopf Bifurcation

Consider the parameterized autonomous system

$$\frac{\mathrm{d}x}{\mathrm{d}t} = F(\mu, x) \tag{A.12}$$

with a continuously differentiable function $F : U \to \mathbb{R}^n$, U an open set of $\mathbb{R}^{n+1}$. Assume that there exists a branch x^*_μ, μ in some interval I, of equilibria such that $(\mu, x^*_\mu) \in U$ and

$$F(\mu, x^*_\mu) = 0 \quad \forall \mu \in I. \tag{A.13}$$

We want the possible values of μ at which there is a (Hopf) bifurcation of periodic solutions from the branch of equilibria. Let

$$\mathcal{J}(\mu) = D_x F(\mu, x^*_\mu)$$

be the Jacobian matrix of $F(\mu, \cdot)$ at x^*_μ. We assume that there exists some $\mu_0 \in \mathbb{R}$, $\omega_0 > 0$ with the following properties:

(1) F is twice continuously differentiable in a neighborhood of $(\mu_0, x^*_{\mu_0})$.

(2) $\mathcal{J}(\mu_0)$ is an invertible matrix and has a pair of simple purely imaginary eigenvalues $\pm \mathrm{i}\omega_0$.

(3) (*Nonresonance*) $\mathcal{J}$ has no eigenvalues of the form $k\mathrm{i}\omega_0$, $k \in \mathbb{Z}$, $k \neq \pm 1$.

(4) (*Transversality*) If $\lambda(\mu) = \alpha(\mu) + \mathrm{i}\beta(\mu)$ is the branch of eigenvalues of $\mathcal{J}(\mu)$ with $\lambda(\mu_0) = \mathrm{i}\omega_0$, assume that $\alpha'(\mu_) \neq 0$.

Remark.

(a) The local existence and differentiability in μ of such a branch of eigenvalues follows from the other assumptions.

(b) It follows from these assumptions and the implicit function theorem that x^*_μ is a twice continuously differentiable function of μ in a neighborhood of μ_0.

(c) In the planar case, $n = 2$, the assumptions (2), ..., (4) can be replaced by

($\diamond_2$) the trace of $\mathcal{J}(\mu_0)$ is 0 and the determinant of $\mathcal{J}(\mu_0)$ is strictly positive;

($\diamond_3$) the derivative of the trace of $\mathcal{J}(\mu)$ at $\mu = \mu_0$ is different from 0.

The following theorem by Hopf holds (for a proof see Ambrosetti, Prodi, 1993, for example).

Theorem A.40. *Let this scenario hold. Then, at $\mu = \mu_0$, a branch of periodic solutions of (A.12) bifurcates with period close to $2\pi/\omega_0$.*

More precisely, there exists an open interval $J \ni 0$ and continuously differentiable functions $\omega(s)$, $\mu(s)$ from J to $\mathbb{R}$ and a family x_s, $s \in J$, of nonconstant periodic solutions of (A.12) such that

(i) $\omega(0) = \omega_0$, $\mu(0) = \mu_0$;

(ii) x_s *has period* $2\pi/\omega(s)$;

(iii) $\sup |x_s(\mathbb{R})| \to 0$ *as* $s \to 0$.

In order to discuss the (orbital) stability of the bifurcating periodic solutions in the plane, one makes a coordinate transformation such that the Jacobian matrix obtains the form

$$\begin{pmatrix} 0 & -\gamma \\ \gamma & 0 \end{pmatrix}$$

(cf. ($\diamond_2$)). Let $F = (F^1, F^2)$ be the vector field associated with (A.12) and $x = (u, v)$, after the transformation, and let F be three times continuously differentiable. Then the stability of the bifurcating periodic orbit is determined by the following number, a (Wiggins, 1990, Section 3.1B, (3.1.107)):

$$\begin{aligned} a = \gamma[F^1_{uuu} + F^1_{uvv} + F^2_{uuv} + F^2_{vvv}] \\ + [F^1_{uv}(F^1_{uu} + F^1_{vv}) - F^2_{uv}(F^2_{uu} + F^2_{vv}) - F^1_{uu}F^2_{uu} + F^1_{vv}F^2_{vv}]. \end{aligned} \tag{A.14}$$

Here the subscripts mean partial derivatives. The partial derivatives need to be evaluated at the bifurcation point, $(\mu_0, x^*_{\mu_0})$. If $a < 0$, the bifurcating periodic orbits are locally asymptotically stable; this case is called *supercritical bifurcation* because one simultaneously has that $\mu(s) > \mu_0$ for all $\mu \in J$. If $a > 0$, the bifurcating orbits are unstable; this case is called *subcritical bifurcation* because one simultaneously has that $\mu(s) < \mu_0$ for all $\mu \in J$.

One can apply this stability criterion to more than two space dimensions by performing a center-manifold reduction.

A.8 Perron–Frobenius Theory of Positive Matrices and Associated Linear Dynamical Systems

A vector $x \in \mathbb{R}^n$ is called *positive*, symbolically $x > 0$, if all components are nonnegative and at least one is positive. It is called *strictly positive*, $x \gg 0$, if all components are positive. A square matrix is called *positive* if all entries are nonnegative numbers and the matrix is not the zero matrix. It is called *quasi-positive* if it is not the zero matrix and all off-diagonal entries are nonnegative numbers. It is called *strictly positive* if all entries are strictly positive. If $n \geqslant 2$, an $n \times n$ matrix $A = (a_{ik})$ is called *irreducible* if the following hold: for any proper nonempty subset P of $\{1, \ldots, m\}$ there are $k \in P$, $j \notin P$ such that $a_{jk} \neq 0$. A 1×1 matrix is called irreducible if it is not the 0 matrix.

Equivalently, A is irreducible if and only if, for all $i, k = 1, \ldots, n$, there exist numbers $j_1, \ldots, j_r \in \{1, \ldots, n\}$ such that $i = j_1, k = j_r$ and $a_{j_l j_{l+1}} \neq 0$ for all $l = 1, \ldots, r-1$. A positive matrix A is irreducible if and only if the matrix exponential e^A is strictly positive.

A positive square matrix A is called *primitive* if one of its powers, A^k, has strictly positive entries. It is easily seen that a positive matrix is primitive if it is irreducible and all entries in its main diagonal are strictly positive (Exercise A.2).

If A is a complex square matrix, a complex number λ is called a *spectral value* of A if the matrix $\lambda \boldsymbol{I} - A$ is singular. The set of spectral values of A is called the *spectrum* of A and is denoted by $\sigma(A)$. For a matrix, a spectral value is an *eigenvalue* and vice versa, i.e., there exists a nonzero vector x, called an eigenvector of A, such that $Ax = \lambda x$. The *spectral radius* of the matrix A, $r(A)$, is defined as

$$r(A) = \max\{|\lambda|;\ \lambda \in \sigma(A)\},$$

while the spectral bound (modulus of stability) of the matrix A, $s(A)$, is defined as

$$s(A) = \max\{\Re\lambda;\ \lambda \in \sigma(A)\}.$$

Obviously, $s(A) \leqslant r(A)$.

Theorem A.41. *Let A be a positive matrix. Then its spectral radius, $r(A)$, is an eigenvalue associated both with a positive eigenvector of A and a positive eigenvector of the transposed matrix A^*. In particular, $s(A) = r(A)$.*

For a proof see Schaefer (1974, Proposition I.2.3).

Theorem A.42. *Let A be a positive matrix and $\mu \geqslant 0$ such that*

$$A^q z \geqslant \mu z$$

for some natural number $q > 0$ and some vector $z = \mathbb{R}^m$, $-z \notin \mathbb{R}^m_+$. Then the spectral radius of A satisfies $r(A) \geqslant \mu^{1/q}$.

Proof. This is the finite-dimensional special case of Theorem 2.5 by Krasnosel'skii (1964). □

Theorem A.43. *Let A be a quasi-positive matrix. Then its spectral bound, $s(A)$, is an eigenvalue of A associated both with a positive eigenvector of A and a positive eigenvector of the transposed matrix A^*. Moreover, if $x > 0$ is a vector and $\mu \in \mathbb{R}$ such that $Ax \geqslant \mu x$, there exists some vector $z > 0$ and some scalar $\lambda \geqslant \mu$ such that $Az = \lambda z$ and, in particular, $s(A) \geqslant \mu$.*

Proof. Since all off-diagonal elements of A are nonnegative, the matrix $A + \nu I$ is positive for some (and then all) sufficiently large $\nu > 0$. Let $\lambda \in \mathbb{C}$ be an eigenvalue of A such that $\Re\lambda = s(A)$. Then there exists a (possibly complex) vector $x \neq 0$ such that $Ax = \lambda x$. So, $(A + \nu)x = (\lambda + \nu)x$. Let $|x| = (|x_1|, \dots, |x_n|)$ be the modulus (or absolute value) of the vector x. Since $A + \nu I$ is a positive matrix, $|\nu + \lambda|\,|x| = |(A + \nu)x| \leqslant (A + \nu)|x|$. By Theorem A.42 and Theorem A.41, there exists some $r \geqslant |\nu + \lambda| \geqslant \nu + s(A)$ and some vector $z > 0$ such that $(A + \nu)z = rz$. So, $Az = (r - \nu)z$. By definition of $s(A)$, $r - \nu \leqslant s(A)$. Together with our previous inequality, $r - \nu = s(A)$ and $s(A)$ is an eigenvalue of A associated with a positive eigenvector. Since $s(A) = s(A^*)$

and A^* is a quasi-positive matrix, we can conclude that $s(A)$ is also associated with a positive eigenvector of A^*.

Now let $Ax \geqslant \mu x$ for some vector $x > 0$ and some $\mu \in \mathbb{R}$. Then $(A+\nu)x \geqslant (\nu+\mu)x$. Since $A+\nu I$ is a positive matrix, by Theorem A.42 and Theorem A.41, there exists some $r \geqslant (\nu+\mu)$ and some vector $z > 0$ such that $(A+\nu)z = rz$. Obviously, $Az = (r-\nu)z$ and $r-\nu \geqslant \mu$. So, we choose $\lambda = r-\nu$. By definition of $s(A)$, $\lambda \leqslant s(A)$ and so $\mu \leqslant s(A)$. □

Theorem A.44. *Let A and D be positive matrices, D diagonal with all diagonal elements being positive. Then* $s(A-D)$ *and* $r(D^{-1}A)-1$ *have the same sign, i.e., these numbers are simultaneously positive, zero, or negative.*

Proof. Let $\lambda = s(A-D)$. Since the off-diagonal elements of $A-D$ are nonnegative, by Theorem A.43 there exists some vector $x > 0$ such that $(A-D)x = \lambda x$. Reorganizing terms, $Ax = Dx + \lambda x$. Since D is an invertible matrix, $D^{-1}Ax = x + \lambda D^{-1}x \geqslant (1+\lambda\epsilon)x$ with ϵ being the reciprocal of the largest of the diagonal elements in D. So, $r(D^{-1}A) \geqslant 1+\lambda\epsilon$. Now let $r = r(D^{-1}A) \geqslant 1$. By Theorem A.41 there exists some vector $x > 0$ such that $D^{-1}Ax = rx$. Reorganizing terms, $Ax = rDx$. So, $(A-D)x = (r-1)Dx \geqslant (r-1)\,\mathrm{d}x$ with d being the smallest diagonal element of D. By Theorem A.43, $s(A-D) \geqslant (r-1)d$. □

Theorem A.44 also holds in the more general case that D is a nonsingular M-matrix (van den Driessche, Watmough, 2002). The continuous and discrete dynamical systems associated with irreducible quasi-positive or even primitive matrices have a strikingly simple large-time behavior. In the following, $\langle x, y\rangle = \sum_{k-1}^{m} x_k y_k$ is the canonical scalar (or inner) product on $\mathbb{R}^m$.

Theorem A.45. *Let A be a quasi-positive irreducible matrix. Then* $s = s(A)$ *is an eigenvalue of both A and* A^* *with strictly positive eigenvectors* v *and* v^* *and* $s(A)$ *is larger than the real parts of all other eigenvalues of A. Furthermore, any nonnegative solution x of the differential equation* $x' = Ax$ *which is not the zero vector satisfies*

$$e^{-s(t-r)}x(t) \to \frac{\langle x(r), v^*\rangle}{\langle v, v^*\rangle}v, \quad t\to\infty, \quad r > 0.$$

Proof. If A is a quasi-positive irreducible matrix, then $A+\nu$ is a positive irreducible matrix for a sufficiently large $\nu > 0$. So, all matrices $e^{tA} = e^{-\nu t}e^{t(A+\nu)}$ are strictly positive and thus form an irreducible uniformly continuous semigroup of compact operators on the Banach lattice $\mathbb{R}^m$. The claim now follows from Theorem 9.11 in Heijmans, de Pagter (1987). □

Remark A.46. Theorem A.45 has significant side effects for an irreducible quasi-positive matrix A, as follows.

(a) Every subspace that is forward invariant under A and contains a positive vector also contains the eigenvector v associated with $s(A)$.

(b) In particular, eigenvalues of A different from $s(A)$ have no positive eigenvector or positive generalized eigenvector.

(c) There are no generalized eigenvectors associated with $s(A)$ and the eigenspace associated with $s(A)$ is one dimensional. In other words, $s(A)$ is a simple eigenvalue.

(d) $s(A) > \Re\lambda$ for all eigenvalues λ of A that are different from $s(A)$.

Proof of (a). Let Y be a subspace of X that is forward invariant under A and let $y_0 \in Y$ be positive. Consider the solution $x' = Ax$ with $x(0) = x_0$. Then $x(t) = e^{tA}x_0 \in Y$ for all $t \geqslant 0$ and

$$v = \lim_{t\to\infty} \frac{\langle v, v^*\rangle}{\langle x_0, v^*\rangle} e^{-s(A)t} x(t), \quad x(t) \in Y.$$

(d) Let λ be an eigenvalue of A that is different from $s(A)$, but with $\Re\lambda = s(A)$. Let x be the solution of $x' = Ax$ with $x(0) = v$ being the eigenvector associated with λ. Then $e^{-s(A)t}x(t) = e^{-i(\Im\lambda)t}v$ does not converge, contradicting the statement in Theorem A.45. The proofs of (b) and (c) are left as an exercise. □

Lemma A.47. *Let A be a quasi-positive irreducible matrix and $Ax \leqslant \lambda x$ or $A^*x \leqslant \lambda x$ with $\lambda \in \mathbb{R}$ and x being a positive vector. Then $s(A) \leqslant \lambda$.*

Proof. We consider $Ax \leqslant \lambda x$; the other case is done similarly. By Theorem A.45, there exists a strictly positive vector v^* such that $A^*v^* = sv^*$ with $s = s(A)$. Then

$$\lambda\langle x, v^*\rangle = \langle \lambda x, v^*\rangle \geqslant \langle Ax, v^*\rangle = \langle x, A^*v^*\rangle = \langle x, sv^*\rangle = s\langle x, v^*\rangle.$$

Since the vector x is positive and the vector v^* strictly positive, $\langle x, v^*\rangle > 0$ and $\lambda \geqslant s = s(A)$ follows by division. □

Remark A.48. Choosing x with $x_j = 1$ for $j = 1, \ldots, m$ in Lemma A.47 provides the estimates

$$s(A) \leqslant \max_{1\leqslant j\leqslant m} \sum_{k=1}^{m} a_{jk}, \qquad s(A) \leqslant \max_{1\leqslant k\leqslant m} \sum_{j=1}^{m} a_{jk}.$$

Since the vector x is strictly positive, it is actually sufficient that A is quasi-positive.

Theorem A.49. *Let A be a primitive matrix with spectral radius $r = r(A) = s(A)$. Then*

$$r^{-k}A^k x \to \frac{\langle x, v^*\rangle}{\langle v, v^*\rangle} v, \quad k \to \infty,$$

where v and v^* are strictly positive eigenvectors of A and A^* associated with the eigenvalue r according to Theorem A.45.

A perhaps more intuitive equivalent formulation is the following one.

Corollary A.50. *Let A be a primitive matrix with spectral radius $r = r(A) = s(A)$. Then, for every positive vector x and for every norm $\|\cdot\|$ on $\mathbb{R}^m$,*

$$\frac{A^k x}{\|A^k x\|} \to \frac{v}{\|v\|}, \quad k \to \infty,$$

where v is a positive eigenvector of A associated with the eigenvalue r according to Theorem A.45.

Proof. Corollary A.50 obviously follows from Theorem A.49 by the continuity of the norm. The converse follows by choosing the norm $\|x\| = \langle |x|, v^*\rangle$, where $|x|$ is the vector $(|x_1|, \dots, |x_n|)$ and v^* is a strictly positive eigenvector of A^* associated with r. □

Theorem A.49 is only valid for primitive matrices. Actually, for positive matrices, primitivity is equivalent to the ergodicity statement in Theorem A.49 (Schaefer, 1974, Chapter I, Proposition 7.3). But mean ergodicity still holds for irreducible matrices, which means that the convergence in Theorem A.49 holds for averages (see the end of Section I.6 in Schaefer (1974)). Notice that the convergence statement in Theorem A.49 implies the convergence statement in Theorem A.51 (Exercise A.2).

Theorem A.51. *Let A be an irreducible positive matrix with spectral radius $r = r(A)$. Then*

$$\frac{1}{k+1}\sum_{j=0}^{k} r^{-j}A^j x \to \frac{\langle x, v^*\rangle}{\langle v, v^*\rangle} v, \quad k \to \infty,$$

where v and v^* are the strictly positive eigenvectors of A and A^* associated with the eigenvalue r according to Theorem A.45.

Noticing that $\langle A^j x, v^*\rangle = r^j\langle x, v^*\rangle$, Theorem A.51 can be reformulated as follows.

Theorem A.52. *Let A be an irreducible positive matrix with spectral radius $r = r(A)$ and x a positive vector. Then*

$$\frac{1}{k+1}\sum_{j=0}^{k} \frac{A^j x}{\langle A^j x, v^*\rangle} \to \frac{v}{\langle v, v^*\rangle}, \quad k \to \infty,$$

where v and v^* are the strictly positive eigenvectors of A and A^* associated with the eigenvalue r according to Theorem A.45.

The following surprising relation holds for the spectral radius of a square matrix A:

$$r(A) = \inf_{n \in \mathbb{N}} \|A^n\|^{1/n} = \lim_{n \to \infty} \|A^n\|^{1/n},$$

where $\|\cdot\|$ is an arbitrary matrix norm (Yosida, 1968, Section VIII.2). This relation can be used to show the following result we used in Section 23.7.

Lemma A.53. *Let A, B be two square matrices of the same size. Then $r(AB) = r(BA)$.*

Proof. We can assume that A and B are not the 0 matrix. It is easily seen by induction that $(AB)^{n+1} = A(BA)^n B$. Let $\|\cdot\|$ be a so-called associated matrix norm, in particular $\|CD\| \leqslant \|C\| \, \|D\|$ for all square matrices C and D. Then $\|(AB)^{n+1}\| \leqslant \|A\| \, \|(BA)^n\| \, \|B\|$. By the relation above, $r(AB) \leqslant r(BA)$. By symmetry, we have equality. □

Exercises

A.1. Show that an irreducible positive matrix is primitive if all entries in the main diagonal are strictly positive.

A.2. Let $(z(\ell))$ be a convergent sequence of vectors in a normed vector space. Show that the averages

$$\frac{1}{\ell + 1} \sum_{l=0}^{\ell} z(l)$$

converge to the same limit as $\ell \to \infty$.

A.3. Let A and B be two square matrices and C the block matrix

$$C = \begin{pmatrix} A & 0 \\ 0 & B \end{pmatrix}.$$

Then $r(C) = \max\{r(A), r(B)\}$.

A.4. Prove Remark A.46 (b) and (c).

Bibliographic Remarks

The infinite-dimensional generalizations of these results are equally powerful (Krasnosel'skii, 1964; Schaefer, 1974; Clément, Heijmans et al., 1987, Chapters 8 and 9; Nagel, 1986, Chapters C-III and C-IV; Clément, Heijmans et al., 1987, Chapter 10; Capasso, 1993, Chapters 5 and 6; Thieme, 1998).

Appendix B

Integration, Integral Equations, and Some Convex Analysis

B.1 The Stieltjes Integral of Regulated Functions

We want to generalize the Riemann integral

$$\int_a^b f(t)g'(t)\,\mathrm{d}t, \quad a, b \in \mathbb{R}, \quad a < b,$$

to functions f and g, where f is not continuous and g is not continuously differentiable. This is motivated by the fact that sojourn functions (Chapter 12) are not necessarily continuous or even differentiable and that optimal vaccination profiles (Chapter 22) are found among the step functions (Section 22.4). A suitable generalization is the Stieltjes integral, which can be introduced at various levels of sophistication: as the Stieltjes integral for regulated functions, the Riemann–Stieltjes integral, and the Lebesgue–Stieltjes integral.

Those who would like to follow the main body of the text without going into any of these theories should know three things.

On a formal level, the Stieltjes integral satisfies

$$\int_a^b f(t)\,\mathrm{d}g(t) = \int_a^b f(t)g'(t)\,\mathrm{d}t, \quad a, b \in \mathbb{R}, \quad a < b, \tag{$\bullet$}$$

if f is continuous and g continuously differentiable on $[a, b]$. The left integral denotes the Stieltjes integral. It also satisfies the integration-by-parts formula

$$\int_a^b f(t)\,\mathrm{d}g(t) = f(b)g(b) - f(a)g(a) - \int_a^b f'(t)g(t)\,\mathrm{d}t \tag{$\bullet\bullet$}$$

if f is continuously differentiable on $[a, b]$ and g is monotone. Notice that, by ($\bullet$), this is the usual integration-by-parts formula if g is continuously differentiable. Conceptually, one should understand that, if f is continuous and g is monotone (nonincreasing or nondecreasing), the Stieltjes integral is the limit of (Riemann) sums

$$\sum_{j=0}^{n} f(s_j)(g(t_{j+1}) - g(t_j)) \tag{$\bullet\bullet\bullet$}$$

formed for partitions $a = t_0 < \cdots < t_{n+1} = b$ and arbitrarily chosen $s_j \in (t_j, t_{j+1})$, as the lengths of these intervals tend to 0.

A rigorous definition of the Stieltjes integral uses the concept of functions of bounded variation and can be based on Riemann sums, on regulated functions, or on measure theory. For the reader who wants a deeper understanding of the Stieltjes theory used in Chapter 12, we want to develop one of the theories in sufficient detail. Since the measure-theoretic approach to integration is beyond the scope of an appendix and since I do not like to manipulate Riemann sums, we develop the Stieltjes integral for regulated functions. In the next section we give a flavor of measure-theoretic integration and, in Toolbox B.4, make the connection to the Lebesgue–Stieltjes integral, because the fundamental law of stage dynamics in Section 13.5, while it can and will also be formulated in terms of the Stieltjes integral, can only be proved with satisfactory generality and convenience using the advanced techniques of Lebesgue–Stieltjes integration. Sections B.3 and B.5 contain elements of convex analysis which fruitfully interact with integration: Jensen's inequality is used in demographics (Chapter 14), and linear optimization on convex sets is employed to find optimal vaccination schedules (Chapter 22). The renewal equation of demographics (a Volterra integral equation of convolution type) is treated in Toolbox B.6.

The Construction of the Stieltjes Integral

A function $f : [a, b] \to \mathbb{R}$ is called a *step function* if there exists a partition $a = t_0 < \cdots < t_{n+1} = b$ of $[a, b]$ such that f is constant on all intervals (t_j, t_{j+1}), $j = 0, \ldots, n$.

A function $f : [a, b] \to \mathbb{R}$ is called *regulated* if there exists a sequence of step functions (f_k) on $[a, b]$ such that $f_k(t) \to f(t)$ for $k \to \infty$ uniformly in $t \in [a, b]$. The following characterization holds (e.g., Hönig, 1975, Chapter I, Theorem 3.1).

Proposition B.1.

(a) *A function $f : [a, b] \to \mathbb{R}$ is regulated if and only if the left limits,*

$$f(t-) = \lim_{t>s\to t} f(s),$$

exist for all $t \in (a, b]$ and the right limits,

$$f(t+) = \lim_{t<s\to t} f(s),$$

exist for all $t \in [a, b)$.

(b) *A regulated function has only countably many discontinuities, i.e., there are only countably many $t \in (a, b)$ such that $f(t+) \neq f(t-)$.*

Notice that sums and products of regulated functions are regulated functions.

A function $g : [a, b] \to \mathbb{R}$ is said to be of *bounded variation* (or *finite variation*) if

$$v_g := \sup\left\{\sum_{j=0}^{n} |g(t_{j+1}) - g(t_j)|\right\} < \infty, \tag{B.1}$$

where the supremum is taken over all partitions $a = t_0 < \cdots < t_{n+1} = b$, $n \in \mathbb{N}$.

We have the following characterization of functions of bounded variation.

Lemma B.2.

(a) *A function $g : [a, b] \to \mathbb{R}$ is of bounded variation if and only if it is the difference of a monotone nondecreasing and a monotone nonincreasing function.*

(b) *A function of bounded variation is regulated.*

A proof of part (a) can be found, for example, in Wade (1995, Corollary 3.12), or in McDonald, Weiss (1999, Theorem 6.3). Part (b) follows from part (a).

Notice that sums and scalar multiples of functions of bounded variation are of bounded variation again.

The *Stieltjes integral over a step function* f for a function g of bounded variation is defined as follows.

Let $a = t_0 < \cdots < t_{n+1} = b$ be a partition such that f is constant on every interval (t_j, t_{j+1}). Pick some points $s_j \in (t_j, t_{j+1})$ and set

$$\int_a^b f(t)\,\mathrm{d}g(t) = \sum_{j=0}^{n} f(s_j)(g(t_{j+1}) - g(t_j)). \tag{B.2}$$

One can easily check that this definition does not depend on the choice of the partition or of the points s_j. In the following it is convenient to introduce the supremum norm

$$\|f\| = \sup\{|f(t)|;\ t \in [a, b]\}. \tag{B.3}$$

Recall that a sequence (f_k) of functions on $[a, b]$ converges toward a function f as $k \to \infty$ uniformly on $[a, b]$, if and only if $\|f_k - f\| \to 0$ as $k \to \infty$.

The following properties of the integral for step functions are easily checked.

Lemma B.3.

(a) *Let $f_1, f_2, g : [a, b] \to \mathbb{R}$, f_1, f_2 step functions and g of bounded variation. Then*

$$\int_a^b (f_1(t) + f_1(t))\,\mathrm{d}g(t) = \int_a^b f_1(t)\,\mathrm{d}g(t) + \int_a^b f_2(t)\,\mathrm{d}g(t).$$

(b) *Let $f, g_1, g_2 : [a, b] \to \mathbb{R}$, f a step function and g_1, g_2 of bounded variation. Then*

$$\int_a^b f(t)\,\mathrm{d}(g_1 + g_2)(t) = \int_a^b f(t)\,\mathrm{d}g_1(t) + \int_a^b f(t)\,\mathrm{d}g_2(t).$$

(c) *Let $f, g : [a, b] \to \mathbb{R}$, f a step function and g of bounded variation. Then, for any $\alpha \in \mathbb{R}$,*

$$\int_a^b \alpha f(t)\,\mathrm{d}g(t) = \alpha \int_a^b f(t)\,\mathrm{d}g(t) = \int_a^b f(t)\,\mathrm{d}(\alpha g)(t).$$

(d) *Let $f, g : [a, b] \to \mathbb{R}$, f a step function and g of bounded variation. Then the following estimate holds:*

$$\left|\int_a^b f(t)\,\mathrm{d}g(t)\right| \leqslant \|f\| \boldsymbol{v}_g.$$

Manipulating the partitions in a suitable way provides the following result.

Lemma B.4. *Let $s \in (a, b)$, $f, g : [a, b] \to \mathbb{R}$, f a step function and g of bounded variation. Then*

$$\int_a^b f(t)\,\mathrm{d}g(t) = \int_a^s f(t)\,\mathrm{d}g(t) + \int_s^b f(t)\,\mathrm{d}g(t).$$

Now let f be a regulated function. Then there exists a sequence of step functions (f_k) such that $\|f_k - f\| \to 0$ as $n \to \infty$. Lemma B.3 (a), (c) and (d) imply that

$$\left|\int_a^b f_k(t)\,\mathrm{d}g(t) - \int_a^b f_l(t)\,\mathrm{d}g(t)\right| \leqslant \|f_k - f_l\| \boldsymbol{v}_g,$$

i.e., the sequence of Stieltjes integrals of step functions is a Cauchy sequence of real numbers and so has a limit, the *Stieltjes integral of the regulated function f*:

$$\int_a^b f(t)\,\mathrm{d}g(t) = \lim_{k\to\infty} \int_a^b f_k(t)\,\mathrm{d}g(t).$$

Lemma B.3 (a), (c) and (d) also imply that the limit does not depend on the choice of the approximating sequence of step functions.

The Properties of the Stieltjes Integral

The Stieltjes integral has the following properties, which easily follow from approximation by step functions.

Proposition B.5.

(a) *Let $f_1, f_2, g : [a,b] \to \mathbb{R}$, f_1, f_2 regulated and g of bounded variation. Then*

$$\int_a^b (f_1(t) + f_1(t))\,\mathrm{d}g(t) = \int_a^b f_1(t)\,\mathrm{d}g(t) + \int_a^b f_2(t)\,\mathrm{d}g(t).$$

(b) *Let $f, g_1, g_2 : [a,b] \to \mathbb{R}$, f regulated and g_1, g_2 of bounded variation. Then*

$$\int_a^b f(t)\,\mathrm{d}(g_1 + g_2)(t) = \int_a^b f(t)\,\mathrm{d}g_1(t) + \int_a^b f(t)\,\mathrm{d}g_2(t).$$

(c) *Let $f, g : [a,b] \to \mathbb{R}$, f a regulated function and g of bounded variation. Then, for any $\alpha \in \mathbb{R}$,*

$$\int_a^b \alpha f(t)\,\mathrm{d}g(t) = \alpha \int_a^b f(t)\,\mathrm{d}g(t) = \int_a^b f(t)\,\mathrm{d}(\alpha g)(t).$$

(d) *Let $f, g : [a,b] \to \mathbb{R}$, f regulated and g of bounded variation. Then*

$$\left| \int_a^b f(t)\,\mathrm{d}g(t) \right| \leqslant \|f\| \boldsymbol{v}_g.$$

(e) *Let $f, g : [a,b] \to \mathbb{R}$, f regulated and g of bounded variation, $s \in [a,b]$. Then*

$$\int_a^b f(t)\,\mathrm{d}g(t) = \int_a^s f(t)\,\mathrm{d}g(t) + \int_s^b f(t)\,\mathrm{d}g(t).$$

There is a special case in which the Stieltjes integral is identical to the familiar Riemann integral.

Theorem B.6. *Let $f, g : [a,b] \to \mathbb{R}$, f regulated and g continuously differentiable on $[a,b]$. Then g is of bounded variation and*

$$\int_a^b f(t)\,\mathrm{d}g(t) = \int_a^b f(t) g'(t)\,\mathrm{d}t,$$

with the second integral being the usual Riemann integral.

Proof. Let $a = t_0 < \cdots < t_{n+1} = b$ be a partition. Then

$$\begin{aligned}\sum_{j=0}^{n} |g(t_{j+1}) - g(t_j)| &= \sum_{j=0}^{n} \left| \int_{t_j}^{t_{j+1}} g'(s)\, \mathrm{d}s \right| \\ &\leqslant \sum_{j=0}^{n} \int_{t_j}^{t_{j+1}} |g'(s)|\, \mathrm{d}s \\ &= \int_a^b |g'(s)|\, \mathrm{d}s,\end{aligned}$$

and we see that g is of bounded variation. We first consider a step function f and a partition as above such that f is constant on every interval (t_j, t_{j+1}). Choose points $s_j \in (t_j, t_{j+1})$. By the fundamental theorem of calculus,

$$\begin{aligned}\int_a^b f(t)\, \mathrm{d}g(t) &= \sum_{j=0}^{n} f(s_j)(g(t_{j+1}) - g(t_j)) \\ &= \sum_{j=0}^{n} f(s_j) \int_{t_j}^{t_{j+1}} g'(s)\, \mathrm{d}s \\ &= \sum_{j=0}^{n} \int_{t_j}^{t_{j+1}} f(s_j) g'(s)\, \mathrm{d}s.\end{aligned}$$

Since $f(s) = f(s_j)$ for all $s \in (t_j, t_{j+1})$,

$$\int_a^b f(t)\, \mathrm{d}g(t) = \sum_{j=0}^{n} \int_{t_j}^{t_{j+1}} f(s) g'(s)\, \mathrm{d}s = \int_a^b f(s) g'(s)\, \mathrm{d}s.$$

Now let f be regulated and $\epsilon > 0$ be arbitrary. Then there exists a step function $\tilde{f}$ such that $\|f - \tilde{f}\| < \epsilon$. From the triangle inequality,

$$\begin{aligned}\left| \int_a^b f(t)\, \mathrm{d}g(t) - \int_a^b f(t) g'(t)\, \mathrm{d}t \right| &\leqslant \left| \int_a^b f(t)\, \mathrm{d}g(t) - \int_a^b \tilde{f}(t)\, \mathrm{d}g(t) \right| \\ &\quad + \left| \int_a^b \tilde{f}(t)\, \mathrm{d}g(t) - \int_a^b \tilde{f}(t) g'(t)\, \mathrm{d}t \right| \\ &\quad + \left| \int_a^b \tilde{f}(t) g'(t)\, \mathrm{d}t - \int_a^b f(t) g'(t)\, \mathrm{d}t \right|.\end{aligned}$$

The second term on the right-hand side of this inequality is 0 because $\tilde{f}$ is a step function. By Proposition B.5 and the properties of the Riemann integral we have

$$\left| \int_a^b f(t)\,\mathrm{d}g(t) - \int_a^b f(t)g'(t)\,\mathrm{d}t \right| \leqslant \|f - \tilde{f}\| \boldsymbol{v}_g + \|(f - \tilde{f})g'\|(b-a)$$
$$\leqslant \epsilon(\boldsymbol{v}_g + \|g'\|(b-a)).$$

Since this estimate holds for every $\epsilon > 0$, we have that

$$\left| \int_a^b f(t)\,\mathrm{d}g(t) - \int_a^b f(t)g'(t)\,\mathrm{d}t \right| \leqslant 0$$

and so is 0 because it is nonnegative. This proves the assertion. □

Convergence theorems are crucial in every integration theory.

Theorem B.7. *Let $f, f_k : [a,b] \to \mathbb{R}$, $k \in \mathbb{N}$, be regulated functions and let $g, g_k : [a,b] \to \mathbb{R}, k \in \mathbb{N}$, be functions of bounded variation. Let $f_k(t) \to f(t)$ as $k \to \infty$ uniformly in $t \in [a,b]$ and $g_k(t) \to g(t)$ as $k \to \infty$, pointwise in $t \in [a,b]$. Finally, let the sequence of variations $(\boldsymbol{v}_{g_k})$ be bounded. Then*

$$\int_a^b f_k(t)\,\mathrm{d}g_k(t) \to \int_a^b f(t)\,\mathrm{d}g(t), \quad k \to \infty.$$

Remark. The assumption that the sequence of variations $(\boldsymbol{v}_{g_k})$ is bounded can be omitted, for it follows from the pointwise convergence of g_k and the uniform boundedness theorem (e.g., McDonald, Weiss, 1999, Theorem 8.2).

Proof. We split the proof into two parts.

Step 1.

$$\int_a^b f_k(t)\,\mathrm{d}g_k(t) - \int_a^b f(t)\,\mathrm{d}g_k(t) \to 0, \quad k \to \infty.$$

Step 2.

$$\int_a^b f(t)\,\mathrm{d}g_k(t) \to \int_a^b f(t)\,\mathrm{d}g(t).$$

Step 1 follows from the following estimate (using Proposition B.5):

$$\left| \int_a^b f_k(t)\,\mathrm{d}g_k(t) - \int_a^b f(t)\,\mathrm{d}g_k(t) \right| = \left| \int_a^b (f_k(t) - f(t))\,\mathrm{d}g_k(t) \right|$$
$$\leqslant \|f_k - f\| \boldsymbol{v}_{g_k} \to 0, \quad k \to \infty,$$

because $\boldsymbol{v}_{g_k}$ is bounded and $\|f_k - f\| \to 0, k \to \infty$.

In order to prove Step 2, we first assume that f is a step function. Choose a partition $a = t_0 < \cdots < t_{n+1} = b$ such that f is constant on all intervals (t_j, t_{j+1}). Let the points s_j be arbitrarily chosen points in (t_j, t_{j+1}). Since $g_k(t) \to g(t)$ as $k \to \infty$ for every $t \in [a, b]$,

$$\begin{aligned}\int_a^b f(t)\,\mathrm{d}g_k(t) &= \sum_{j=0}^{n} f(s_j)(g_k(t_{j+1}) - g_k(t_j)) \\ &\xrightarrow{k\to\infty} \sum_{j=0}^{n} f(s_j)(g(t_{j+1}) - g(t_j)) \\ &= \int_a^b f(t)\,\mathrm{d}g(t).\end{aligned}$$

Now let f be regulated. For any $\epsilon > 0$, there exists a step function $\tilde{f}$ such that $\|f - \tilde{f}\| < \epsilon$. By the triangle inequality,

$$\begin{aligned}&\left|\int_a^b f(t)\,\mathrm{d}g_k(t) - \int_a^b f(t)\,\mathrm{d}g(t)\right| \\ &\quad\leqslant \left|\int_a^b f(t)\,\mathrm{d}g_k(t) - \int_a^b \tilde{f}(t)\,\mathrm{d}g_k(t)\right| \\ &\qquad + \left|\int_a^b \tilde{f}(t)\,\mathrm{d}g_k(t) - \int_a^b \tilde{f}(t)\,\mathrm{d}g(t)\right| \\ &\qquad + \left|\int_a^b \tilde{f}(t)\,\mathrm{d}g(t) - \int_a^b f(t)\,\mathrm{d}g(t)\right| \\ &\quad\leqslant \|f - \tilde{f}\|(\boldsymbol{v}_{g_k} + \boldsymbol{v}_g) + \left|\int_a^b \tilde{f}(t)\,\mathrm{d}g_k(t) - \int_a^b \tilde{f}(t)\,\mathrm{d}g(t)\right| \\ &\quad\leqslant \epsilon\Big(\sup_l \boldsymbol{v}_{g_l} + \boldsymbol{v}_g\Big) + \left|\int_a^b \tilde{f}(t)\,\mathrm{d}g_k(t) - \int_a^b \tilde{f}(t)\,\mathrm{d}g(t)\right|.\end{aligned}$$

The last expression converges to 0 as $k \to \infty$ because $\tilde{f}$ is a step function. This implies

$$\limsup_{k\to\infty}\left|\int_a^b f(t)\,\mathrm{d}g_k(t) - \int_a^b f(t)\,\mathrm{d}g(t)\right| \leqslant \epsilon\Big(\sup_l \boldsymbol{v}_{g_l} + \boldsymbol{v}_g\Big).$$

Since this relation holds for every $\epsilon > 0$,

$$\limsup_{k\to\infty}\left|\int_a^b f(t)\,\mathrm{d}g_k(t) - \int_a^b f(t)\,\mathrm{d}g(t)\right| = 0.$$

This provides the statement in Step 2. □

Every continuous function is a regulated function, which follows from Proposition B.1, but can also be easily seen directly. If the integrand is continuous, the Stieltjes integral can be approximated by arbitrary Riemann–Stieltjes sums.

Theorem B.8. *Let $f, g : [a, b] \to \mathbb{R}$, f continuous, g of bounded variation. Then, for every $\epsilon > 0$, there exists some $\delta > 0$ such that the following holds.*

If $a = t_0 < \cdots < t_{n+1} = b$ is an arbitrary partition with $t_{j+1} - t_j < \delta$ for $j = 0, \ldots, n$ and s_j are arbitrary points in $[t_j, t_{j+1}]$, then

$$\left| \int_a^b f(t)\,\mathrm{d}g(t) - \sum_{j=1}^{n} f(s_j)(g(t_{j+1}) - g(t_j)) \right| < \epsilon.$$

Proof. Since the continuous function f is uniformly continuous on the compact interval $[a, b]$, there exists some $\delta > 0$ such that

$$|f(s) - f(r)| < \frac{\epsilon}{2\boldsymbol{v}_g} \quad \text{whenever } |r - s| < \delta.$$

Let $a = t_0 < \cdots < t_{n+1} = b$ be an arbitrary partition with $t_{j+1} - t_j < \delta$ for $j = 0, \ldots, n$ and let the points s_j be arbitrary points in $[t_j, t_{j+1}]$. Let $\tilde{f}$ be the step function that is constant on (t_j, t_{j+1}) taking the values $f(s_j)$. Then

$$\int_a^b \tilde{f}(t)\,\mathrm{d}g(t) = \sum_{j=1}^{n} f(s_j)(g(t_{j+1}) - g(t_j)).$$

The way we have chosen the partition implies

$$|f(t) - \tilde{f}(t)| < \frac{\epsilon}{2\boldsymbol{v}_g} \quad \forall t \in [a, b].$$

By Proposition B.5,

$$\left| \int_a^b f(t)\,\mathrm{d}g(t) - \int_a^b \tilde{f}(t)\,\mathrm{d}g(t) \right| \leqslant \|f - \tilde{f}\|\boldsymbol{v}_g \leqslant \tfrac{1}{2}\epsilon.$$

This implies the assertion. □

Lemma B.9. *If $g : [a, b] \to \mathbb{R}$ is of bounded variation, then there exists a sequence of step functions g_k such that $\|g_k - g\| \to 0$ as $k \to \infty$ and $\boldsymbol{v}_{g_k} \leqslant \boldsymbol{v}_g$.*

Proof. There exists a sequence of step functions $\tilde{g}_k$ such that $\|\tilde{g}_k - g\| \to 0$ as $k \to \infty$. Fix k for a moment. There exists a partition $a = t_0 < \cdots < t_{n+1} = b$ such that $\tilde{g}_k$ is constant on (t_j, t_{j+1}). We define another sequence of step functions g_k as follows,

$$\begin{aligned} g_k(t_j) &= g(t_j), \quad j = 0, \ldots, n+1, \\ g_k(t) &= g(s_j), \quad t \in (t_j, t_{j+1}), \quad j = 0, \ldots, n, \end{aligned}$$

where the points s_j are arbitrarily chosen points in (t_j, t_{j+1}). Since $\tilde{g}_k$ is constant on (t_j, t_{j+1}), we have for all $t \in (t_j, t_{j+1})$,

$$|g_k(t) - g(t)| \leqslant |g(s_j) - \tilde{g}_k(s_j)| + |\tilde{g}_k(t) - g(t)| \leqslant 2\|g - \tilde{g}_k\|.$$

Since this holds for all j,

$$\|g_k - g\| \leqslant 2\|\tilde{g}_k - g\| \to 0, \quad k \to \infty.$$

By construction, the functions g_k are step functions and

$$\boldsymbol{v}_{g_k} \leqslant \boldsymbol{v}_g. \qquad \square$$

The following integration-by-parts formula for Stieltjes integrals generalizes the usual integration-by-parts formula (see Theorem B.6).

Theorem B.10. *Let $f, g : [a, b] \to \mathbb{R}$ be of bounded variation and have no common discontinuities and assume that either f or g is continuous both at a and b. Then*

$$\int_a^b f(t)\,\mathrm{d}g(t) - \int_a^b g(t)\,\mathrm{d}f(t) = f(b)g(b) - f(a)g(a).$$

Proof. By the symmetry of the integration-by-parts formula we can assume that f is continuous at a and b.

Step 1. f, g are step functions.

Choose a partition $a = t_1 < \cdots < t_{n+1} = b$ such that g is constant on all intervals (t_j, t_{j+1}), $j = 1, \ldots, n$. Since f and g have no common discontinuities, we can, possibly after adding points to the partition, find $s_j \in (t_j, t_{j+1})$, $j = 1, \ldots, n$, such that f is constant on $(s_j, s_j + 1)$ for $j = 1, \ldots, n-1$ and on $[0, s_1)$ and $(s_n, b]$. Set $s_0 = 0$ and $s_{n+1} = b$. Then f is constant on (s_j, s_{j+1}) for $j = 0, \ldots, n$. By definition of the Stieltjes integral for step functions (equation (B.2)),

$$\begin{aligned}
\int_a^b & g(t)\,\mathrm{d}f(t) \\
&= \sum_{j=1}^{n} g(s_j)(f(t_{j+1}) - f(t_j)) \\
&= \sum_{j=1}^{n} g(s_j)f(t_{j+1}) - \sum_{j=0}^{n-1} g(s_{j+1})f(t_{j+1}) \\
&= \sum_{j=0}^{n} g(s_j)f(t_{j+1}) - g(a)f(a) - \sum_{j=0}^{n} g(s_{j+1})f(t_{j+1}) + g(b)f(b) \\
&= -\sum_{j=0}^{n} f(t_{j+1})(g(s_{j+1}) - g(s_j)) + f(b)g(b) - f(a)g(a).
\end{aligned}$$

$a = s_0 < \cdots < s_{n+1} = b$ is a partition of $[a, b]$ and $s_j < t_{j+1} < s_{j+1}$ for $j = 1, \ldots, n-1$. Since f is constant on $[s_0, s_1)$ and on $(s_n, s_{n+1}]$, it does not matter that $s_0 = a = t_1$ and $s_{n+1} = b = t_{n+1}$; so

$$\int_a^b g(t)\,\mathrm{d}f(t) = -\int_a^b f(t)\,\mathrm{d}g(t) + f(b)g(b) - f(a)g(a).$$

Step 2. f regulated, g a step function.

We uniformly approximate f by step functions f_k. Since f and g have no discontinuities in common and g has only finitely many discontinuities and f is continuous at a and b, one can modify every f_k such that it has no discontinuities in common with g and is continuous at a and b. The integration-by-parts formula follows from Step 1 and Theorem B.7.

Step 3. f and g of bounded variation.

In particular, f is regulated. By Lemma B.9, we uniformly approximate g by step functions g_k with variation $\boldsymbol{v}_{g_k} \leqslant \boldsymbol{v}_g$. Since f and g have no common discontinuities and f has only countably many discontinuities, we can modify each g_k to have no common discontinuities with f, possibly enlarging $\boldsymbol{v}_{g_k}$, but satisfying $\boldsymbol{v}_{g_k} \leqslant \boldsymbol{v}_g + 1$. The integration-by-parts formula follows from Step 2 and Theorem B.7. □

Theorems B.6 and B.10 have the following integration-by-parts formula as an immediate consequence.

Corollary B.11. *Let $f, g : [a, b] \to \mathbb{R}$, f continuously differentiable on $[a, b]$, g of bounded variation. Then*

$$\int_a^b f(t)\,\mathrm{d}g(t) = f(b)g(b) - f(a)g(a) - \int_a^b f'(t)g(t)\,\mathrm{d}t,$$

with the last integral being the usual Riemann integral.

We conclude with a result for the sign of a Stieltjes integral.

Theorem B.12. *Let $f, g : [a, b] \to \mathbb{R}$, f regulated and nonnegative, g nondecreasing. Then*

$$\int_a^b f(t)\,\mathrm{d}g(t) \geqslant 0.$$

Proof. By Lemma B.2, g is of bounded variation. If f is a step function, this follows from formula (B.2). If f is regulated, it is the uniform limit of nonnegative step functions and the Stieltjes integral is nonnegative as the limit of a nonnegative sequence. □

As an application of Theorem B.12 we consider Laplace–Stieltjes transforms. Let $g : [0, \infty) \to [0, \infty)$, $g(0) = 0$ be nondecreasing and exponentially bounded, i.e., $g(t) \leqslant M\mathrm{e}^{at}$ for appropriate positive numbers $M, a > 0$. Then

$$\check{g}(\lambda) := \int_0^\infty \mathrm{e}^{-\lambda t}\, \mathrm{d}g(t) = \lim_{b\to\infty} \int_0^b \mathrm{e}^{-\lambda t}\, \mathrm{d}g(t)$$

exists for $\lambda > a$ and

$$\check{g}(\lambda) = \lambda \int_0^\infty \mathrm{e}^{-\lambda t} g(t)\, \mathrm{d}t =: \lambda \hat{g}(\lambda),$$

with $\hat{g}$ denoting the Laplace transform. One can prove that $\check{g}$ is infinitely often differentiable and that the derivatives are obtained by formally differentiating with respect to λ under the integral sign:

$$\frac{\mathrm{d}^n}{\mathrm{d}\lambda^n} \check{g}(\lambda) = (-1)^n \int_0^\infty t^n \mathrm{e}^{-\lambda t}\, \mathrm{d}g(t), \quad \lambda > a.$$

It follows from Theorem B.12 that $\check{g}$ is completely monotonic, i.e.,

$$(-1)^n \frac{\mathrm{d}^n}{\mathrm{d}\lambda^n} \check{g}(\lambda) \geqslant 0, \quad \lambda > a.$$

The converse is true as well.

Theorem B.13 (Bernstein). *Let $h : (a, \infty) \to \mathbb{R}$. Then h is completely monotonic (infinitely often differentiable and $(-1)f^{(n)}$ nonnegative) if and only if it is the Laplace–Stieltjes transform of an nondecreasing function g on $[0, \infty)$, $h(\lambda) = \check{g}(\lambda)$. g is uniquely determined by h if $g(0) = 0$ and g is right continuous on $(0, \infty)$.*

Exercises

B.1. Let $g : [a, b] \to \mathbb{R}$ be continuous and of bounded variation. Then there exists a sequence of continuously differentiable functions $g_k : [a, b] \to \mathbb{R}$ which converge to g as $k \to \infty$ uniformly on $[a, b]$ and $\boldsymbol{v}_{g_k} \leqslant \boldsymbol{v}_g$.

B.2. Let $g : [a, b] \to \mathbb{R}$ and $f : g([a, b]) \to \mathbb{R}$. Assume that one of the following two assumptions holds:

(i) f, g continuous, g of bounded variation; or

(ii) f regulated and g monotone.

Then we can integrate by substitution,

$$\int_{g(a)}^{g(b)} f(x)\, \mathrm{d}x = \int_a^b f(g(t))\, \mathrm{d}g(t).$$

B.3. Let $g : [a, b] \to (0, \infty)$ be continuous and of bounded variation. Then

$$\ln g(b) - \ln g(a) = \int_a^b \frac{\mathrm{d}g(t)}{g(t)}.$$

The continuity of g can be dropped if g is monotone.

B.4. Let $g : [a, b] \to \mathbb{R}$ and let $f, h : g([a, b]) \to \mathbb{R}$ be such that g strictly increasing, h of bounded variation and f regulated on their respective domains. Then we have the change of variables formula

$$\int_{g(a)}^{g(b)} f(x)\,\mathrm{d}h(x) = \int_0^t f(g(t))\,\mathrm{d}(h \circ g)(t).$$

Hint: approximate f by step functions and notice that if $a = t_0 < \cdots < t_{n+1} = b$ is a partition of $[a, b]$, so is $g(a) = x_0) < \cdots < x_{n+1} = g(b)$, with $x_j = g(t_j)$ a partition of $[g(a), g(b)] = g([a, b])$. If f is a step function, so is the composition $f \circ g$. If f is regulated, so is $f \circ g$. If h is of bounded variation, so is $h \circ g$.

B.2 Some Elements from Measure Theory

True integration comfort (convergence theorems, changing the order of integration, support of a distribution) can only be achieved in the framework of measure theory. In Chapter 12 we consider sojourn distributions $\mathcal{G} : [0, \infty) \to [0, 1]$ with $\mathcal{G}(a)$ being the probability of having left a certain stage a time units after having entered it. In Chapter 22 we consider vaccination distributions $W : [0, \infty) \to [0, 1]$, with $W(a)$ being the probability of having been vaccinated up to age a.

In other words, we can interpret $\mathcal{G}(a)$ and $W(a)$ as being the probability of leaving the stage, or being vaccinated, in the (stage) age-interval from 0 to a. This interpretation is ambiguous, because we do not know whether it refers to the interval $[0, a)$ or the interval $[0, a]$. We will address this ambiguity in Toolbox B.4; here we want to extend the concept of the probability of being vaccinated at some age a in a certain set B from intervals of the above form to more general sets B.

This leads us to a concept of a (probability) measure m. Let

$$\Omega = [0, \infty)$$

(or, more generally, a locally compact space which is the union of countably many compact sets).

Let $\mathcal{B}$ be a collection of subsets of Ω.

$\mathcal{B}$ is called a σ-algebra on Ω if the following conditions are satisfied:

(i) $\emptyset, \Omega \in \mathcal{B}$;

(ii) if $B \in \mathcal{B}$, then $\Omega \setminus B \in \mathcal{B}$;

(iii) if $B_n, n \in \mathbb{N}$, is a countable collection of sets in $\mathcal{B}$, then $\bigcup_{n\in\mathbb{N}} B_n \in \mathcal{B}$.

A function $m : \mathcal{B} \to \mathbb{R}$ is called a *measure* on $\mathcal{B}$ if the following hold:

(i) $m(\emptyset) = 0$;

(ii) m is countably additive: if $B_n, n \in \mathbb{N}$, is a countable collection of pairwise disjoint sets, then

$$m\left(\biguplus_{n\in\mathbb{N}} B_n\right) = \sum_{n\in\mathbb{N}} m(B_n).$$

The symbol $\uplus$ denotes a disjoint union. Notice that our measures are assumed to be finite. A measure is called *nonnegative* if it takes nonnegative values. A nonnegative measure m on $\mathcal{B}$ is called positive if $m(\Omega) > 0$. Notice that, for a nonnegative measure m and sets $B_1, B_2 \in \mathcal{B}$ with $B_1 \subseteq B_2$, we have

$$m(B_2) = m(B_1 \uplus (B_2 \setminus B_1)) = m(B_1) + m(B_2 \setminus B_1) \geqslant m(B_1).$$

Furthermore, if m is a nonnegative measure and B_n, $n \in \mathbb{N}$, is a countable collection of sets in $\mathcal{B}$, then

$$m\left(\bigcup_{n\in\mathbb{N}} B_n\right) \leqslant \sum_{n=1}^{\infty} m(B_n),$$

with the series on the right-hand side possibly being infinite.

The set of measures on $\mathcal{B}$ becomes a real vector space by setting

$$(\alpha_1 m_1 + \alpha_2 m_2)(B) = \alpha_1 m_1(B) + \alpha_2 m_2(B), \quad B \in \mathcal{B}, \quad \alpha_1, \alpha_2 \in \mathbb{R},$$

for two measures m_1, m_2. One can make the space of measures a Banach space using the *total variation* of a measure as the norm (McDonald, Weiss, 1999, Section 6.7). The topology induced by the total variation is called the strong topology.

It looks as if an obvious choice for $\mathcal{B}$ is the power set of Ω, but it turns out that it is not a fruitful one. The correct choice is the smallest σ-algebra that contains all intervals $[0, a]$ (i.e., the intersection of all σ-algebras that contain these intervals—or, equivalently, all compact sets; it can easily be shown that this is a σ-algebra again). The sets in this σ-algebra are called the *Borel sets* in Ω. One can show that open and closed subsets of Ω are Borel sets. Measures on $\mathcal{B}$ are called *Borel measures* on Ω.

A nonnegative Borel measure on Ω is called *regular* if, for every $\epsilon > 0$ and every Borel set B in Ω, there exist subsets C and U of Ω such that C is compact in Ω, U is open in Ω,

$$C \subseteq B \subseteq U, \quad m(U \setminus B) < \epsilon, \quad m(B \setminus C) < \epsilon.$$

One can extend the notion of a regular measure from nonnegative measures to measures. If the topology of Ω is induced by a metric, as is the case for $\Omega = [0, \infty)$, then every Borel measure is regular.

The regular Borel measures on Ω again form a Banach space which is denoted by $M(\mathcal{B})$ or $M(\Omega)$.

Obvious examples for measures are *Dirac* or *point measures*, δ_a, $a \in \Omega$:

$$\delta_a(B) = \begin{cases} 1, & a \in B, \\ 0, & a \notin B. \end{cases}$$

Other examples are Lebesgue–Stieltjes measures, which will be considered in Toolbox B.4.

A function $f : \Omega \to \mathbb{R}$ is called *Borel measurable* if the pre-images $f^{-1}(J)$ of intervals J are Borel sets. One can show that continuous functions and regulated functions are Borel measurable. Endowed with the supremum norm, the space of bounded Borel measurable functions from Ω to $\mathbb{R}$ becomes a Banach space.

If f is a bounded Borel measurable function, we can define the integral of f with respect to the measure m,

$$\int_\Omega f \, \mathrm{d}m = \int_\Omega f(a)m(\mathrm{d}a).$$

Let $\epsilon > 0$. Choose a finite number of disjoint intervals $I_1, \dots, I_n$ of length less than ϵ such that

$$f(\Omega) \subseteq \bigcup_{j=1}^{n} I_j.$$

Furthermore, we choose elements $b_j \in I_j$, $j = 1, \dots, n$, and define

$$S(f, m, \epsilon) = \sum_{j=1}^{n} b_j m(f^{-1}(I_j)).$$

One can show that $S(f, m, \epsilon)$ converges as $\epsilon \to 0$ and that the limit does not depend on the choice of intervals I_j and elements $b_j \in I_j$. In particular, by restricting to intervals that have nonempty intersection with $f(\Omega)$ we can choose $b_j \in f(\Omega) \cap I_j$.

If $m = \delta_b$ is a Dirac measure, then

$$\int_\Omega f \, \mathrm{d}\delta_b = f(b).$$

This definition of an integral also works for set functions that are only finitely additive; countable additivity allows us to extend the integral to Borel measurable functions that are not bounded. (See the bibliographic remarks for references.)

We present the following example for a convergence theorem which we use in Chapter 16.

Lebesgue–Fatou lemma. *Let m be a nonnegative Borel measure on Ω and $f_j : \Omega \to \mathbb{R}$, $j \in J$, $J = \mathbb{N}$ or $J = (\alpha, \infty)$, a sequence or family of Borel measurable functions. If $J = (\alpha, \infty)$, assume that $f_j(a)$ depends continuously on j for every $a \in \Omega$. Then the following hold:*

(a) *If there exists a Borel measurable function g with $f_j \geqslant g$ on Ω for all $j \in J$ and $\int_\Omega g \, \mathrm{d}m$ is defined and finite, then*

$$\int_\Omega \liminf_{j\to\infty} f_j \, \mathrm{d}m \leqslant \liminf_{j\to\infty} \int_\Omega f_j \, \mathrm{d}m.$$

(b) *If there exists a Borel measurable function g with $f_j \leqslant g$ on Ω for all $j \in J$ and $\int_\Omega g \, \mathrm{d}m$ is defined and finite, then*

$$\limsup_{j\to\infty} \int_\Omega f_j \, \mathrm{d}m \leqslant \int_\Omega \limsup_{j\to\infty} f_j \, \mathrm{d}m.$$

Proof. If $J = \mathbb{N}$, this is the usual Lebesgue–Fatou lemma (Yosida, 1968, Section 0.3). Let $J = (\alpha, \infty)$. For $k \in \mathbb{N}$, set

$$g_k(a) = \inf\{f_j(a);\ j \geqslant \alpha + k\}, \quad a \in \Omega.$$

Since $f_j(a) \geqslant g(a)$ for all $j \in J$, g_k is defined and

$$g(a) \leqslant g_k(a) \leqslant f_k(a) \quad \forall a \in \Omega, \quad k \in \mathbb{N}.$$

Since $f_j(a)$ depends continuously on j, we have

$$g_k(a) = \inf\{f_j(a);\ j \geqslant \alpha + k,\ j \in \mathbb{Q}\},$$

and g_k is Borel measurable because the supremum is taken over a countable set (Hewitt, Stromberg, 1969, Theorem 11.12). Furthermore, $g_k(a)$ is nondecreasing in k and

$$\lim_{k\to\infty} g_k(a) = \liminf_{j\to\infty} f_j(a), \quad a \in \Omega.$$

By Beppo Levi's theorem of monotone convergence (Hewitt, Stromberg, 1969, 12.22; McDonald, Weiss, 1999, Theorem 4.6),

$$\begin{aligned}\int_\Omega \liminf_{j\to\infty} f_j \,\mathrm{d}m &= \lim_{k\to\infty} \int_\Omega g_k \,\mathrm{d}m\\ &= \liminf_{k\to\infty} \int_\Omega g_k \,\mathrm{d}m \leqslant \liminf_{k\to\infty} \int_\Omega f_k \,\mathrm{d}m.\end{aligned}$$

The statement for lim sup follows by using that $-\limsup_{j\to\infty} f_j(a) = \liminf_{j\to\infty}(-f_j(a))$. □

An important subspace of the bounded Borel measurable functions is $C_0(\Omega)$, the space of continuous functions on Ω which converge to 0 as $a \to \infty$.

The Banach space $C_0(\Omega)$ induces a topology on $M(\mathcal{B})$, called the weak* topology, which makes $M(\mathcal{B})$ a complete locally convex Hausdorff space. Convergence in this topology can be characterized by

$$m_j \to m \quad \iff \quad \int_\Omega f \,\mathrm{d}m_j \to \int_\Omega f \,\mathrm{d}m \quad \forall f \in C_0(\Omega).$$

In particular, for fixed $f \in C_0(\Omega)$, the integral $\int_\Omega f \,\mathrm{d}m$ is a continuous real-valued function of $m \in M(\mathcal{B})$ with respect to the weak* topology on $M(\mathcal{B})$.

The strong and the weak* topology have a fruitful, Heine–Borel type, interaction. A set in $M(\mathcal{B})$ that is bounded in the strong topology has compact closure in the weak* topology. We only need this result (a consequence of the Alaoglu–Bourbaki theorem and the fact that $M(\mathcal{B})$ is the dual space of $C_0(\Omega)$ (see McDonald, Weiss, 1999, Theorems 10.12 and 9.16)) in the following special case.

Proposition B.14. *Let K be a weakly* closed subset of $M(\mathcal{B})$ and $c > 0$ a constant such that*

$$0 \leqslant m(B) \leqslant c \quad \forall B \in \mathcal{B}, \quad m \in K.$$

Then K is weakly compact.*

Here we have used the terms *weakly* closed* and *weakly* compact* as shorthands for *closed in the weak* topology* and *compact in the weak* topology.*

Another reason to consider the measure-theoretic integral rather than the Stieltjes integral in the previous section consists of measures having a more convenient notion of support than distribution functions.

Definition B.15. Let m be a nonnegative Borel measure on Ω. Let B_m be the union of all (relatively) open sets A in Ω such that $m(A) = 0$. Then

$$\mathrm{supp}(m) = \Omega \setminus B_m$$

is called *the support* of m.

If m is the zero measure, the support of m is the empty set. The set B_m in Definition B.15 is (relatively) open as a union of (relatively) open sets.

Lemma B.16. *The support of a nonnegative Borel measure m on Ω is a closed set. Furthermore, the following hold:*

(a) *If K is a subset of $\Omega \setminus \operatorname{supp}(m)$, which is compact in Ω, then $m(K) = 0$.*

(b) *If m is a regular Borel measure, then*

$$m(\Omega \setminus \operatorname{supp}(m)) = 0.$$

Proof. (a) Let K be a subset of B_m that is compact in Ω. By definition of B_m there exist finitely many sets $A_1, \ldots, A_n$ which are open in Ω such that

$$K \subseteq \bigcup_{j=1}^{n} A_j, \quad m(A_j) = 0.$$

Hence

$$m(K) \leqslant \sum_{j=1}^{n} m(A_j) = 0.$$

(b) Now assume that m is a regular Borel measure. Let $\epsilon > 0$. Then there exists a subset K of B_m that is compact in Ω such that

$$m(B_m \setminus K) < \epsilon.$$

Hence, by part (a),

$$m(B_m) \leqslant m(B_m \setminus K) + m(K) < \epsilon.$$

Since this holds for every $\epsilon > 0$, $m(B_m) = 0$. □

Lemma B.17. *Let m be a positive Borel measure on $\Omega = [0, \infty)$. Then $0 \in \operatorname{supp}(m)$ if and only if $m([0, \epsilon)) > 0$ for every $\epsilon > 0$. Furthermore, $a > 0$ is in the support of m if and only if $m(a - \epsilon, a + \epsilon) > 0$ for every $\epsilon \in (0, a)$.*

Proof. We restrict our proof to $a > 0$. The case $a = 0$ is proved in a similar way. Let B_m be the set in Definition B.15. Equivalently, to the original statement, we prove that $a \in B_m$ if and only if there exists some $\epsilon > 0$ such that $m((a - \epsilon, a + \epsilon)) = 0$.

Let $a \in B_m$. By definition of B_m, there exists a subset A of Ω that is open in Ω such that $a \in A$ and $m(A) = 0$. Then there exists some $\epsilon > 0$ such that $(a - \epsilon, a + \epsilon) \subseteq A$. Hence $m(a - \epsilon, a + \epsilon)) \leqslant m(A) = 0$.

Now let $m((a - \epsilon, a + \epsilon)) = 0$ for some $\epsilon \in (0, a)$. Since $(a - \epsilon, a + \epsilon)$ is an open subset of Ω, we have $(a - \epsilon, a + \epsilon) \subseteq B_m$ by definition of B_m. In particular, $a \in B_m$. □

Proposition B.18. *Assume that m is a positive Borel measure on $\Omega = [0, \infty)$ and that the support of m contains n different points $a_1, \dots, a_n$ in Ω. Then*

$$m = \sum_{j=1}^{n} s_j m_j$$

with numbers $s_j > 0$ and regular positive Borel measures m_j such that

$$\sum_{j=1}^{n} s_j = 1, \quad m_j(\Omega) = m(\Omega), \quad j = 1, \dots, n,$$

and the measures $m_1, \dots, m_n$ form an orthogonal family of measures, i.e., there exist pairwise disjoint subsets $\Omega_1, \dots, \Omega_n$ of Ω such that

$$m_j(\Omega_k) = \begin{cases} m(\Omega), & j = k, \\ 0, & j \neq k. \end{cases}$$

Proof. Since a_j is in the support of m, by Lemma B.17,

$$m((a_j - \epsilon, a_j + \epsilon) \cap \Omega) > 0$$

for every $\epsilon > 0$. We choose $\epsilon > 0$ so small that the intervals $(a_j - \epsilon, a_j + \epsilon)$ are pairwise disjoint. Set

$$\Omega_j = (a_j - \epsilon, a_j + \epsilon) \cap \Omega, \quad j = 1, \dots, n-1, \qquad \Omega_n = \Omega \setminus \left(\biguplus_{j=1}^{n-1} \Omega_j \right).$$

Then $m(\Omega_j) > 0$ for all $j = 1, \dots, n$, and Ω is the disjoint union of $\Omega_1, \dots, \Omega_n$. Set

$$s_j = \frac{m(\Omega_j)}{m(\Omega)}$$

and

$$m_j(B) = m(B \cap \Omega_j) \frac{1}{s_j}, \quad B \in \mathcal{B}.$$

Then m_j is a regular Borel measure on Ω for $j = 1, \dots, n$, and

$$\sum_{j=1}^{n} s_j = \frac{1}{m(\Omega)} \sum_{j=1}^{n} m(\Omega_j) = \frac{1}{m(\Omega)} m\left(\biguplus_{j=1}^{n} \Omega_j \right) = 1,$$

and

$$m_j(\Omega) = \frac{m(\Omega_j)}{s_j} = m(\Omega).$$

By construction, the measures $m_1, \dots, m_n$ form an orthogonal family of measures. $\square$

B.3 Some Elements from Convex Analysis

Let X be a vector space. If $x, y \in X$ and $0 \leqslant s \leqslant 1$, then $z = sx + (1-s)y$ is called the *convex combination* of x and y. If $0 < s < 1$, then z is called a *proper convex combination* of x and y. Geometrically, the convex combinations of x and y form the line connecting x and y. A subset K of X is called *convex* if, for any points $x, y \in K$, all convex combinations are in K as well. Geometrically, this means that every finite line with endpoints in K completely lies in K.

Definition B.19. Let K be a convex set in X.

(a) An element $z \in K$ is called an *extreme point* of K if $K \setminus \{z\}$ is still a convex set, i.e., if we still have a convex set after the element z is removed. In other words, a point $z \in K$ is an extreme point if it is not a proper convex combination of two other points in K; formally,

$$\text{if } z = sy + (1-s)x \text{ with } x, y \in K \text{ and } 0 < s < 1, \text{ then } x = y = z.$$

(b) A subset E of K is called an *extreme subset* of K if a proper convex combination $z = sy + (1-s)x$ of two points $x, y \in K$ is in E only if both y and x are in E. A convex extreme subset of K is called a *face* of K.

Definition B.20. Let K be a convex subset of X. A functional $f : K \to \mathbb{R}$ is called *concave* if

$$f(sx + (1-s)y) \geqslant sf(x) + (1-s)f(y) \quad \forall x, y \in K, \quad s \in [0, 1].$$

Proposition B.21. *Let K be a nonempty convex subset of X and f a concave functional on K. Assume that*

$$\alpha := \inf f(K) > -\infty.$$

Then $E = f^{-1}(\{\alpha\})$ is empty or an extreme subset of K.

Proof. Assume that E is not empty. In order to show that E is an extreme subset of K, let $z = sx + (1-s)y$ be an element in E that is a proper convex combination of two elements $x, y \in K$. In other words, let $x, y \in K$ and $0 < s < 1$ such that $f(z) = \alpha$ for $z = sx + (1-s)y$. We have $f(x) \geqslant \alpha$ and $f(y) \geqslant \alpha$ by definition of α. Assume that $x \notin E$. Then $f(x) > \alpha$ and, since f is concave,

$$f(z) \geqslant sf(x) + (1-s)f(y) > s\alpha + (1-s)\alpha = \alpha,$$

a contradiction. Similarly, one sees that $y \in E$. □

We now turn to locally convex Hausdorff spaces X, Banach spaces, for example. The following result is the first part of the proof of the Krein–Milman theorem (see the proof of Theorem 10.1.2 (a) in Edwards (1965, 1995)).

Proposition B.22 (Krein–Milman). *Let K be a nonempty compact convex subset of a locally convex Hausdorff space X, and let E be a nonempty closed extreme subset of K. Then E contains an extreme point of K.*

We have seen in Section 22.4 that the problem of optimal vaccination schedules can be formulated as an extreme-value problem for an affine function on a convex set.

Theorem B.23 (fundamental theorem of convex optimization). *Let K be a nonempty compact convex subset of a locally convex Hausdorff space X and $f : K \to \mathbb{R}$ be concave and continuous. Then there exists an extreme point z of K such that*

$$f(z) = \inf f(K).$$

Proof. Let $\alpha = \inf f(K)$. Since K is compact and f is continuous, $\alpha > -\infty$ and

$$E = \{x \in K;\ f(x) = \alpha\}$$

is nonempty and closed. Since f is concave and K is convex, E is an extreme subset of K by Proposition B.21. By Proposition B.22, E contains an extreme point, z, of K, and $f(z) = \alpha$ by definition of E. □

We have seen in the last section that the regular Borel measures, $M(\Omega)$, with the weak* topology, form a locally convex Hausdorff space in which strongly bounded subsets are compact. Let C_j, $j = 1, \ldots, n$, be functionals on $M(\Omega)$, given by

$$C_j(m) - C_j(0) = \int_\Omega g_j(a)m(\mathrm{d}a),$$

with functions $g_j \in C_0(\Omega)$. By Proposition B.14, the set of nonnegative Borel measures m on Ω with $m(\Omega) \leqslant c_0$ is weakly* compact. Since $C_j(m)$ is a weakly* continuous function of m, the set of measures with $C_j(m) \leqslant c_j$ is a weakly* closed set. As the intersection of a weakly* closed set and a weakly* compact set, the set of measures described by the following constraints is weakly* compact (and convex):

$$C_i(m) \leqslant c_i, \quad i = 1, \ldots, n, \qquad m(\Omega) \leqslant c_0, \quad m \geqslant 0. \tag{B.4}$$

Proposition B.24. *If a measure in the set (B.4) is an extreme point of this set, then its support contains at most $n + 1$ points.*

Proof. Let us assume that the support of m contains at least $n+2$ different points. By Proposition B.18, we find regular positive Borel measures $m_1, \ldots, m_{n+2}$ such that

$$m = \sum_{j=1}^{n+2} s_j m_j, \quad s_j > 0, \quad \sum_{j=1}^{n+2} s_j = 1, \quad m_j(\Omega) = m(\Omega).$$

For $\epsilon = (\epsilon_1, \ldots, \epsilon_{n+2}) \in \mathbb{R}^{n+2}$, we set

$$m(\epsilon) = \sum_{j=1}^{n+2} (s_j + \epsilon_j) m_j.$$

Obviously,

$$m = \tfrac{1}{2} m(\epsilon) + \tfrac{1}{2} m(-\epsilon).$$

Our goal is to find $\epsilon \in \mathbb{R}^{n+2}$, $\epsilon \neq 0$ such that $m(\epsilon)$ and $m(-\epsilon)$ satisfy the constraints (B.4). Then m will be a proper convex combination of two elements in the convex set described by (B.4) and is not an extreme point of this set.

In order to achieve our goal, we try to find $\epsilon \in \mathbb{R}^{n+2}$, $\epsilon \neq 0$, such that

$$s_j \pm \epsilon_j > 0, \quad \sum_{j=1}^{n+2} (s_j + \epsilon_j) = 1, \quad \sum_{j=1}^{n+2} (s_j - \epsilon_j) = 1,$$
$$C_i(m(\pm\epsilon)) = C_i(m), \quad i = 1, \ldots, n.$$

This amounts to solving the linear system

$$\sum_{j=1}^{n+2} \epsilon_j = 0, \quad \sum_{j=1}^{n} \epsilon_j C_i(m_j) = 0, \quad i = 1, \ldots, n,$$

under the constraints $s_j - |\epsilon_j| > 0$. Since the linear system is underdetermined ($n+1$ equations for $n+2$ unknowns) and homogeneous, it always has a nonzero solution $\epsilon = (\epsilon_1, \ldots, \epsilon_{n+2})$. Multiplying this solution by a small positive scalar, we also satisfy the inequalities $s_j - \epsilon_j > 0$. Finally, we have $m(\epsilon) \neq m(-\epsilon)$. Indeed, equality would imply that $\sum_{j=1}^{n+2} \epsilon_j m_j = 0$, in contradiction to the fact that the measures $m_1, \ldots, m_{n+2}$ form an orthogonal family of measures.

As pointed out before, this implies that m is a convex combination of two different elements in the convex set described by (B.4) and is not an extreme point of this set. □

Theorem B.25. *Let the functional R on $M(\Omega)$ be given by*

$$R(m) - R(0) = \int_\Omega g(a) m(\mathrm{d}a)$$

with some $g \in C_0(\Omega)$. Then $R(m)$ takes its minimum (maximum) under the side conditions (B.4) at a measure which is the convex combination of at most $n + 1$ Dirac measures.

Proof. By Theorem B.23, the minimum is taken at some regular nonnegative Borel measure m that is an extreme point in the convex set described by (B.4). By Proposition B.24, the support of m contains $n + 1$ points at most, and m is the linear combination of no more than $n + 1$ Dirac measures by Lemma B.16 (b). □

The problem of vaccinating at optimal ages, considered in Section 22.4, has two constraints: the natural constraint that the maximal fraction which can be vaccinated is 1, and the economic constraint of a cost ceiling. So, the optimizing measure is the convex combination of at most two Dirac measures, i.e., optimal vaccinations are performed at one or two specific ages.

B.4 Lebesgue–Stieltjes Integration

The relation between sojourn functions and vaccination distributions on the one hand and regular Borel measures on the other is given by the following result (e.g., McDonald, Weiss, 1999, Theorem 4.13).

Lemma B.26. *Let $g : [0, \infty) \to \mathbb{R}$ be nondecreasing and bounded. Then there exists a uniquely determined nonnegative regular Borel measure m_g on $[0, \infty)$ such that*

$$m_g([0, a)) = m_g([0, a]) = g(a) - g(0) \qquad \text{whenever } a \geqslant 0 \text{ and } g \text{ is continuous at } a.$$

An analogous result holds if g is nonincreasing and bounded, and a nonpositive measure $m_g = -m_{-g}$ is obtained.

The measure m_g is called the Lebesgue–Stieltjes measure associated with g and satisfies

$$m_g([0, a]) = g(a+) - g(0) = \lim_{b \searrow a} g(b) - g(0), \quad a \geqslant 0,$$

$$m_g([0, a)) = g(a-) - g(0), \quad a > 0.$$

This clarifies the problem of whether, for a vaccination distribution W, $W(a)$ is the probability of being vaccinated in the interval $[0, a)$ or the interval $[0, a]$. The first is $W(a-)$ and the second is $W(a+)$.

If $g : [0, \infty) \to \mathbb{R}$ is a monotone function, in the following $\mathrm{d}g(s)$ denotes Stieltjes integration as defined in Toolbox B.1, while $g(\mathrm{d}s)$ denotes the measure-theoretic integral using the Lebesgue–Stieltjes measure uniquely associated with g. In the following lemma we explore the case in which the two integrals are equal.

Lemma B.27. *Let $u : [0, a] \to \mathbb{R}$ be regulated and g be monotone and bounded and continuous at a, and let u and g have no common discontinuities. Then the Stieltjes and Lebesgue–Stieltjes integrals are equal:*

$$\int_{[0,a)} u(s) g(\mathrm{d}s) = \int_0^a u(s)\, \mathrm{d}g(s).$$

If u and g have common discontinuities, the two integrals can be different.

Proof. Let $\epsilon > 0$ and v be a step function such that $|u(s) - v(s)| < \epsilon$ for all $s \in [a, b]$. Since g has at most countably many discontinuities which are all different from the discontinuities of u, one can modify v in such a way that the discontinuities of v are different from the discontinuities of g. First we show that the Stieltjes and Lebesgue–Stieltjes integrals are the same for the step function v. Let v be constant on the intervals (a_j, a_{j+1}), $0 = a_0 < \cdots < a_{n+1} = a$, such that g is continuous at $a_1, \dots, a_{n+1}$. Let $s_j \in (a_j, a_{j+1})$. By Lemma B.26,

$$\begin{aligned}\int_{[0,a)} v(s)g(\mathrm{d}s) &= v(s_0)(g(a_1) - g(a_0)) + \sum_{j=1}^{n} v(s_j)m_g((a_j, a_{j+1}))\\ &= \sum_{j=1}^{n} v(s_j)(g(a_{j+1}) - g(a_j)) = \int_a^b v(s)\,\mathrm{d}g(s).\end{aligned}$$

By the triangle inequality,

$$\begin{aligned}&\left|\int_{[0,a)} u(s)g(\mathrm{d}s) - \int_0^a u(s)\,\mathrm{d}g(s)\right|\\ &\qquad \leqslant \left|\int_{[0,a)} (u(s) - v(s))g(\mathrm{d}s)\right| + \left|\int_0^a (u(s) - v(s))\,\mathrm{d}g(s)\right|\\ &\qquad \leqslant \epsilon(m_g([0, a)) + |g(a) - g(0)|) = 2\epsilon|g(a) - g(0)|.\end{aligned}$$

Since this holds for every $\epsilon > 0$, the Stieltjes integral and the Lebesgue–Stieltjes integral are equal. □

A Dirac or point measure concentrated at $a \geqslant 0$ is associated with a step function with one jump, namely at a. So, we have the following reformulation of Theorem B.25.

Theorem B.28. *Let the functionals R and $C_1, \dots, C_n$ be given by Lebesgue–Stieltjes integrals (or, equivalently, improper Stieltjes integrals)*

$$R(W) - R(0) = \int_{[0,\infty)} g(a)W(\mathrm{d}a), \qquad C_j(W) - C(0) = \int_{[0,\infty)} g_j(a)W(\mathrm{d}s)$$

with functions $g, g_j \in C_0([0, \infty)$. Then $R(W)$ takes its minimum (maximum) under the side conditions

$$C_i(W) \leqslant c_i, \quad i = 1, \dots, n, \qquad |W(a) - W(0)| \leqslant c_0 \quad \forall a \geqslant 0, \quad W \geqslant 0,$$

at a monotone step function W with at most $n + 1$ discontinuities.

If W is a vaccination distribution as in Chapter 22, it represents a vaccination strategy with vaccinations at $n + 1$ different ages at most.

Convolutions and Stieltjes Convolutions

In the following we prove the *fundamental balance equation of stage dynamics* (Section 13.5) for the case where the sojourn functions, here represented by P, are not differentiable. The proof uses Lebesgue–Stieltjes integrals, but the results are also formulated in terms of Stieltjes integrals.

Lemma B.29. *Let u be integrable on $[0, \tau]$ and let P be nonincreasing. Then the convolution*

$$(u * P)(r) = \int_0^r u(r-s)P(s)\,\mathrm{d}s, \quad 0 \leqslant r \leqslant \tau,$$

is absolutely continuous and, outside a set of Lebesgue measure 0,

$$(u * P)'(r) = u(r)P(0) + \int_{[0,r)} u(r-s)P(\mathrm{d}s).$$

If u is regulated, outside a set of Lebesgue measure 0,

$$(u * P)'(r) = u(r)P(0) + \int_0^r u(r-s)\,\mathrm{d}P(s).$$

$u * P$ is called the *convolution* of u and P, while

$$(u \star P)(r) = \int_{[0,r)} u(r-s)P(\mathrm{d}s)$$

is called the *Stieltjes convolution* of u and P.

Proof.

$$\begin{aligned}
\int_0^t \left(\int_{[0,r)} u(r-s)P(\mathrm{d}s) \right) \mathrm{d}r &= \int_{[0,t)} \left(\int_s^t u(r-s)\,\mathrm{d}r \right) P(\mathrm{d}s) \\
&= \int_{[0,t)} \left(\int_0^{t-s} u(r)\,\mathrm{d}r \right) P(\mathrm{d}s) \\
&= \int_0^t u(r) \left(\int_{[0,t-r)} P(\mathrm{d}s) \right) \mathrm{d}t \\
&= \int_0^t u(r)(P(0) - P(t-r))\,\mathrm{d}t \\
&= P(0) \int_0^t u(r)\,\mathrm{d}r - (u * P)(t).
\end{aligned}$$

In the middle of this chain of equalities we have used Fubini's theorem (Hewitt, Stromberg, 1969, (21.13); McDonald, Weiss, 1999, Theorem 4.17) and the result from Lemma B.26 that $\int_{[0,a)} P(\mathrm{d}s) = P(a) - P(0)$ for almost all $a \geqslant 0$. Strictly speaking, the chain must be read from bottom to top, because the existence of the first integral is not a priori clear, while the existence of the last integrals, if also taken over the absolute values, follows from the assumptions.

The equality of $(u * P)'(t)$ with the Stieltjes integral follows from Lemma B.27, because there are at most countably many points r such that P is not continuous at r and the functions $s \mapsto u(r-s)$ and P have a common discontinuity. Indeed, the set of r for which the second happens is contained in $\bigcup_{s \in \mathrm{Dis}(P)}(s - \mathrm{Dis}(u))$, which is countable as a countable union of countable sets, where $\mathrm{Dis}(u)$ and $\mathrm{Dis}(P)$ denote the set of points at which u and P, respectively, are not continuous. □

Lemma B.30. *Let $u \in L^1_+[0,\infty)$ and $P : [0,\infty) \to [0,\infty)$ nonincreasing, $P(a) \to 0$ as $a \to \infty$, then the function w,*

$$w(t) = \int_t^\infty u(s-t)P(s)\,\mathrm{d}s, \quad 0 \leqslant r \leqslant \tau,$$

is absolutely continuous and, outside a set of Lebesgue measure 0,

$$w'(t) = \int_{[t,\infty)} u(s-t)P(\mathrm{d}s).$$

If u is regulated, outside a set of Lebesgue measure 0,

$$w'(t) = \int_0^\infty u(s-t)\,\mathrm{d}P(s).$$

In particular, w is nonincreasing and

$$w(t) \leqslant w(0) = \int_0^\infty u(s)P(s)\,\mathrm{d}s,$$

$$\int_0^\infty |w'(t)|\,\mathrm{d}t \leqslant w(0).$$

Proof. Similarly, as in the proof of Lemma B.29,

$$\begin{aligned}
\int_r^\infty \left(\int_{[t,\infty)} u(s-t)P(\mathrm{d}s)\right)\mathrm{d}t &= \int_{[r,\infty)} \left(\int_r^s u(s-t)\,\mathrm{d}t\right)P(\mathrm{d}s)\\
&= \int_{[r,\infty)} \left(\int_0^{s-r} u(t)\,\mathrm{d}t\right)P(\mathrm{d}s)\\
&= \int_0^\infty u(t)\left(\int_{[t+r,\infty)} P(\mathrm{d}s)\right)\mathrm{d}t\\
&= \int_0^\infty u(t)P(r+t)\,\mathrm{d}t\\
&= \int_r^\infty u(t-r)P(t)\,\mathrm{d}t.
\end{aligned}$$

That w' and the Stieltjes integral are equal almost everywhere follows from Lemma B.17 in a way similar to the proof of Lemma B.29. □

The next results follow from Lemmas B.29 and B.30 and Theorem (21.67) in Hewitt, Stromberg (1969).

Proposition B.31. *Let $u \in L^1_+[0,\infty)$ and $P, Q : [0,\infty) \to [0,\infty)$ non-increasing, $P(a)Q(a) \to 0$ as $a \to \infty$.*

(a) *Then v, defined by*

$$v(t) = \int_0^t u(t-s)Q(s)P(s)\,\mathrm{d}s,$$

is absolutely continuous and, outside a set of Lebesgue measure 0,

$$v'(t) = u(t)P(0)Q(0) + \int_{[0,t)} u(t-s)Q(s-)P(\mathrm{d}s) + \int_{[0,t)} u(t-s)P(s+)Q(\mathrm{d}s)$$

with $P(0-) = P(0)$.

(b) *Furthermore, w, defined by*

$$w(t) = \int_t^\infty u(s-t)P(s)Q(s)\,\mathrm{d}s,$$

is absolutely continuous and, outside a set of Lebesgue measure 0,

$$w'(t) = \int_{[t,\infty)} u(s-t)[P(s+)Q(\mathrm{d}s) + Q(s-)P(\mathrm{d}s)].$$

Proof. Define

$$P_0(a) = P(a)Q(a), \qquad P_1(a) = \int_{[0,a]} P(s+)Q(\mathrm{d}s),$$

$$P_2(a) = \int_{[0,a]} Q(s-)P(\mathrm{d}s).$$

By Theorem (21.67) in Hewitt, Stromberg (1969),

$$P_0(a+) - P_0(0) = P_1(a) + P_2(a).$$

This implies that $m_0 = m_1 + m_2$, where the m_j are the Lebesgue–Stieltjes measures associated with P_j. The assertion now follows from Lemmas B.29 and B.30. □

Corollary B.32. *Let $u \in L^1_+[0,\infty)$ and $P, Q : [0,\infty) \to [0,\infty)$ non-increasing, $P(a)Q(a) \to 0$ as $a \to \infty$. Assume that P and Q have no points of discontinuity in common.*

(a) *Then v, defined by*

$$v(t) = \int_0^t u(t-s)Q(s)P(s)\,\mathrm{d}s,$$

is absolutely continuous and, outside a set of Lebesgue measure 0,

$$v'(t) = u(t)P(0)Q(0) + \int_{[0,t)} u(t-s)Q(s)P(\mathrm{d}s) + \int_{[0,t)} u(t-s)P(s)Q(\mathrm{d}s).$$

(b) *Furthermore, w, defined by*

$$w(t) = \int_t^\infty u(s-t)P(s)Q(s)\,\mathrm{d}s,$$

is absolutely continuous and, outside a set of Lebesgue measure 0,

$$w'(t) = \int_{[t,\infty)} u(s-t)[P(s)Q(\mathrm{d}s) + Q(s)P(\mathrm{d}s)].$$

Proof. P and Q are discontinuous at countably many points only. Since P and Q have no points of discontinuity in common, the points of discontinuity of P form a set of measure 0 for the Lebesgue–Stieltjes measure associated with Q and vice versa. The assertion now follows from Proposition B.31. □

Corollary B.33. *Let $u \in L^1_+[0,\infty)$ and $P, \mathcal{F} : [0,\infty) \to [0,\infty)$ non-increasing, $P(a)Q(a) \to 0$ as $a \to \infty$, $\mu > 0$. Assume that P and Q have no points of discontinuity in common.*

Then v, defined by

$$v(t) = \int_0^t u(t-s)\mathrm{e}^{-\mu s}\mathcal{F}(s)P(s)\,\mathrm{d}s,$$

is absolutely continuous and, outside a set of Lebesgue measure 0,

$$\begin{aligned} v'(t) &= u(t)P(0)\mathcal{F}(0) - \mu v(t) \\ &\quad + \int_{[0,t)} u(t-s)\mathrm{e}^{-\mu s}[\mathcal{F}(s)P(\mathrm{d}s) + P(s)\mathcal{F}(\mathrm{d}s)] \\ &\leqslant u(t)P(0)\mathcal{F}(0) - \mu v(t) \\ &\quad + \int_{[0,t)} u(t-s)\mathrm{e}^{-\mu s}\mathcal{F}(s)P(\mathrm{d}s). \end{aligned}$$

Furthermore, w, defined by

$$w(t) = \int_t^\infty u(s-t)\mathrm{e}^{-\mu t} P(s)\mathcal{F}(s)\,\mathrm{d}s,$$

is absolutely continuous and, outside a set of Lebesgue measure 0,

$$\begin{aligned} w'(t) &= -\mu w(t) + \mathrm{e}^{-\mu t}\int_{[t,\infty)} u(s-t)[\mathcal{F}(s)P(\mathrm{d}s) + P(s)\mathcal{F}(\mathrm{d}s)] \\ &\leqslant -\mu w(t) + \mathrm{e}^{-\mu t}\int_{[t,\infty)} u(s-t)\mathcal{F}(s)P(\mathrm{d}s). \end{aligned}$$

Again, in Corollaries B.32 and B.33, the derivatives can be expressed by Stieltjes integrals rather than Lebesgue–Stieltjes integrals if u is also assumed to be regulated.

Exercises

B.5. Consider the age-structured SEIR model with vaccination. Remember the two extreme strategies represented by the distributions

$$W_0(0) = 0, \qquad W_0(a) = 1, \quad a > 1,$$
$$W_\infty(a) = 0, \quad a \geqslant 0.$$

The strategy represented by W_0 consists of vaccinating everybody immediately after birth. With the second strategy, nobody is ever vaccinated (see Section 22.4).

Find the measures that correspond to W_0 and W_∞ via Lemma B.26.

Answer. W_0 is represented by the Dirac measure δ_0,

$$\delta_0(B) = \begin{cases} 1, & 0 \in B, \\ 0, & 0 \notin B, \end{cases}$$

for every Borel set B in $[0,\infty)$. Indeed, approximation by Riemann sums provides

$$\int_0^\infty f(a)W_0(\mathrm{d}a) = f(0) = \int_{[0,\infty)} f\,\mathrm{d}\delta_0$$

for all $f \in C_0[0,\infty)$.

W_∞ is represented by the zero measure.

B.6. For a vaccination distribution W (Section 22.4) and $\alpha \in (0,1)$ define

$$f_\alpha(W) = \alpha\frac{\mathcal{R}_0(W)}{\mathcal{R}_0(W_\infty)} + (1-\alpha)\frac{C(W)}{C(W_0)}.$$

Show that

(a) for each $\alpha \in (0, 1)$ there exists a vaccination strategy W_α that minimizes f_α;

(b) W_α can be chosen as a one-age strategy where all vaccinations occur exactly at the same age;

(c)

$$C(W_\alpha) \leqslant \min\left\{1, \frac{\alpha}{1-\alpha}\right\} C(W_0),$$

$$\mathcal{R}_0(W_\alpha) \leqslant \min\left\{1, \frac{1}{\alpha} - 1\right\} \mathcal{R}_0(W_\infty).$$

Answer. (a) From (22.13) we notice that

$$f_\alpha(W) = \frac{\alpha}{\mathcal{R}_0} f(0) + \int_\Omega h_\alpha \, \mathrm{d}m$$

with

$$h_\alpha = \frac{1-\alpha}{C(W_0)} g - \frac{\alpha}{\mathcal{R}_0(W_\infty)} f$$

being a function in $C_0(\Omega)$ and m being a regular nonnegative Borel measure on $\Omega = [0, \infty)$, $m(\Omega) \leqslant 1$.

Let K be the set of nonnegative regular Borel measures m on $\Omega = [0, \infty)$ with $m(\Omega) \leqslant 1$. K is convex and compact in the weak* topology (see Proposition B.14). One easily sees from the representation of f_α above that f_α is affine and weakly* continuous. By the fundamental theorem of convex optimization, f_α takes its minimum on K, and it takes the minimum at an extreme point of K (though it may also take the minimum in points that are not extreme points).

(b) Let m be the extreme point in K in which f_α takes its minimum. Suppose that the support of m contains more than one point, i.e., at least two points. By Proposition B.18,

$$m = s_1 m_1 + s_2 m_2,$$

with numbers $s_j > 0$ and regular positive Borel measures m_j such that

$$s_1 + s_2 = 1, \qquad m_j(\Omega) = m(\Omega), \quad j = 1, 2.$$

In this case, m_1 and m_2 are both elements of K, and m is not an extreme point of K because it is the proper convex combination of two elements in K.

(c) Since $\mathcal{R}_0(W_0) = 0$ and $C(W_\infty) = 0$, we have

$$f_\alpha(W_\alpha) \leqslant f_\alpha(W_0) = 1 - \alpha, \qquad f_\alpha(W_\alpha) \leqslant f_\alpha(W_\infty) = \alpha.$$

Hence

$$f_\alpha(W_\alpha) \leqslant \min\{\alpha, 1-\alpha\}.$$

Furthermore,

$$f_\alpha(W_\alpha) \geqslant \alpha \frac{\mathcal{R}_0(W_\alpha)}{\mathcal{R}_0(W_\infty)}, \qquad f_\alpha(W_\alpha) \geqslant (1-\alpha)\frac{C(W_\alpha)}{C(W_0)}.$$

Thus

$$\mathcal{R}_0(W_\alpha) \leqslant \frac{1}{\alpha}\mathcal{R}_0(W_\infty) \min\{\alpha, 1-\alpha\}$$

and

$$C(W_\alpha) \leqslant \frac{1}{1-\alpha}\mathcal{R}_0(W_\infty) \min\{\alpha, 1-\alpha\}.$$

This implies the assertion.

B.5 Jensen's Inequality and Related Material

The following are special cases of Jensen's inequality (Hewitt, Stromberg, 1969, (13.34) (d)). Recall that a real function φ defined on an interval I is called *convex* if

$$\varphi(tx + (1-t)y) \leqslant t\varphi(x) + (1-t)\varphi(y) \quad \forall x, y \in I, \quad t \in [0,1].$$

φ is called *concave* if the inequality is reversed. The formulation of the subsequent theorem is in terms of Borel measurable functions and Lebesgue (Stieltjes) integrals, but could be in terms of regulated functions and Stieltjes integrals instead.

Theorem B.34. *Let I be a closed interval, and let $\varphi : I \to \mathbb{R}$ be convex. Furthermore, let $-\infty \leqslant a < b \leqslant \infty$ and $f : [a,b] \to I$, $g : [a,b] \to [0,\infty)$ be Borel measurable and bounded, $0 < \int_a^b g(x)\,\mathrm{d}x < \infty$. Then*

$$\frac{\int_a^b \varphi(f(x))g(x)\,\mathrm{d}x}{\int_a^b g(x)\,\mathrm{d}x} \geqslant \varphi\left(\frac{\int_a^b f(x)g(x)\,\mathrm{d}x}{\int_a^b g(x)\,\mathrm{d}x}\right).$$

Theorem B.35. *Let I be a closed interval, and let $\varphi : I \to \mathbb{R}$ be convex. Furthermore, let $-\infty \leqslant a < b \leqslant \infty$ and let $f : [a,b] \to I$ be Borel measurable, $g : [a,b] \to \mathbb{R}$ be increasing, $g(b) - g(a) > 0$. Then*

$$\frac{\int_a^b \varphi(f(x))g(\mathrm{d}x)}{g(b)-g(a)} \geqslant \varphi\left(\frac{\int_a^b f(x)\,\mathrm{d}g(x)}{g(b)-g(a)}\right).$$

If φ is concave rather than convex, then the inequalities in Theorems B.34 and B.35 are reversed.

If no information about the convexity of φ is available, one can still get an approximate result in the case in which f is the identity and φ and g operate on a different scale.

Proposition B.36. *Let $-\infty \leqslant a < b \leqslant \infty$ and $\varphi, g : [a, b] \to [0, \infty]$ be nonnegative and integrable. Let φ_0 and g_0 be the normalizations of φ and g,*

$$\varphi_0(s) = \varphi(sM_\varphi), \qquad g_0(s) = g(sM_g),$$

with

$$M_\varphi = \int_a^b \varphi(t)\,\mathrm{d}t, \qquad M_g = \int_a^b g(t)\,\mathrm{d}t.$$

Finally, assume that φ_0 is twice continuously differentiable on (a, b) with

$$|\varphi_0''(s)| \leqslant \Lambda \quad \forall s \in (a, b).$$

Then

$$\int_a^b \varphi(t)g(t)\,\mathrm{d}t = M_g\varphi(E_g) + R_{\varphi,g}$$

with

$$E_g = \frac{\int_a^b tg(t)\,\mathrm{d}t}{M_g} = M_g \int_{a_0}^{b_0} sg_0(s)\,\mathrm{d}s,$$

$a_0 = a/M_g$, $b_0 = b/M_g$, and

$$|R_{\varphi,g}| \leqslant \frac{\Lambda}{2}\left(\frac{M_g}{M_\varphi}\right)^2 \left(\int_{a_0}^{b_0} s^2 g_0(s)\,\mathrm{d}s - \frac{E_g}{M_g}\right) \leqslant \frac{\Lambda}{2}\left(\frac{M_g}{M_\varphi}\right)^2 \frac{\int_a^b t^2 g(t)\,\mathrm{d}t}{M_g^3}.$$

This shows that $\int_a^b \varphi(t)g(t)\,\mathrm{d}t \approx M_g\varphi(E_g)$ if M_g/M_φ is very small and Λ is not large. Notice that the approximation is an equality if φ is linear on (a, b).

Proof. Substituting $t = sM_g$ and setting $a_0 = a/M_g$, $b_0 = b/M_g$,

$$\begin{aligned}\int_a^b \varphi(t)g(t)\,\mathrm{d}t - M_g\varphi(E_g) &= M_g \int_{a_0}^{b_0} \varphi(sM_g)g_0(s)\,\mathrm{d}s - M_g\varphi(E_g)\\ &= M_g \int_{a_0}^{b_0} [\varphi(sM_g) - \varphi(E_g)]g_0(s)\,\mathrm{d}s.\end{aligned}$$

In the last step we have used that $\int_{a_0}^{b_0} g_0(s)\,\mathrm{d}s = 1$. By the integral form of Taylor's theorem (Kirkwood, 1995, Section 8-3),

$$\begin{aligned}\int_{a_0}^{b_0} [\varphi(sM_g) - \varphi(E_g)]g_0(s)\,\mathrm{d}s \\ &= \int_{a_0}^{b_0} \varphi'(E_g)(M_g s - E_g)g_0(s)\,\mathrm{d}s \\ &\quad + \int_{a_0}^{b_0}\left(\int_{E_g}^{sM_g}(sM_g - r)\varphi''(r)\,\mathrm{d}r\right)g_0(s)\,\mathrm{d}s \\ &= \varphi'(E_g)\left(M_g\int_{a_0}^{b_0} sg_0(s)\,\mathrm{d}s - E_g\right) \\ &\quad + M_g\int_{a_0}^{b_0}\left(\int_{E_g/M_g}^{s}(s-u)\varphi''(uM_g)\,\mathrm{d}r\right)g_0(s)\,\mathrm{d}s.\end{aligned}$$

The first term on the right-hand side of the equation is zero if

$$E_g = M_g\int_{a_0}^{b_0} sg_0(s)\,\mathrm{d}s = \frac{\int_a^b sg(s)\,\mathrm{d}s}{M_g}.$$

Then

$$R_{\varphi,g} = M_g^2\int_{a_0}^{b_0}\left(\int_{E_g/M_g}^{s}(s-u)\varphi_0''\left(\frac{uM_g}{M_\varphi}\right)\frac{1}{M_\varphi^2}\,\mathrm{d}r\right)g_0(s)\,\mathrm{d}s.$$

Let us look at the case where $a_0 \leqslant E_g/M_g \leqslant b_0$. Splitting the outer integral at E_g/M_g and using the estimate for φ_0'',

$$\begin{aligned}|R_{\varphi,g}|\left(\frac{M_\varphi}{M_g}\right)^2\frac{1}{\Lambda} &\leqslant \int_{a_0}^{E_g/M_g}\left(\int_{s}^{E_g/M_g}(u-s)\,\mathrm{d}u\right)g_0(s)\,\mathrm{d}s \\ &\quad + \int_{E_g/M_g}^{b_0}\left(\int_{E_g/M_g}^{s}(s-u)\,\mathrm{d}u\right)g_0(s)\,\mathrm{d}s \\ &= \int_{a_0}^{E_g/M_g}\frac{1}{2}\left[\left(\frac{E_g}{M_g}\right) - s\right]^2 g_0(s)\,\mathrm{d}s \\ &\quad + \int_{E_g/M_g}^{b_0}\frac{1}{2}\left[s - \left(\frac{E_g}{M_g}\right)\right]^2 g_0(s)\,\mathrm{d}s \\ &= \frac{1}{2}\int_{a_0}^{b_0}\left[s - \left(\frac{E_g}{M_g}\right)\right]^2 g_0(s)\,\mathrm{d}s \\ &= \frac{1}{2}\left(\int_{a_0}^{b_0} s^2 g_0(s) - \left[\frac{E_g}{M_g}\right]^2\right).\end{aligned}$$

In the other two cases we obtain the same result. □

B.6 Volterra Integral Equations

In Section 14.1 we derive the renewal equation for the population birth rate B of an age-structured population:

$$B(t) = \int_0^t \beta(a)B(t-a)\mathcal{F}(a)\,\mathrm{d}a + B_0(t),$$
$$B_0(t) = \int_0^\infty \beta(a+t)\mathcal{F}(a+t)\frac{u_0(a)}{\mathcal{F}(a)}\,\mathrm{d}a, \quad t \geqslant 0.$$

Here $u_0 \in L_1[0,c]$ with $L_1[0,c]$ denoting the space of Lebesgue integrable functions on $[0,c]$ (or rather their equivalence classes) and $c \in (0,\infty]$ being the maximum life span such that the survival function $\mathcal{F}$ satisfies $\mathcal{F}(a) > 0$ for $a \in [0,c)$ and $\mathcal{F}(a) = 0$ for $a > c$. We use the convention $u_0(a)/\mathcal{F}(a) := 0$ for $a \geqslant c$. We assume that $\beta \in L_1[0,c]$. We want to show that the renewal equation has a unique solution in $L_{1,\mathrm{loc}}[0,\infty)$, the space of functions that are integrable on every finite interval in $[0,\infty)$. We first notice that $B_0 \in L_1[0,\infty)$. Indeed, by Tonelli's theorem (e.g., McDonald, Weiss, 1999, Theorem 4.16),

$$\int_0^\infty |B_0(t)|\,\mathrm{d}t \leqslant \int_0^c \left(\int_0^\infty |\beta(a+t)|\frac{\mathcal{F}(a+t)}{\mathcal{F}(a)}\,\mathrm{d}t\right)|u_0(a)|\,\mathrm{d}a.$$

Since $\mathcal{F}$ is nonincreasing,

$$\int_0^\infty |B_0(t)|\,\mathrm{d}t \leqslant \int_0^c \left(\int_a^c |\beta(t)|\,\mathrm{d}t\right)|u_0(a)|\,\mathrm{d}a$$
$$\leqslant \int_0^c |\beta(t)|\,\mathrm{d}t \int_0^c |u_0(a)|\,\mathrm{d}a.$$

Setting $k(a) = \beta(a)\mathcal{F}(a)$, the renewal equation becomes the linear Volterra integral equation (of convolution form)

$$B(t) = \int_0^t B(t-a)k(a)\,\mathrm{d}a + B_0(t), \quad t \geqslant 0,$$

with $B_0, k \in L_1[0,\infty)$.

Existence and Uniqueness of Solutions

The nonlinear demographic model in Chapter 16 leads to a nonlinear convolution equation

$$B(t) = \int_0^t f(B(t-a))k(a)\,\mathrm{d}a + B_0(t), \quad t \geqslant 0. \qquad \text{(VIE)}$$

For simplicity, we consider the case where $f : \mathbb{R} \to \mathbb{R}$ is globally Lipschitz continuous, i.e.,

$$|f(x) - f(y)| \leqslant \Lambda|x - y| \quad \forall x, y \in \mathbb{R},$$

with Lipschitz constant Λ.

Theorem B.37. *Let $B_0, k \in L_{1,\mathrm{loc}}[0, \infty)$ and f be globally Lipschitz continuous. Then there exists a unique solution of (VIE) on $[0, \infty)$.*

Proof. We want to apply the contraction mapping principle (Banach's fixed-point theorem) (see, for example, McDonald, Weiss, 1999, Theorem 8.3). To this end, we choose an arbitrary number $r \in (0, \infty)$ and endow $X = L_1[0, r]$ with the exponentially weighted norm,

$$\|B\|_\lambda = \int_0^r \mathrm{e}^{-\lambda t}|B(t)|\,\mathrm{d}t, \quad \lambda \geqslant 0.$$

Since

$$\|B\|_\lambda \leqslant \|B\|_0 \leqslant \mathrm{e}^{\lambda r}\|B\|_\lambda,$$

this norm is equivalent to the standard norm of $L_1[0, r]$ and so makes $X = L_1[0, r]$ a Banach space, in particular a complete metric space. We define

$$\Phi(B)(t) = \int_0^t f(B(t - a))k(a)\,\mathrm{d}a + B_0(t), \quad t \in [0, r], \quad B \in L_1[0, r].$$

Notice that

$$|\Phi(0)(t)| \leqslant |f(0)| \int_0^t |k(a)|\,\mathrm{d}a + |B_0(t)|,$$

so $\Phi(0) \in L_1[0, r]$. Furthermore, by the Lipschitz continuity of f,

$$|\Phi(B)(t) - \Phi(\tilde{B})(t)| \leqslant \int_0^t \Lambda|B(t - a) - \tilde{B}(t - a)|\,|k(a)|\,\mathrm{d}a.$$

So,

$$\int_0^r \mathrm{e}^{-\lambda t}|\Phi(B)(t) - \Phi(\tilde{B})(t)|\,\mathrm{d}t$$
$$\leqslant \Lambda \int_0^r \left(\int_0^t \mathrm{e}^{-\lambda(t-a)}|B(t - a) - \tilde{B}(t - a)|\mathrm{e}^{-\lambda a}|k(a)|\,\mathrm{d}a \right) \mathrm{d}t.$$

Changing the order of integration (Tonelli's theorem (McDonald, Weiss, 1999, Theorem 4.16)),

$$\begin{aligned}
&\int_0^r e^{-\lambda t}|\Phi(B)(t) - \Phi(\tilde{B})(t)|\,dt \\
&\quad \leqslant \Lambda \int_0^r \left(\int_a^r e^{-\lambda(t-a)}|B(t-a) - \tilde{B}(t-a)|\,dt \right) e^{-\lambda a}|k(a)|\,da \\
&\quad = \Lambda \int_0^r \left(\int_0^{r-a} e^{-\lambda s}|B(s) - \tilde{B}(s)|\,ds \right) e^{-\lambda a}|k(a)|\,da \leqslant \Lambda \|B - \tilde{B}\|_\lambda \|k\|_\lambda.
\end{aligned}$$

This shows that Φ maps $L_1[0, r]$ into itself and

$$\|\Phi(B) - \Phi(\tilde{B})\|_\lambda \leqslant \Lambda \|B - \tilde{B}\|_\lambda \|k\|_\lambda.$$

So, Φ is Lipschitz continuous with Lipschitz constant $\Lambda\|k\|_\lambda$. By Lebesgue's theorem of dominated convergence (e.g., McDonald, Weiss, 1999, Theorem 4.9), we have $\|k\|_\lambda \to 0$ as $\lambda \to \infty$; so, choosing λ large enough, we achieve that $\Lambda\|k\|_\lambda < 1$, i.e., Φ becomes a strict contraction on $L_1[0, r]$ with $\|\cdot\|_\lambda$. The contraction mapping theorem implies that Φ has a unique fixed point, B_r, in $L_1[0, r]$. Since $r > 0$ was arbitrary and every fixed point of Φ is a solution of our convolution equation and vice versa, we obtain a unique integrable solution of our convolution equation on every interval $[0, r]$.

If $t \geqslant 0$, we set $B(t) = B_r(t)$ for some $r > t$. Notice that every solution of the convolution equation on some interval $[0, r]$ is also a solution of the convolution equation on $[0, s]$ for every $s \in [0, r]$. So, by uniqueness, this definition does not depend on the choice of r, and we obtain a locally integrable solution B of the convolution equation on $[0, \infty)$. Since its restriction to $[0, r]$ is a solution on $[0, r]$, it is uniquely determined. □

Large-Time Behavior of Solutions

We now consider the scalar integral equation

$$\left.\begin{aligned}
x(t) = x_\diamond(t) + \int_{[0,t]} f(x(t-s))P(ds), \quad t \geqslant 0, \\
x_\diamond(t) \to 0, \quad t \to \infty,
\end{aligned}\right\} \tag{IE}$$

where f is a continuous real function and P a probability measure on $[0, \infty)$, i.e., P is a nonnegative Borel measure with $P([0, \infty)) = 1$, and where $x_\diamond$ is a given function. To make sense, this equation tacitly assumes that $x, x_\diamond$: $[0, \infty) \to \mathbb{R}$ are Borel measurable and that x takes values in the domain of f. We are interested in the large-time behavior of a solution x, so we consider

$$x_\infty = \liminf_{t\to\infty} x(t), \qquad x^\infty = \limsup_{t\to\infty} x(t).$$

Lemma B.38. *Let $f : I \to \mathbb{R}$ be a continuous function defined on a closed interval I. If $x : [0, \infty) \to I$ is a bounded solution of (IE), then $[x_\infty, x^\infty] \subseteq f([x_\infty, x^\infty])$ and the interval $[x_\infty, x^\infty]$ contains a fixed point of f.*

Proof. Choose a sequence $t_n \to \infty, n \to \infty$, such that $x(t_n) \to x_\infty$ as $n \to \infty$. By the Lebesgue–Fatou lemma (cf. Toolbox B.2),

$$\int_{[0,\infty)} \liminf_{n\to\infty} f(x(t_n - s))P(\mathrm{d}s) \leqslant x_\infty \leqslant x^\infty$$
$$\leqslant \int_{[0,\infty)} \limsup_{n\to\infty} f(x(t_n - s))P(\mathrm{d}s).$$

Since, for every $s \in [0, \infty)$,

$$x_\infty \leqslant \liminf_{n\to\infty} x(t_n - s) \leqslant \limsup_{n\to\infty} x(t_n - s) \leqslant x^\infty$$

and f is continuous,

$$\liminf_{n\to\infty} f(x(t_n - s)), \limsup_{n\to\infty} f(x_n(t - s)) \in f([x_\infty, x^\infty]).$$

By the intermediate value theorem, $f([x_\infty, x^\infty])$ is an interval which, moreover, is compact, i.e.,

$$f([x_\infty, x^\infty]) = [r, s]$$

with $r, s \in \mathbb{R}$. Since P is a probability measure,

$$x_\infty \geqslant \int_{[0,\infty)} rP(\mathrm{d}s) = r,$$

and similarly $x^\infty \leqslant s$. This implies

$$[x_\infty, x^\infty] \subseteq [r, s] = f([x_\infty, x^\infty]).$$

Finally, $x_\infty = f(z)$ for some $z \in [x_\infty, x^\infty]$, so $f(z) \leqslant z$ for this z. Similarly, $f(y) \geqslant y$ for some $y \in [x_\infty, x^\infty]$. By the intermediate value theorem, there exists some x^* between z and y such that $f(x^*) = x^*$, in particular $x^* \in [x_\infty, x^\infty]$. □

Proposition B.39. *Let $f : I \to \mathbb{R}$ be a continuous function on a closed interval I mapping an interval $[a_0, b_0]$ in I into itself. Let $x : [0, \infty) \to I$ be a bounded solution of (IE) such that $a_0 \leqslant x_\infty < x^\infty \leqslant b_0$.*

Then

- *f has at least two fixed points in $[a_0, b_0]$, or*
- *f has at least one fixed point x^* and the iterate f^2 has at least one fixed point y^* such that $a_0 \leqslant y^* < x^* < f(y^*) \leqslant b_0$ (with $f(y^*)$ being a fixed point of f^2 as well).*

Proof. By Lemma B.38 and the assumptions of this proposition,

$$[x_\infty, x^\infty] \subseteq [a_0, b_0] \cap f([x_\infty, x^\infty]).$$

So, we are in the situation of Proposition 9.3 and the assertion follows. □

Theorem B.40. *Let $f : [0, \infty) \to [0, \infty)$ be continuous and $f(x) > 0$ for all $x > 0$. Assume that f has a unique fixed point $x^* > 0$ (with 0 possibly being another fixed point) and that f^2 has no fixed point in $(0, \infty)$ other than x^*. Furthermore, assume that there exist x_1 and x_2 such that*

$$0 < x_1 < x^* < x_2 < \infty, \qquad f(x_1) > x_1, \qquad f(x_2) < x_2.$$

Then all nonnegative solutions x of (IE) which are bounded and bounded away from 0 satisfy

$$x(t) \to x^*, \quad t \to \infty.$$

Proof. Suppose the statement is wrong. Then there is a solution x of (IE) such that $0 < x_\infty < x^\infty < \infty$. By Proposition 9.5, we find a_0 and b_0 such that $0 < a_0 < x_\infty < x^\infty < b_0 < \infty$ and f maps the interval $[a_0, b_0]$ into itself. By Proposition B.39, f has two positive fixed points or f^2 has a positive fixed point different from x^*, a contradiction. □

B.7 Critical and Regular Values of a Function

Let I be an interval with endpoints $a < b$, let $\breve{I}$ denote the open interval (a, b) and $f : I \to \mathbb{R}$ be continuous. f is called *piecewise continuously differentiable* if there exists a partition $a = t_0 < \cdots < t_k = b$ such that f is continuously differentiable on all intervals (t_{j-1}, t_j), $j = 1, \ldots, k$.

$t \in I$ is called a *critical point* of f if one of the following holds:

(i) $t \in I \setminus \breve{I}$.

(ii) $t \in \breve{I}$, and f is not differentiable at t.

(iii) $t \in \breve{I}$, f is differentiable at t, but f' is not continuous at t.

(iv) $t \in \breve{I}$, f is differentiable at t, f' is continuous at t, and $f'(t) = 0$.

A number $r \in \mathbb{R}$ is called a *critical value* of f if there exists a critical point $t \in I$ such that $r = f(t)$. r is called a *regular value* of f if it is not a critical value of f.

Proposition B.41. *Let I be a bounded closed interval, and let $f : I \to \mathbb{R}$ be continuous and piecewise continuously differentiable. Then the set of critical values of f is closed and has Lebesgue measure 0. So the set of regular values is dense in $\mathbb{R}$ and open.*

Proof. Let $r \in \mathbb{R}$ be the limit of critical values r_n. Then there exist critical points $t_n \in I$ such that $f(t_n) \to r$ as $n \to \infty$. After choosing a subsequence, $t_n \to t \in I$ as $n \to \infty$. Since f is continuous, $r = f(t)$. Without loss of generality we can assume that f is continuously differentiable at t. Since f is piecewise continuously differentiable, we can also assume that f is continuously differentiable at all t_n. Since all t_n are critical points, $f'(t_n) = 0$. Since f' is continuous at t, $f'(t) = 0$ and t is a critical point and r a critical value.

Now choose a partition $a = t_0 < \cdots < t_k = b$ such that f is continuously differentiable on all subintervals (t_{j-1}, t_j). If r is a critical value, either it is contained in the finite set $\{f(t_1), \ldots, f(t_k)\}$ or it is a critical value of the restriction of f to one of the subintervals (t_{j-1}, t_j). It is a well-known special case of Sard's lemma that the critical values of a continuously differentiable function on an open set in $\mathbb{R}$ form a set of Lebesgue measure 0 (see, for example, Deimling, 1985, Proposition 1.4). So, the critical values of f are contained in the finite union of sets of Lebesgue measure 0 and so form a set of Lebesgue measure 0. □

If I is an unbounded interval and $f : I \to \mathbb{R}$ is continuous, then f is called *piecewise continuously differentiable* if the restriction of f to any bounded subinterval is piecewise differentiable.

Theorem B.42. *Let I be a nonempty interval, $f : I \to \mathbb{R}$ continuous and piecewise continuously differentiable. Then the critical values of f form a set that has Lebesgue measure* 0. *The set of regular values contains a dense set that is the countable intersection of open sets.*

Proof. We can write I as the countable union of bounded closed intervals I_j. Let f_j denote the restriction of f to I_j. Then the set of critical values of f is contained in the union of the sets of critical values of f_j and so in a countable union of closed sets with Lebesgue measure 0. The countable union also has Lebesgue measure 0. The set of regular values contains the complement of this countable union, and this complement is dense and the countable intersection of open sets. □

Bibliographic Remarks

There are three standard approaches to the Stieltjes integral: the Stieltjes integral for regulated functions; the Riemann–Stieltjes integral; and the Lebesgue–Stieltjes integral. The first has been presented here in great detail, because it appears to be technically the easiest and has far-reaching generalizations to vector-valued functions (Diekmann, Gyllenberg, Thieme, 1993, 1995), which, to my taste, are less technically involved than the generalization of the Riemann–Stieltjes integral to vector-valued functions (Hönig, 1975). For the Riemann–Stieltjes integral we also refer to Kirkwood (1989, 1995) and Wade (1995), and for the Lebesgue and Riemann–Stieltjes integral to Hewitt, Stromberg (1969). In the long run, the Lebesgue–Stieltjes integral (see also McDonald, Weiss, 1999) is the more useful tool, but it involves the elaborate concepts of measure theory (Berberian, 1965; Folland, 1984; Hewitt, Stromberg, 1969; Aliprantis, Burkinshaw, 1998; McDonald, Weiss, 1999). Our Toolbox B.3 on convex analysis contains only a part of the Krein–Milman theorem. For the full theorem and its applications see Edwards (1965, 1995) and McDonald, Weiss (1999). More on (in particular nonlinear) optimization can be found in Zeidler (1985). A proof of Bernstein's theorem (Theorem B.13) can be found in Widder (1971, p. 155). Bernstein's theorem has been extended to ordered Banach spaces by Arendt (1994) (see also Thieme, 1996; Arendt et al., 2001). A compendium on Volterra integral equations is the book by Gripenberg, Londen, Staffans (1990). Stieltjes convolutions (Section B.4) can be generalized to Banach spaces (Thieme, Voßeler, 2002).

Appendix C

Some MAPLE Worksheets with Comments for Part 1

The subsequent MAPLE worksheets have been used to produce some of the figures in Part 1. They contain both the commands one needs to type (with explanations) and the MAPLE replies which appear on the screen. I consulted the book by Kofler (1997).

C.1 Fitting the Growth of the World Population (Figure 3.1)

This MAPLE worksheet leads to be construction of Figure 3.1 in Chapter 3, linear and exponential fits to the growth of the world population from 1950 to 1985 according to data from Keyfitz, Flieger (1990, p. 105). We first let MAPLE forget assignments for variables from previous work.

```
restart;
```

We take a look at a graphical display of the data. To this end we make lists of the population sizes in billions and of the corresponding years.

```
dataN:=[2.515 312, 2.751 559, 3.019 376, 3.335 927,
        3.697 918, 4.079 753, 4.450 210, 4.853 848];
```

$$dataN := [2.515\,312, 2.751\,559, 3.019\,376, 3.335\,927, 3.697\,918, 4.079\,753, 4.450\,210, 4.853\,848]$$

```
datat:=[1950, 1955, 1960, 1965, 1970, 1975, 1980, 1985];
```

$$datat := [1950,\ 1955,\ 1960,\ 1965,\ 1970,\ 1975,\ 1980,\ 1985]$$

In order to plot the data, we combine the two lists into one.

```
data:=zip((x,y)->[x,y], datat, dataN);
```

$$data := [[1950, 2.515\,312],[1955, 2.751\,559],[1960, 3.019\,376], [1965, 3.335\,927],[1970, 3.697\,918],[1975, 4.079\,753], [1980, 4.450\,210],[1985, 4.853\,848]]$$

Since we want to use the plot of the source data several times, we give it a name.

```
p:=plot(data, style=point, color=black):
```

We call the plot subroutine and plot.

```
with(plots):
display(p);
```

We also want to look at the natural logarithms of the population sizes. The "map" command allows us to obtain the logarithms of the whole list in one go.

```
datalnN:=map(ln, dataN);
```

$$datalnN := [.922\,396\,851\,5,\ 1.012\,167\,660,\ 1.105\,050\,188,\ 1.204\,750\,602,\ 1.307\,769\,959,\ 1.406\,036\,447,\ 1.492\,951\,286,\ 1.579\,771\,792]$$

```
dataln:=zip((x,y)->[x,y], datat, datalnN);
```

$$dataln := [[1950,\ .922\,396\,851\,5],[1955,\ 1.012\,167\,660],[1960,\ 1.105\,050\,188],[1965,\ 1.204\,750\,602],[1970,\ 1.307\,769\,959],[1975,\ 1.406\,036\,447],[1980,\ 1.492\,951\,286],[1985,\ 1.579\,771\,792]]$$

```
pln:=plot(dataln, style=point, color=black):
display(pln);
```

In order to fit a line through the original population data, we summon the statistics subprogram and the fit subroutine.

```
with(stats):
```

Warning, these names have been redefined: anova, describe, fit, importdata, random, statevalf, statplots, transform

```
with(fit):
```

Warning, the names leastmediansquare and leastsquare have been redefined

We ignore the warnings. Before we fit, we shift the time such that 1950 becomes 0.

```
datax:=[0, 5, 10, 15, 20, 25, 30, 35];
```

$$datax := [0,\ 5,\ 10,\ 15,\ 20,\ 25,\ 30,\ 35]$$

We first fit a line to the original population data and plot the result.

```
leastsquare[[x,y], y=a0+a1*x, a0, a1]([datax, dataN]);
```

$$y = 2.404\,399\,167 + .067\,633\,640\,48x$$

```
f:=rhs(%);
```

$$f := 2.404\,399\,167 + .067\,633\,640\,48x$$

```
f1:=unapply(f,x);
```

$$f1 := x{->}\ 2.404\,399\,167 + .067\,633\,640\,48x$$

We prepare the plot with the linear fit.

```
p1:=plot([1950+x, f1(x), x=-5..40], color=black, title=` linear fit ` ):
```

The command "1950+x" in square brackets reverses the shift in time. We plot the original data and the linear fit together.

```
display([p,p1]);
```

This gives us the left-hand side of Figure 3.1. Next we fit a line to the logarithms of the population data and plot the results.

```
leastsquare[[x,y], y=b0+b1*x, b0,b1]([datax, datalnN]);
```

$$y = .920\,048\,476\,7 + .019\,075\,049\,76x$$

In order to get the exponential fit to the original data, we give a name to this expression and then exponentiate it.

```
g:=rhs(%);
```

$$g := .920\,048\,476\,7 + .019\,075\,049\,76x$$

```
g1:=exp(g);
```

$$g1 := \exp(.920\,048\,476\,7 + .019\,075\,049\,76x)$$

We prepare the plot of the exponential fit. Again the command "1950+x" reverses the shift in time.

```
p2:=plot([1950+x, g1, x=-5..40], color=black, title=`exponential
fit` ):
```

We plot the original data and the exponential fit.

```
display([p,p2]);
```

This provides the right-hand side of Figure 3.1. To obtain the Malthusian parameter, we differentiate g with respect to x.

```
r:=diff(g,x);
```

$$r := .019\,075\,049\,76$$

We calculate the doubling time in years.

```
double:=ln(2)/r;
```

$$double := 52.424\,502\,82\ln(2)$$

```
double:=evalf(%);
```

$$double := 36.337\,896\,32$$

C.2 Periodic Modulation of Exponential Growth in Closed Populations (Figures 3.2 and 3.3)

We first erase prior variable assignments from the MAPLE memory.

```
restart;
```

We introduce the per capita birth rate.

```
f:=(1+cos(2*Pi*t))^8;
```

$$f := (1 + \cos(2\pi t))^8$$

We calculate the time average over one period.

```
fav:=int(f,t=0..1);
```

$$fav := \frac{6435}{128}$$

We divide by the time average to normalize the per capita birth rate to have time average 1.

```
f0:=f/fav;
```

$$f0 := \frac{128}{6435}(1 + \cos(2\pi t))^8$$

We introduce the normalized per capita death rate.

```
g:=1+.5*sin(2*Pi*t-Pi/2);
```

$$g := 1 - .5\cos(2\pi t)$$

We plot the normalized per capita birth and death rates and obtain Figure 3.2A.

```
p1:=plot([f0,g], t=0..3, title=`Per capita birth and death rates`,
        labels=[`time`,` `], axes=frame, color=black):
p2:=textplot([.5,1.7,`death`]):
p3:=textplot([1.2,4.9,`birth`]):
with(plots):
display([p1,p2,p3]);
```

We work towards the immigration rate.

```
k:=(cos(Pi*t))^8;
```

$$k := \cos(\pi t)^8$$

We introduce the immigration rate.

```
m:=2+10*k/(1+20*k);
```

$$m := 2 + \frac{10\cos(\pi t)^8}{1 + 20\cos(\pi t)^8}$$

We plot the immigration rate and obtain Figure 3.2B.

```
p4:=plot(m, t=0..3, x=0..3, title=` Immigration rate` ,
        labels=[` time` ,` ` ], color=black):
display(p4);
```

We introduce the intrinsic growth rates we use in the cases A, B, C in Figure 3.3. We write r0 for ρ_{A}, rp for ρ_{B} and rm for ρ_{C}.

```
r0:=unapply(f0-g,t);
```

$$r0 := t \to \frac{128}{6435}(1+\cos(2\pi t))^8 - 1 + .5\cos(2\pi t)$$

```
rp:=unapply(f0-.7*g,t);
```

$$rp := t \to \frac{128}{6435}(1+\cos(2\pi t))^8 - .7 + .35\cos(2\pi t)$$

```
rm:=unapply(f0-1.9*g,t);
```

$$rm := t \to \frac{128}{6435}(1+\cos(2\pi t))^8 - 1.9 + .95\cos(2\pi t)$$

We introduce the initial data: time=0, population size=1.

```
start:=[0,1];
```

$$start := [0, 1]$$

We introduce the differential equations for the various intrinsic growth rates, prepare their plots and plot them.

```
eq0:=diff(y(t),t)=r0(t)*y(t);
```

$$eq0 := \frac{\partial}{\partial t}y(t) = \left(\frac{128}{6435}(1+\cos(2\pi t))^8 - 1 + .5\cos(2\pi t)\right)y(t)$$

The command "DEplot" only works in the DEtools subroutine, which we summon with the next command.

```
with(DEtools):
p0:=DEplot(eq0,[y(t)], t=0..10, y=0..1.5, stepsize=0.01, {start},
        arrows=NONE, thickness=1, linecolor=black,
        axes=frame, title=` A` ):
```

The command "eq0" specifies the equation, "[y(t)]" the variable we want to plot, "t=0..10" the time interval, "y=0..1.5" the window in the y-direction; "stepsize=0.01" specifies the steps for the numerical procedure, and {start} calls the initial data we have specified before. The command "arrows=NONE" keeps the direction field from showing, "thickness=1" gives us the thinnest possible line of the solution graph, "linecolor=black" makes the solution graph black, "axes=frame" moves the axes away from the graph, and "title=` A` " provides A in the picture.

```
with(plots):
```

Warning, the name changecoords has been redefined

```
display([p0]);
eqp:=diff(y(t),t)=rp(t)*y(t);
```

$$eqp := \frac{\partial}{\partial t} y(t) = \left(\frac{128}{6435}(1+\cos(2\pi t))^8 - .7 + .35\cos(2\pi t) \right) y(t)$$

```
pp:=DEplot(eqp, [y(t)], t=0..10, stepsize=0.01, {start},
        arrows=NONE, thickness=1, linecolor=black,
        axes=frame, title=` B` ):
display(pp);
eqm:=diff(y(t),t)=rm(t)*y(t);
```

$$eqm := \frac{\partial}{\partial t} y(t) = \left(\frac{128}{6435}(1+\cos(2\pi t))^8 - 1.9 + .95\cos(2\pi t) \right) y(t)$$

```
pm:=DEplot(eqm, [y(t)], t=0..10, stepsize=0.01, {start},
        arrows=NONE, thickness=1, linecolor=black,
        axes=frame, title=` C` ):
display(pm);
```

C.3 Fitting Sigmoid Population-Growth Curves (Figures 6.1 and 6.2)

In Section 6.2 we fitted the logistic equation and the Smith equation to the population-growth data of the bacterium *Escheria coli* from McKendrick, Kesava (1911). Here are detailed MAPLE worksheets showing how this can be done.

We first take a look at a graphical display of the data. To this end we make a list of the times at which measurements were taken (in hours). But first we erase possible previous work from the MAPLE memory.

```
restart;
datax:=[0, 0.5, 1, 2, 3, 4, 5, 6, 7, 8];
```

$$datax := [0,\ .5,\ 1,\ 2,\ 3,\ 4,\ 5,\ 6,\ 7,\ 8]$$

We also make a list of the number of bacteria that were counted in the tube (in millions).

```
datay:=[.176, .280, .60Ϝ, 3.87, 28.2, 74.2, 127., 150., 149., 154.];
```

$$datay := [.176, .280, .608, 3.87, 28.2, 74.2, 127., 150., 149., 154.]$$

In order to plot the data, we combine the two lists into one.

```
data:=zip((x,y)->[x,y], datax, datay);
```

$$data := [[0, .176],[.5, .280],[1, .608],[2, 3.87],[3, 28.2], [4, 74.2],[5, 127.],[6, 150.],[7, 149.],[8, 154.]]$$

Before we plot we need to alert MAPLE that we intend to do so.

```
with(plots):
```

Warning, the name changecoords has been redefined

Since we want to use the plot of the source data several times, we give it a name.

```
p1:=plot(data,x=-1..9, y=-10..165, style=point, labels=[` time` ,
        `` ], title=` A` ):
```

Finally, we have a look at the data.

```
display(p1);
```

We obtain a figure that only contains the dotted parts of Figure 6.1. In order to check whether there is a reasonable chance of being able to fit the data using any of the population curves introduced in Chapter 5, we graph the natural logarithms of the population numbers against time. MAPLE's "map" command allows us to do this for the whole data list at once.

```
datalny:=map(ln, datay);
```

$$datalny := [-1.737\,271\,284, -1.272\,965\,676, -.497\,580\,397\,0, 1.353\,254\,507, 3.339\,321\,978, 4.306\,764\,150, 4.844\,187\,086, 5.010\,635\,294, 5.003\,946\,306, 5.036\,952\,602]$$

```
dataln:=zip((x,y)->[x,y], datax, datalny);
```

$$dataln := [[0, -1.737\,271\,284], [.5, -1.272\,965\,676], [1, -.497\,580\,397\,0], [2, 1.353\,254\,507], [3, 3.339\,321\,978], [4, 4.306\,764\,150], [5, 4.844\,187\,086], [6, 5.010\,635\,294], [7, 5.003\,946\,306], [8, 5.036\,952\,602]]$$

We plot the logarithmic data.

```
plot(dataln, style=point);
```

The curve we obtain is essentially concave, though it shows a slight convexity at the beginning and the end. This plot is not shown in the text.

Fitting the Logistic Equation to *Escheria coli* Data

Since the present versions of MAPLE have a problem fitting parameters that are located in the argument of a nonlinear function, we need to make a reasonable guess for the carrying capacity K. Apparently, it should be slightly larger than 154. For the logistic equation to be a good fit, the inflection point should be $\frac{1}{2}K$. Now the data points are too scarce to locate the inflection point. For lack of a better idea, we fit a cubic through the data points and determine the inflection points of the cubic. To this end we activate the statistics subprogram.

```
with(stats):
```

Warning, these names have been redefined: anova, describe, fit, importdata, random, statevalf, statplots, transform

```
with(fit);
```

Warning, the names leastmediansquare and leastsquare have been redefined [leastmediansquare, leastsquare]

```
leastsquare[[x,y], y=a0+a1*x+a2*x^2+a3*x^3,
            {a0,a1,a2,a3}]([datax,datay]);
```

$$y = 5.728\,781\,064 - 27.999\,477\,12x + 17.061\,844\,65x^2 - 1.414\,103\,320x^3$$

For further processing, we give the right-hand side a name.

```
g:=rhs(%);
```

$$g := 5.728\,781\,064 - 27.999\,477\,12x + 17.061\,844\,65x^2 - 1.414\,103\,320x^3$$

We take a look at how well the cubic fits the data.

```
p2:=plot(g,x=0..8):

display([p1,p2]);
```

This looks rather gruesome, but this was to be expected. To find the inflection point of the cubic, we look for the zero of its second derivative and substitute it back into the cubic.

```
g2:=diff(g,x$2);
```

$$g2 := 34.123\,689\,30 - 8.484\,619\,920x$$

```
x1:=fsolve(g2=0,x);
```

$$x1 := 4.021\,828\,865$$

```
L:=subs(x=x1,g);
```

L:=77.104 455 80

$2L \approx 154.2$ is slightly bigger than 154, so we may try $K = 154.2$. If this does not give a good fit, we can modify it in a second attempt. We should be careful, however, to set up the MAPLE worksheet in such a way that we can easily repeat the procedure.

```
K:=154.2;
```

$$K := 154.2$$

It is convenient to normalize the population data with respect to the carrying capacity.

```
f:=y/K;
```

$$f := .006\,485\,084\,306y$$

```
f1:=unapply(f,y);
```

$$f1 := y \to -.006\,485\,084\,306y$$

```
datay1:=map(f1, datay);
```

$$datay1 := [.001\,141\,374\,838, .001\,815\,823\,606, .003\,942\,931\,258, .025\,097\,276\,26, .182\,879\,377\,4, .481\,193\,255\,5, .823\,605\,706\,9, .972\,762\,645\,9, .966\,277\,561\,6, .998\,702\,983\,1]$$

We want to determine an optimal fit of the parameters k and ρ in the equation $k + \rho t = \ln y - \ln(1 - y)$. We need to transform our data correspondingly.

```
g:=ln(z)-ln(1-z);
```

$$g := \ln(z) - \ln(1 - z)$$

```
g1:=unapply(g,z);
```

$$g1 := z \to \ln(z) - \ln(1 - z)$$

```
datay2:=map(g1,datay1);
```

$$datay2 := [-6.774\,379\,718, -6.309\,398\,663, -5.531\,880\,133, -3.659\,578\,370, -1.496\,959\,929, -.075\,262\,484\,6, 1.540\,970\,113, 3.575\,550\,768, 3.355\,287\,680, 6.646\,390\,485]$$

These data, when plotted against time, should form a straight line. Let us check whether they do.

```
data2:=zip((x,y)->[x,y], datax, datay2);
```

$$data2 := [[0,-6.774\,379\,718], [.5,-6.309\,398\,663], [1,-5.531\,880\,133],[2,-3.659\,578\,370], [3,-1.496\,959\,929],[4,-.075\,262\,484\,6], [5,1.540\,970\,113],[6,3.575\,550\,768], [7,3.355\,287\,680], [8,6.646\,390\,485]]$$

```
p4:=plot(data2, style=point):

display(p4);
```

The curve we get (not shown in the text) is not too far from a straight line; we see, however, that we should discard the last-but-one data point before we do our fit.

```
datax1:=datax[1..8];
```

$$datax1 := [0, .5, 1, 2, 3, 4, 5, 6]$$

```
datax2:=[op(datax1),8];
```

$$datax2 := [0, .5, 1, 2, 3, 4, 5, 6, 8]$$

```
datay3:=datay2[1..8];
```

$$datay3 := [-6.774\,379\,718,-6.309\,398\,663,-5.531\,880\,133, -3.659\,578\,370,-1.496\,959\,929, -.075\,262\,484\,6, 1.540\,970\,113,3.575\,550\,768]$$

```
datay4:=datay2[10];
```

$$datay4 := 6.646\,390\,485$$

```
datay5:=[op(datay3), datay4];
```

$$datay5 := [-6.774\,379\,718,-6.309\,398\,663,-5.531\,880\,133, -3.659\,578\,370,-1.496\,959\,929, -.075\,262\,484\,6, 1.540\,970\,113,3.575\,550\,768,6.646\,390\,485]$$

Before we start our fitting exercise, we activate the statistics subprogram.

```
with(stats):
with(fit):
leastsquare[[x,y], y=a0+a1*x, {a0,a1}] ([datax2, datay5]);
```

$$y = -7.004\,380\,749 + 1.727\,284\,027x$$

We read from this answer that $k \approx -7.004\,380\,749$ and $\rho \approx 1.727\,284\,027$. For further processing we give a name to the right-hand side of the previous expression.

```
h:=rhs(%);
```

$$h := -7.004\,380\,749 + 1.727\,284\,027x$$

We plot this line in order to see how well it fits our transformed data.

```
p5:=plot(h,x=-1..10):
display([p4,p5]);
```

We transform back to the original variables.

```
y:=solve(h=g, z);
```

$$y := \frac{\exp(-7.004\,380\,749 + 1.727\,284\,027x)}{1 + \exp(-7.004\,380\,749 + 1.727\,284\,027x)}$$

```
M:=K*y;
```

$$M := K\frac{\exp(-7.004\,380\,749 + 1.727\,284\,027x)}{1 + \exp(-7.004\,380\,749 + 1.727\,284\,027x)}$$

We compare this graph to the original data.

```
p8:=plot(M,x=-1..10):
display([p1,p8]);
```

This yields the left-hand side of Figure 6.1.

Fitting Smith's Equation to *Escheria coli* Data

We start a new MAPLE worksheet. We use the same data lists as before, but without the last-but-one entry.

```
restart;
datax:=[0, 0.5, 1, 2, 3, 4, 5, 6, 8];
```

$$datax := [0, .5, 1, 2, 3, 4, 5, 6, 8]$$

```
datay:=[.176, .280, .608, 3.87, 28.2, 74.2, 127., 150., 154.];
```

$$datay := [.176, .280, .608, 3.87, 28.2, 74.2, 127., 150., 154.]$$

We choose the carrying capacity as in the fit of the logistic equation.

```
K:=154.2;
```

$$K := 154.2$$

We prepare to normalize the data with respect to the carrying capacity.

```
f:=y/K;
```

$$f := .006\,485\,084\,306y$$

```
f1:=unapply(f,y);
```

$$f1 := y \rightarrow .006\,485\,084\,306y$$

Now we can perform the normalization.

```
datay1:=map(f1, datay);
```

$$\begin{aligned} datay1 := [&.001\,141\,374\,838, .001\,815\,823\,606,\\ &.003\,942\,931\,258, .025\,097\,276\,26,\\ &.182\,879\,377\,4, .481\,193\,255\,5, .823\,605\,706\,9,\\ &.972\,762\,645\,9, .998\,702\,983\,1] \end{aligned}$$

We prepare the least-squares fit of the equation $\ln y = k + rx + a\ln(1 - y)$.

```
datay2:=map(ln,datay1);
```

$$\begin{aligned} datay2 := [&-6.775\,521\,745, -6.311\,216\,137, -5.535\,830\,858,\\ &-3.684\,995\,954, -1.698\,928\,483, -.731\,486\,311\,0,\\ &-.194\,063\,374\,6, -.027\,615\,167\,05,\\ &-.001\,297\,858\,754] \end{aligned}$$

```
h:=ln(1-z);
```

$$h := \ln(1 - z)$$

```
h1:=unapply(h,z);
```

$$h1 := z \rightarrow \ln(1 - z)$$

```
dataz:=map(h1,datay1);
```

$$\begin{aligned} dataz := [&-.001\,142\,026\,664, -.001\,817\,474\,206,\\ &-.003\,950\,725\,147, -.025\,417\,583\,53,\\ &-.201\,968\,554\,1, -.656\,223\,826\,4, -1.735\,033\,488,\\ &-3.603\,165\,935, -6.647\,688\,344] \end{aligned}$$

We activate the statistics subprogram.

```
with(stats):
```

Warning, these names have been redefined: anova, describe, fit, importdata, random, statevalf, statplots, transform

```
with(fit):
```

Warning, the names leastmediansquare and leastsquare have been redefined

Ignoring the warning, we perform the least-squares fit.

```
leastsquare[[x,z,y], y=a0+a1*x+a2*z, { a0,a1, a2 }]
            ([datax, dataz, datay2]);
```

$$y = -7.057\,657\,711 + 1.765\,405\,962x + 1.050\,099\,717z$$

We extract the constants from the equation.

```
F:=rhs(%);
```

$$F := -7.057\,657\,711 + 1.765\,405\,962x + 1.050\,099\,717z$$

```
F1:=unapply(F,(x,z));
```

$$F1 := (x, z) \to -7.057\,657\,711 + 1.765\,405\,962x + 1.050\,099\,717z$$

```
k:=F1(0,0);
```

$$k := -7.057\,657\,711$$

```
r:=diff(F,x);
```

$$r := 1.765\,405\,962$$

```
a:=diff(F,z);
```

$$a := 1.050\,099\,717$$

We return to our equation.

```
eq:=k+r*x=-(ln(1-(M/K)))*a+ln(M/K);
```

$$eq := -7.057\,657\,711 + 1.765\,405\,962x = -1.050\,099\,717\ln(1 - .006\,485\,084\,306M) + \ln(.006\,485\,084\,306M)$$

MAPLE cannot give us a symbolic solution for M, but can give us one for x.

```
G:=solve(eq,x);
```

$$G := 3.997\,753\,414 \\ -.594\,820\,534\,0 \ln(1. - .006\,485\,084\,306M) \\ +.566\,441\,952\,5 \ln(.006\,485\,084\,306M)$$

We get ready to make a parametric plot to display the graph.

```
G1:=unapply(G,M);
```

$$G1 := M \to 3.997\,753\,414 \\ -.594\,820\,534\,0 \ln(1. - .006\,485\,084\,306M) \\ +.566\,441\,952\,5 \ln(.006\,485\,084\,306M)$$

This is the parametric plot.

```
with(plots):
p4:=plot([G(M),M, M=-10..170]):
```

To compare this curve with the original data, we extend the data sets from above.

```
datax1:=datax[1..8];
```

$$datax1 := [0, .5, 1, 2, 3, 4, 5, 6]$$

```
datax2:=[op(datax1), 7, 8];
```

$$datax2 := [0, .5, 1, 2, 3, 4, 5, 6, 7, 8]$$

```
datay3:=datay[1..8];
```

$$datay3 := [.176, .280, .608, 3.87, 28.2, 74.2, 127., 150.]$$

```
datay4:=[op(datay3),149,154];
```

$$datay4 := [.176,.280,.608,3.87,28.2,74.2,127.,150., \\ 149,154]$$

```
data4:=zip((x,y)->[x,y], datax2, datay4);
```

$$data4 := [[0,.176],[.5,.280],[1,.608],[2,3.87],[3,28.2], \\ [4,74.2],[5,127.],[6,150.],[7,149],[8,154]]$$

We plot the original data.

```
p3:=plot(data4, t=-1..10, N=-10..160, style=point, labels=[`time`,
      ``], title=`B`):
```

We show the original data and the fitted curve in one figure.

```
display([p3,p4]);
```

This yields the right-hand side of Figure 6.1.

C.4 Fitting Bernoulli's Equation to Population Data of Sweden (Figure 6.3)

We start by making a MAPLE list for the years for which data are available in Keyfitz, Flieger (1990).

```
restart;

datax1:=[1950,1955,1960,1965,1967,1970,1975,1980,1985];
```

$$datax1 := [1950,1955,1960,1965,1967,1970,1975,\\ 1980,1985]$$

We also make a MAPLE list for the associated population numbers in millions.

```
datay1:=[7.017, 7.262, 7.480, 7.734, 7.868, 8.043, 8.193,
    8.310, 8.350];
```

$$datay1 := [7.017, 7.262, 7.480, 7.734, 7.868, 8.043, 8.193, 8.310,\\ 8.350]$$

To plot these data, we combine them in one MAPLE list using the "zip" command.

```
data1:=zip((x,y)->[x,y], datax1, datay1);
```

$$data1 := [[1950,7.017],[1955,7.262],[1960,7.480],\\ [1965,7.734],[1967,7.868],[1970,8.043],\\ [1975,8.193],[1980,8.310], [1985,8.350]]$$

Before we plot, we call up the plot program.

```
with(plots):
```

We could plot directly, but later we will combine this plot with the fit, so we first give the plot a name.

```
p1:=plot(data1, style=point):
```

Now we finally want to see the plot in order to guess the values of K and L.

```
display(p1);
```

Looking at the graph we guess the values for K and L and calculate the value of K/L.

```
K:=8.4;
```

$$K := 8.4$$

```
L:=7.8;
```

$$L := 7.8$$

```
K/L;
```

$$1.076923077$$

Next we determine $\theta = w$.

```
w:=fsolve(1=(1+x)*(L/K)^x, x=1..100);
```

$$w := 54.098\,363\,52$$

Without adding "x=1..100" to the "fsolve" command, MAPLE would give us 0 as an answer.

Following the transformation of the Bernoulli equation to the logistic equation, we transform the data in two steps. First we normalize the population numbers with respect to the carrying capacity.

```
f:=(y/K);
```

$$f := .119\,047\,619\,0y$$

```
f1:=unapply(f, y);
```

$$f1 := y \rightarrow .119\,047\,619\,0y$$

The "map" command allows us to transform a list in one go.

```
datay2:=map(f1, datay1);
```

$$datay2 := [.835\,357\,142\,5, .864\,523\,809\,2, .890\,476\,190\,1, .920\,714\,285\,3, .936\,666\,666\,3, .957\,499\,999\,6, .975\,357\,142\,5, .989\,285\,713\,9, .994\,047\,618\,6]$$

We complete the transformation as follows.

```
g:=y^w;
```

$$g := y^{54.098\,363\,52}$$

```
g1:=unapply(g, y);
```

$$g1 := y \rightarrow y^{54.098\,363\,52}$$

```
datay3:=map(g1, datay2);
```

$$datay3 := [.000\,059\,348\,99222, .000\,379\,959\,1523, .001\,882\,129\,495, .011\,460\,796\,39, .029\,026\,757\,04, .095\,419\,908\,78, .259\,281\,903\,7, .558\,359\,175\,8, .723\,991\,021\,2]$$

Before we make our least-squares estimate, we need to shift the time and make yet another transformation of the dependent variable.

```
datax2:=[0,5,10,15,17,20,25,30,35];
```

$$datax2 := [0, 5, 10, 15, 17, 20, 25, 30, 35]$$

```
h:=ln(y/(1-y));
```

$$h := \ln\left(\frac{y}{1-y}\right)$$

```
h1:=unapply(h, y);
```

$$h1 := y \rightarrow \ln\left(\frac{y}{1-y}\right)$$

```
datay4:=map(h1, datay3);
```

$$datay4 := [-9.732\,016\,067, -7.875\,066\,773, -6.273\,467\,530, -4.457\,296\,100, -3.510\,080\,851, -2.249\,183\,604, -1.049\,704\,215, .234\,505\,502\,4, .964\,345\,593\,4]$$

We also need to activate the statistics subroutine.

```
with(stats):
```

Warning, these names have been redefined: anova, describe, fit, importdata, random, statevalf, statplots, transform

```
with(fit):
```

Warning, the names leastmediansquare and leastsquare have been redefined

Ignoring the warnings we go ahead with our least-squares estimate.

```
leastsquare[[x,z], z=a0+a1*x, {a0,a1}] ([datax2, datay4]);
```

$$z = -9.305\,258\,210 + .317\,193\,374\,8x$$

We could calculate y by hand, but we would rather let MAPLE do the work for us.

```
h2:=rhs(%);
```

$$h2 := -9.305\,258\,210 + .317\,193\,374\,8x$$

```
y:=solve(h=h2, y);
```

$$y := \frac{e^{(-9.305\,258\,210+.317\,193\,374\,8x)}}{e^{(-9.305\,258\,210+.317\,193\,374\,8x)} + 1.}$$

We transform back to the original dependent variable N.

```
N:=K*y^(1/w);
```

$$N := 8.4\left(\frac{e^{(-9.305\,258\,210+.317\,193\,374\,8x)}}{e^{(-9.305\,258\,210+.317\,193\,374\,8x)}+1.}\right)^{.018\,484\,847\,51}$$

The return shift in time is accomplished by a two-dimensional "plot" command. We want to show values from 1900 to 2010. Before that we must change the function type.

```
G:=unapply(N, x);
```

$$G := x \to 8.4\left(\frac{e^{(-9.305\,258\,210+.317\,193\,374\,8x)}}{e^{(-9.305\,258\,210+.317\,193\,374\,8x)}+1.}\right)^{.018\,484\,847\,51}$$

```
p5:=plot([x+1950, G(x), x=-50..60]):
display([p1, p5]);
```

This provides Figure 6.3.

C.5 Illustrating the Allee Effect (Figures 7.2–7.4)

First we will illustrate that, in the model of Chapter 7, it can depend on the initial conditions whether or not the population becomes extinct. We define the two nontrivial equilibria.

```
restart;
x1:=1;
```

$$x1 := 1$$

```
x2:=2;
```

$$x2 := 2$$

The parameter M is related to the equilibria as follows.

```
M:=1+x1+x2;
```

$$M := 4$$

We define the differential equation.

```
de:=diff(x(t),t)=x(t)/(M*(1+x(t)))*(x2-x(t))*(x(t)-x1);
```

$$de := \frac{d}{dt}x(t) = 1/4\frac{x(t)(2-x(t))(x(t)-1)}{1+x(t)}$$

We activate the differential equations subprogram.

```
with(DEtools):
```

The following command provides Figure 7.2.

```
phaseportrait(de, [x(t)], t=-4..25, [[x(0)=0.9], [x(0)=1.2],
                    [x(0)=2.5]], stepsize=0.05, dirgrid=[10,15],
                    linecolor=black);
```

Next we want to demonstrate that a temporary increase in the extra mortality experienced during mate search from predators can drive a population that would otherwise survive into extinction, provided that the increase lasts long enough.

```
restart:
```

```
M:=4;
```

$$M := 4$$

We calculate the threshold for η, $\eta_\diamond$.

```
thr:=((M+1)/ 2)^2/M;
```

$$thr := \frac{25}{16}$$

```
thr:=evalf(%);
```

$$thr := 1.562\,500\,000$$

We define the function η_0 from which we will obtain two versions with different time scales.

```
f:=x->piecewise(x<-1,0,x<1,1+cos(Pi*x),0):
```

```
f(x);
```

$$\begin{cases} 0 & x < -1 \\ 1 + \cos(\pi x) & x < 1 \\ 0 & \text{otherwise} \end{cases}$$

```
g:=1.5+0.1*f(x);
```

$$g := 1.5 + .1 \left(\begin{cases} 0 & x < -1 \\ 1 + \cos(\pi x) & x < 1 \\ 0 & \text{otherwise} \end{cases} \right)$$

```
g0:=unapply(g,x);
```

$$g0 := x \to 1.5 + .1 \left(\begin{cases} 0 & x < -1 \\ 1 + \cos(\pi x) & x < 1 \\ 0 & \text{otherwise} \end{cases} \right)$$

We define η_1 and η_2, which are called b_1 and b_2, here.

```
b1:=g0(0.08*t);
```

$$b1 := 1.5 + .1 \left(\begin{cases} 0 & .08t < -1 \\ 1 + \cos(.08\pi t) & .08t < 1 \\ 0 & \text{otherwise} \end{cases} \right)$$

```
b2:=g0(0.1*t);
```

$$b2 := 1.5 + .1 \left(\begin{cases} 0 & .1t < -1 \\ 1 + \cos(.1\pi t) & .1t < 1 \\ 0 & \text{otherwise} \end{cases} \right)$$

We plot Figure 7.3.

```
with(plots):

p1:=plot(b1, t=-21..21,1..2, color=black, axes=frame,
        thickness=2, title=` A` , labels=[` time` , ` `]):

p2:=plot(b2, t=-21..21,1..2, color=black, axes=frame,
        thickness=3, title=` B` , labels=[` time` , ` `]):

display(p1); display(p2);
```

We define the differential equations and give names to the plots of their solutions.

```
de1:=diff(x(t),t)=(x(t)/(1+x(t)))*((1-(x(t)/M))*(1+x(t))-b1);
```

$$de1 := \frac{\partial}{\partial t} x(t)$$

$$= \frac{x(t)}{1 + x(t)} \left((1 - \tfrac{1}{4} x(t))(1 + x(t)) - 1.5 \right.$$

$$\left. - .1 \left(\begin{cases} 0 & .08t < -1 \\ 1 + \cos(.08\pi t) & .08t < 1 \\ 0 & \text{otherwise} \end{cases} \right) \right)$$

```
sol1:=dsolve({de1, x(-20)=2.01}, x(t), numeric);
```

$sol1 := \mathbf{proc}(rkf45\_x)...\mathbf{end\ proc}$

```
with(plots):
p3:=odeplot(sol1, [t,x(t)],-15..35,color=black):
de2:=diff(x(t),t)=(x(t)/(1+x(t))) * ((1-(x(t)/M))*(1+x(t))-b2):
sol2:=dsolve({de2, x(-20)=2.01}, x(t), numeric);
```

$sol2 := \mathbf{proc}(rkf45\_x)...\mathbf{end\ proc}$

```
p4:=odeplot(sol2, [t,x(t)],-15..35, color=black, axes=frame,
            labels=[`time`, ``], title=`two populations under
            different extra mortalities`):
p5:=textplot([30,.2,`A`]):
p6:=textplot([30,2,`B`]):
```

Finally, we obtain Figure 7.3.

```
display([p3,p4,p5,p6]);
```

C.6 Dynamics of an Aquatic Population Interacting with a Polluted Environment (Figure 10.3E)

This work sheet was used to prepare the phase-plane plot in Figure 10.3E. We first erase previous work from the MAPLE memory.

```
restart;
```

We define the parameter a.

```
a:=0.25;
```

$a := .25$

We define the parameter σ.

```
s:=.2;
```

$s := .2$

We introduce the nonlinearity β.

```
b:=(max(0,1-y))/(1+x);
```

$$b := \frac{\max(0, 1 - y)}{1 + x}$$

We make preparations to determine the boundary equilibria.

```
f:=a-y * (s+b);
```

$$f := .25 - y\left(.2 + \frac{\max(0, 1 - y)}{1 + x}\right)$$

We change the function specification.

```
f1:=unapply(f,x,y);
```

$$f1 := (x, y) \to .25 - y\left(.2 + \frac{\max(0, 1 - y)}{1 + x}\right)$$

In order to calculate the boundary equilibria, we set $x = 0$.

```
f0:=f1(0,y);
```

$$f0 := .25 - y(.2 + \max(0, 1 - y))$$

We solve for the y-coordinates of the boundary equilibria.

```
y1:=fsolve(f0=0,y=0..0.5);
```

$$y1 := .268\,337\,521\,0$$

```
y2:=fsolve(f0=0,y=0.5..1);
```

$$y2 := .931\,662\,479\,0$$

```
y3:=fsolve(f0=0,y=1..2);
```

$$y3 := 1.250\,000\,000$$

We make preparations for determining the interior equilibrium. We choose its y-value, y^*, which we call Y.

```
Y:=.74;
```

$$Y := .74$$

We calculate the associate value for μ.

```
m:=a/Y-s;
```

$$m := .137\,837\,837\,8$$

We change the function specification of β before we calculate the interior equilibrium.

```
b1:=unapply(b,x,y);
```

$$b1 := (x, y) \to \frac{\max(0, 1 - y)}{1 + x}$$

We have specified the y-value of the interior equilibrium before, $y^* = Y$, and we now substitute it into the previous expression.

```
b0:=b1(x,Y);
```

$$b0 := .26\frac{1}{1+x}$$

We solve for the biomass component of the interior equilibrium.

```
M0:=fsolve(b0=m,x);
```

$$M0 := .8862745103$$

We get ready to prepare a plot for the interior equilibrium.

```
data0:=[[M0,Y]];
```

$$data0 := [[.8862745103, .74]]$$

We do not plot the interior equilibrium at this point, but only give the plot a name.

```
p0:=plot(data0, style=point, symbol=box, symbolsize=20,
        color=black):
```

We prepare plots for the boundary equilibria. Sinks will be denoted by boxes, sources by circles, and saddles by diamonds.

```
data3:=[[0,y3]];
```

$$data3 := [[0, 1.250000000]]$$

```
p3:=plot(data3, style=point, symbol=box, symbolsize=20,
        color=black):
```

```
data1:=[[0,y1]];
```

$$data1 := [[0, .2683375210]]$$

```
p1:=plot(data1, style=point, symbol=diamond, symbolsize=30,
        color=black):
```

```
data2:=[[0,y2]];
```

$$data2 := [[0, .9316624790]]$$

```
p2:=plot(data2, style=point, symbol=diamond, symbolsize=30,
        color=black):
```

We now turn to solving the differential equations. We introduce the right-hand side of the differential equation for M. We use x for M.

```
g:=x * (b-m);
```

$$g := x\left(\frac{\max(0, 1-y)}{1+x} - .137\,837\,837\,8\right)$$

We change the function specification for g.

```
g1:=unapply(g,x,y);
```

$$g1 := (x, y) \to x\left(\frac{\max(0, 1-y)}{1+x} - .137\,837\,837\,8\right)$$

We tell MAPLE that we want to do differential equations.

```
with(DEtools):
```

We specify the differential equation for the population biomass.

```
de1:=diff(x(t),t)=g1(x(t),y(t));
```

$$de1 := \frac{\partial}{\partial t}x(t) = x(t)\left(\frac{\max(0, 1-y(t))}{1+x(t)} - .137\,837\,837\,8\right)$$

We specify the differential equation for the body burden.

```
de2:=diff(y(t),t)=f1(x(t),y(t));
```

$$de2 := \frac{\partial}{\partial t}y(t) = .25 - y(t)\left(.2 + \frac{\max(0, 1-y(t))}{1+x(t)}\right)$$

We will make two phase-plane plots which we will overlay. The first plot will only present solutions in forward time, the second in both forward and backward time. This is due to the fact that the first quadrant is only forward invariant, but not backward invariant. So, the initial data for the second plot must be chosen in such a way that the solutions stay in the first quadrant also in backward time.

We specify four triples of initial data for the first plot. In each triple $[\cdot, \cdot, \cdot]$, the first component is the initial value for time t, the second component is the initial value for x, and the third component is the initial value for y.

```
start:=[0,.2,0], [0,1.5,1], [0,1.5,y3], [0,1.5,1.3];
```

$$start := [0,.2,0], [0,1.5,1], [0,1.5,1.250\,000\,000], [0,1.5,1.3]$$

We prepare the first plot.

```
p4:=DEplot([de1,de2], [x(t),y(t)], 0..100, {start}, stepsize=0.1,
        linecolor=black, thickness=1, dirgrid=[15,15],
        arrows=MEDIUM, title=` E` , axes=frame):
```

We have used the "DEplot" command to get both the various orbits and the direction field. We calculate the solutions from $t = 0$ to $t = 100$. The "dirgrid" command specifies the number of arrows used in the direction field and the "arrow" command the size of the arrows. The "title" command makes the letter E show in the plot. The "axes" command lets the axes move a little away from the actual picture.

Warning, the name changecoords has been redefined

We specify the initial data for the second phase-plane plot.

```
start2:=[0,.02,.9],[0,.2,1],[0,.7,.8];
```

$$start2 := [0, .02, .9], [0, .2, 1], [0, .7, .8]$$

We prepare the second phase-plane plot.

```
p6:=DEplot([de1,de2], [x(t),y(t)], -500..500, {start2},
            stepsize=0.1, linecolor=black, thickness=1,
            arrows=NONE):
```

As it turns out, we need many steps to get a clear picture about the large-time behavior of the solutions. This calculation is quite time consuming, and it is advisable to have some other work that one can do while MAPLE is busy. We have therefore put all the specifications into the first phase-plane plot because this takes less time and is better suited for experimenting with the various options. Finally, we overlay the plots of the four equilibria and the two phase-plane plots.

```
with(plots):
display([p0,p1,p2,p3,p4,p6]);
```

References

Aiello, W. G., Freedman, H. I., and Wu, J., 1992, Analysis of a model representing stage-structured population growth with state-dependent time delay, *SIAM J. Appl. Math.* **52**, 855–869.

Aliprantis, C. D., and Burkinshaw, O., 1998, *Principles of real analysis*, 3rd edn, Academic Press, San Diego, CA.

Allee, W. C., 1931, *Animal aggregations*, University of Chicago Press, Chicago, IL.

Allee, W. C., 1951, *Cooperation among animals*, Henry Schuman, New York.

Ambrosetti, A., and Prodi, G., 1993, *A primer of nonlinear analysis*, Cambridge University Press, Cambridge.

Andersen, T., 1997, *Pelagic nutrient cycles. Herbivores as sources and sinks*, Springer, Berlin, Heidelberg.

Anderson, R. M. (ed.), 1982a, *Population dynamics of infectious diseases. Theory and applications*, Chapman and Hall, London.

Anderson, R. M., 1982b, Transmission dynamics and control of infectious disease agents, *Population biology of infectious diseases* (ed. R. M. Anderson, R. M. May), pp. 1–12, Life Sciences Research Report, no. 25, Springer, Berlin.

Anderson, R. M., and May, R. M. (eds), 1982a, *Population biology of infectious diseases*, Dahlem Konferenzen, Life Sciences Research Report, no. 25, Springer, Berlin.

Anderson, R. M., and May, R. M., 1982b, Directly transmitted infectious diseases: control by vaccination, *Science* **215**, 1053–1060.

Anderson, R. M., and May, R. M., 1991, *Infectious diseases of humans. Dynamics and control*, Oxford University Press, Oxford.

Andreasen, V., 1995, Instability in an SIR-model with age-dependent susceptibility, *Mathematical population dynamics: analysis of heterogeneity. Volume one: theory of epidemics* (ed. O. Arino, D. Axelrod, M. Kimmel, M. Langlais), pp. 3–14, Wuerz, Winnipeg.

Anita, S., Iannelli, M., Kim Mi, Y., and Park, E., 1998, Optimal harvesting for periodic age-dependent population dynamics, *SIAM J. Appl. Math.* **58**, 1648–1666.

Anita, S., 2000, *Analysis and control of age-dependent population dynamics*, Kluwer Academic Publishers, Dordrecht.

Arendt, W., 1994, Vector-valued versions of some representation theorems in real analysis, *Functional analysis* (ed. K. D. Bierstedt, A. Pietsch, W. M. Ruess, D. Vogt), pp. 33–50, Marcel Dekker, New York.

Arendt, W., Batty, C. J. K., Hieber, M., and Neubrander, F., 2001, *Vector-valued Laplace transforms and Cauchy problems*, Birkhäuser, Basel.

Arino, O., Axelrod, D., and Kimmel, M. (eds), 1997, *Advances in mathematical population dynamics—molecules, cells and man*, World Scientific, Singapore.

Arino, O., Axelrod, D., Kimmel, M., and Langlais, M. (eds), 1995, *Mathematical population dynamics: analysis of heterogeneity*, vol. 1, *Theory of epidemics*, Wuerz, Winnipeg.

Aron, J. L., 1989, Simple versus complex epidemiological models, *Applied mathematical ecology* (ed. S. A. Levin, T. G. Hallam, L. J. Gross), pp. 176–192, Springer, Berlin, Heidelberg.

Aron, J. L., and Schwartz, I. B., 1984a, Seasonality and period-doubling bifurcations in an epidemic model, *J. Theor. Biol.* **110**, 665–679.

Aron, J. L., and Schwartz, I. B., 1984b, Some new directions for research in epidemic models, *IMA J. Math. Appl. Med. Biol.* **1**, 267–276.

Bailey, N. T. J., 1975, *The mathematical theory of infectious diseases and its applications*, Griffin, High Wycombe.

Barbalat, I., 1959, Systèmes d'équations différentielles d'oscillations nonlinéaires, *Rev. Roumaine Math. Pures Appl.* **4**, 267–270.

Barbu, V., Iannelli, M., and Martcheva, M., 2001, On the controllability of the Lotka–McKendrick model of population dynamics, *J. Math. Analysis Appl.* **253**, 142–165.

Barner, K., 2001, Das Leben Fermats, *Mitteilungen der DMV*, Issue 3 (2001), pp. 12–26, Deutsche Mathematiker-Vereinigung.

Bartlett, M. S., 1957, Measles periodicity and community size, *J. R. Statist. Soc.* A **120**, 48–70.

Bartlett, M. S., 1960, The critical community size for measles in the United States, *J. R. Statist. Soc.* A **123**, 37–44.

Bazykin, A. D., 1998, *Nonlinear dynamics of interacting populations*, World Scientific, Singapore.

Bell, E. T., 1937, *Men of mathematics*, reprinted in 1986 by Simon & Schuster.

Benaïm, M., and Hirsch, M. W., 1996, Asymptotic pseudotrajectories and chain recurrent flows, with applications, *J. Dyn. Diff. Eqns* **8**, 141–176.

Berberian, S. K., 1965, *Measure and integration*, Chelsea Publishing Co., New York.

Berec, L., Boukal, D. S., and Berec, M., 2002, Linking the Allee effect, sexual reproduction, and temperature-dependent sex determination via spatial dynamics, *Am. Naturalist* **157**, 217–230.

Bernoulli, D., 1766, Essai d'une nouvelle analyse de la mortalité causée par la petite vérole, *Mém. Math. Phys. Acad. Roy. Sci. Paris* **1**.

Beverton, R. J. H., and Holt, S. J., 1957, On the dynamics of exploited fish populations, *Fishery Investigations* Series 2, vol. 19, HMSO.

Boukal, D. S., and Berec, L., 2002, Single species models of the Allee effect: extinction boundaries, sex ratios and mate encounters, *J. Theor. Biol.* **218**, 375–394.

Boyce, W. E., and DiPrima, R. C., 1992, *Elementary differential equations and boundary value problems*, 5th edn, Wiley, New York.

Brauer, F., and Castillo-Chavez, C., 2001, *Mathematical models in population biology and epidemiology*, Springer, New York.

Brauer, F., and van den Driessche, P., 2002, Some directions for mathematical epidemiology, *Dynamical systems and their applications to biology* (ed. S. Ruan, G. Wolkowicz, and J. Wu), pp. 95–112, Fields Institute Communications, American Mathematical Society, Providence, RI.

Bürger, R., 2000, *The mathematical theory of selection, recombination, and mutation*, Wiley, Chichester.

Busenberg, S. N., and Castillo-Chavez, C., 1991, A general solution of the problem of mixing of subpopulations and its application to risk- and age-structured epidemic models for the spread of AIDS, *IMA J. Math. Appl. Med. Biol.* **8**, 1–29.

Busenberg, S., and Cooke, K., 1993, *Vertically transmitted diseases. Models and control*, Springer, Berlin, Heidelberg.

Butler, G., Freedman, H. I., and Waltman, P., 1986, Uniformly persistent systems, *Proc. Am. Math. Soc.* **96**, 425–430.

Butler, G., and Waltman, P., 1986, Persistence in dynamical systems, *J. Diff. Eqns* **63**, 255–263.

Capasso, V., 1993, *Mathematical structures of epidemic systems*, Lecture Notes in Biomathematics, vol. 97, Springer, Berlin.

Castillo-Chavez, C., with Blower, S., van den Driessche, P., Kirschner, D., and Yakubu, A.-A. (eds), 2002a, *Mathematical approaches for emerging and reemerging infectious diseases: an introduction*, Springer, New York.

Castillo-Chavez, C., with Blower, S., van den Driessche, P., Kirschner, D., and Yakubu, A.-A. (eds), 2002b, *Mathematical approaches for emerging and reemerging infectious diseases: models, methods and theory*, Springer, New York.

Castillo-Chavez, C., Cooke, K. L., Huang, W., and Levin, S. A., 1989, On the role of long incubation periods in the dynamics of acquired immunodeficiency syndrome (AIDS). Part 1: single population models, *J. Math. Biol.* **27**, 373–398.

Castillo-Chavez, C., and Thieme, H. R., 1995, Asymptotically autonomous epidemic models, *Mathematical population dynamics: analysis of heterogeneity. Volume one: theory of epidemics* (ed. O. Arino, D. Axelrod, M. Kimmel, M. Langlais), pp. 33–50, Wuerz, Winnipeg.

Castillo-Chavez, C., Velasco-Hernandez, J. X., and Fridman, S., 1994, Modeling contact structures in biology, *Frontiers in mathematical biology* (ed. S. A. Levin), pp. 454–491, Lecture Notes in Biomathematics, vol. 100, Springer, Berlin.

Caswell, H., 1989, 2001, *Matrix population models: construction, analysis, and interpretation*, Sinauer Associates, Sunderland.

Cha, Y., Iannelli, M., and Milner, F. A., 1998, Existence and uniqueness of endemic states for the age-structured S-I-R epidemic model, *Math. Biosci.* **150**, 177–190.

Cha, Y., Iannelli, M., and Milner, F. A., 2000, Stability change of an epidemic model, *Dynamic Syst. Appl.* **9**, 361–376.

Chance, J. K., 1989, *Conquest of the Sierra. Spaniards and Indians in colonial Oaxaca*, University of Oklahoma Press, Norman, OK.

Charlesworth, B., 1980, 1994, *Evolution in age-structured populations*, 1st and 2nd edns, Cambridge University Press, Cambridge.

Clark, C. W., 1976, 1990, *Mathematical bioeconomics*, Wiley, New York.

Clément, Ph., Heijmans, H. J. A. M., Angenent, S., van Duijn, C. J., and de Pagter, B., 1987, *One-parameter semigroups*, North-Holland, Amsterdam.

Cliff, A. D., Haggett, P., and Ord, J. K., 1986, *Spatial aspects of influenza epidemics*, Pion Ltd, London.

Coale, A. J., 1972, *The growth and structure of human populations*, Princeton University Press, Princeton, NJ.

Coale, A. J., and Trussell, T. J., 1974, Model fertility tables: variations in the age structure of childbearing in human populations, *Population Index* **40**, 186–192 (reprinted in Smith, Keyfitz (eds), 1977, pp. 323–331).

Coddington, E. A., and Levinson, N., 1955, *Theory of ordinary differential equations*, McGraw-Hill, New York.

Collins, J. P., 1979, Intrapopulation variation in the body size at metamorphosis and timing of metamorphosis in the bullfrog, *Rana castesbeiana*, *Ecology* **60**, 738–749.

Collins, J. P., and Cheek, J. E., 1983, Effect of food and density on development of typical and cannibalistic salamander larvae in *Ambystoma tigrinum nebulosum*, *Amer. Zool.* **23**, 77–84.

Collins, J. P., Zerba, K. E., and Sredl, M. J., 1993, Shaping intraspecific variation: development, ecology and the evolution of morphology and life history variation in tiger salamanders, *Genetica* **89**, 167–183.

Cooke, K. L., 1967, Functional differential equations: some models and perturbation problems, *Differential equations and dynamical systems* (ed. J. K. Hale, J. P. LaSalle), pp. 167–183, Academic Press, New York.

Cooke, K. L., van den Driessche, P., and Zou, X, 1999, Interaction of maturation delay and nonlinear birth in population and epidemic models, *J. Math. Biol.* **39**, 332–352 (Erratum, 2002, *J. Math. Biol.* **45**, 470).

Coppel, W. A., 1965, *Stability and asymptotic behaviour of differential equations*, Heath Mathematical Monographs.

Costantino, R. F., Cushing, J. M., Dennis, B., and Desharnais, R. A., 1995, Experimentally induced transitions in the dynamics behaviour of insect populations, *Nature* **375**, 227–230.

Cox, D. R., 1962, *Renewal theory*, Methuen, London.

Crow, E. L., and Shimizu, K. (eds), 1988, *Lognormal distributions, theory and applications*, Marcel Dekker, New York.

Cull, P., 1986, Local and global stability for population models, *Biological Cybernetics* **54**, 141–149.

Cushing, J. M., 1998, *An introduction to structured population dynamics*, CBMS-NSF Regional Conference Series in Applied Mathematics, vol. 71, SIAM, Philadelphia, PA

Cushing, J. M., Dennis, B., Desharnais, R. A., and Costantino, F. R., 1996, An interdisciplinary approach to understanding nonlinear ecological dynamics, *Ecological Modeling* **91**, 111–119.

Deimling, K., 1985, *Nonlinear functional analysis*, Springer, Berlin.

Dennis, B., 1989, Allee effects: population growth, critical density, and the chance of extinction, *Nat. Resource Modeling* **3**, 481–538.

Dennis, B., Desharnais, R. A., Cushing, J. M., and Costantino, R. F., 1995, Nonlinear demographic dynamics: mathematical models, statistical methods, and biological experiments, *Ecol. Monogr.* **65**, 261–281.

Dennis, B., and Patil, G. P., 1984, The gamma distribution and weighted multinomial gamma distributions as models of population abundance, *Math. Biosci.* **68**, 187–212.

de Roos, A. M., 1997, A gentle approach to physiologically structured population models, *Structured population models in marine, freshwater, and terrestrial systems* (ed. S. Tuljapurkar, H. Caswell), pp. 119–204, Chapman and Hall, New York.

de Roos, A. M., Persson, L., and Thieme, H. R., 2003, Emergent Allee effects in top predators feeding on structured prey population, *Proc. R. Soc. Lond.* B **270**, 611–618.

Desowitz, R. S., 1981, *New Guinea tapeworms and Jewish grandmothers. Tales of parasites and people*, Avon, New York.

Desowitz, R. S., 1997, *Who gave Pinta to the Santa Maria? Torrid diseases in a temperate world*, W. W. Norton & Co., New York.

Devaney, R. L., 1987, *An introduction to chaotic dynamical systems*, Addison-Wesley, Harlow.

Dieckmann, U., Metz, J. A. J., Sabelis, M. W., and Sigmund, K. (eds), 2002, *Adaptive dynamics of infectious diseases: in pursuit of virulence management*, Cambridge University Press, Cambridge.

Diekmann, O., Dietz, K., and Heesterbeek, J. A. P., 1991, The basic reproduction ratio for sexually transmitted diseases, I, Theoretical considerations, *Math. Biosci.* **107**, 325–339.

Diekmann, O., Gyllenberg, M., Huang, H., Kirkilionis, M., Metz, J. A. J., and Thieme, H. R., 2001, On the formulation and analysis of general deterministic structured population models, II, Nonlinear theory, *J. Math. Biol.* **43**, 157–189.

Diekmann, O., Gyllenberg, M., Metz, J. A. J., and Thieme, H. R., 1998, On the formulations and analysis of general deterministic structured population models, I, Linear Theory, *J. Math. Biol.* **36**, 349–388.

Diekmann, O., Gyllenberg, M., and Thieme, H. R., 1993, Perturbing semigroups by solving Stieltjes renewal equations, *Diff. Int. Eqns* **6**, 155–181.

Diekmann, O., Gyllenberg, M., and Thieme, H. R., 1995a, Perturbing evolutionary systems by step responses and cumulative outputs, *Diff. Int. Eqns* **8**, 1205–1244.

Diekmann, O., and Heesterbeek, J. A. P., 2000, *Mathematical epidemiology of infectious diseases; model building, analysis and interpretation*, Wiley, Chichester.

Diekmann, O., Heesterbeek, J. A. P., and Metz, J. A. J., 1990, On the definition and the computation of the basic reproduction ratio $\mathcal{R}_0$ in models for infectious diseases in heterogeneous populations, *J. Math. Biol.* **28**, 365–382.

Diekmann, O., and Kretzschmar, M., 1991, Patterns in the effects of infectious diseases on population growth, *J. Math. Biol.* **29**, 539–570.

Diekmann, O., Metz, J. A. J., and Heesterbeek, J. A. P., 1995b, The legacy of Kermack and McKendrick, *Epidemic models: their structure and relation to data* (ed. D. Mollison), pp. 95–115, Cambridge University Press, Cambridge.

Diekmann, O., Nisbet, R. M., Gurney, W. S. C., and van den Bosch, F., 1986, Simple mathematical models for cannibalism: a critique and a new approach, *Math. Biosci.* **78**, 21–46.

Diekmann, O., van Gils, S. A., Verduyn Lunel, S. M., and Walther, H.-O., 1995c, *Delay equations. Functional, complex and nonlinear analysis*, Springer, New York.

Dietz, K., 1975, Transmission and control of arbovirus diseases, *Epidemiology* (ed. D. Ludwig, K. L. Cooke), pp. 104–119, SIAM, Philadelphia, PA.

Dietz, K., 1976, The incidence of infectious diseases under the influence of seasonal fluctuations, *Mathematical models in medicine* (ed. J. Berger, W. Bühler, R. Repges, P. Tautu), pp. 1–15, Lecture Notes in Biomathematics, vol. 11, Springer, Heidelberg.

Dietz, K., 1982, Overall population patterns in the transmission cycle of infectious disease agents, *Population biology of infectious diseases* (ed. R. M. Anderson, R. M. May), pp. 1–12, Life Sciences Research Report, no. 25, Springer, Berlin.

Dietz, K., 1988, The first epidemic model: a historical note on P. D. En'ko, *Austral. J. Statist.* **30A**, 56–65.

Dietz, K., 1995, Some problems in the theory of infectious disease transmission and control, *Epidemic models. Their structure and relation to data* (ed. D. Mollison), pp. 3–16, Cambridge University Press, Cambridge.

Dietz, K., and Heesterbeek, J. A. P., 2002, Daniel Bernoulli's epidemiological model revisited, *Math. Biosci.* **180**, 1–21.

Dietz, K., and Schenzle, D., 1985, Mathematical models for infectious disease statistics, *A celebration of statistics*, The ISI Centenary Volume (ed. A. C. Atkinson, S. E. Fienberg), Springer, New York.

Dobson, A., and Grenfell, B. T. (eds), 1995, *Ecology of infectious diseases in natural populations*, Cambridge University Press, Cambridge.

Doedel, E., 1981, AUTO: a program for the automatic bifurcation analysis of autonomous systems, *Cong. Num.* **30**, 265–284.

Edelstein-Keshet, L., 1988, *Mathematical models in biology*, McGraw-Hill, New York.

Edwards, C. H., Jr., and Penny, D. E., 1988, *Elementary linear algebra*, Prentice Hall, Upper Saddle River, NJ.

Edwards, R. E., 1965, 1995, *Functional analysis*, Holt, Rinehart and Winston, New York (1965) and Dover, New York (1995).

Evans, M., Hastings, N., and Peacock, B., 2000, *Statistical distributions*, 3rd edn, Wiley, New York.

Farkas, M., 2001, *Dynamical models in biology*, Academic Press, San Diego, CA.

Farrington, C. P., Kanaan, M. N., and Gay, N. J., 2001, Estimation of the basic reproduction number for infectious diseases from age-stratified serological survey data (with discussion), *J. Royal Statist. Soc.* C; *Appl. Statistics* **50**, 251–283.

Feller, W., 1941, On the integral equation of renewal theory, *Ann. Math. Statistics* **12**, 243–267 (reprinted in Smith, Keyfitz (eds), 1977, pp. 131–156).

Feller, W., 1966, *An introduction to probability theory and its applications*, vol. II, Wiley, New York.

Feng, Z., 1999, Stability of periodic solutions in an epidemiological model, *Bifurcation theory and its numerical analysis* (ed. Z. Chen, S. N. Chow and K. Li), pp. 69–78, Springer, Singapore, New York.

Feng, Z., and Thieme, H. R., 1995, Recurrent outbreaks of childhood diseases revisited: the impact of isolation, *Math. Biosci.* **128**, 93–130.

Feng, Z., and Thieme, H. R., 2000a, Endemic models with arbitrarily distributed periods of infection, I, General theory, *SIAM J. Appl. Math.* **61**, 803–833.

Feng, Z., and Thieme, H. R., 2000b, Endemic models with arbitrarily distributed periods of infection, II, Fast disease dynamics and permanent recovery, *SIAM J. Appl. Math.* **61**, 983–1012.

Fiedler, B., 1986, Global Hopf bifurcation for Volterra integral equations, *SIAM J. Math. Analysis* **17**, 911–932.

Fiedler, B., and Mallet-Paret, J., 1989, A Poincaré–Bendixson theorem for scalar reaction diffusion equations, *Arch. Ration. Mech. Analysis* **107**, 325–345.

Folland, G. B., 1984, *Real analysis. Modern techniques and their applications*, Wiley, New York.

Fonda, A., 1988, Uniformly persistent semidynamical systems, *Proc. Am. Math. Soc.* **104**, 111–116.

Fox, L. R., 1975, Cannibalism in natural populations, *A. Rev. Ecol. Syst.* **6**, 87–106.

Frauenthal, J. C., 1986, Analysis of age-structure models, *Mathematical ecology. An introduction* (ed. T. G. Hallam, S. A. Levin), pp. 117–147, Springer, Berlin.

Freedman, H. I., and Moson, P., 1990, Persistence definitions and their connections, *Proc. Am. Math. Soc.* **109**, 1025–1032.

Gao, L., Mena-Lorca, J., and Hethcote, H. W., 1995, Four SEI endemic models with periodicity and separatrices, *Math. Biosci.* **128**, 157–184.

Garrett, L., 1994, *The coming plague*, Penguin, New York.

Gilpin, M. E., and Ayala, F. J., 1973, Global models of growth and competition, *Proc. Nat. Acad. Sci. USA* **70**, 3590–3593.

Gompertz, B., 1825, On the nature of the function expressive of the law of human mortality, and on a new mode of determining the value of life contingencies. In a letter to Francis Batley, esq., *Phil. Trans. R. Soc. Lond.* **115**, 513–585.

Goodall, E. W., 1931, Incubation period of measles (letter to editor), *Brit. Med. J.* **1**, 73–74.

Greenhalgh, D., 1990, Vaccination compaigns for common childhood diseases, *Math. Biosci.* **100**, 201–240.

Greenhalgh, D., and Dietz, K., 1994, Some bounds on estimates for reproductive ratios from the age-specific force of infection, *Math. Biosci.* **124**, 9–57.

Grenfell, B. T., and Dobson, A. P. (eds), 1995, *Ecology of infectious diseases in natural populations*, Cambridge University Press, Cambridge.

Gripenberg, G., Londen, S-O., and Staffans, O., 1990, *Volterra integral and functional equations*, Cambridge University Press, Cambridge.

Grossman, Z., 1980, Oscillatory phenomena in a model of infectious diseases, *Theor. Pop. Biol.* **18**, 204–243.

Grossman, Z., Gumowski, I., and Dietz, K., 1977, The incidence of infectious diseases under the influence of seasonal fluctuations—analytic approach, *Nonlinear systems and applications to life sciences*, pp. 525–546, Academic Press, San Diego, CA.

Gurney, W. S., and Nisbet, R. M., 1998, *Ecological dynamics*, Oxford University Press, Oxford.

Gyllenberg, M., Hanski, I., and Lindström, T., 1997, Continuous versus discrete single species population models with adjustable reproductive strategies, *Bull. Math. Biol.* **59**, 679–705.

Hadeler, K. P., and Castillo-Chavez, C., 1995, A core group model for disease transmission, *Math. Biosci.* **128**, 41–55.

Hadeler, K. P., and Dietz, K., 1984, Population dynamics of killing parasites which reproduce in the host, *J. Math. Biol.* **21**, 45–65.

Hadeler, K. P., and Müller, J., 1996, Optimal vaccination patterns in age-structured populations, II, Optimal strategies, *Model for infectious human diseases: their structure and relation to data* (ed. V. Isham, G. Medley), pp. 102–114, Cambridge University Press, Cambridge.

Hadeler, K. P., and Tomiuk, J., 1977, Periodic solutions of difference-differential equations, *Arch. Ration. Mech. Analysis* **65**, 87–96.

Hahn, W., 1967, *Stability of motion*, Springer, Berlin.

Hale, J. K., 1977, *Theory of functional differential equations*, Springer, New York.

Hale, J. K., 1980, *Ordinary differential equations*, Krieger, Malabar.

Hale, J. K., 1988, *Asymptotic behavior of dissipative systems*, AMS, Providence, RI.

Hale, J. K., and Waltman, P., 1989, Persistence in infinite-dimensional systems, *SIAM J. Math. Analysis* **20**, 388–395.

Hallam, T. G., 1986, Population dynamics in a homogeneous environment, *Mathematical ecology* (ed. T. G. Hallam, S. A. Levin), pp. 61–94, Biomathematics, vol. 17, Springer, Berlin.

Hallam, T. G., and De Luna, J. L., 1984, Effects of toxicants on populations: a qualitative approach, III, Environmental and food chain pathways, *J. Theor. Biol.* **109**, 411–429.

Hallam, T. G., Lassiter, R. R., and Kooijman, S. A. L. M., 1989, Effects of toxicants on aquatic populations, *Applied mathematical ecology* (ed. S. A. Levin, T. G. Hallam, L. J. Gross), pp. 352–382, Springer, Berlin.

Hassell, M. P., 1975, Density dependence in single-species populations, *J. Animal Ecol.* **44**, 283–295.

Hassell, M. P., Anderson, R. C., Cohen, J. E., Cvjetanović, B., Dobson, A. P., Gill, D. E., Holmes, J. C., May, R. M., McKeown, T., Pereira, M. S., and Tyrell, D. A. J., 1982, Impact of infectious diseases on host populations. Group report, *Population biology of infectious diseases* (ed. R. M. Anderson, R. M. May), pp. 15–35, Dahlem Konferenzen, Life Sciences Research Report, no. 25, Springer, Berlin.

Hassett, M. J., and Stewart, D. G., 1999, *Probability for risk management*, ACTEX, Winsted.

Heesterbeek, J. A. P., 2002, A brief history of $\mathcal{R}_0$ and a recipe for its calculation, *Acta Biotheoretica* **50**, 189–204.

Heesterbeek, J. A. P., and Metz, J. A. J., 1993, The saturating contact rate in marriage and epidemic models, *J. Math. Biol.* **31**, 529–539.

Heijmans, H. J. A. M., and de Pagter, B., 1987, Asymptotic behavior, *One-parameter semigroups* (ed. Ph. Clément, H. J. A. M. Heijmans, S. Angenent, C. J. van Duijn, B. de Pagter), pp. 213–233, North-Holland, Amsterdam.

Hethcote, H. W., 1976, Qualitative analysis for communicable disease models, *Math. Biosci.* **28**, 335–356.

Hethcote, H. W., 1989a, Three basic epidemiological models, *Applied mathematical ecology* (ed. S. A. Levin, T. G. Hallam, L. J. Gross), pp. 119–144, Biomathematics, vol. 18, Springer, Berlin.

Hethcote, H. W., 1989b, Rubella, *Applied mathematical ecology* (ed. S. A. Levin, T. G. Hallam, L. J. Gross), pp. 212–234, Biomathematics, vol. 18, Springer, Berlin.

Hethcote, H. W., 1994, A thousand and one epidemic models, *Frontiers in mathematical biology* (ed. S. A. Levin), pp. 504–515, Lecture Notes in Biomathematics, vol. 100, Springer, Berlin.

Hethcote, H. W., 1995, Modeling heterogeneous mixing in infectious disease dynamics, *Models for infectious human diseases. Their structure and relation to data* (ed. V. Isham, G. Medley), Cambridge University Press, Cambridge.

Hethcote, H. W., 2000, The mathematics of infectious disease, *SIAM Rev.* **42**, 599–653.

Hethcote, H. W., and Levin, S. A., 1989, Periodicity in epidemiological models, *Applied mathematical ecology* (ed. S. A. Levin, T. G. Hallam, L. J. Gross), pp. 193–211, Biomathematics, vol. 18, Springer, Berlin.

Hethcote, H. W., Stech, H. W., and van den Driessche, P., 1981, Nonlinear oscillations in epidemic models, *SIAM J. Appl. Math.* **40**, 1–9.

Hethcote, H. W., and Thieme, H. R., 1985, Stability of the endemic equilibrium in epidemic models with subpopulations, *Math. Biosci.* **75**, 205–227.

Hethcote, H. W., and Van Ark, J. W., 1992, *Modeling HIV transmission and AIDS in the United States*, Lecture Notes in Biomathematics, vol. 95, Springer, Berlin.

Hethcote, H. W., and Yorke, J. A., 1984, *Gonorrhea: transmission dynamics and control*, Lecture Notes in Biomathematics, vol. 56, Springer, Berlin.

Hethcote, H. W., Zhien, M., and Shengbing, L., 2002, Effects of quarantine in six endemic models for infectious diseases, *Math. Biosci.* **180**, 141–160.

Hewitt, E., and Stromberg, K., 1969, *Real and abstract analysis*, Springer, Berlin.

Heyward, W. L., and Curran, J. W., 1988, The epidemiology of AIDS in the US, *Scient. Amer.* **259** (4), 72–81.

Hirsch, M. W., 1990, Systems of differential equations which are competitive or cooperative, IV, Structural stability in three dimensional systems, *SIAM J. Math. Analysis* **21**, 1225–1234.

Hirsch, W. M., Hanisch, H., and Gabriel, J.-P., 1985, Differential equation models for some parasitic infections: methods for the study of asymptotic behavior, *Commun. Pure Appl. Math.* **38**, 733–753.

Hirsch, M. W., and Smale, S., 1974, *Differential equations, dynamical systems, and linear algebra*, Academic Press, San Diego, CA.

Hofbauer, J., Mallet-Paret, J., and Smith, H. L., 1991, Stable periodic solutions for the hypercycle system, *J. Dyn. Diff. Eqns* **3**, 423–436.

Hofbauer, J., and Sigmund, K., 1988, *The theory of evolution and dynamical systems*, Cambridge University Press, Cambridge.

Hofbauer, J., and Sigmund, K., 1998, *Evolutionary games and population dynamics*, Cambridge University Press, Cambridge.

Holling, C. S., 1965, The functional response of predators to prey density and its role in mimicry and population regulation, *Mem. Ent. Soc. Canada* **45**, 1–60.

Holling, C. S., 1966, The functional response of invertebrate predators to prey density, *Mem. Entom. Soc. Can.* **48**.

Hönig, C. S., 1975, *Volterra Stieltjes-integral equations*. North-Holland/American Elsevier, Amsterdam.

Hoppensteadt, F. C., 1974, Asymptotic stability in singular perturbation problems, II, Problems having matched asymptotic expansion solutions, *J. Diff. Eqns* **15**, 510–521.

Hoppensteadt, F. C., 1975, *Mathematical theories of populations: demographics, genetics and epidemics*, Regional Conference Series in Applied Mathematics, vol. 20, SIAM, Philadelphia, PA.

Hoppensteadt, F. C., 1982, *Mathematical methods of population biology*, Cambridge University Press, New York.

Hoppensteadt, F. C., and Waltman, P., 1970, A problem in the theory of epidemics, *Math. Biosci.* **9**, 71–91.

Hoppensteadt, F. C., and Waltman, P., 1971, A problem in the theory of epidemics, II, *Math. Biosci.* **12**, 133–145.

Horn, M. A., Simonett, G., and Webb, G. F. (eds), 1998, *Mathematical models in medical and health science*, Vanderbilt University Press, Nashville, TN.

Huang, W., Cooke, K. L., and Castillo-Chavez, C., 1992, Stability and bifurcation for a multiple-group model for the dynamics of HIV/AIDS transmission, *SIAM J. Appl. Math.* **52**, 835–854.

Hutchinson, G. E., 1978, *An introduction to population ecology*, Yale University Press, New Haven, CT and London.

Hutson, V., 1984, A theorem on average Lyapunov functions, *Monatsh. Math.* **98**, 267–275.

Hutson, V., and Schmitt, K., 1992, Permanence in dynamical systems, *Math. Biosci.* **111**, 1–71.

Iannelli, M., 1995, *Mathematical theory of age-structured population dynamics*, Applied Mathematics Monographs (CNR), vol. 7, Giardini Editori e Stampatori, Pisa.

Impagliazzo, J., 1985, *Deterministic aspects of mathematical demography*, Springer, Berlin.

Inaba, H., 1989, Weak ergodicity of population evolution processes, *Math. Biosci.* **96**, 195–219.

Inaba, H., 1990, Threshold and stability results for an age-structured epidemic model, *J. Math. Biol.* **28**, 411–434.

Isham, V., and Medley, G. (eds), 1996, *Models for infectious human diseases: their structure and relation to data*, Cambridge University Press, Cambridge.

Ivanov, A. F., and Sharkovsky, A. N., 1991, Oscillations in singularly perturbed delay equations, *Dynamics reported*, vol. I (ed. C. K. R. T. Jones, U. Kirchgraber, H. O. Walther), pp. 164–224, Springer, Berlin.

Jacquez, J. A., Simon, C. P., and Koopman, J., 1995, Core groups and R_0s for subgroups in heterogeneous SIS and SI models, *Epidemic models: their structure and relation to data* (ed. D. Mollison), pp. 279–301, Cambridge University Press, Cambridge.

Johnson, N. L., and Kotz, S., 1970, *Continuous univariate distributions, I*, Houghton Mifflin, Boston, MA.

Jordan, A. M., 1986, *Trypanosomiasis control and African rural development*, Longman, Harlow.

Kato, T., 1976, *Perturbation theory for linear operators*, Springer, Berlin.

Keeling, M. J., and Grenfell, B. T., 1998, Effect of variability in infectious period on the persistence and spatial spread of infectious diseases, *Math. Biosci.* **147**, 207–226.

Kendall, M. G., and Stuart, A., 1958, *The advanced theory of statistics*, 3 volumes, Hafner Publishing Company, New York.

Kermack, W. O., and McKendrick, A. G., 1927–1939, Contributions to the mathematical theory of epidemics: Part I (1927) *Proc. R. Soc. Lond.* A **115**, 700–721 (reprinted in *Bull. Math Biol.* **53** (1991), 33–55); Part II (1932) *Proc. R. Soc. Lond.* A **138**, 55–85 (reprinted in *Bull. Math Biol.* **53** (1991), 57–87); Part III (1933) *Proc. R. Soc. Lond.* A **141**, 94–122 (reprinted in *Bull. Math Biol.* **53** (1991), 89–118); Part IV (1937) *J. Hyg. Camb.* **37**, 172–187; Part V (1939) *J. Hyg. Camb.* **39**, 271–288.

Keyfitz, N., 1977, *Introduction to the mathematics of population with revisions*, Addison-Wesley, Reading.

Keyfitz, N., and Flieger, W., 1971, *Population. Facts and methods of demography*, Freeman and Co., San Francisco, CA.

Keyfitz, N., and Flieger, W., 1990, *World population growth and aging*, The University of Chicago Press, Chicago, IL.

Kim, Y. J., and Aron, J. L., 1989, On the equality of average age and average expectation of remaining life in a stationary population, *SIAM Rev.* **31**, 110–113.

Kingsland, S. E., 1982, The refractory model: the logistic curve and the history of population ecology, *Q. Rev. Biol.* **57**, 29–52.

Kingsland, S. E., 1995, *Modeling nature. Episodes in the history of population ecology*, 2nd edn, The University of Chicago Press, Chicago, IL.

Kirkilionis, M., Krömker, S., Rannacher, R., and Tomi, F. (eds), 2003, *Trends in nonlinear analysis*, Springer, Berlin, Heidelberg.

Kirkwood, J. R., 1989, 1995, *An introduction to analysis*, PWS Publishing Co., Boston, MA.

Koch, M. G., 1987, *AIDS—Vom Molekül zur Pandemie*, Spektrum der Wissenschaft, Heidelberg.

Kofler, M., 1997, *Maple. An introduction and reference*, Addison-Wesley, Harlow.

Kooijman, S. A. L. M., 1986, Population dynamics on the basis of budgets, *The dynamics of physiologically structured populations* (ed. J. A. J. Metz, O. Diekmann), pp. 267–297, Lecture Notes in Biomathematics, vol. 68, Springer, Berlin.

Kooijman, S. A. L. M., 1993, 2000, *Dynamic energy budgets in biological systems*, Cambridge University Press, Cambridge.

Kooijman, S. A. L. M., and Metz, J. A. J., 1984, On the dynamics of chemically stressed populations: the deduction of population consequences from effects on individuals, *Ecotox. Env. Saf.* **8**, 254–274.

Kot, M., 2001, *Elements of mathematical ecology*, Cambridge University Press, Cambridge.

Krasnosel'skii, M. A., 1964, *Positive solutions of operator equations*, Noordhoff, Groningen.

Krause, U., 1997, Positive nonlinear difference equations, *Nonlinear Analysis TMA* **30**, 301–308.

Kribs-Zaleta, C. M., and Martcheva, M., 2002, Vaccination strategies and backward bifurcation in an age-since-infection structured model, *Math. Biosci.* **177–178**, 317–322.

Kribs-Zaleta, C. M., and Velasco-Hernández, J. X., 2000, A simple vaccination model with multiple endemic states, *Math. Biosci.* **164**, 183–201.

Kuang, Y., 1990, Global stability of Gause-type predator–prey systems, *J. Math. Biol.* **28**, 463–474.

Kuang, Y., 1993, *Delay differential equations with applications in population dynamics*, Academic Press, San Diego, CA.

Lajmanovich, A., and Yorke, J. A., 1976, A deterministic model for gonorrhea in a nonhomogeneous population, *Math. Biosci.* **28**, 221–236.

Levin, S. A., 1995, Grouping in population models, *Epidemic models: their structure and relation to data* (ed. D. Mollison), pp. 271–278, Cambridge University Press, Cambridge.

Levin, S. A., Hallam, T. G., and Gross, L. J. (eds), 1989, *Applied mathematical ecology*, Springer, Berlin.

Li, M. Y., and Muldowney, J. S., 1995, Global stability for the SEIR model in epidemiology, *Math. Biosci.* **125**, 155–164.

Li, M. Y., Muldowney, J. S., and van den Driessche, P., 1999, Global stability of SEIRS models in epidemiology, *Can. Appl. Math. Quart.* **7**, 409–425.

Li, M. Y., Smith, H. L., and Wang, L., 2001, Global dynamics of an SEIR epidemic model with vertical transmission, *SIAM J. Appl. Math.* **62**, 58–69.

Lin, C. C., and Segel, L. A., 1974, *Mathematics applied to deterministic problems in the natural sciences*, Macmillan (republished in 1988, SIAM, Philadelphia, PA).

Lin, X., 1991, On the uniqueness of endemic equilibria of an HIV/AIDS transmission model for a heterogeneous population, *J. Math. Biol.* **29**, 779–790.

Lin, X., and So, J. W.-H., 1993, Global stability of the endemic equilibrium and uniform persistence in epidemic models with subpopulations, *J. Austral. Math. Soc.* B **34**, 282–295.

Liu, W.-M., 1993, Dose-dependent latent period and periodicity of infectious diseases, *J. Math. Biol.* **31**, 487–494.

Liu, W.-M., 1995, Models for recurrent outbreaks of infectious diseases, *Mathematical population dynamics: analysis of heterogeneity*, vol. 1, *Theory of epidemics* (ed. O. Arino, D. Axelrod, M. Kimmel, M. Langlais), pp. 119–130, Wuerz, Winnipeg.

Liu, W.-M., Hethcote, H. W., and Levin, S. A., 1987, Dynamical behavior of epidemiological models with nonlinear incidence rates, *J. Math. Biol.* **25**, 359–380.

Loeb, M. L. G., Collins, J. P., and Maret, T. J., 1994, The role of prey in controlling expression of a trophic polymorphism in *Ambystoma tigrinum nebulosum*, *Functional Ecology* **8**, 151–158.

London, W. P., and Yorke, J. A., 1973a, Recurrent outbreaks of measles, chickenpox and mumps, I, Seasonal variation in contact rates, *American J. Epidem.* **98**, 453–468.

London, W. P., and Yorke, J. A., 1973b, Recurrent outbreaks of measles, chickenpox and mumps, II, Systematic differences in contact rates and stochastic effects, *American J. Epidem.* **98**, 468–482.

Lotka, A. L., 1907, Relation between birth rates and death rates, *Science* **26**, 21–22 (reprinted in Smith, Keyfitz (eds), 1977, pp. 93–95).

Lotka, A. J., 1922, The stability of the normal age distribution, *Proc. Nat. Acad. Sci.* **8**, 339–345 (reprinted in Smith, Keyfitz (eds), 1977, pp. 101–107).

McDonald, J. N., and Weiss, N. A., 1999, *A course in real analysis*, Academic Press, San Diego, CA.

McGeary, J., 2001, Death stalks a continent, *Time* **157**(6), 36–45.

McKendrick, A. G., 1926, Applications of mathematics to medical problems, *Proc. Edinb. Math. Soc.* **14**, 98–130.

McKendrick, A. G., and Kesava Pai, M., 1911, The rate of multiplication of microorganisms: a mathematical study, *Proc. R. Soc. Edinb.* **31**, 649–655.

McNeill, W. H., 1976, *Plagues and people*, Anchor-Doubleday, New York.

Makeham, W. M., 1867, On the law of mortality, *J. Inst. Actuaries* **13**, 325–367.

Mallet-Paret, J., and Sell, G. R., 1996, The Poincaré–Bendixson theorem for monotone cyclic feedback systems with delay, *J. Diff. Eqns* **125**, 441–489.

Mallet-Paret, J., and Smith, H. L., 1990, The Poincaré–Bendixson theorem for monotone cyclic feedback systems, *J. Dyn. Diff. Eqns* **2**, 367–421.

Malthus, T., 1798, *An essay on the principle of population*, Harmondsworth, Middlesex (republished by Penguin, New York, in 1970).

Mann, J. M., Chin, J., Piot, P., and Quinn, Th., 1988, The international epidemiology of AIDS, *Scient. Amer.* **259** (4), 82–89.

Maret, T. J., and Collins, J. P., 1996, Effect of prey vulnerability on population size structure of a gape-limited predator, *Ecology* **77**, 320–324.

Marks, G., and Beatty, W. K., 1976, *Epidemics*, Charles Scribner's Sons, New York.

Martcheva, M., 1999, Exponential growth in age-structured two-sex populations, *Math. Biosci.* **157**, 1–22.

Martcheva, M., and Thieme, H. R., n.d., Progression age enhanced backward bifurcation in an epidemic model with super-infection, *J. Math. Biol.*, to appear.

Martelli, M., Cooke, K., Cumberbatch, E., Tang, B., and Thieme, H. R. (eds), 1996, *Differential equations and applications to biology and industry*, World Scientific, Singapore.

May, R. M., 1982, Introduction, *Population biology of infectious diseases* (ed. R. M. Anderson, R. M. May), pp. 1–12, Life Sciences Research Report, no. 25, Springer, Berlin.

May, R. M., and Anderson, R. M., 1989, The transmission dynamics of human immunodeficiency virus (HIV), *Applied mathematical ecology* (ed. S. A. Levin, T. G. Hallam, L. J. Gross), Biomathematics, vol. 18, Springer, Berlin.

Maynard Smith, J., 1974, *Models in ecology*, Cambridge University Press, Cambridge.

Merriam-Webster's Collegiate Dictionary, 1994, *Merriam-Webster's Collegiate Dictionary.*

Merrill, S. J., and Murphy, B. M., 2002, Detecting autocatalytic dynamics in data modeled by a compartmental model, *Math. Biosci.* **180**, 255–262.

Mesterton-Gibbons, M., 1989, *A concrete approach to mathematical modelling*, Addison-Wesley, Redwood City, CA.

Metz, J. A. J., and Diekmann, O., 1986, *The dynamics of physiologically structured populations*, Lecture Notes in Biomathematics, vol. 68, Springer, Berlin.

Metz, J. A. J., and van den Bosch, F., 1995, Velocities of epidemic spread, *Epidemic models: their structure and relation to data* (ed. D. Mollison), pp. 150–186, Cambridge University Press, Cambridge.

Michaelis, L., and Menten, M. I., 1913, Die Kinetik der Invertinwirkung, *Biochem. Z.* **49**, 333–369.

Mischaikow, K., Smith, H. L., and Thieme, H. R., 1995, Asymptotically autonomous semiflows, chain recurrence and Lyapunov functions, *Trans. Am. Math. Soc.* **347**, 1669–1685.

Mode, C. J., and Sleeman, C. K., 2000, *Stochastic processes in epidemiology: HIV/AIDS, other infectious diseases, and computers*, World Scientific, Singapore, Rivers Edge, NJ.

Mollison, D. (ed.), 1995, *Epidemic models: their structure and relation*, Cambridge University Press, Cambridge.

Monod, J., 1942, *Recherches sur la croissance des cultures bacteriennes*, Herman, Paris.

Monod, J., 1950, La technique de culture continue, théorie et applications, *Annales l'Institut Pasteur* **79**, 390–401.

Muldowney, J. S., 1990, Compound matrices and ordinary differential equations, *Rocky Mountain J. Math.* **20**, 857–872.

Müller, J., 1998, Optimal vaccination patterns in age structured populations, *SIAM J. Appl. Math.* **59**, 222–241.

Murray, J. D., 1989, *Mathematical biology*, Springer, Berlin, Heidelberg.

Nagel, R. (ed.), 1986, *One-parameter semigroups of positive operators*, Lecture Notes in Mathematics, vol. 1184, Springer, Berlin.

Nåsell, I., 1985, *Hybrid models of tropical infections*, Lecture Notes in Biomathematics, vol. 59, Springer, Berlin.

Nåsell, I., 2002, Endemicity, persistence, and quasi-stationarity, *Mathematical approaches for emerging and reemerging infectious diseases: models, an introduction* (ed. C. Castillo-Chavez, with S. Blower, P. van den Driessche, D. Kirschner, A.-A. Yakubu), Springer, New York.

Nåsell, I., 2002, Measles outbreaks are not chaotic, *Mathematical approaches for emerging and reemerging infectious diseases: models, methods, and theory* (ed. C. Castillo-Chavez, with S. Blower, P. van den Driessche, D. Kirschner, A.-A. Yakubo), Springer, New York.

Navarova, N., and Thieme, H. R., 1998, Remarks on an environmental control problem, *Mathematical models in medical and health sciences* (ed. M. A. Horn, G. Simonett, G. F. Webb), pp. 267–279, Vanderbilt University Press, Nashville, TN.

Nesemann, T., 1999, A nonlinear extension of the Coale–Lopez theorem, *Positivity* **3**, 135–148.

Nisbet, R. M., 1997, Delay-differential equations for structured populations, *Structured population models in marine, freshwater, and terrestrial systems* (ed. S. Tuljapurkar, H. Caswell), pp. 90–118, Chapman and Hall, New York.

Nisbet, R. M., and Gurney, W. S. C., 1982, *Modelling fluctuating populations*, Wiley, Chichester.

Nisbet, R. M., Muller, E. B., Brooks, A. J., and Hosseini, P., 1997, Models relating individual and population response to contaminant, *Environmental Modeling Assessment* **2**, 7–12.

Nisbet, R. M., and Onyiah, L. C., 1994, Population dynamic consequences of competition within and between age classes, *J. Math. Biol.* **32**, 329–344.

Norton, H. T. J., 1928, Natural selection and Mendelian variation, *Proc. Lond. Math. Soc.* **28**, 1–45.

Nowak, M. A., and May, R. A., 2000, *Virus dynamics: mathematical principles of immunology and virology*, Oxford University Press, Oxford.

Nussbaum, R. D., 1990, Some nonlinear weak ergodic theorems, *SIAM J. Math. Analysis* **21**, 436–460.

Olsen, L. F., Truty, G. L., and Schaffer, W. M., 1988, Oscillations and chaos in epidemics: a nonlinear dynamic study of six childhood diseases in Copenhagen, Denmark, *Theor. Pop. Biol.* **33**, 344–370.

Patel, J. K., Kapadia, C. H., and Owen, D. B., 1976, *Handbook of statistical distributions*, Marcel Dekker, New York.

Peters, C. J., 1997, *Virus hunter. Thirty years of battling hot viruses around the world*, Anchor Books, Double Day, New York.

Pfennig, D. W., Sherman, P. W., and Collins, J. P., 1994, Kin recognition and cannibalism in polyphenic salamanders, *Behav. Ecol.* **5**, 225–232.

Poirier, R., 1962, *Die 15 Weltwunder*, Deutsche Buchgemeinschaft.

Polis, G. A., 1981, The evolution and dynamics of intraspecific predation, *A. Rev. Ecol. Syst.* **12**, 225–251.

Preston, S. H., Heuveline, P., and Guillot, M., 2001, *Demography. Measuring and modeling population processes*, Blackwell.

Preston, G., 1995, *The hot zone*, Anchor Books, New York.

Pugliese, A., 1990, Population models for diseases with no recovery, *J. Math. Biol.* **28**, 65–82.

Rass, L., Radcliffe, J., 2003, *Spatial deterministic epidemics*, AMS, Providence, RI.

Razzell, P. E., 1974, An interpretation of the modern rise of populations in Europe—a critique, *Pop. Studies* **28**, 5–17.

Ricker, W. E., 1954, Stock and recruitment, *J. Fish. Res. Board Canada* **11**, 559–623.

Ricker, W. E., 1975, Computation and interpretation of biological studies of fish populations, *Bull. Fish. Res. Bd. Canada* **191**.

Robinson, C., 1995, *Dynamical systems*, CRC Press, Boca Raton, FL.

Roughgarden, J., 1979, *Theory of population genetics and evolutionary ecology: an introduction*, Macmillan, New York.

Royama, T., 1992, *Analytical population dynamics*, Chapman and Hall.

Ruan, S., and Zhao, X.-Q., 1999, Persistence and extinction in two species reaction-diffusion systems with delays, *J. Diff. Eqns* **156**, 71–92.

Sartwell, P. E., 1950, The distribution of incubation periods of infectious diseases, *Am. J. Hyg.* **51**, 310–318.

Sartwell, P. E., 1966, The incubation period and the dynamics of infectious disease, *Am. J. Epid.* **83**, 204–318.

Schaefer, H. H., 1974, *Banach lattices and positive operators*, Springer, Berlin.

Schaffer, W. M., 1985, Can nonlinear dynamics help us infer mechanisms in ecology and epidemiology?, *IMA J. Math. Appl. Med. Biol.* **2**, 221–252.

Schaffer, W. M., and Kot, M., 1985, Nearly one dimensional dynamics in an epidemic, *J. Theor. Biol.* **112**, 403–427.

Schenzle, D., 1984, An age-structured model of pre- and postvaccination measles transmission, *IMA J. Math. Appl. Med. Biol.* **1**, 169–191.

Schwartz, I. B., and Smith, H. L., 1983, Infinite subharmonic bifurcations in an SEIR epidemic model, *J. Math. Biol.* **18**, 233–253.

Scott, M. E., and Smith, G. (eds), 1994, *Parasitic and infectious diseases. Epidemiology and ecology*, Academic Press, San Diego, CA.

Scott, S., and Duncan, C. J., 1998, *Human demography and disease*, Cambridge University Press, Cambridge.

Sell, G. R., and You, Y., 2002, *Dynamics of evolutionary equations*, Springer, New York.

Sharkovskii, A. N., 1964, Coexistence of cycles of a continuous map of a line into itself, *Ukrainian Math. J.* **16**, 61–71.

Sharpe, F. R., and Lotka, A. J., 1911, A problem in age-distribution, *Phil. Mag.* (6) **21**, 435–438.

Sigmund, K., 1993, *Games of life*, Oxford University Press (hard cover), Penguin (paperback).

Simon, C. P., and Jacquez, J. A., 1992, Reproduction numbers and the stability of equilibria of SI models for heterogeneous populations, *SIAM J. Appl. Math.* **22**, 541–576.

Simpson, H. N., 1980, *Invisible armies. The impact of disease on American history*, The Bobbs-Merrill Company, Indianapolis, IN.

Smith, D., and Keyfitz, N. (eds), 1977, *Mathematical demography*, Springer, Heidelberg.

Smith, F. E., 1963, Population dynamics in *Daphnia magna* and a new model for population growth, *Ecology* **44**, 651–663.

Smith, H. L., 1983a, Subharmonic bifurcation in an SIR epidemic model, *J. Math. Biol.* **17**, 163–177.

Smith, H. L., 1983b, Multiple stable subharmonics for a periodic epidemic model, *J. Math. Biol.* **17**, 179–190.

Smith, H. L., 1995, *Monotone dynamical systems. An introduction to the theory of competitive and cooperative systems*, AMS, Providence, RI.

Smith, H. L., and Waltman, P., 1995, *Theory of the chemostat. Dynamics of microbial competition*, Cambridge University Press, Cambridge.

Stech, H. W., and Williams, M., 1981, Stability in a class of cyclic epidemic models with delay, *J. Math. Biol.* **11**, 95–103.

Stillerman, M., and Thalhimer, W., 1944, Attack rate and incubation period of measles, *Am. J. Dis. Child.* **67**, 15.

Takáč, P., 1996, Convergence in the part metric for discrete dynamical systems in ordered topological cones, *Nonlinear Analysis TMA* **26**, 1753–1777.

Takeuchi, Y., 1996, *Global dynamical properties of Lotka–Volterra systems*, World Scientific, Singapore.

Tan, W.-Y., 2000, *Stochastic modeling of AIDS epidemiology and HIV pathogenesis*, World Scientific, Singapore.

Tang, B., and Kuang, Y., 1996, Permanence in Kolmogorov-type systems of nonautonomous functional differential equations, *JMAA* **197**, 427–447.

Tang, B., Sitomer, A., and Jackson, T., 1997, Population dynamics and competition in chemostat models with adaptive nutrient uptake, *J. Math. Biol.* **35**, 453–479.

Thieme, H. R., 1983, Local stability in epidemic models for heterogeneous populations, *Mathematics in biology and medicine* (ed. V. Capasso, E. Grosso, S. L. Paveri-Fontana), pp. 185–189, Lecture Notes in Biomathematics, vol. 57, Springer, Berlin.

Thieme, H. R., 1985, Renewal theorems for some mathematical models in epidemiology, *J. Integ. Eqns* **8**, 185–216.

Thieme, H. R., 1988a, Asymptotic proportionality (weak ergodicity) and conditional asymptotic equality of solutions to time-heterogeneous sublinear difference and differential equations, *J. Diff. Eqns* **73**, 237–268.

Thieme, H. R., 1988b, Well-posedness of physiologically structured population models for *Daphnia magna*, *J. Math. Biol.* **26**, 299–317.

Thieme, H. R., 1991, Stability change of the endemic equilibrium in age-structured models for the spread of S-I-R type infectious diseases, *Differential equations models in biology, epidemiology and ecology* (ed. S. Busenberg, M. Martelli), pp. 139–158, Lecture Notes in Biomathematics, vol. 92, Springer, Berlin.

Thieme, H. R., 1992, Convergence results and a Poincaré–Bendixson trichotomy for asymptotically autonomous differential equations, *J. Math. Biol.* **30**, 755–763.

Thieme, H. R., 1993, Persistence under relaxed point-dissipativity (with applications to an epidemic model), *SIAM J. Math. Analysis* **24**, 407–435.

Thieme, H. R., 1994a, Asymptotically autonomous differential equations in the plane, *Rocky Mountain J. Math.* **24**, 351–380.

Thieme, H. R., 1994b, Asymptotically autonomous differential equations in the plane, II, Stricter Poincaré–Bendixson type results, *Diff. Integral Eqns* **7**, 1625–1640.

Thieme, H. R., 1996, Positive perturbations of dual and integrated semigroups, *Adv. Math. Sci. Appl.* **6**, 445–507.

Thieme, H. R., 1998, Positive perturbation of operator semigroups: growth bounds, essential compactness, and asynchronous exponential growth, *Discr. Cont. Dyn. Syst.* **4**, 735–764.

Thieme, H. R., 2000, Uniform persistence and permanence for non-autonomous semiflows in population biology, *Math. Biosci.* **166**, 173–201.

Thieme, H. R., 2002, The transition through stages with arbitrary length distributions, and applications in epidemics, *Mathematical approaches for emerging and reemerging infectious diseases: models, methods and theory* (ed. C. Castillo-Chavez, S. Blower, P. van den Driessche, D. Kirschner, A.-A. Yakubu), pp. 45–84, Springer, New York.

Thieme, H. R., and Voßeler, H., 2002, A Stieltjes type convolution for integrated semigroups of strong bounded variation and L_p solutions to the abstract Cauchy problem, *J. Diff. Integ. Eqns* **15**, 1171–1218.

Thieme, H. R., and Yang, J., 2000, On the complex formation approach in modeling predator prey relations, mating and sexual disease transmission, *Proceedings of Nonlinear Differential Equations, University of Miami* (ed. R. S. Cantrell, C. Cosner), *Electron. J. Diff. Eqns Conf.* **05**, 255–283.

Thomas, D. M., Snell, T. W., and Jaffar, S. M., 1996, A control problem in a polluted environment, *Math. Biosci.* **133**, 139–163.

Tikhonov, A. N., Vasil'eva, A. B., and Sveshnikov, A. G., 1985, *Differential equations*, Springer, Berlin, Heidelberg.

Trenerry, C. F., 1926, *The origin and early history of insurance*, P. S. King & Son, London.

Tuljapurkar, S., and Caswell, H. (eds), 1997, *Structured population models in marine, freshwater, and terrestrial systems*, Chapman and Hall, New York.

van den Bosch, F., and Diekmann, O., 1986, Interactions between egg-eating predator and prey: the effect of the functional response and of age structure, *IMA J. Math. Appl. Med. Biol.* **3**, 53–69.

van den Driessche, P., and Watmough, J., 2002, Reproduction numbers and sub-threshold endemic equilibria for compartmental models of disease transmission, *Math. Biosci.* **180**, 29–48.

Verhulst, P.-F., 1845, Recherches mathématiques sur la loi d'accroissement de la population, *Nouveaux Mémoires de l'Académie Royale des Sciences et Belles-Lettres de Bruxelles* **18**, 3–38.

von Bertalanffy, L., 1934, Untersuchungen über die Gesetzlichkeit des Wachstums, Teil I, Allgemeine Grundlagen der Theorie: mathematische und physiologische Gesetzlichkeiten des Wachstums bei Wassertieren, *Arch. Entwicklungsmech. Org.* **131**, 613–652.

VonFoerster, H., 1959, Some remarks on changing populations, *The kinetics of cellular proliferation* (ed. F. Stohlman), Grune and Strutton, New York.

Wade, W. R., 1995, *An introduction to analysis*, Prentice Hall, Upper Saddle River, NJ.

Waltman, P., 1991, A brief survey of persistence in dynamical systems, *Delay differential equations and dynamical systems* (ed. S. Busenberg, M. Martelli), pp. 31–40, Lecture Notes in Mathematics, vol. 1475, Springer.

Webb, G. F., 1984, A semigroup proof of the Sharpe–Lotka theorem, *Infinite dimensional systems* (ed. F. Kappel, W. Schappacher), pp. 254–268, Lecture Notes in Mathematics, vol. 1076, Springer, Berlin.

Webb, G. F., 1985, *Theory of nonlinear age-dependent population dynamics*, Marcel Dekker, New York.

Widder, D. V., 1971, *An introduction to transform theory*, Academic Press, San Diego, CA.

Wiggins, S., 1990, *Introduction to applied nonlinear dynamical systems and chaos*, Springer, New York.

Wilbur, H. M., and Collins, J. P., 1973, Ecological aspects of amphibian metamorphosis, *Science* **182**, 1305–1314.

Wills, C., 1996, *Yellow fever, black goddess*, Addison-Wesley.

Wu, L.-I., and Feng, Z., 2000, Homoclinic bifurcation in an SIQR model for childhood diseases, *J. Diff. Eqns* **168**, 150–167.

Yosida, K., 1968, *Functional analysis*, 2nd edn, Springer, Berlin.

Zeidler, E., 1985, *Nonlinear functional analysis and its applications, III, Variational methods and optimization*, Springer, Berlin, Heidelberg.

Zhao, X.-Q., 1995, Uniform persistence and periodic coexistence states in infinite-dimensional periodic semiflows with applications, *Canad. Appl. Math. Quart.* **3**, 473–495.

Zhao, X.-Q., 2001, Uniform persistence in processes with application to nonautonomous competitive models, *J. Math. Analysis Appl.* **258**, 87–101.

Zhu, Hsiu-Rong, 1991 *The existence of stable periodic orbits for systems of three dimensional differential equations that are competitive*, Ph.D. thesis, Arizona State University.

Zhu, Hsiu-Rong, and Smith, H. L., 1994, Stable periodic orbits for a class of three dimensional competitive systems, *J. Diff. Eqns* **110**, 143–156

Index

age at death
  average, 263, 264, 266
  expected, 29, 187, 224, 380
age at reproduction (child bearing)
  average, 259, 260
  expected, 256, 259, 260, 267, 269
age profile
  final, 251, 252
  large-time, 251
Allee effect, 52, 65, 72, 74, 75, 104, 107, 147, 149, 181, 510
almost periodic, 31
asymmetric logistic equation, 39, 104
asymptotic equality, xiv, 75, 251, 252, 271
average age
  at death, *see* age at death, average
  at infection, xv, 341, 366, 371–373, 375, 381
  at stage exit, 224
  of a population, 185, 219, 221, 262–267
  of a stage, 166, 167, 188, 201, 206, 330, 337, 368, 378, 380
average duration of susceptibility, xv, xvi, 330, 331, 341, 366, 371, 373, 378–381
average expectation of remaining sojourn (duration), 195–198, 202, 204, 205, 207, 208, 210, 220, 339
average intrinsic growth rate, 16, 18, 28
average (mean) length (duration) of the infectivity period, 295, 303, 368, 369, 380
average per capita mortality rate of a population, 264

balanced exponential growth, 242–244, 251, 255, 262–265, 267, 269–271
Bendixson criterion, 159, 160, 427
Bernoulli difference equation (discrete Bernoulli equation), 81–83
Bernoulli differential equation (continuous Bernoulli equation), 39, 48, 49, 53, 56, 59–61, 63, 78 100, 508
Beverton–Holt difference equation (discrete Beverton–Holt equation), xiv, 81, 83, 95, 100–105, 520
Beverton–Holt differential equation (continuous Beverton–Holt equation), xiv, xvi, 37–45, 48, 53, 56–62, 65, 74, 110, 133, 134, 230, 235
bifurcation
  Hopf, *see* Hopf bifurcation
  subcritical, 146, 164–167, 178–180, 399, 417, 446
  supercritical, 164–167, 176, 179
birth rate
  per capita, 8, 13, 17, 20, 21, 28, 29, 39, 42, 46, 56, 60, 151, 168, 179, 231, 234, 243, 262, 265, 318, 338, 496
  population, 7, 219, 221, 222, 239, 240, 273, 486
bistability, bistable, 138, 145, 147, 159, 179, 180
body burden, 108–113, 116, 133, 134, 138–140, 145, 147–149, 516
Borel
  measurable, 360, 467–469, 483, 488
  measure, 466–471, 473–475, 482, 488
  set, 206, 466, 467, 481
breakpoint, 71

cannibalism, xvi, 40, 42, 43, 46, 83, 84, 230–235, 237
carrying capacity, 38, 39, 44, 46, 47, 53, 55, 57–59, 66, 67, 69, 500, 501, 504, 508

central epoch (of an epidemic), 302
change of variables, 233, 234, 465
chaos, 100, 105, 237, 252
characteristic equation, 242, 255, 256, 259, 327, 328, 352, 434, 435
competitive system, 338
concave, 34, 49, 54, 57, 70, 135, 174, 200, 209, 472, 473, 483, 484, 500
contact probabilities, 386, 417
contact rate, 293, 305–307, 342, 392, 417
convergence, 85, 87, 91–93, 245, 277, 337, 370, 428–433, 465
  dominated, 246, 350, 379, 488
  in Abel average, 246, 247
  monotone, 469
  of averages, 408, 450
  pointwise, 205, 245, 459
  uniform, 233, 370, 459
  weak*, 469
convex
  function, 54, 234, 266
  Hausdorff space, 469, 472, 473
  optimization, xvii, 473
  set, 454, 472–474
cooperative system, 427
critical point, 69, 425, 426, 433, 490, 491
critical value, 171, 490, 491
cumulants, 257, 271

Daphnia, 104, 151, 209
determinant, 126, 127, 155, 158, 160, 163, 171, 173, 174, 326, 435, 445
discretization, 100, 101, 105
distribution
  exponential, *see* exponential distribution
  log-normal, 202–205, 210
  normal, 202, 203, 205
divergence, 137–139, 143, 159, 172, 426, 427, 443
domain of attraction (stable set), 93, 136, 137, 143, 147, 176, 178, 443
doubling time, 14, 28, 29, 259, 261, 262, 495
Duhamel's principle, 20
Dulac function, 131, 134, 427
Dulac–Bendixson criterion, 138, 140, 179, 426

E. coli, 58
eigenvalue, 126–130, 134, 139, 144, 158, 162, 326, 327, 393, 404, 407, 411–413, 428, 434, 435, 445, 447–451
eigenvector, 164, 407, 411–413, 434, 447–451
  Perron–Frobenius, 407
  probability, 407, 408
emigration rate
  per capita, 10, 11, 20, 21, 28
  population, 7
environment, xiv, 14, 17, 33, 53, 104, 107–109, 114, 117, 151, 239, 244, 289, 513,
  periodic, 14–17
equilibrium
  boundary, 108, 120–123, 125–132, 135–138, 140–149, 514, 515
  disease-free, 321–326, 332, 334, 335, 338, 341, 358, 381, 389–391, 399, 401, 404–406, 408, 417
  endemic, 127, 322, 324, 326, 327, 329–332, 335–338, 341, 349–351, 366, 370, 371, 373, 380, 381, 390–399, 417, 442
  extinction, xiv, 120, 135, 147, 148, 155, 158, 159
  interior, 121–142, 149, 157, 159 163, 166, 168, 170–172, 175, 514, 515
  stable, 71, 72, 136, 160, 180, 434
  strongly endemic, 384, 390–399, 417
  survival, 121, 138, 145, 147, 148
  unstable, 127, 134, 180
ergodicity
  mean, *see* mean ergodicity
  strong, 251, 270
  weak, 251, 271
exit rate, 12, 199, 201, 202, 208, 209, 213, 221–224, 226, 230
expectation
  of life (life expectancy), *see* life expectancy
  of remaining life, xiv, 190, 191, 194, 195, 202, 207–210, 221, 245, 256, 262, 267, 269, 271, 307, 360, 363, 375, 376

of unvaccinated life, 367, 371, 375
expected age
at child bearing (reproduction), *see* age at reproduction (child bearing), expected
at death, *see* age at death, expected
at infection, 330, 331, 380
at stage exit, 187, 224
expected remaining sojourn (life), 190–201, 207–209, 220, 360, 363, 375, 376
expected reproductive age, 256, 259, 268
exponential distribution, 205, 235–237, 339
exponential growth, 14, 15, 241–244, 251, 255, 256, 262–271, 381, 496
extinction, xiv, 3, 70, 108, 120, 123, 135, 136, 145, 147–149, 155–160, 332, 334, 345, 348, 351–354, 399, 417, 511

feedback loop, 107, 117–120
final replacement ratio, 301
final size of an epidemic (of the susceptible, removed population), xiv, 296, 297, 300, 301, 314–316
fixed point, 84–94, 96–103, 261, 274, 277, 300, 301, 393, 394, 443, 488–490
Floquet
exponent, 16
multiplier, 16
fluctuation, 14, 114, 115, 121, 122, 157, 334, 339, 428, 441
Fourier series, 25–27
Frauenthal's rule of thumb, 267, 269, 270
fundamental
balance equation of population (stage) dynamics, 7, 13, 47, 222, 224, 454, 477
mode of oscillation, 27
solution, 14, 16, 20, 23
theorem of calculus, 9, 14, 24, 25, 193, 313, 458
theorem of convex optimization, 473, 482

generalist predator, 65, 68, 74, 107
Gompertz
equation, 37, 47–49, 52, 53, 56, 59, 61, 62
law, 194, 208, 210, 525

half-value time, 331
heteroclinic orbit, 130, 138, 146–148
homoclinic orbit, 130, 160, 172, 176, 179, 426
Hopf bifurcation, xv, 139, 145, 146, 149, 164–167, 173, 174, 176, 178–180

immigration rate, 7, 12, 18, 20, 21, 27–30, 217, 496, 497
immunity (period), (immune period), 185, 284, 285, 290, 291, 317–319, 327, 331, 337, 338, 341
incidence, 291, 293, 305, 312, 316, 339
incubation period, 185, 202, 205, 209, 290, 291
infection age, 311, 312
expected at removal, 311
infection rate, 288, 293, 294, 303, 315, 339, 366, 373
inflection point, 38, 49, 58, 135, 310, 500
infectious (infective) period (period of infectivity), 185, 205, 206, 290, 291, 303, 317, 323, 327, 330, 337–339, 344, 357, 368, 369, 380, 381
integrable at 0, 55, 56, 62
integrable at ∞, 55–57, 62
integrating factor, xv, 19, 21, 22
integration by parts, 21, 188, 189, 228, 229, 277, 376, 453, 462, 463
integration by substitution, 464
interepidemic period, xv, 317, 318, 330, 339
intermediate value theorem, xv, 51, 53, 85–91, 98, 119, 121, 124, 128, 144, 242, 350, 363, 489
interstage competition, 153, 154, 159, 168
intrastage competition, 152, 159, 180
intrinsic growth rate, 16–18, 28, 497

invariant, 123, 126, 129, 136, 138, 140, 141, 156, 441, 442, 443
  forward, 123, 129, 136, 137, 336, 337, 442, 443, 449, 516
irreducible, 391–394, 401, 403, 406–410, 413, 446, 448–451

Jacobian
  determinant (the Jacobian), 139, 161, 171, 173, 428
  matrix, 125, 126, 139, 158–160, 163, 164, 167, 171–173, 175, 428, 444–446
Jensen's inequality, 256, 259, 260, 266, 454, 483

kurtosis (coefficient of excess), 190, 205, 257

Laplace transform, 234, 240–242, 246, 248, 253, 256, 257, 355, 402, 464
Laplace–Stieltjes transform, 233, 234, 236, 464
latency, 57, 202, 317
latent period (period of latency), 186, 205, 206, 290, 291, 311, 315, 317–319, 321, 324, 330, 337
least-squares fit, 35, 47, 58, 504, 505
Lebesgue–Fatou lemma, 314, 353, 354, 468, 489
Lebesgue–Stieltjes integral, 345, 453, 454, 475–477, 481, 491, 492
Leibnitz, 37, 82, 100, 101, 104
life expectancy, 9, 10, 29, 41, 43, 46, 60, 66–69, 113, 187, 202, 207, 208, 263, 266, 327, 369, 377, 378, 381
limiting system, 118, 386
linear dose response, 110, 113, 121, 140, 143, 144
logarithmic convexity, 57
logistic equation, 38, 39, 43, 44, 47, 57, 58, 60, 83, 100–105, 110, 498, 500, 504, 508

maintenance cost, 33, 34, 170,
Malthus, 14
Malthusian parameter, 14, 16, 241–245, 247, 255, 256, 259–266, 268, 271, 495
mass action, 40, 42, 68, 109, 111, 231
matrix, 35, 44, 155, 325, 326, 338, 392, 393, 403, 404, 409, 410, 414, 415, 434, 435, 444, 451
  diagonal, 158, 164, 408, 435, 446–448, 451
  irreducible, 391–393, 403, 407–410, 413, 446, 448–451
  positive, xv, 446–451
  primitive, 408, 416, 446, 448, 450, 451
  quasi-positive, xv, 418, 446–449
  replacement, *see* replacement matrix
  second generation, *see* replacement matrix
  strictly positive, 394, 408, 446, 448, 451
  transposed, 407, 447
mean ergodicity, 407, 408, 450
mean sojourn time, 10, 11, 66, 166, 187, 188, 190, 191, 197, 201, 202, 204, 207, 208, 218, 219, 230, 323, 405
mean value theorem, 11, 53, 77, 91, 93, 94, 174, 428, 429, 431, 432
measurable, 360, 467–469, 483, 488
measure
  Borel, *see* Borel measure
  nonnegative, 466, 467
  probability, 206, 274, 488, 489
  regular, 467
metered model, 105
minimal set, 160, 338, 442, 443
mixing, 382, 417, 418
  separable, 341, 344–346, 349, 381, 411–415, 417
  separable extra-group, 412–414
modulus of stability (spectral bound), 447
moments, 189–190, 203, 204, 234, 256, 257
mortality (death) rate
  per capita, 8, 10, 13, 14, 17, 29, 37, 39, 42, 46, 56, 60, 65, 71, 81, 83, 103, 151, 152, 170, 179, 180, 199–202, 209, 221, 227, 230, 250, 251, 264, 265, 275, 323, 338, 342, 385, 537
  population, 7, 8, 222

next-generation matrix, 406, 417
  see also replacement matrix
node, 126–128, 136–138, 143–146, 178, 327
nonhomogeneous linear differential equation, 23

open loop, 107
optimal strategy, 362–364
optimization, xv, xvii, 361, 365, 376, 454, 473, 482, 492

period-two point, 85
periodic, xiv, 14–18, 21, 23–28, 34, 38, 81, 98, 100
  almost, 31
  modulation, 15, 17, 23, 28
  orbit, 130–141, 146–148, 158–160, 163–168, 172, 176–180, 338, 425–428, 443, 444, 446
  quasi-periodic, *see* quasi-periodic
Perron–Frobenius
  theory, xv, 407
  vector, 407
persistence, xiv, xvi, xviii, 78, 124, 156, 157, 159, 181, 279, 332, 334, 336, 345, 348, 351, 354, 383, 399, 401, 403, 417, 436–442
Poincaré–Bendixson
  dichotomy, 131, 425
  property, 338
  theorem, 157, 160, 163, 166, 168, 179
  theory, 107, 108, 130, 424
  trichotomy, 129, 130, 136, 138, 426
prevalence, 289–291, 317, 330
primitive, 408, 416, 446, 448, 450, 451
probability of dying during a stage, 12, 228, 230
probability of lifetime emigration, 11–12
probability of surviving a stage, 12, 156, 172, 175, 276, 278, 323, 324

quasi-periodic, 30, 31
quasi-steady-state, xvi, 40, 42, 43, 66, 67, 69, 111, 113

regular value, 128–130, 171, 490, 491
regulated function, xv, 453, 454, 456, 457, 459, 461, 467, 483, 491
removal rate, 294, 314,
removed, 293, 296, 297, 302, 306, 311, 313, 315, 383, 394
renewal equation, 240, 454, 486
renewal theorem, 244, 246, 250, 251, 266, 270, 316
replacement (second generation) matrix, 383, 406–411, 413–416
replacement ratio, 294, 295, 299, 302, 351, 367
  basic, xiv–xvi, 322, 323, 325, 332–334, 338, 351, 367, 375, 383, 386, 404, 406, 407, 409–412, 414–416
  initial, 295, 297, 301–303, 308, 314
  net, 341, 351–354, 358–363, 365–367, 375, 378, 382
reproduction ratio, 156
  basic, 29, 68, 69, 156, 243, 256, 259, 262, 266, 267, 269, 271, 324
resource–consumer model, 40, 111
Ricker difference equation (discrete Ricker equation), 81, 83–85, 96–100, 102, 103, 105, 532
Ricker differential equation (continuous Ricker equation), 37, 39, 40, 45–49, 53, 56, 60–62, 76, 78, 83, 167, 171, 174, 230, 235, 278
Routh–Hurwitz criterion, 327, 395, 435

saddle, 128–131, 136–139, 144–147, 161, 162, 171, 174–180, 515
semiflow, xvi, 116, 120, 124, 125, 336, 403, 436–443
semistable, 72
separation of variables, xv, 9, 14
separatrix, 71
sex ratio, 8, 65, 66
sigmoid population growth, 49, 53, 57, 69, 104, 105, 233
sink, 126, 129–131, 138, 146, 147, 161, 162, 171, 175, 177–179
skewness, 190, 204

Smith's equation, 39, 44, 48, 49, 56, 58, 62, 104, 498, 503
sojourn
  function, 186, 188–190, 193, 197, 203, 205–210, 226, 231, 275, 311, 314, 453, 475, 477
    de Moivre type, 208
    Pareto type, 208
  time, 10–12, 66, 166, 185–191, 197, 201, 202, 204–208, 218, 219, 230, 323, 405
solution
  complementary, 22, 27
  general, 15, 22, 23, 26
  particular, 22, 26
  persistent, 23, 125
  transient, 123
source, 126, 127, 130, 131, 134, 136, 138, 139, 145, 146, 148, 161, 162, 171, 175, 177–180, 287, 339, 340, 382, 493, 499
spectral bound, $s(A)$, 447–450
spectral radius, $r(A)$, 382, 383, 393, 394, 400, 403, 404, 406, 407, 409, 410, 414, 415, 417, 447, 450, 451
spectral value, 447
stable age distribution, *see* age profile, large-time
stable manifold, 128, 131, 136–139, 145, 146, 176–178, 180
stable set (domain of attraction), *see* domain of attraction (stable set)
stage, 12, 40, 43, 45, 66, 107, 108, 293, 383–385
  age, 211–216
  duration, xiii, 186, 187, 189, 196–198, 202–204, 207, 217–219, 221, 224, 226, 234, 237
  structure, xiii–xvi, 151, 152, 157, 181, 418
  transition, xiii, xv, 151, 157–160, 162, 167, 168, 171, 179, 180, 185, 385
standard deviation, 202, 203, 256
step function, 194, 197, 364, 381, 453–456, 458, 460–463, 465, 476
Stieltjes
  convolution, 477–481, 492
  integral, xv, 188, 189, 223, 228, 231, 232, 345, 359, 368, 453–464, 469, 475–479, 481, 483, 491
superposition, principle of, 22
support (of a measure), 469–471, 473, 474, 482
survival function (survivorship), 185, 186, 193, 195, 197, 207, 209, 226, 229–231, 239, 240, 252, 262–264, 275, 344, 352, 374, 486

threshold
  condition, 120, 121, 130, 141, 156, 417
  phenomenon, 301
  theorem (for epidemics), 295, 297, 307–309, 314–316
time average, 16, 18, 26–28, 77, 496
  asymptotic, 17, 18, 21, 22, 30, 31, 35
  for periodic intrinsic growth rate, 16
time scale, 40, 43, 46, 66, 107, 232, 283, 365, 366, 368, 369, 511
toxicant, xiv, 107–110, 113–117, 136, 148, 149
trace (of a matrix), 126, 127, 158–160, 163, 167, 171–175, 326, 435, 445

undetermined coefficients, methods of, 22, 23
unstable manifold, 128, 129, 131, 136–138, 145–147, 176

vaccination, 384, 385, 399, 417
  cost, 358–365, 375, 376, 475
  distribution, 359–362, 377, 465, 475, 476, 481
  profile, 344, 351, 359, 375, 453
  schedule, 193, 359, 366, 454, 473
  strategy, xiii, xv, 341, 358, 359, 361–366, 375, 377, 381, 382, 476, 482
variance, xiv, 189, 196–198, 202, 204, 206, 207, 256, 257, 259, 262, 268

variation (of a function)
bounded, 224, 454–465
finite, 455
variation (of a measure) (total variation), 466
variation of constants (parameters) formula, 20, 21, 24
Verhulst equation, 37–39, 43, 44, 48, 49, 53, 56, 62, 74, 81, 83, 100, 101, 104
Volterra integral equation, 240, 273, 454, 486, 492
von Bertalanffy equation (model), 33–35, 56, 61, 62, 104, 535

watershed equilibrium, 71, 72
weak* topology, 469, 473, 482
weakly* closed, 469, 473
weakly* compact, 469, 473

www.ingramcontent.com/pod-product-compliance
Ingram Content Group UK Ltd.
Pitfield, Milton Keynes, MK11 3LW, UK
UKHW022248121224
452420UK00005B/359

9 780691 092911